W9-AYL-982

THE HISTORY OF THE TELESCOPE

THE HISTORY OF THE TELESCOPE

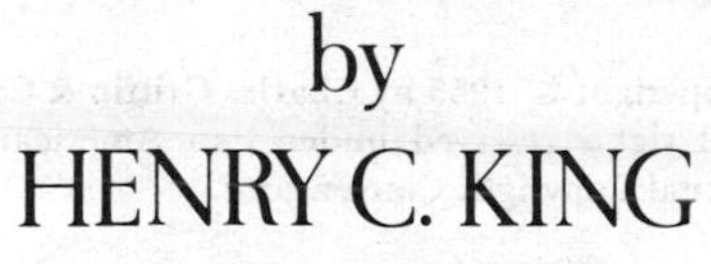

by
HENRY C. KING

Director, McLaughlin Planetarium,
Toronto, Ontario, Canada

With a Foreword by
SIR HAROLD SPENCER JONES
Astronomer Royal 1933–1955

Dover Publications, Inc.
New York

Copyright © 1955 by Charles Griffin & Company Ltd.
All rights reserved under Pan American and International Copyright Conventions.

Published in Canada by General Publishing Company, Ltd., 30 Lesmill Road, Don Mills, Toronto, Ontario.
Published in the United Kingdom by Constable and Company, Ltd., 10 Orange Street, London WC2H 7EG.

This Dover edition, first published in 1979, is an unabridged and unaltered republication of the work originally published in 1955 by Charles Griffin & Company Ltd. The Dover edition is published by special arrangement with Charles Griffin & Company Ltd., 5a Crendon St., High Wycombe, Bucks. HP13 6LE England.

International Standard Book Number: 0-486-23893-8
Library of Congress Catalog Card Number: 79-87811

Manufactured in the United States of America
Dover Publications, Inc.
180 Varick Street
New York, N.Y. 10014

TO

MARY

What crowd is this? what have we here!
we must not pass it by;
A Telescope upon its frame,
and pointed to the sky.

William Wordsworth

FOREWORD

By Sir Harold Spencer Jones, F.R.S., Astronomer Royal

For many years Dr H. C. King has been engaged in collecting material for a comprehensive history of the telescope and of its evolution from its invention to its present culmination in the great 200-inch Hale reflector on Palomar Mountain, California. This work, now completed, fills a notable gap in astronomical literature. Its compilation has entailed much research and the large number of references to original sources of information add greatly to its value.

There has been much controversy over the question of when and by whom the telescope was invented. As Dr King mentions, most historians have concluded that it was invented in Holland in the year 1608. Various claims for earlier invention have been made, amongst them a claim in favour of Giambattista della Porta about the year 1586. Dr King discounts the evidence in favour of Porta on the ground that he was an unreliable and imaginative writer; though there is truth in this, Porta, when writing in 1586 to the Cardinal d'Este to announce his *Magia Naturalis*, did state explicitly that he could make telescopes (*occhiali*) by means of which a man could be recognized at a distance of several miles.

Zacharias Jansen of Middelburg is amongst those for whom claims to the invention of the telescope have been made, which are discounted—and I think rightly—by Dr King. M. C. de Waard, the historian of the telescope, discovered a note written in 1634 by Isaac Beeckman, a friend of Descartes, in which he mentioned that he was told by Johannes Jansen (son of Zacharias) that his father had made his first telescope in the year 1604, after the model of an Italian telescope, on which was written *anno 1590*. The evidence seems to me to favour the conclusion that the first telescope was made in Italy *circa* 1590, and possibly by Porta. One or more of the early Italian telescopes probably found their way to Holland, where telescopes were made by more than one person, either as a result of seeing the Italian telescope or from hearing some description of it.

Whether this is so or not, by 1610 the invention was becoming widely known and its application to astronomical observations by Galileo, Simon Marius and others, soon bore rich fruit in the discovery of sunspots, of the four major satellites of Jupiter, of mountains on the moon, and of the phases of Venus. The changing position from day to day of spots on the sun proved that the sun was rotating, and the observation of the satellites of Jupiter showed that they were revolving around the parent planet.

The opponents of the Copernican system had found it convenient to assert that it provided a useful mathematical representation of the motions in the solar system but that the earth did not really revolve round the sun. The discoveries of Galileo provided the strongest evidence in favour of the truth of the views of Copernicus and did more than anything else to ensure their eventual acceptance.

To overcome the optical defects of the first telescopes, longer and longer telescopes were made. It has always seemed to me remarkable that these long telescopes, with focal lengths of up to 200 feet and with their unsteady methods of support, could be used satisfactorily in the open air, and even more remarkable that important discoveries were made with them. The observers must have been men of great patience and perseverance.

The invention by Isaac Newton of the reflecting telescope provided a form of instrument that was compact and free from the aberrations of the refracting telescope with a single-lens objective. The early mirrors were made of speculum metal and the casting, figuring, and polishing of these mirrors introduced many problems. It was not until after the invention of the achromatic lens about the middle of the eighteenth century that the refracting telescope came back into favour. Its development then depended upon improvements in the making of optical glass, with which the names of Guinand and of Fraunhofer are associated. Improvements in the form of mounting, and the introduction of a clock drive for turning the polar axis at the sidereal rate, resulted in greater convenience in observation. I have endeavoured without much success to discover who first employed a clock-driven equatorial for regular astronomical observation. This is a question which Dr King has not discussed.

The large reflectors of William Herschel and Lord Rosse, with speculum metal mirrors, were an interesting development in the attempt to secure greater light-gathering power. But the difficulties of casting large specula and the frequent repolishing that they required caused attention for a time to revert to the refracting telescope, culminating in the 40-inch refractor of the Yerkes Observatory.

The discovery by Liebig of a simple chemical process by which a thin, highly reflecting film of silver could be deposited on a glass surface, the application of this discovery to glass mirrors, and the methods of making and testing such mirrors, due primarily to Foucault in the sixties of last century, paved the way for the development of the modern large reflecting telescope.

In Dr King's book all these developments are described. His book is, however, much more than a mere history of the evolution of the modern telescope. Much information of interest is given about the many craftsmen and instrument-makers who were concerned in the successive developments, and about the associated advances in astronomy.

The book is one that I can heartily commend. I am sure that it will be warmly welcomed not only for its inherent interest but also as a source of information on all matters relating to the evolution of the telescope.

ROYAL GREENWICH OBSERVATORY
HERSTMONCEUX CASTLE
SUSSEX
January, 1955

PREFACE

So far as the author is aware, this is the first attempt to give the history of the telescope the prominence it deserves. It is true that, in most books on the history of astronomy, references are made to certain instruments important for their size or for the discoveries made with them. Robert Grant in the middle of the last century, and Agnes Clerke at its close, both adopted this treatment and included chapters in which they outlined the evolution of the telescope. Later historians, however, have repeated rather than extended this information. Authors of the few books dealing with the theory and construction of telescopes have sometimes spared a few pages to their history, which only the French authors A. Danjon and A. Couder have thought worth extending to three chapters. J. N. Lockyer's *Stargazing* gave a fairly consecutive history of the telescope, but this book was written nearly seventy years ago. In this century, G. E. Pendray popularized the telescope's history in his *Men, Mirrors, and Stars*, and D. O. Woodbury in 1940, and Helen Wright in 1953, told the story of the construction of the 200-inch Hale telescope. After we exhaust the information provided by these and a few other books, we embark on a seemingly endless quest in the large field of published and unpublished scientific literature.

When we think of the telescope we are at once reminded of its long and brilliant association with astronomy. But for this, our present knowledge of the heavens would be little in advance of that of the Renaissance astronomers, who thought that the solar system extended no further than Saturn's orbit and that stars just visible to the naked eye marked the outer limits of the sidereal universe. Modern astronomy, in fact, dates from 1610, when Galileo Galilei directed one of the first telescopes skywards and at once extended the frontiers of the observable domain nearly a hundredfold. Since then the process has continued, owing to the use of larger and more powerful instruments and to the rise of photography and spectrography in the last century.

The telescope and its equipment having thus dictated the course of observational astronomy, the reader must expect to find, in the following pages, repeated references to astronomers and astronomy. For this reason also, instrumental improvements receive more attention than the discoveries made through them. The description of Sir William Herschel's 20-foot reflector, for example, occupies more space than his theory of stellar distribution, and far greater importance is attached to the introduction of silvering processes than to the discovery of, say, the planet Neptune. For the same reason, instrument-makers and opticians take precedence over astronomers, except when the latter happened to assist directly the telescope's progress. Craftsmen like Graham, Bird, and the Dollonds were as important in their particular sphere as were Flamsteed, Halley, and Maskelyne in theirs. The fact that the former made telescopes and other instruments for a living, worked in shops like their lesser optical brethren, and were usually self-educated men of humble origin, was sufficient to subordinate them to professional astronomers, generally university-trained and having high social and scientific responsibilities. Yet the three craftsmen already mentioned, together with others like Ramsden, Fraunhofer, the Clarks, and Brashear, rose by their technical knowledge and practical skill to join ranks with the leading scientific men of their age.

Several hundred telescopes of all sizes, types, and degrees of usefulness are described in the text. The descriptions generally refer to the appearance of each instrument at the time of its construction or over a particular period in its career. These descriptions may no longer be true in some instances. Changes in an observatory's programme of work require changes in observational procedure. New items of equipment are added, a mounting is likely to require modification, a better telescope drive and control-system might be fitted, perhaps major optical alterations are undertaken. Hence some older telescopes have acquired new mountings or new optical elements, have had their optics refigured, or have been fitted with more modern attachments. Others have been dismantled, are now in museums, or cannot be traced. In some cases these changes are mentioned in the text and the known fate of an old and interesting telescope is recorded. For knowledge of the precise condition and whereabouts of others, further research is necessary—research which would have delayed publication by several years.

In its design, a modern astronomical telescope embodies the results of many years' research in glass-technology, optics, instrument-making, practical astronomy, and engineering. The history of the telescope consequently covers an extensive field, one made larger by the early application of optical sights to graduated instruments. Many modern observatory instruments which now incorporate a telescope are, in principle, of great antiquity. In view of this last fact, the author has thought it advisable to outline the methods and results of pre-telescopic observation. Indeed, it is not until we compare these pre-telescopic results with those obtained with the telescope that we begin to appreciate how greatly that instrument revolutionized human thought.

Nearly a hundred years ago the application of spectroscopes to the astronomical telescope rejuvenated physical astronomy. Chapter XIV outlines the early growth of modern astrophysics from the instrumental side; its incorporation paves the way for a better appreciation of modern techniques and achievements.

While the term *speculum* refers to a mirror of any suitable material, many writers prefer to call metal mirrors *specula* and reserve the term *mirrors* for silver-on-glass reflectors. The reader will find that we have used the words *speculum* and *mirror* indiscriminately in early chapters for a reflector of the well-known tin–copper alloy, speculum metal. The change from speculum metal to silver-on-glass mirrors took place in the fifties of the last century, after which time we meet with very few metal mirrors. When such appear, as in the 48-inch Melbourne telescope, the text will leave no doubt as to the type of reflecting surface employed. The words *object-glass* and *objective* are similarly interchanged; again, the text makes it clear as to whether the glass was a single lens, visual achromatic, photovisual or photographic. The time-honoured method of estimating object-glass quality by the ability to separate close double stars has been adhered to, for in many cases no other record is available. As the reader probably knows, the performance of a glass on such objects is governed largely by its aperture, as Dawes demonstrated about a century ago. Wherever possible, information concerning optical performance on more suitable test-objects is given, but here again it is difficult to say whether the tests were of the observer's visual acuity, of local atmospheric conditions (the 'seeing'), or of the quality of the telescope. Test results, therefore, are given for what they are worth and mainly for completeness of the record. When the test-objects are double stars, their separations at the time of observation are stated.

Of the many articles and books consulted in the preparation of this work, the author wishes to acknowledge his indebtedness to papers by D. Baxandall, Prof. Moritz von Rohr, R. S. Clay and T. H. Court which appeared in *The Transactions of the Optical Society*. From the study of these original sources came the incentive to create the more detailed picture now placed before the reader. J. L. E. Dreyer's now classic *Planetary Systems from Thales to Kepler* (1906) and *Tycho Brahe* (1890), also his *A Short Account of Sir William Herschel's Life and Work* (1912), have been freely consulted. The value of the Dollond narrative was enhanced when Mr H. W. Robinson, Librarian of the Royal Society Library, introduced the author to Ramsden's important letter. Full use of the contents of this letter has been made in the text. The account of Fraunhofer's work at Benediktbeuern rests on W. H. S. Chance's paper to the Physical Society on *The Optical Glassworks at Benediktbeuern*, while Mr E. Wilfred Taylor generously provided information about the activities of Cooke, Troughton, and the Simms family. In lieu of a comprehensive biography of Dr G. E. Hale, the author turned to the director's *Annual Reports of the Mount Wilson Observatory* and Hale's own *Study of Stellar Evolution*, one of the decennial publications of the University of Chicago. The account of twentieth-century American work was built up from articles in *Popular Astronomy* and *The Astrophysical Journal*. Of particular value were the astronomical summaries which appeared in the former journal during 1945 and 1946. Finally, the author's task was lightened considerably by the lists of telescopes published by P. Stroobant and astronomers of the Royal Observatory of Belgium (1931), A. Danjon and A. Couder (1935), and G. Z. Dimitroff and J. G. Baker 1946).

The *Philosophical Transactions of the Royal Society*, *Memoirs* and *Monthly Notices of the Royal Astronomical Society*, *Journal of the British Astronomical Association* and *The Observatory* have provided valuable material, and the author wishes to express his gratitude to the librarians of the Royal Society, Royal Astronomical Society, British Astronomical Association, and Science Museum for permission to consult these and other sources at his leisure. The monthly periodical *The Telescope*, published at Harvard Observatory until 1941, when it was merged with *The Sky* to form *Sky and Telescope*, together with *Nature* and the Scientific American's *Amateur Telescope Making* were invaluable in writing of the more recent development of the telescope. The accounts of Gregory, Short, Bird, Ramsden, and Tulley are developments of the author's articles on the history of optics published in the periods 1938 to 1939 and 1950 to 1953 in *The Optician*; so also is the story of the Dollonds which first appeared in 1938 in *The Dioptric Review*. He wishes to thank both the Hatton Press and the British Optical Association for permission to reproduce most of this published information.

Modern telescope-makers have generously provided material and illustrations. The author wishes to thank in this respect the Sir Howard Grubb, Parsons Company, the Warner and Swasey Company, the Perkin-Elmer Corporation, C. Épry and Jacquelin of Paris and the late J. W. Fecker. To Mr J. Evershed, F.R.S., and the late Mrs Evershed, also to Mr F. J. Hargreaves, the author owes a special debt of gratitude. During the difficult days of war they gave freely of their time to read and criticize the original layout and text. Dr W. H. Steavenson also assisted in the removal of obscurities, and his advice was always freely given. At a later date, and at the suggestion of Dr Roy K. Marshall, the manuscript was revised and enlarged and began to adopt its present form. In 1951 it was approved by the University of London for the award of the Ph.D. degree. Grateful acknowledgement is due to Mr N. J. Goodman, editor of the British Astronomical Association and formerly with Messrs Charles Griffin & Co. Ltd, for his earnest cooperation and invaluable advice during the printing stages; also to Mr and Mrs C. A. Federer and Dr J. Ashbrook, editors of *Sky and Telescope*, who read all proofs and suggested many considerable improvements in content and presentation. While every effort has been made to maintain a high degree of accuracy in the text, errors have doubtless occurred. The author will always be pleased to receive notices of any additions or corrections from his readers.

SLOUGH, BUCKINGHAMSHIRE — H.C.K.

June, 1955

CONTENTS

THE HISTORY OF THE TELESCOPE

CHAPTER I

It is necessary that the genuine astronomer should not, like Hesiod and others such as Hesiod, confine himself to a knowledge of the risings and settings of the constellations: he ought to be likewise acquainted with the circuits of the seven planets, and of the eighth celestial sphere.

PLATO

ASTRONOMICAL observation has its roots deep in remote antiquity. To primitive man, the apparent regularity underlying the celestial panorama was the only sure way of noting the passage of time. The growth of communal life, the cultivation of crops, the celebration of feasts and festivals, all necessitated the adoption of a reliable time-scale. We picture the Mesopotamian agriculturalist and shepherd of 3000 B.C. glancing at the position of the sun or the direction and length of his shadow to judge the time of day. We imagine the astronomer-priests of Babylon grouping the brighter stars into constellations, and recording on clay tablets the waxing and waning of the moon, the occurrence of eclipses, and the risings and settings of planets.

The Babylonians favoured the sexagesimal notation in their arithmetic and readily chose a year of 360 days. They counted thirty days from one new moon to the next. Twelve such lunar months gave a year. The sun, however, requires just over 365 days to circuit the heavens, so that, to correlate better the lunar and solar calendars, the Babylonians had occasionally to interpolate extra months.[1] The Egyptians adopted a year of 365 days and we have evidence of the extra five days as early as about 2500 B.C.[2] Even so, the relation was still imperfect, and later generations included a leap year once every fourth year.

Man passed from the observation of the changing shadows of buildings and pyramids to the construction of simple sun-dials at a very early date. The earliest known shadow-clock is an Egyptian one of the tenth to eighth century B.C., when it was probably in regular use.[3] The first sun-dials no doubt consisted of a vertical post surrounded by steps or graduations on stone. The Egyptians were content with temporal hours, that is, with hours of unequal length according to the season of the year. The length of an hour of the day thus differed from that of the night (except at the equinoxes), an arrangement good enough for regulating the civil life of the period. Water-clocks existed as early as 1400 B.C.[4] and were supplemented at night by sighting bright stars as they passed behind walls or buildings. An alternative method was to sight the stars with two plumb-lines suspended in the plane of the meridian, an arrangement known as the *merkhet*[5] (Fig. 1).

The Egyptians also constructed calendars, based on the risings and settings of certain bright stars,[6] that is, by counting the number of days between successive first appearances of the same star in the east just before sunrise, or successive last appearances in the west just after sunset. For this purpose they often used Sirius,

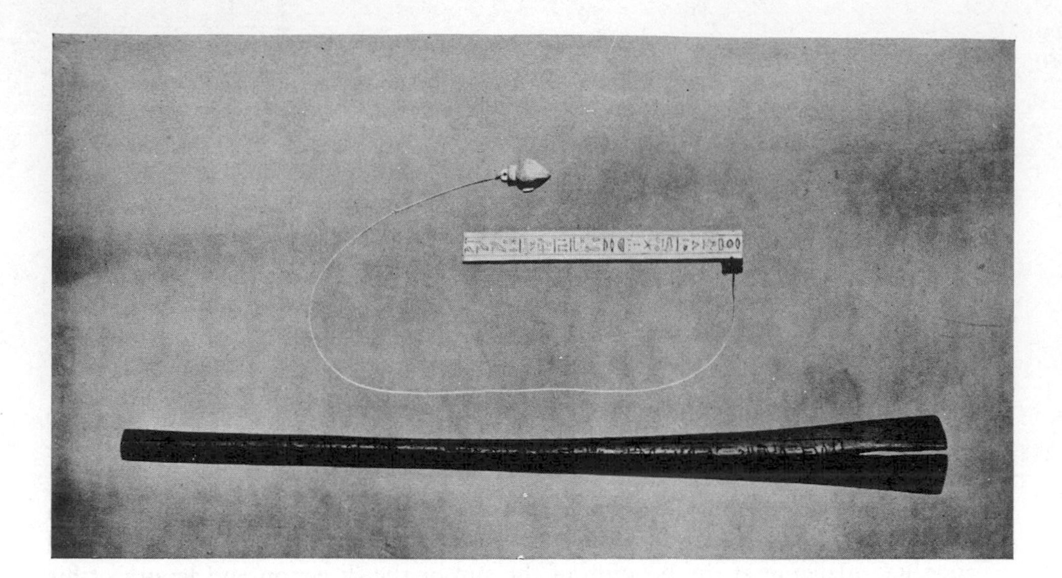

Fig. 1—The merkhet

An ancient Egyptian astronomical instrument

(Science Museum, London. British Crown copyright)

the brightest star in the sky, for it was visible on the eastern horizon at dawn when all others had paled into invisibility. The interval between two of these *heliacal* risings just before the sun was, consequently, another way of estimating the length of the solar year.

The Egyptians were astrologers rather than astronomers. If we marvel at their craftsmanship and organizing ability, we should remember that trial-and-error methods extended over many centuries do not constitute science as it is now understood. One looks in vain for the spirit of inquiry, the fertility of thought so evident in Greek times. Thousands of bodies were disembowelled prior to embalming, yet no science of anatomy resulted. Practical skill in glazes and glassmaking, in the mounting of real and imitation jewellery, and in the making and working of alloys, reached high levels, but no system of chemistry arose. Pyramids and temples with their obelisks dotted the Nile valley, but there was no engineering science. Lever, wedge, and inclined plane were employed in the erection of pyramids, as was also the brute force of countless gangs of slaves. The latter afforded good cover for the technical deficiencies of ' engineer ' and overseer.[7]

Both the Egyptian and Babylonian cosmologies were influenced by current religious cults. Both were fancifully childish. The Egyptian world took the shape of a rectangular box, with Egypt in the middle and the sky a flat or domed ceiling supported by columns or high mountains.[8] The Nile, believed to surround the then known world, continued its course in the heavens as a celestial river—the Milky Way. The stars were lamps hung from the sky by ropes and among them roamed numerous deities and spirits. The sun-god Ra travelled across the sky in a boat, accompanied by an adoring throng of the hosts of heaven. Reference

Fig. 2—The use of the merkhet

The observer S, with his assistant N, is laying down a meridian line by observation of the Pole Star. The observer N_1 is taking time observations by noting when certain stars pass above the eyes, elbows, etc., of his assistant S_1.

(By courtesy of the Director of the Science Museum, London)

is made in the Books of the Dead to a ladder which stretched from earth to heaven whereby spirits could climb to the abodes of the blessed.[9] The Egyptian religion, inconsistent to a degree, overloaded with a hierarchy of gods and dominated by the fear of eternal death, effectively stifled rational thought.

To recover observational astronomy we must turn to the fertile crescent of Mesopotamia. From the mounds where once stood Uruk, Borsippa, and Sippara, archaeologists have unearthed clay tablets which bear astronomical tables.[10] These tablets date to the last centuries of the pre-Christian era and mention the astronomers Kidinnu and Naburiannu. Rows of numbers witness to a good knowledge of the positions and periods of the sun, moon, and planets. The tables imply no conception of motion in circular orbits in space. They are purely numerical, and reflect a purely numerical outlook, yet they tell of the apparently erratic motion of the planets among the stars with their stationary points, advances, and retrogressions. The tables enshrine no speculative system, but bespeak careful observation.[11]

When we turn to the ancient Greeks, in particular to the Ionian School, we witness the results of imagination unsupported by observation. In seeking a rational description of phenomena, Thales of Miletus in the sixth century B.C. dispensed with the capricious gods and heroes of Homeric legend. He considered the earth to be a flat disk, a living organism, nourished and sustained by water as if by a seminal fluid.[12] To Thales, who probably had Nile floods and fertility in mind, everything sprang from water, the self-generating essence of the world. Pliny says that Thales was the first to investigate the cause of eclipses, while Herodotus says that he (Thales) predicted a solar eclipse, the occurrence of which, during a battle between the Lydians and Medes, gained him much reputation among his contemporaries.[13] This prediction, however, was at best a rough one. Thales gave as the limit of time the very year in which the eclipse took place, so that its occurrence at the time of a battle was a lucky coincidence. Thales had a rough knowledge of geometry. He was apparently aware, for instance, of the properties of similar triangles, and calculated the heights of mountains by measuring the lengths of their shadows and comparing these with the shadow cast by a vertical stick of known height.[14] He is also said to have ' marked ' the stars of the ' Little Bear ' and, like the Phoenician navigators, used it to find the pole.[15]

Thales' disciple, Anaximander, imagined the earth to be a short cylinder, placed at the centre of the world, and without tendency to fall in any particular direction. Through the spherical heavens ran tubes of fire. Holes in the tubes allowed the fire to escape and gave rise to the heavenly bodies. The moon had an orifice of variable aperture while solar and lunar eclipses were caused by temporary obstructions in the tubes.[16] Anaximenes, another pre-Socratic philosopher, held that the earth floats on air, and appears to have taught that the stars are fixed, stud-like, to a rotating solid sphere.[17]

Another school of thought, established by Pythagoras in the sixth century B.C. in Southern Italy, made the world a product of harmonious relationships. Numbers took the place of the atoms—introduced by Democritus and Epicurus—and cosmology became infused with mathematics. Speculation again ran riot. Philolaus, a Pythagorean and a contemporary of Socrates, imagined that the spherical earth moved once in twenty-four hours in an orbit around a ' central fire '. Sun, moon, and planets also rotated about this ' fire ' which was invisible since it remained on the uninhabited side of the earth.[18] To bring the number of bodies encircling this central fire to the mystical number ten, Philolaus postulated the existence of a ' counter-earth ' which kept pace with the earth and so remained invisible.[19] This strange system did at least account for the daily motion of the sun and stars, and provided an all-important geometrical harmony so cherished by the Greeks. The anthropocentric step of shifting the earth to the centre of the system or, the same thing, of placing the ' central fire ' within the earth, was taken by Heraclides of Pontus in the fourth century B.C.[20] By giving the spherical earth a rotation about its axis once in twenty-four hours Heraclides, and apparently the Pythagoreans Hicetus[21] and Ecphantus[22], gave the system greater geometrical simplicity. Heraclides went further. He maintained that, while the sun revolved about the earth in its own orbit, Mercury and Venus both revolved about the sun.[23] This distinct advance is mentioned by several ancient writers and was known to Copernicus in the fifteenth century A.D., but it never became popular with the Greeks.

In the fifth and fourth centuries B.C., Plato and Aristotle were content with a spherical earth, mainly because they considered the sphere the most perfect of geometrical solids. The earth was in the centre of the world because this was its natural place; it did not rotate because rotation was a property of the far more perfect celestial sphere. Plato favoured a finite, spherical universe, with the outer firmament rotating once in twenty-four hours and the planets, each a divine luminary, rotating in the opposite direction in circular orbits, in order from the earth—Moon, Sun, Venus, Mercury, Mars, Jupiter, and Saturn.[24] Plato was aware that this animistic and over-simplified system did not account for the erratic motions of the planets as recorded by the Babylonians. An attempt to remedy this defect was made by Eudoxus who evolved a remarkable system of twenty-seven geocentric spheres. Four spheres were assigned to each planet. These rotated in such periods, on axes suitably inclined, that the combined movements produced in the planet a motion which corresponded to that observed.[25] The system was a masterpiece of applied geometry, although Eudoxus recognized that it was a purely mathematical device.[26] Not so Aristotle. He pictured physical crystalline spheres, out in space, to which he attached spherical stars and planets.[27] The outermost sphere of fixed stars, or the *Primum Mobile*, rotated at such speed that it imparted its motion to all the planetary spheres which it enclosed. This meant that twenty-six of the Eudoxan spheres shared in this motion, whereas Eudoxan theory required only the outer planetary spheres to be so affected. Aristotle, perforce, added a set of unrolling or neutralizing spheres and ended with a system of fifty-five physical homocentric spheres.[28]

Aristotle pictured an eternal, spherical universe beyond which there was no void, no time, and no place. The tiny earth stood motionless at the centre of a large company of transparent spheres beyond which was the divine and unchanging sphere of the fixed stars, directly motivated by God. Like Plato and the Pythagoreans, Aristotle postulated a spherical earth, but gave good practical reasons for adopting this shape. Travellers had noticed that the altitudes of southern stars decreased as they travelled northwards. Stars visible from Egypt were invisible from Asia Minor. Aristotle says that this fact, together with the circular outline of the earth's shadow on the moon during an eclipse of the latter, suggests that the earth is both spherical and a mere speck compared with the vault of heaven.[29] This implies, of course, the material insignificance of man. Above the earth's surface are the layers of the last three of the four elements, water, air, and fire—each in its proper and natural place. Beneath the moon's orbit all is change and decay, fit stage for such transitory phenomena as comets and meteors. Above the moon is the pure, eternal and unchanging realm of *aether*, wherein the planets, and the spheres in which they are embedded, strive to acquire, but never reach, the perfect circular motion of the *Primum Mobile*.

Aristotle placed little reliance on observation and, from insecure and often false premises, swept on magnificently to great generalizations. Yet his discoveries in biology, based on direct personal observation, show that he was no pure speculator like Plato. It is unfortunate that he overlooked the observational basis of astronomy, the more readily to pursue deductive reasoning in cosmology. Aristotelian philosophy and cosmology dominated men's minds until the seventeenth century, a domination made stronger by their earlier infusion into inflexible Christian creeds. Little wonder, therefore, that when Renaissance devotees of experimental science met such combined opposition at the hands of the Church, Aristotle was soundly abused as the cause of all the trouble.

We now leave Athens for Alexandria, where, about 300 B.C., Ptolemy Soter founded a university and library. Here, Greek philosophy took on a more practical aspect and gave birth to Greek astronomy as a separate science. Here, Aristillus and Timocharis[30] observed and recorded the positions of the brighter stars; here, Eratosthenes and Hipparchus founded observational astronomy, Herophilus and Erasistratus made dissections and founded anatomy. Here, such great mathematicians as Euclid, Archimedes, and Apollonius gave a crowning glory to Greek geometry. Here, the practical arts and crafts of Egypt combined with Greek metaphysics to produce *chemia*, the beginning of alchemy.

To Aristarchus of Samos in the third century B.C. we owe propositions on the measurement of the sizes and distances of the sun and moon.[31] These are sound in principle, but require observations made to a degree of accuracy unattainable in those days. Aristarchus considered the sun–moon–earth triangle when the moon was at the half phase and, by observation, found that the angular distance between the sun and the moon was then 29/30ths of a right angle. He deduced from this, by geometrical considerations, that the sun is nineteen times more distant than the moon, an estimate over twenty times too small. According to Archimedes, Aristarchus believed that the earth and planets move in orbits about the sun, that is, he entertained the heliocentric theory which Copernicus reformulated in the sixteenth century.[32]

An important deviation from the usual run of sun-dials, gnomons, and clepsydrae was the *staff of Archimedes*, the earliest known Greek sighting instrument. It was used in the third century B.C. and may have been in use long before.[33] It consisted of a small, flat disk attached to a wooden staff, to one end of which the observer applied his eye while moving the disk until it appeared to cover the sun—an observation made, no doubt, when the sun was thinly veiled by clouds. He then noted on a scale the distance of the disk from his eye, deriving, in this case, a measure of the sun's apparent diameter.

Eratosthenes, friend of Archimedes and custodian of the great library at Alexandria, is remembered for his attempt to measure the size of the earth. He noticed that, at the summer solstice, the sun cast no shadows at Syene (Aswan), but that vertical rods at Alexandria cast short shadows.[34] On the assumption that the sun is at a distance great compared with the earth's radius, Eratosthenes measured, in effect, the zenith distance of the sun at Alexandria; this came out at 1/50th of a great circle. Knowing the distance between Alexandria and Syene by direct pacing, he calculated the circumference of the earth to be 250,000 stadia—a figure which he seems later to have corrected to 252,000.[35] Our assessment of the accuracy of this result depends on the value we ascribe to the stadium; if we take the Egyptian value, the earth's radius, according to Eratosthenes, is 3925 miles—a good result.

The last and greatest astronomer of the pre-Christian era was Hipparchus, of the second century B.C. He spent some years at Alexandria, but observed mostly from Bithynia in Asia Minor. He was a careful and exact astronomer and used graduated measuring instruments. He refers to two equatorial armillaries, standing in the square portico of the museum at Alexandria, which Eratosthenes may have used.[36] Each consisted of a bronze ring, mounted in the plane of the celestial equator, and was used for determining the times of the equinoxes. On these dates, the shadow of the upper half of the ring fell upon the inner surface of the lower half.

Hipparchus was familiar with a simple form of armillary sphere, an instrument which perhaps Eratosthenes had introduced. By taking the altitude of the sun at the summer and winter solstices, the times when it is farthest north and south of the celestial equator, Eratosthenes was able to fix the position of the latter and to measure the tilt of the earth's axis—the obliquity of the ecliptic. For this observation he may have used a gnomon, but it is more likely that he used a solstitial armillary.[37] The latter took the form of two concentric rings fixed in the plane of the meridian, one of which revolved inside the other. The external ring was divided into degrees; the inner ring carried two diametrically opposed stops. When the sun crossed the meridian, the inner ring was moved until the shadow of the upper stop fell upon the lower. An index, referred to the outer scale, indicated the meridian altitude of the sun.

Hipparchus' observations of planetary positions with simple armillaries[38] revealed the shortcomings of the Eudoxan theory. He adopted, instead, a system of epicycles, first suggested by Apollonius, the Alexandrian mathematician. In this system,[39] each planet moves on a circle (the epicycle), the centre of which describes a circular path (the deferent) round the earth. A combination of suitable motions accounted for the stationary points and retrogressions of the planets, but only up to a point—Hipparchus predicted that even this system did not quite fit the phenomena. He noticed the irregular motion of the sun and determined the (unequal) lengths of the seasons, which he ascribed to the sun's eccentric orbit about the earth.[40] He thus took the bold step of placing the earth slightly out of centre of the sun's orbit, but conceded that this was a purely geometrical expedient.

Hipparchus was the first observational astronomer to display the modern spirit. To increase the accuracy of his observations he constructed graduated instruments; to discuss his results the better he invented spherical trigonometry;[41] to account for phenomena he indulged in no sweeping physical generalizations. Convinced of the essential orderliness of nature, he sought always the most geometrically simple hypothesis compatible with observation. He investigated the major inequalities of the moon's motion, measured the moon's parallax, discovered the precession of the equinoxes, and recorded the positions and brightnesses of many stars. The practical nature of this work contrasts with the speculative philosophy of the Athenian school of only 150 years previously, and certainly reaches more fruitful levels.

The observational work of Hipparchus was continued and re-discussed about A.D. 150 by Ptolemy, who was the last great astronomer of antiquity. Ptolemy drew freely from the observational data and geometrical ideas of Hipparchus, to which he added his own. He accepted the epicyclic theory, investigated afresh its more outstanding deficiencies, and made it the framework of a masterly geometrical representation of the cosmos. He retained the idea of eccentric circles by making the deferents eccentric to the earth, and he introduced a new concept, the *punctum equans*, or point outside the centre of the deferent, with regard to which the angular motion was uniform.[42] He thus violated the cherished notion of uniform motion about the physical earth, but saved the situation mathematically. The earth was at the centre of the universe, and did not rotate. If it rotated, he argued, objects on its surface would be hurled away like mud from a spinning wheel. In general, however, he was little interested in physical theory—in the 'why' of phenomena. He considered his scheme no more than a clever geometrical hypothesis, the best possible way of describing and predicting the movements of the sun, moon, and planets. To Ptolemy, the goal of geometry was the fitting of

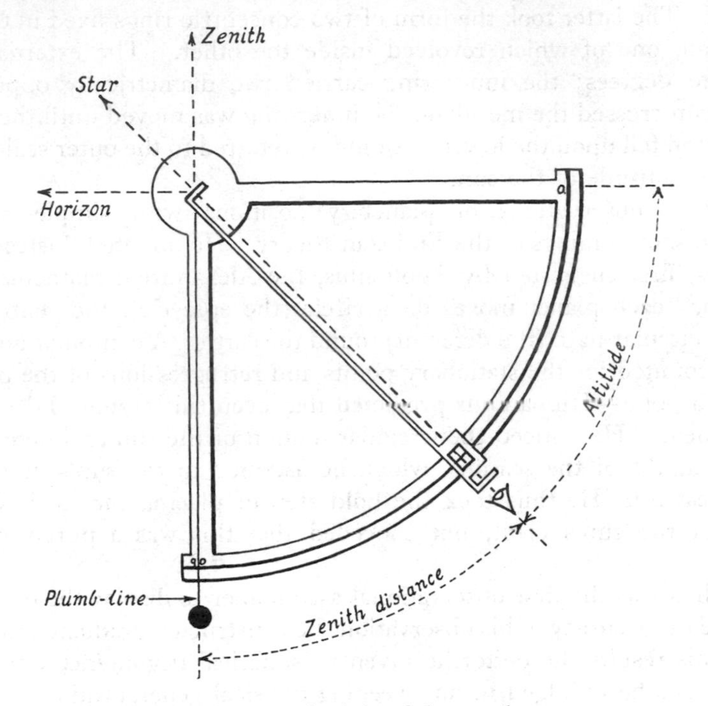

Fig. 3—The principle of the quadrant

uniform and circular motion to the cosmos, so he was justly proud of his achievement.[43]

Ptolemy suggested that quadrants (Fig. 3) should replace complete circles on the plea that an arc could be graduated into smaller and more legible divisions. To prove this, he designed a quadrant[44] with which to measure the meridian altitude of the sun and which, once set in the meridian, could be adjusted for verticality by means of a plumb-line. Ptolemy made no mention of a proposed method for setting the instrument in the meridian, neither does he appear to have made a quadrant of this type.[45] The Arabians, however, adopted it, and their successors so altered and improved its design that, to the end of the eighteenth century, it was a standard observatory instrument.

Ptolemy also mentioned another instrument, the *triquetrum*, often called *Ptolemy's rules* (Fig. 4), designed to overcome the difficulty of graduating arcs and circles.[46] It consisted of a vertical post to which were hinged two rods or arms, the distance between the hinges being graduated and made equal to the length of the upper arm. The latter carried sights, and with the post formed two sides of an isosceles triangle. The third side, the base, was also graduated, and its reading, referred to a table of chords, gave the apical angle or zenith distance of the object under observation.

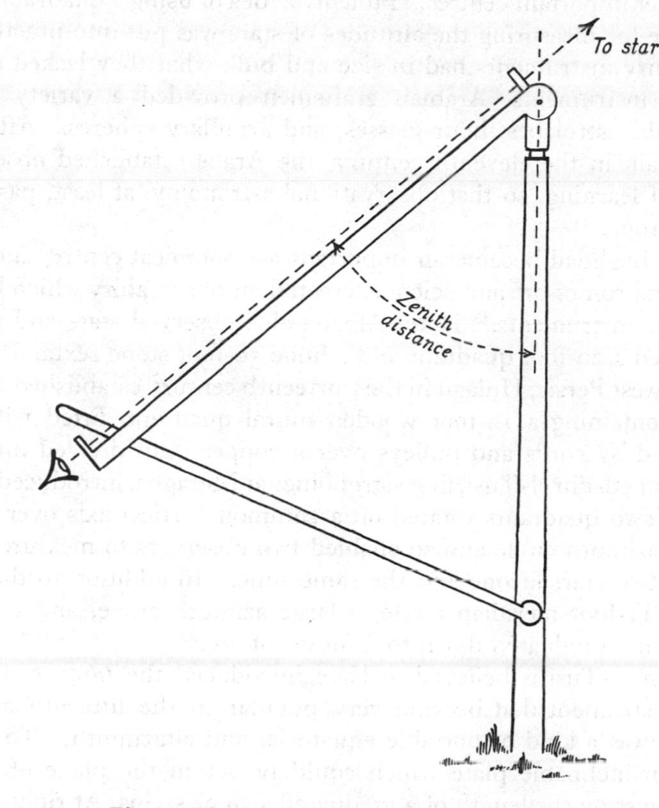

Fig. 4—The triquetrum or 'Ptolemy's rules'

Ptolemy was apparently the first to map, by latitude and longitude, important towns around the Mediterranean,[47] but his figures are far from accurate owing to the absence of reliable astronomical data. For only a few places had latitudes been obtained by means of vertical gnomons; for the rest he relied on the statements of travellers and mariners. Owing to this difficulty, and to the fact that his work was not translated and made known generally until the fifteenth century, Ptolemy's latitude–longitude system remained in obscurity for over a thousand years.

With the sack of Alexandria in A.D. 389, part of the university library and museum was destroyed. Many Greek philosophers fled to escape Christian persecution, and the scientific tradition vanished from Alexandria. For nearly three centuries the shadows of the old school lurked in the city of unrest. Then, in A.D. 632, the followers of Mahomet began the greatest crusade in history. Within ten years they had overrun Egypt, Palestine, and the East. Alexandria fell in 641. The museum was destroyed, its library burnt; most of the Greek philosophers there fled to Constantinople.

The Arabs encouraged learning. They translated Greek scientific works and gave special attention to Ptolemy's *Mathematical Symposium* or *Almagest*, as his great work was called. Observatories were established at Baghdad, Cairo, Damascus, and at other important centres. Ptolemy's idea of using a quadrant instead of a complete circle for measuring the altitudes of stars was put into practice, and impressive masonry instruments had in size and bulk what they lacked in accuracy. Besides fixed instruments, Arabian craftsmen provided a variety of portable quadrants, dials, astrolabes, hour-glasses, and armillary spheres. After they had conquered Spain in the eleventh century, the Arabs established observatories at new centres of learning, so that observational astronomy, at least, passed without break into Europe.

The city of Baghdad became an important astronomical centre, and here Kalif al Mumun, a patron of art and science, erected an observatory which he furnished with Ptolemaic instruments.[48] Here, Albategnius observed stars and planets, and Abul Wefa used a 20-foot quadrant and a huge 56-foot stone sextant.[49] At Meragha, in north-west Persia, Hulagu in the thirteenth century established an ambitious observatory containing a 12-foot wooden mural quadrant, fitted with a sighted alidade* moved by cords and pulleys over a copper limb divided into degrees.[50] Here also, Nasir ed-din el-Tusi, first astronomer at Meragha, introduced the azimuth quadrant.[51] Two quadrants rotated on a common vertical axis over a graduated horizontal or azimuth circle and so enabled two observers to measure the angular separation of two stars at one and the same time. In addition to the quadrants, there was an 11-foot meridian circle, a large azimuth circle, and armillae fitted with alidades and graduated down to minutes of arc.[52]

Nasir ed-din el-Tusi is believed to have introduced the *turquet* or *torquetum*[53] (Fig. 5), an instrument that became very popular in the fifteenth and sixteenth centuries. It was a kind of portable equatorial and altazimuth. To a base plate was hinged an inclinable plate which could be set in the plane of the celestial equator by adjusting the length of a graduated arm or stylus. At right angles to the inclinable plate was a polar axis carrying two circles. A movable alidade indicated declinations on the upper circle, while the equatorial circle, in the plane of the inclinable plate, indicated right ascensions. Regiomontanus ascribes this instrument to Geber[54] (twelfth century), whose theodolite was of similar construction, but with a vertical moving alidade instead of a full vertical circle and alidade. In any case, the torquetum appears to have been in regular use in the fourteenth and fifteenth centuries.

The last Arabian observatory of importance was that which Ulugh Beigh established at Samarkand in the fifteenth century. Here he erected large masonry

* alidade—an Arabic name for an index or ruler which is movable about the centre of a graduated arc or circle, carries the sights, and indicates the number of degrees marked on the limb of the instrument.

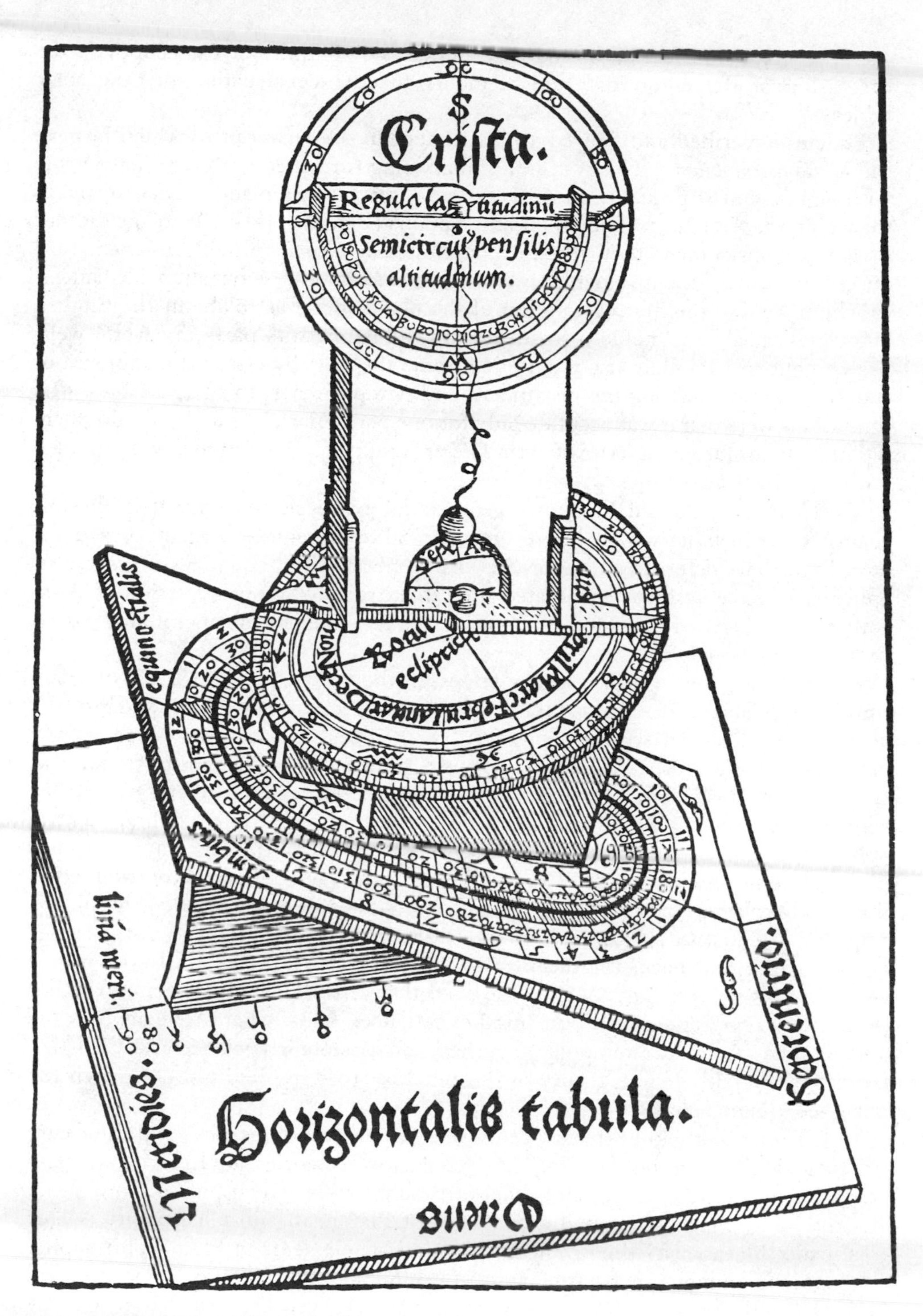

Fig. 5—Regiomontanus' torquetum, ***circa*** **1470**

(*Science Museum, London. British Crown copyright*)

quadrants, one 180 feet high, and determined the obliquity of the ecliptic,[55] the precession of the equinoxes, and elements for the construction of new solar tables.[56]

Ptolemy described and figured an instrument in the *Almagest* to which he gave the name *astrolabium*. This was an armillary ring furnished with a diametric rule with sights, and it probably formed the basis for the flat planispheric astrolabes for which the Arabian craftsmen became famous (Fig. 6). If the Arabians did not invent this instrument they certainly developed it into a veritable mathematical jewel—an instrument invaluable in surveying and for safe navigation. Chaucer, to whom we owe the first description of the planispheric astrolabe in the English language, called it a ' noble instrument ' and illustrated its parts in several well-drawn figures.[57] Its main use was to determine the time by observations of sun or stars merely by adjusting the instrument and without resort to special tables. Its limitations were that it was accurate only for one particular place and for one epoch. To remedy the former, a series of extra *tablets* for appropriate latitudes was provided with each instrument.

The Arabians adopted Ptolemy's geometrical synthesis but gave it a physical reality for which it was ill fitted—geometrical representation again passed for actuality. The deferent and epicyclic circles became the equators of crystalline spheres, and the system degenerated into an extremely complex version of Aristotle's cosmology. In this form, alongside the Aristotelian scheme, it was passed on to the western world.

Spain during the twelfth century witnessed the translation into Latin of many important Arabic texts. Among these were Arabic versions of Aristotle and Ptolemy and the cosmological speculations of Alfraganus and Albategnius based on the *Almagest*. Here arose a school of Moslem philosophers who took Aristotle as the infallible master of all scientific knowledge. Averroes and Alpetragius of this school rejected the epicycles of Ptolemy and re-introduced a system of homocentric spheres.[58] But it was not easy to dispense with Ptolemy in this way. Given a physical interpretation, his system and the mechanics of so many eccentric revolving crystal spheres became incredibly complex. The fact that homocentric spheres required a planet to remain at the same distance from the earth at all times was alone a great drawback, for the planets' apparent brightnesses are by no means constant. There was no denying the fact that the Ptolemaic system, whether material or representational, best fitted experience. Thus, when Albertus Magnus and Thomas Aquinas attempted a synthesis of Aristotelian science and Christian theology, the fairly logical theory of the universe that resulted did not match the observed phenomena.

In the fourteenth and fifteenth centuries, the pursuit of learning and the desire to reconcile the rival theories led many scholars to seek original Greek texts. Plato was studied direct, and another school of thought arose which placed Plato above Aristotle and which attempted a synthesis of Platonism and Christianity. This was made the easier by the mystical and animistic elements in Plato's philosophy. Pythagorean number mysticism, animistic motions, and belief in a geometrical harmony and unity underlying phenomena characterized the neo-Platonists. By the sixteenth century there was a distinct clash between the conflicting cosmologies

Fig 6—Copy of an Arabian astrolabe, A.D. 1067

(*Science Museum, London. British Crown copyright*)

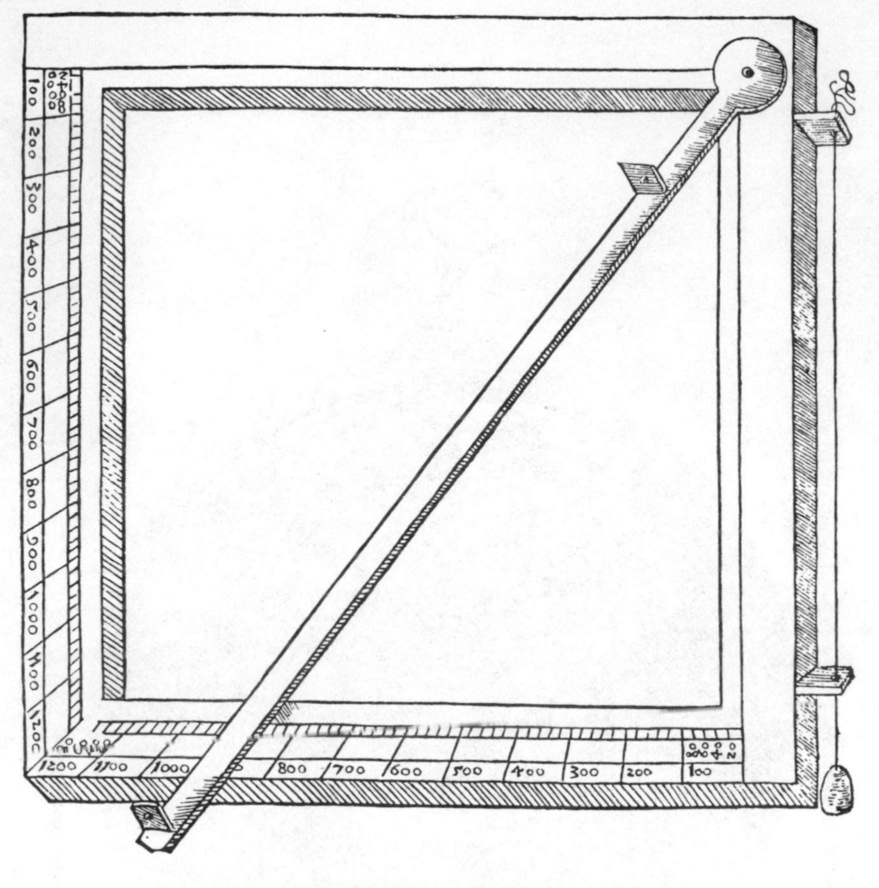

Fig. 7—Purbach's geometrical quadrant

(*Science Museum, London. British Crown copyright*)

of Aristotle and Plato.[59] In the struggle, Ptolemy remained supreme as the only writer whose system (for those times) represented the observed facts.

In the middle of the thirteenth century, King Alphonso of Castille supervised the computation of tables based on Ptolemy's theory of planetary motion and the observations of the Arabs. These *Alphonsine Tables* form the first state-sponsored astronomical ephemerides published for general use. In the fifteenth century, Purbach and his pupil Johannes Müller, better known as Regiomontanus, found with their instruments that the places derived from these tables differed considerably from the actual positions in the sky. Regiomontanus later settled in Nürnberg, where he erected an observatory and, in collaboration with a wealthy citizen named Walther, compiled improved ephemerides and nautical almanacs.[60] For their observations they used the best instruments that the Nürnberg craftsmen could make, including some of the first weight-driven clocks. According to Regiomontanus, there were five main types of instrument—the cross-staff and Jacob's staff (modifications of Ptolemy's triquetrum), Purbach's geometrical square (based on the quadrant—see Fig. 7), the regula (for taking altitudes of the sun and moon) and the torquetum.[61] Yet, with all these instruments, he could only estimate that the comet of 1472 could not have a parallax greater than 6 degrees.[62]

During the Middle Ages, when the union between Christianity and Aristotelianism was complete, freethinkers like Roger Bacon, Leonardo da Vinci, and Nicolas of Cusa, entertained neo-Platonic ideas. In the sixteenth-century revival of neo-Platonism due to the direct study of Plato's works, one of those caught up in the rising school of thought was Novara, professor of mathematics at Bologna. To Novara, the cumbrous Ptolemaic system violated the notion of a harmoniously simple universe. We imagine that he told his friend and pupil, Nicholas Copernicus, as much. Yet, prior to his visit to Italy in 1496, Copernicus had already embraced neo-Platonic opinions. When he returned to his native Poland he was all the more convinced that Ptolemy's solution was not the best one.[63]

To recognize the moulding influence of neo-Platonism on the ideas of Copernicus is to appreciate that he was not particularly interested in the truth or falsity of the earth's motion. His aim was, rather, to place the earth and to move it so as best to describe the observed apparent motions of sun, moon, and planets.[64] To Copernicus, the universe was fundamentally geometrical and essentially simple. Its nature was such that it could be represented in a way far simpler than that detailed in the *Almagest*. The circle was both simple and perfect; circular orbits in space must come into any good scheme. This is pure Platonism—to admit it is to reject the rank materialism of Aristotelian cosmology. Copernicus took this step in 1506 when, in letters to friends, he outlined his new theory. By 1540, this outline was fairly well known and, in *De Revolutionibus*, published in 1543 and dedicated to Pope Paul III, the outline blossomed into maturity.

Copernicus believed in the motion of a physical earth round a physical sun. Osiander, who saw the work through the printing, thought fit, in view of so revolutionary an opinion, to add a preface wherein he emphasized the arbitrary nature of heliocentric motions.[65] Copernicus acknowledged that Philolaus had considered the earth's motion about some centre, as also had Hicetus and Ecphantus, so that there was every justification why he should entertain the idea.[66] Considerably encouraged and comforted by this reflection, Copernicus now asked where he could best place the sun. Straightway he assumed its immobility and transferred its annual motion to the earth.

> In this most beautiful temple [he writes][67] could we place this luminary in any better position from which he can illuminate the whole at once? . . . So the Sun sits as on a royal throne ruling his children the planets which circle round him.

These and similar arguments form but a small part of *De Revolutionibus*. The rest of the work reflects ancient thought—Ptolemy reorientated. Copernicus added the results of his own 8-foot triquetrum observations to those of Ptolemy,[68] but overlooked the importance of accurate observations over long periods of time. So great was his belief in and respect for the observations of the ancients that he even used Ptolemy's value of 3 minutes of arc for the sun's horizontal parallax,[69] a figure over twenty times too great. Yet the observations he did possess required that the sun be shifted from the centre of the system and that its distinguished

' royal throne ' be replaced by a point.[70] He clung also to the old idea that the earth is fixed to a revolving radius, and had to introduce a special third motion—an annual conical motion of the earth's axis.[71] He rejected Ptolemy's *punctum equans*, but added so many epicycles that his exposition reads more like a copy of Ptolemy than something new. To save the phenomena, his system required thirty-four epicycles,[72] less than half the number required by Ptolemy's successors.

For a hundred years after the death of Regiomontanus, in 1476, little advance was made in knowledge of the positions of the fixed stars. Astronomical instruments were crude, but not incapable of showing that planetary tables could not predict positions with reasonable accuracy. Their unsuitability for precise measurement was abundantly evident when, in 1572, a ' new ' star blazed out in Cassiopeia and astronomers tried to measure its parallax. Schuler at Wittenberg, using an old wooden quadrant, at first found a displacement of 19 minutes of arc, but subsequent measures with a new and large triquetrum revealed no sensible parallax. Hagecius, physician to the Emperor of Bohemia, obtained values of 7′ to 16′;[73] Camerarius at Frankfurt on Oder found a parallax of 12′ which he later changed to $4\frac{1}{2}$′, and Nolthius, a German writer, obtained the absurd result of 39′.[74] Tycho Brahe (Fig. 8) measured the star's position with a cross-staff and then obtained a different result with a sextant. Only Mästlin obtained consistent results, and he dispensed with instruments altogether. He picked out stars in line with the nova and, with the assistance of a piece of thread, used them as reference stations for checking its position.[75] He concluded that the nova had no parallax and that it was situated among the fixed stars whose distance, Copernicus had suggested, was very great.

To Tycho Brahe, these discordant observations provided further evidence of the backward state of astronomical observation. He became interested in astronomy after observing a conjunction of Jupiter and Saturn, the cause, he thought, of the great plague in 1563.[76] He found that planetary tables were in error by several days, and there and then decided to make their improvement his prime concern. The opportunity came in 1576, when the King of Denmark, Frederick II, granted him the island of Hveen (between Copenhagen and Elsinore) and sufficient money to build and maintain an observatory. Tycho called the observatory *Uraniborg* (Fig. 9). In its scope and plan it was similar to a modern establishment, with its own garden, living-rooms, printing-presses, workshops, library, and four different buildings filled with a varied array of instruments.

Tycho saw that one way of improving observations would be to construct instruments of great size, a plan which he put into practice when he was only twenty-two by designing a 19-foot quadrant for his friend, Paul Hainzel.[77] This great instrument, constructed by the best craftsmen in Augsburg, was so heavy that it took twenty men to erect it on a hill in Hainzel's estate at Göggingen. The massive body was made almost entirely of oak beams, with the exception of iron braces and a brass strip along the limb divided into 5400 minute spaces. The entire quadrant rotated on a large oak pillar set in masonry and was made accessible for low-altitude stars by steps. When not in use, it was covered by a roof, but this and its massive proportions did not save it from destruction during a violent storm some five years later.[78]

Fig. 8—Tycho Brahe

(By courtesy of the Director of the Science Museum, London)

When Tycho planned Uraniborg he dismissed the idea of using the clumsy and heavy torqueta of his predecessors in favour of sextants, quadrants, and various armillae in iron and brass. These and many others he described in his *Astronomiae Instauratae Mechanica*, published in 1598 and now very rare. His instruments were nearly all constructed on a large scale and were open to many sources of error, but we are bound to marvel at the results that he achieved with them. They were decidedly superior to any of the Arabian instruments, with the result

Fig. 9—Tycho Brahe's observatory, Uraniborg

(By courtesy of the Director of the Science Museum, London)

that Tycho was able to reduce the likely error in the measurement of the separation of two stars from $\pm$ 10 minutes to less than $\pm$ 1 minute of arc.[79]

Many of Tycho's instruments were divided by transversals, a method introduced by Homelius, one of the professors at Leipzig University. Homelius, however, was not the real author, and the idea may have originated with Purbach or Regiomontanus, or even with Richard Chanzler, an English instrument-maker who, by 1573, had applied transversals to cross-staff divisions.[80] Whoever the inventor, Tycho was the first to apply transversals to large instruments and the first to devise a way of averaging out their errors.

The subdivision of small spaces by transversals was a simple operation. The arc was divided into minutes or similar intervals by the method of continual bisection and, between these, were placed ten subdivisions along the diagonal drawn from the inner end of one minute mark to the outer end of the next (Fig. 10). The diagonal scale was easier to make than equal arc subdivisions, while the small 10-minute spaces were large enough to be fairly easily read.

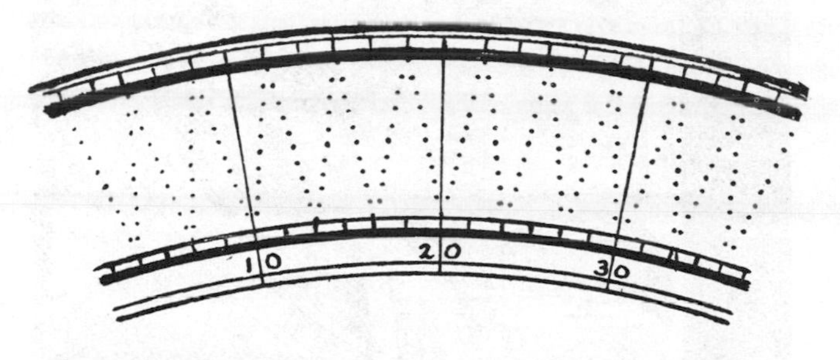

Fig. 10—Portion of an arc of small radius divided by the method of transversals

It will be seen that this method is the forerunner of the modern diagonal scale

The largest and most impressive instrument at Uraniborg was the great mural quadrant in the main building (Fig. 11). This consisted of a brass arc, 5 inches broad and nearly 7 feet in radius, fastened to a meridian wall and divided by transversals. Two sliding pointers on the arc indicated meridian altitudes and enabled two observers to sight two stars at the same time through a hollow brass cylinder fixed at the centre of curvature.[81]

Another quadrant, of only 16 inches radius, was divided according to the plan of Pedro Nunez, professor of mathematics at Coimbra and contemporaneous with Tycho.[82] Within the external arc, divided into 90 degrees, were forty-four concentric arcs divided into 89, 88, 87, etc., equal divisions, down to the innermost with only 46. When the edge of the index or alidade rested between any two particular half-degree divisions on the limb, it also ran over a division on one of the concentric arcs and so indicated what fractional part of 30 minutes ought to be added to the limb reading. This ingenious method had one big disadvantage—the difficulty of dividing a 90° arc into an odd number of divisions like eighty-nine. So far as we know, Tycho had only one instrument so graduated.

Tycho used his large quadrant for determining time and for checking the rates of his clocks. Conversely, if he knew the right ascension of any one star and could rely on his clocks, he could determine the right ascensions of others by noting the times of their upper transits over the meridian. Unfortunately, the clocks of his day were unreliable and an error of only 4 minutes of time introduced an error in space of a whole degree. No clock at that time could be relied on to less than 15 minutes in a day, and an error of an hour was not uncommon.

Fig. 11—Tycho Brahe's mural quadrant

Small wonder that Tycho's clocks had only one hand, the hour hand! Minute hands were not added to clocks until after the middle of the seventeenth century. Yet, had the clocks had a constant rate, Tycho could have applied a correction when he reduced the observations; but even the rate varied over short periods owing to temperature changes, air currents, and the varying position of the driving weight. The escapement, the old weighted *foliot* type, was unreliable and, in those days, it was impossible to make teeth so as to give the wheelwork a uniform motion. Tycho's friend, the Landgrave of Hesse-Cassel, relied on clocks when he determined the right ascension of the 1572 nova and other bright stars, with the result that his observations were in error, in some instances, by as much as 2 degrees.[83]

In the absence of a good timekeeper, Tycho relied on altazimuth quadrants and equatorial armillae for the determination of hour angles. Hipparchus had used the moon as an intermediate link for deducing the longitude of stars from that of the sun. Tycho used Venus, with her smaller diameter and parallax, and thence found the position of α Arietis, a bright star near the vernal equinox. He then proceeded to measure the distance of other stars from α Arietis until he had eight

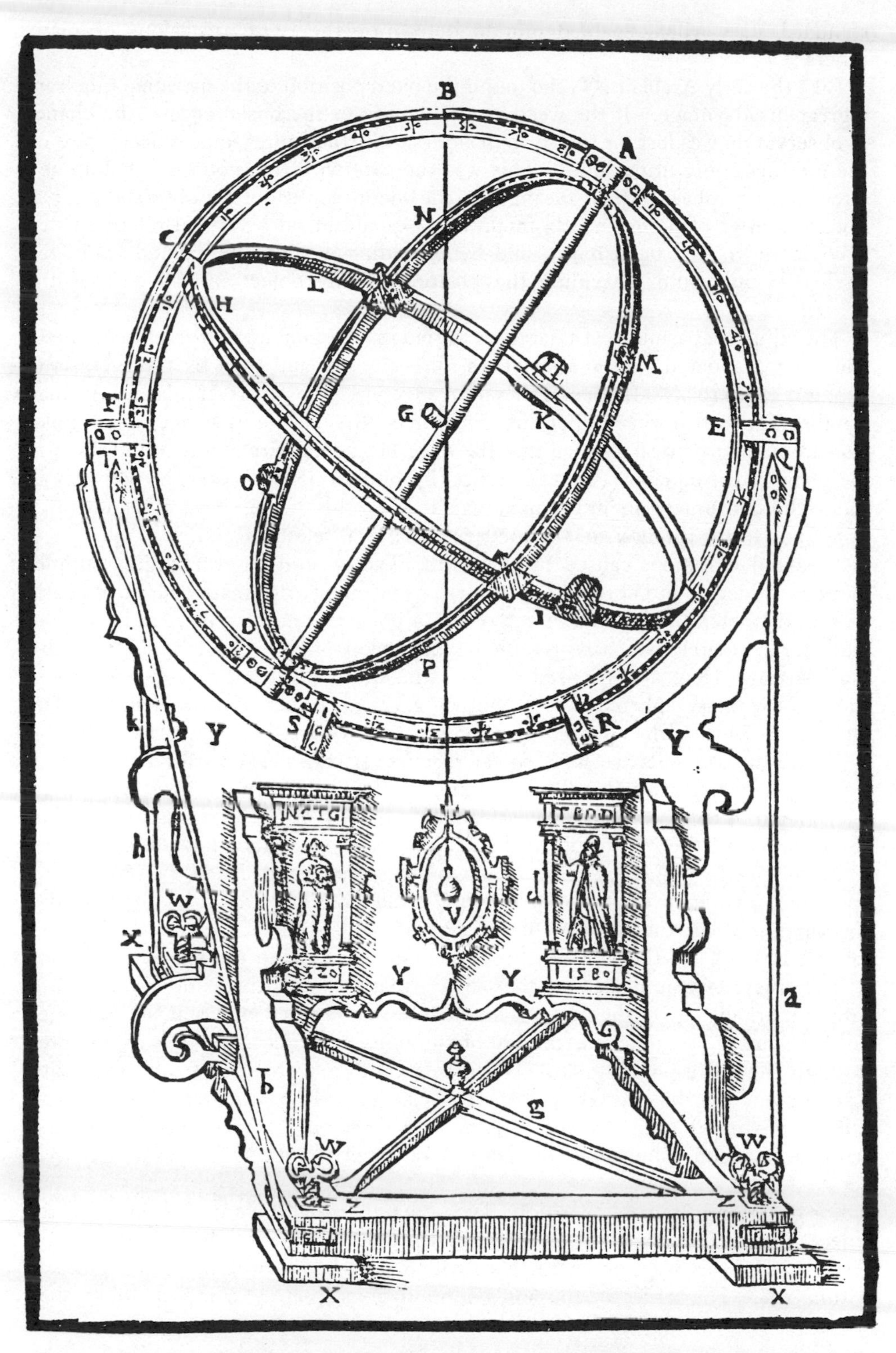

Fig. 12—Tycho Brahe's equatorial armillary, 1580

(*Science Museum, London. British Crown copyright*)

standard stars which enabled him to determine the right ascensions of many others.[84]

Like the early Arabians, Tycho found the fixed position of the meridian quadrant a great disadvantage. If the weather was cloudy or the observer late, the chance of observation was lost for twenty-four hours. Furthermore, some objects were on the meridian only in daylight. This was the case with the moon just before and after the new phase, which meant that the moon could not be observed over at least a quarter of its orbit. By mounting a quadrant on a vertical pillar, so that it could be rotated in azimuth, and by providing it with a horizontal graduated circle, Tycho could determine the position of any object in any part of the heavens.

The equatorial armillae at Uraniborg varied in size from $4\frac{1}{2}$ to $9\frac{1}{2}$ feet in diameter and were graduated, in some instances, direct to 10′ and then by transversals to 15″ and even 10″.[85] In their general arrangement they were similar to Arabian armillae, but with this important difference—the circles revolved on a polar instead of on a vertical axis, so that the once horizontal circle now found itself in the plane of the equator (Fig. 12). Once Tycho had sighted a star, he had only to rotate the instrument about the polar axis to keep the star in view. This idea of a polar axis forms the basis of the modern equatorial telescope.

To avoid vibrations caused by the wind, Tycho built Stjerneborg, a supplementary underground observatory.[86] Here he installed a declination circle of nearly 10-foot diameter. This could be rotated on a polar axis and swept round a 12-foot diameter semicircle which represented the portion of the celestial equator above the horizon. Here also he used a brass altazimuth quadrant 7 feet in radius, a zodiacal armillary sphere and, in another crypt, a sextant with alidades $5\frac{1}{2}$ feet long for measuring the apparent distances between stars.

All the instruments we have so far discussed were fitted with some form of sighting device. Just before Tycho's time, this took the form of two pinnules, one at each end of the alidade. They were merely brass plates with a small hole in the centre and had the disadvantage that, if the holes were too small, the observer could not see faint stars. A larger hole, on the other hand, made the observation uncertain owing to what Tycho termed *parallax*. He therefore introduced a special pinnule at the eye end of the alidade, consisting of four adjustable slits arranged parallel to the sides of the brass plate and thus in the form of a square. At the object end he placed another square plate, exactly the same in size as the square marked out by the slits. The alidade was pointed to a star and adjusted until the latter just touched the side of the object pinnule when it was observed through the corresponding slit. When this condition obtained for all four slits, Tycho knew that he had reduced the parallax, or *error of collimation* as it is now called, to a minimum. In this way, and by correcting any residual collimation error, he was able to improve the accuracy of his observations from 2′ to less than 30″.[87]

For observing the sun's position, Tycho made a small hole in the centre of the object pinnule and moved the alidade until the light fell on a small circle on the eye pinnule. In this way, and with his divided circles, he read the sun's place to within 10″. His tables of the sun's motion consequently compare more than favourably with the Alphonsine and Copernican tables, which were as much as 15′ to 20′ in error.[88]

Tycho rejected the Copernican theory for an earth-centred system—he regarded any motion of the sluggish and heavy earth as contrary to physical principles. About the earth revolved the moon and, further out, the sun, about which revolved the five known planets. Just beyond the orbit of Saturn rotated the sphere of fixed stars, for Tycho regarded as wasteful the vast space which Copernicus placed between the solar system and the stars. He tried in vain to detect the parallax of the brighter and, as he thought, nearer stars—a result which he interpreted as evidence for a stationary earth.[89] His system won few adherents, however, and he did not work it out in detail. In no way did it interfere with the gradual acceptance of the Copernican model. Rather did his rejection of solid crystalline planetary spheres and unchanging heavens contribute to the overthrow of Aristotelian-Ptolemaic cosmology.

Tycho had an imperious and fiery disposition which made him many enemies among the nobles of Denmark, especially those who envied him his large estate and high salary. After his benefactor's death, these individuals naturally counselled the young King Christian against his father's astronomer, already in some disfavour for having ill-treated one of his tenants. After the coronation of King Christian in 1596, Tycho found his allowances gradually curtailed, even to the pension, so that he felt, as never before, that both he and his work were unwanted. Eventually, his troubles became so acute that he was happy to leave Hveen, taking with him servants, students, instruments, and printing-press. After a short stay in Copenhagen, he left Denmark for ever, seeking safe domicile in Rostock, Wandsbeck, and Wittenburg until, worn out in body and spirit, he settled in Prague.

Tycho continued to hold Hveen until his death in 1601 but he took little interest in the repair and maintenance of his former home. Its gradual decay was hastened by his successors, who appear to have pulled down some parts in order to provide building material for new dwelling-houses. By 1652, there was scarcely any trace of the original buildings. In 1671, Picard arrived to determine the geographical position of Uraniborg for the French Academy, and found that only the foundations remained. Of Stjerneborg there was nothing save a slight hollow in the ground.[90]* Thus perished, in less than a century, the finest observatory of olden times, a residence fit for a prince, tastefully decorated and richly endowed, the home of one of the greatest astronomers and the finest instruments of pre-telescopic astronomy.

* Stjerneborg was recently excavated and the site is now protected by a modern building.

REFERENCES

1 Heath, T. L., 1932, *Greek Astronomy*, pp. xvi–xvii.

2 Lewis, G. C., 1862, *An Historical Survey of the Astronomy of the Ancients*, pp. 279–281.

3 Ward, F. A. B., 1937, *Time Measurement*, London, Science Museum publication, p. 17. The instrument is preserved in the Neues Museum, Berlin, while a copy is shown in the Science Museum, London. *Vide* plate I of *op. cit.*

4 *Ibid.*

5 Borchardt, L., *Zeits. für Ägyptische Sprache und Altertumskunde*, **37**, pp. 10–17, 1899. The Science Museum, London, has a copy of a sixth-century B.C. merkhet which was in the Royal Museum, Berlin.

[6] Lewis, *op. cit.*, p. 281.
[7] Bell, E. T., 1945, *The Development of Mathematics*, pp. 27, 42.
[8] Dreyer, J. L. E., 1906, *History of the planetary systems from Thales to Kepler*, pp. 3–4.
[9] Wallis Budge, E. A., 1929, *Book of the Dead*, British Museum publication, p. 40.
[10] Pannekoek, A., *Pop. Astr.*, **55**, pp. 423–438, 1947.
[11] *Ibid.*, pp. 436–437.
[12] Collingwood, R. G., 1945, *The Idea of Nature*, p. 31.
[13] Heath, T. L., 1913, *Aristarchus of Samos*, pp. 13–18.
[14] *Ibid.*, p. 13.
[15] *Ibid.*, p. 23.
[16] Dreyer, *op. cit.*, pp. 14–15. Heath, Ref. 13, pp. 31–38. Heath (p. 38) says that Anaximander may have been the first to introduce the gnomon in Greece.
[17] Dreyer, *op. cit.*, p. 16.
[18] *Ibid.*, pp. 41–42.
[19] *Ibid.*, pp. 43–44.
[20] Heath, Ref. 13, pp. 250–251.
[21] *Ibid.*, pp. 187–188.
[22] *Ibid.*, p. 251.
[23] *Ibid.*, p. 255.
[24] *Ibid.*, p. 157.
[25] Dreyer, *op. cit.*, pp. 95–102, gives summary of Schiaparelli's classical paper on the Eudoxan system (*Publ. R. Osserv. di Brera in Milano*, **9**, 1875).
[26] *Ibid.*, p. 107; Heath, Ref. 13, p. 217.
[27] Heath, Ref. 13, p. 217.
[28] *Ibid.*, pp. 219–220.
[29] *Ibid.*, p. 236.
[30] Grant, R., 1852, *History of Physical Astronomy*, p. 435.
[31] Heath, Ref. 13, pp. 299–414.
[32] *Ibid.*, p. 302.
[33] Gunther, R. T., 1923, *Early Science at Oxford*, ii, p. 9.
[34] Dreyer, *op. cit.*, pp. 174–175; Heath, Ref. 13, p. 339.
[35] Heath, Ref. 13, p. 339.
[36] Grant, *op. cit.*, p. 436.
[37] *Ibid.*, p. 436.
[38] Dreyer, J. L. E., 1890, *Tycho Brahe*, p. 315.
[39] Dreyer, Ref. 8, pp. 151 ff.
[40] *Ibid.*, pp. 162–163.
[41] Cajori, F., 1924, *A History of Mathematics*, pp. 46–47. Bell, E. T., 1945, *The Development of Mathematics*, p. 102.
[42] Dreyer, Ref. 8, pp. 197 ff.
[43] *Ibid.*, p. 196. Pannekoek stresses this outlook in *Pop. Astr.*, **55**, pp. 474–476, 1947.
[44] Dreyer, Ref. 38, p. 320; Grant, *op. cit.*, p. 440.
[45] Delambre, J. B. J., 1817, *Histoire de l'Astronomie Ancienne*, ii, p. 75.
[46] Gunther, *op. cit.*, p. 14.
[47] Ptolemy, C., *Geographicae enarrationis libri octo*, 1562 edition.
[48] Dreyer, Ref. 8, pp. 245–246.
[49] Dreyer, Ref. 38, p. 326; Gunther, *op. cit.*, p. 154.
[50] Dreyer, Ref. 8, p. 320.
[51] *Ibid.*, p. 321.
[52] Gunther, *op. cit.*, p. 22.
[53] *Ibid.*, p. 35.
[54] *Ibid.*, p. 35.
[55] *Ibid.*, p. 21.
[56] Dreyer, Ref. 8, p. 248.
[57] Chaucer, G., 1391, *Treatise on the Astrolabe. Vide* Gunther, *op. cit.*, pp. 202–204.
[58] Johnson, F. R., 1937, *Astronomical Thought in Renaissance England*, pp. 57–61.
[59] *Ibid.*, p. 63.
[60] Gunther, *op. cit.*, p. 67; Dreyer, Ref. 38, pp. 4–5.
[61] Pearson, W., 1829, *Introduction to Practical Astronomy*, ii, p. 90. Gunther, *op. cit.*, p. 69.
[62] Dreyer, Ref. 8, p. 365.
[63] Burtt, E. A., 1925, *The Metaphysical Foundations of Modern Physical Science*, pp. 42–44.
[64] *Ibid.*, p. 39 ff.
[65] Dreyer, Ref. 8, p. 319.
[66] *De Revolutionibus*, Preface, cited in *Occasional Notes of the R.A.S.*, **2**, p. 5, 1947.
[67] *De Revolutionibus*, i, cited as in Ref. 66, p. 19.
[68] Dreyer, Ref. 38, p. 125.
[69] *Ibid.*, p. 339.
[70] Dreyer, Ref. 8, p. 331.
[71] *Ibid.*, pp. 328–330.
[72] *Ibid.*, p. 343.
[73] Dreyer, Ref. 38, p. 58.
[74] *Ibid.*, p. 60.
[75] *Ibid.*, p. 59.
[76] *Ibid.*, pp. 18–19.
[77] *Ibid.*, p. 31.
[78] *Ibid.*, pp. 31–32.
[79] *Ibid.*, pp. 356–358, 387–388.
[80] *Ibid.*, p. 330.
[81] *Ibid.*, pp. 100–101.
[82] *Ibid.*, pp. 102, 329.
[83] *Ibid.*, p. 57.
[84] *Ibid.*, pp. 349–354.
[85] *Ibid.*, pp. 317–318.
[86] *Ibid.*, p. 103 ff.
[87] *Ibid.*, pp. 331–332.
[88] *Ibid.*, p. 334.
[89] *Ibid.*, p. 168.
[90] *Ibid.*, pp. 358, 374, 377.

CHAPTER II

> If an object is placed in a dense spherical medium of which the curved surface is turned towards the eye and is between the eye and the centre of the sphere, the object will appear magnified.
>
> ALHAZEN

IN COMPARISON with their skill in the manufacture and working of glass, the ancients had a limited knowledge of optics. The apparent bending of sticks dipped in water must have been known even to primitive man who, with the rest of mankind up to about the twelfth century A.D., probably considered it a property inherent in all liquids.

Before the manufacture of glass, gem-stones, including transparent rock-crystal, were fashioned with convex surfaces, and those who made them must have been familiar with their magnifying or distorting powers. Certain crude lenses which archaeologists have unearthed in Crete and along the coast of Asia Minor are thought to date back to 2000 B.C. They are of rock-crystal and bear traces of having been polished on a lapidary's wheel.[1] Whether these were used as ornaments, votive eyes, or as aids to vision we cannot say. The quartz lens discovered by Sir John Layard in the ruins of Nimroud is of true lentoid form, but the surfaces are poor and the glass would have been of little value as an aid to vision.[2]

The lenses mentioned by the Greeks appear to have been globular rather than lenticular in shape. The one referred to by Aristophanes in 424 B.C., for example, was only a glass globe filled with water.[3] In common with concave mirrors, these water spheres were used more as burning-glasses than as aids to vision. Seneca, it is true, remarks that Aristophanes' globe could be used to read letters 'however small and dim', but elsewhere refers more particularly to its burning properties.[4] Euclid writes in similar vein, and Archimedes is said to have used large mirrors to fire and so destroy an invading Roman fleet.[5] It seems strange, after Seneca's remarks, that so many centuries had to elapse before glassworkers took the step from flasks of water to glass lenses of the same or similar shape. The explanation would appear to be that magnification was thought to be a property inherent in the medium rather than in the shape of its bounding surfaces. Perhaps the aberrations of the first 'lenses' so masked their magnifying properties that their makers found they could see better without than with them.

Of the origin of glass we know little save that it appeared in Egypt about 3500 B.C. and that the Phoenicians were the first to manufacture it on a large scale.

> It is said [writes Pliny the elder][6] that some Phoenician merchants, having landed on the shores of the river Belus, were preparing their meal, and not finding suitable stones for raising their saucepans, they used lumps of natron, contained in their cargo, for the purpose. When the natron was exposed to the action of fire, it melted into the sand lying on the banks of the river, and they saw transparent streams of some unknown liquid trickling over the ground; this was the origin of glass.

Fig. 13—Title-page of the 1535 Nürnberg edition of Vitello's treatise on optics

It shows the various phenomena of refraction, also magnification by means of a concave mirror

(British Optical Association)

Whether this was the case or not, the Phoenicians became as famous for their glassware as for their Tyrian purple, and their skill and experience with the former proved invaluable to the Byzantine traders and, ultimately, to the Venetians.

Euclid, in the third century B.C., was the first to write about the refraction and reflection of light and to mention that light travelled in straight lines.[7] Ptolemy's knowledge was equally restricted and is just a summary of the ideas of his predecessors. Alhazen, an Arabian writer of the tenth to the eleventh centuries A.D., was the first to experiment with different media in the hope of finding a working theory of reflection and refraction. He correctly explained the apparent change in the shape of the sun and moon as they approach the horizon and showed the necessity of allowing for atmospheric refraction in astronomical observations. He was also aware of the magnifying effects of spherical glass segments, but did not mention using them as aids to vision.[8] In *Opticae Thesaurus Alhazeni Aribis*, the 1572 Latin translation of Alhazen's works, there is also reference to plane, spherical, and parabolic mirrors.

Alhazen's disciple, Vitello, a Pole who spent most of his life in Italy, also wrote a book on optics (Fig. 13)[9] and tried to establish the laws of refraction. He showed that the scintillation of stars is due to moving air currents and pointed out that the effect is intensified when stars are viewed through running water.[10]

At Oxford, Vitello's contemporary, Roger Bacon (1210–1294), went to great pains and expense to study the effects of mirrors and plano-convex lenses. Bacon was a Franciscan friar well in advance, for learning and originality, of the times in which he lived. Realizing that experiment should form the basis of theory, he spent a small fortune making researches in a variety of subjects. He introduced gunpowder into England, commented on the possibilities of heavier-than-air flight, and discussed the effects of lenses.

In the *Opus Majus*, as Bacon's most important work is called, are passages in which he describes the magnifying properties of what purports to be a convex lens. In one part, for example, he writes that 'If the letters of a book or any minute objects, be viewed through a lesser segment of a sphere of glass or crystal, whose plane base is laid upon them, they will appear far better and larger.'[11] Robert Smith, who gives this translation from the Latin, reads '*superimpositi literis*' for Bacon's '*suppositi literis*' and infers that Bacon laid the glass on the letters. It is likely, however, that he held the glass before his eye and then viewed the letters—as we should use a hand magnifier. In either case, Bacon seems to have appreciated the use of the glass as an aid to vision.

Of greater interest is Bacon's description of the magnification of distant objects by different glasses. The effects of a single convex lens depend on its focal length, its distance from the eye, and the eyesight of the observer. Bacon appears to have investigated the first two factors, for he writes:[12]

> . . . we can give such figures to transparent bodies, and dispose them in such order with respect to the eye and the objects, that the rays shall be refracted and bent towards any place we please; so that we shall see the object near at hand or at a distance, under any angle we please. And thus from an incredible distance we may read the smallest letters, and may number the smallest particles of dust and sand, by reason of the greatness of the angle under which we may see them; . . . the sun, moon, and stars may be made to descend hither in appearance, and to be visible over the heads of our enemies, and many things of the like sort, which persons unacquainted with such things would refuse to believe.

In an unscientific and superstitious age, it is not surprising that these and other statements, together with rumours of Bacon's numerous scientific activities, exaggerated in the first instance and then by repetition, evoked the envy of his colleagues. The latter had little difficulty in convincing their superiors that Bacon was a dangerous necromancer and, on this pretext, he was imprisoned in his cell for ten years, to be released only a few years before his death in 1294.[13]

Although Bacon recognized the usefulness of lenses as aids to vision, he can hardly be said to have invented spectacles, by which we mean two lenses, one for each eye, mounted in a hand-frame. When we try to establish the date of this invention, an important step in the history of the telescope, we are again confronted with inadequate evidence. Two Italians are usually quoted—Alexandro della Spina, a Dominican monk of Pisa, and his friend Salvino d'Armati of Florence. We know little of either, save that Spina died in 1313 and that Armati's tombstone bears the frequently quoted inscription:[14]

> Here lies Salvino degli Armato of the Armati of Florence.
> Inventor of Spectacles. God pardon his sins. A.D.1317.

Further evidence is provided by Giordano de Rivalto, a monk of Pisa. In a sermon which he gave in 1305, Rivalto stated that it was barely twenty years since spectacles were invented and added that he had both seen and spoken to the man who made them.[15] Another manuscript, dated 1299 and quoted by W. Molyneux in his *Dioptrica Nova* (1692),[16] contains the reference: ' I find myself so pressed by age, that I can neither read nor write without those glasses they call spectacles, lately invented, to the great advantage of poor old men when their sight grows weak.' A further manuscript, in the library of St Katherine at Pisa, ascribes the invention to della Spina, ' a modest and good man ',[17] and leaves no doubt that spectacles first appeared, in Italy at any rate, sometime between 1285 and the close of the thirteenth century.

The first Italian spectacle-makers were fortunate in having access to the well-established glassworks around Venice, an industry handed down to the Venetians by the Byzantine traders. As the manufacture of convex spectacles increased, so the news of their miraculous effects passed to other countries and, by the early sixteenth century, when concave glasses appeared, there were spectacle-making centres in both Holland and Germany.

In Oxford, the tradition of Bacon's lens experiments lasted for many years, even to the middle of the sixteenth century, when Robert Recorde referred to them in a book on applied geometry:[18]

> . . . many thynges seme impossible to bee doen, which by arte maie verie well bee wrought. And when thei bee wrought, and the reason thereof not understande, then saie the vulgare people, that those thynges are dooen by Negromancie. And hereof came it that Frier Bacon was accompted so greate a Negromancier, whiche never used that arte (by any coniecture that I can finde) but was in Geometrie, and other Mathematicall sciences so experte, that he could doe by them suche thynges, as were wonderfull in the sight of moste people. Great talke there is of a glasse that he made in Oxforde, in whiche men might se thinges that wer doen in other places, and that was iudged to bee doen by power of evill spirites. But I knowe the reason of it to bee good and naturall, and to be wrought by Geometrie (sith perspective is a parte off it) and to stande as well with reason, as to see your face in common glasse.

Leonard Digges of Oxford, a contemporary of Recorde, also profited by reading Bacon's papers. His son Thomas, in the preface to his father's *Pantometria* (1571), says:[19]

> My father by his continual paynfull practises, assisted with demonstrations Mathematicall, was able, and sundrie times hath, by proportionall Glasses duely situate in convenient angles, not onely discovered things farre off, read letters, numbred peeces of money with the very coyne and superscription thereof, cast by some of his friends of purpose uppon Downes in the open fieldes, but also seven myles off declared what hath been doon at that instante in private places.

Whilst it is possible that Digges used a form of refracting telescope for these observations, it is more likely that he copied Bacon and had only a single convex lens before his eye. There is no direct mention of a combination of two lenses; the term ' perspective glasses ' was then applied to any single lens or mirror. Given a single lens at a certain distance before his eye, together with a small amount of hypermetropia, Digges, like Bacon, would have seen distant objects magnified and the right way up. Neither Digges nor any other investigator of this period mentions a tube for holding the lens or lenses. For observing stars a tubeless telescope is satisfactory, but in daytime the absence of a tube allows extraneous light to pass through the eye-lens and so fog the image.

Leonard Digges was familiar with a primitive reflecting instrument which seems to have consisted of a mirror or mirrors and a lens.[20]

> By concave and convex mirrors of circular [spherical] and parabolic forms, or by paires of them placed at due angles, and using the aid of transparent glasses which may break, or unite, the images produced by the reflection of the mirrors, there may be represented a whole region; also any part of it may be augmented so that a small object may be discerned as plainly as if it were close to the observer, though it may be as far distant as the eye can descrie.

Here, for the first time, we have indications of an instrument which we may call a reflecting telescope. For the first time also, the writer nearly divulges the way in which he saw distant objects magnified.

Another Elizabethan writer, the versatile Dr Dee, tutor to Thomas Digges, makes a few passing references to lenses in his preface to an edition of Euclid (1575). He refers to Bacon's manuscripts as the principal source of his information.

> A commander of an army [he writes][21] may wonderfully help himself by perspective Glasses; In whiche (I trust) our posterity will prove more skillful and expert, and to greater purposes, than in these days, can (almost) be credited to be possible.

The references here and elsewhere to the use of ' perspective glasses ' in military operations is interesting. The same motives inspired Dutch inventors early in the seventeenth century.

By far the best account of the effects of convex lenses is given by William Bourne who, in 1585, wrote on optics at the request of Lord Burghley.[22] Bourne describes glasses for improving both near and distant vision, and says that some lenses called ' perspective glasses ' could be made up to twelve inches across.[23] Like smaller lenses ' commonly called spectacle glasses ', they magnified distant objects according to their distance from the eye and the ' grynding of the Glasse, both in his diameter, and thicknes in the middle, and thinnes towards the sydes.' Bourne's eyes were undoubtedly hypermetropic for, by moving the glass further and further away, distant objects became ' of marvellous bigness.' With his eye at the focus of the lens he could ' discerne nothinge thorowe the glasse; But like a myst, or water.' Beyond the focus, he saw the distant object ' reversed and turned the contrary way.'[24]

In the same treatise, Bourne sets out to describe the effects of two different glasses, ' The one concave with a foyle, uppon the hylly [i.e. convex] side, and the other grounde and polisshed smoothe, the thickest in the myddle, and thinnest towardes the edges or sydes.' Unfortunately, he thinks that Dee and Thomas Digges are better qualified to speak about such things by reason of their learning and ' better tyme to practyse those matters.' He is confident, however, that the two glasses will 'shew the thinge of a marvellous largeness, in a manner uncredable to bee beleeved of the common people.'[25] 'So that those things' he concludes, 'that Mr. Thomas Digges hathe written that his father (Leonard) hathe done, may be accomplisshed very well, withowte any dowbyte of the matter: But that the greatest impediment ys, that yow can not beholde, and see, but the smaller quantity at a tyme.'[26]

Bourne backed his concave and convex mirrors with reflecting lead foil. Convex mirrors showed the face smaller and smaller as he increased their curvature. Concaves gave views of distant objects and could be used singly or in combination with other glasses. For grinding all these he used iron tools, turned to the required concavity or convexity. The glass blank was ground on them with the aid of sand until its figure became the same as that of the tool but of the opposite curvature.

About the same time as Bourne wrote his treatise, Giambattista Della Porta of Naples published his *Magia Naturalis*, in the 1589 edition of which he made the following statement:[27]

> By means of a concave glass you will see distant objects small but clear; with a convex glass, near objects magnified but dim. If you know how to combine them exactly you will see both distant and near objects larger than they would otherwise appear and very distinct.

The first impression on reading this is that Porta had hit on the Galilean combination. Porta was, however, an unreliable and imaginative writer and, on this occasion, was probably out to impress his readers. His earlier remarks on single lenses and their effects are both vague and exaggerated. Kepler, about twenty years later, found them so wrapped in mystery as to be quite unintelligible. Porta, no doubt, picked up some of his ideas from others, for he was vice-president of the Accademia dei Lincei, or Academy of Lynxes, and his house at Naples was a favourite meeting-place for men of science and learning.

Most historians make Holland the country of the telescope's origin and 1608 the year of its birth. Here, for the first time, the documentary evidence is fairly clear and complete, for Van Swinden in the 1820s went to great pains to collect data from the official Hague records, data which Moll published in 1831[28] and which have been supplemented by the researches of later historians. Yet there is still doubt as to the name of the inventor, for the idea seems to have germinated in several minds at once. Many writers follow tradition and hand the merit unconditionally to Hans Lippershey, an obscure spectacle-maker of Middleburg in Zeeland. Huygens calls him an 'illiterate mechanick', but Lippershey's subsequent actions showed that he possessed both practical skill and business acumen. The story goes, and it has several versions, that two children were playing in Lippershey's shop with some lenses and noticed that, by holding two of them in a certain position, the weather-vane of the nearby church appeared much larger. Lippershey at once tried this out for himself and then improved it by mounting the lenses in a tube. Some accounts say that an apprentice held the lenses, others that Lippershey was alone at the time or that he copied the idea from another optician. Some say he used a convex lens together with a concave, others that both lenses were convex and that he saw the steeple upside-down. Suffice it to say, Lippershey made a telescope and lost no time in exploiting its financial possibilities. He wrote first to the Zeeland States and from them obtained letters of recommendation to Prince Maurice of Nassau and the States-General.

At this time, Holland was engaged in a fierce struggle for her independence against the armies of Phillip II of Spain. At so critical a period, no person would be more interested in this valuable addition to military strategy than Prince Maurice, patron of science and head of the Belgian army. The States-General was the governing body of the Netherlands, among whose documents Van Swinden found the following entry, dated October 2, 1608:[29]

> On the petition of Hans Lippershey, a native of Wesel, an inhabitant of Middleburg, spectacle-maker, inventor of an instrument for seeing at a distance, as was proved to the States, praying that the said instrument might be kept secret, and that a privilege for thirty years might be granted to him, by which everybody might be prohibited from imitating these instruments, or else grant to him an annual pension, in order to enable him to make these instruments for the utility of this country alone, without selling any to foreign kings and princes. It was resolved, that some of the Assembly do form a committee, which shall communicate with the petitioner about his said invention, and enquire of him whether it would not be possible to improve upon it, so as to enable one to look through it with both eyes; and further, to enquire what remuneration would satisfy him. And due report being made, it will be laid in deliberation, whether it is expedient to grant the petitioner a remuneration or a privilege.

The suggestion at this early stage that the telescope be made in binocular form seems as unnecessary as the stipulation, two days later, that the object-glasses be made of rock-crystal.[30] The States-General no doubt thought that, for military work, a binocular would be more useful than a single spy-glass. The choice of rock-crystal, a material difficult to select and work, was probably dictated by the scarcity of good optical glass.

Lippershey proved equal to these demands and his first telescope, a monocular, was tested by the committee from a tower on Prince Maurice's palace and declared 'likely to be of utility to the State.' This thrifty body then succeeded in persuading Lippershey to reduce his price for a binocular from 1000 to 900 florins (about £75), promising 300 florins payable immediately and an order for two more should the first prove satisfactory. The States would then deliberate whether 'a privilege or an annual pension' should be granted.[31]

Meanwhile, James Metius or, more correctly, Jacob Adriaanzoon, a native of Alkmaar, sent a petition to the States-General in which he declared that he had made a telescope quite equal in power to Lippershey's, although with inferior materials. He said that he had probed into the secrets of glassmaking, had studied the effects of lenses—with a little encouragement he would make an instrument superior to Lippershey's.[32] But the States-General, having gone so far with his rival, was in no mood to grant Metius' petition on so slender a pretext. It vaguely promised to consider his claim if he could improve his workmanship. This proposal Metius apparently took as a rebuff; he was of eccentric and jealous disposition and made no further effort to press his claim. He refused to show his telescope even to his friends and made sure that his successors gained nothing, by causing all his tools to be destroyed at his death.[33] It is said that he once allowed Prince Maurice to look through the tube. His brother Adrian Metius (who never looked through it) spoke well of its performance and claimed for his brother the discoveries made by Galileo.[34] Adrian Metius was one of Tycho's many pupils. He died in 1635 as professor of astronomy at Frankfurt, and it was he who, on account of his devotion to mathematics, earned the surname Metius, afterwards assumed by the whole family.

The third claimant at this time was Zacharias Jansen, another Middleburg spectacle-maker whose son, Hans, used to play in his boyhood with William Boreel. Boreel later became French ambassador to the Dutch States and, in this capacity, visited the town of his birth in 1655 to find out who actually invented the telescope.[35] He began by searching for all who had known Lippershey and Jansen and who should have been in a position to give fairly reliable evidence. This was not the case, for much of the information was second-hand and told of events fifty years earlier. Hans Jansen said that his father invented the telescope in 1590 and used it to look at the moon and stars, while his sister gave either 1611 or 1619 as the date of invention.[36] Boreel concluded from this that Jansen's telescope appeared about 1610, and that Lippershey gained precedence only because he obtained the information 'from a stranger'. It is far more probable, however, that Zacharias Jansen made, not the first telescope, but the first compound microscope, to which Boreel also refers. According to the latter, Zacharias sent this instrument to Prince Maurice and to Albert, archduke of Austria, while, in 1619, Boreel saw the original instrument for himself.

> When I was an envoy in England in 1619 [he writes][37] Cornelius Drebel of Holland, a man aware of many secrets of Nature, mathematical tutor to James I and known to me, showed me the very instrument, which the Archduke had given to Drebel, viz. the one of Zacharias himself; nor was it (as they are now shown) with a short tube, but the tube was almost a foot and a half long, made of gilt brass two inches in diameter supported by three dolphins of brass. The base was an ebony disc, on which were placed the minute objects which we looked at from above enlarged almost miraculously.

This statement is, to a large extent, confirmed by Huygens, who says that Drebel came across the microscope of Jansen in 1621.

The evidence for Jansen appeared in a book published by Pierre Borel, a French physician, and was based on Boreel's investigations.[38] Subsequent researches have, however, placed Zacharias Jansen in a rather different light. It appears that he was a suspicious character who, besides spectacle-making, coined false Spanish money so as to ruin the enemy's credit. Unfortunately, he grew so interested in this art that he continued to coin money for some years after peace was declared, and was finally convicted for forgery. His previous services to his country delayed the execution of the sentence—immersion in boiling oil—a sentence which he wisely frustrated by quitting the country.[39]

Owing to the clamours of Metius and Jansen and to the fact that 'many other persons had a knowledge of the invention', the States-General finally declined to grant a patent to Lippershey. This decision was fortunate inasmuch as news of the instrument was no longer confined to Middleburg or the Netherlands, but passed quickly to Germany, France, and Italy. Thus, while in 1608 the French ambassador at the Hague was negotiating for a telescope to send to Henry IV, early the next year telescopes were on sale in Paris.[40] At the Frankfurt fair in the autumn of 1609, an itinerant Belgian showed a telescope which magnified ' several times ' to a friend of Simon Marius, the astronomer.[41] In May, *Dutch trunks*, *perspectives* or *cylinders* as they were then called, appeared in Milan, a little later in Venice and Padua and, by the end of the year, they were being made in London.[42]

REFERENCES

1 *Trans. Brit. Optical Assoc.*, **39**, p. 7, 1937.

2 Perrot and Chipiez, *Histoire de l'Art dans l'Antiquité*, ii, p. 718. *Vide* also *Optician*, **95**, p. 184, 1938. This lens is in the British Museum, London.

3 Aristophanes, *The Clouds*, Act 2, Scene 1, performed 424 B.C.

4 *Ibid.*

5 Molyneux, W., 1692, *Dioptrica Nova*, p. 252. Smith, R., 1738, *A Compleat System of Opticks*, ii, Remarks, p. 11.

6 Pliny, *Natural History*, lib. xxxvi, cap 26.

7 Wilde, 1838, *Geschichte der Optik*, **i**, p. 10. Hutton, C., 1795, *A Mathematical and Philosophical Dictionary*, ii, p. 175.

8 *Opticae Thesaurus Alhazeni Aribis*, 1572, pp. 44–45.

9 *Vitellionis . . . de Natura, Ratione, et Proienctione, radiorum visus, Luminum, Colorum atque Formarum*, 1535.

10 Hutton, *op. cit.*, p. 176.

11 Smith, *op. cit.*, Remarks, p. 14, quotes *Opus Majus*, 1733, Jebb edition, p. 352.

12 Grant, R., 1852, *History of Physical Astronomy*, pp. 515–516, cites *Opus Majus*, 1733, Jebb edition, p. 357.

13 Hutton, *op. cit.*, i, pp. 179–180.

14 Gasson, W., quotes in *The Optician*, **96**, p. 316, 1939.

15 Molyneux, *op. cit.*, p. 255.

16 *Ibid.*, p. 255.

17 *Ibid.*, p. 254.

18 Recorde, R., 1551, *Pathway to Knowledge*, Preface, cited by Baxandall, D., *Trans. Opt. Soc.*, **24**, p. 305, 1923.

19 *Pantometria*, 1571 edition, Preface. *Vide* Gunther, R. T., 1923, *Early Science at Oxford*, ii, p. 290.

20 *Pantometria*, 1571; Gunther, *op. cit.*, pp. 289–290.

21 *Elements Euclide*, 1575, Preface. *Vide* Gunther, *op. cit.*, p. 291, and Baxandall, *op. cit.*, p. 305.

22 But not published until 1839 in Halliwell, J. O., *Rara Mathematica.*

23 Baxandall, *op. cit.*, p. 306.

24 *Ibid.*, p. 306.

25 Gunther, *op. cit.*, p. 292.

26 *Ibid.*, p. 293; Baxandall, *op. cit.*, p. 307.

27 Danjon, A., and Couder, A., 1935, *Lunettes et Télescopes*, p. 591. Also Baxandall, *op. cit.*, p. 305.

28 Moll, G., *Journal Royal Institution*, **1**, pp. 319, 483, 1831.

29 *Ibid.*, p. 324.

30 *Ibid.*, p. 325.

31 *Ibid.*, p. 326.

32 *Ibid.*, p. 323.

33 *Ibid.*, pp. 321-322.

34 Doberck, W., *Observatory*, **2**, p. 367, 1879.

35 Baxandall, *op. cit.*, p. 307.

36 *Ibid.*, p. 308.

37 Clay, R. S., and Court, T. H., 1932, *History of the Microscope*, p. 9.

38 Borel, P., 1655, *De vero telescopii inventore. Vide* also Moll, *op. cit.*, pp. 330–331, 483–485; Doberck, *op. cit.*, pp. 368–369.

39 Danjon and Couder, *op. cit.*, p. 593.

40 *Ibid.*, p. 596, citing Richer, J., 1611, *Mercure François*, p. 338.

41 *Ibid.*, p. 594; Moll, *op. cit.*, pp. 486–487.

42 Rigaud, S. P., 1832, *Miscellaneous works and correspondence of James Bradley*. *Vide* section *Account of Harriot's Astronomical Papers*, pp. 23, 26, 46. Moll, *op. cit.*, p. 495.

CHAPTER III

Oh! when will there be an end put to the new observations and discoveries of this admirable instrument?

GALILEO GALILEI

GALILEO GALILEI first heard of Lippershey's invention in May, 1609. At the time he was forty-five years of age, professor of mathematics at Padua and in some disfavour with his professional colleagues because of his outspoken anti-Aristotelian views. He was staying in Venice when he heard the rumour that 'a Dutchman had constructed a telescope, by the aid of which visible objects, although at a great distance from the eye of the observer, were seen distinctly as if near.'[1] Some denied and others confirmed this rumour but, a few days later, Galileo

> . . . received confirmation of the report in a letter written from Paris by a noble Frenchman, Jacques Badovere, which finally determined me to give myself up first to inquire into the principle of the telescope, and then to consider the means by which I might compass the invention of a similar instrument, which a little while after I succeeded in doing, through deep study of the theory of Refraction.

At Padua, and within twenty-four hours of his arrival there, Galileo made his first telescope (Fig. 15):[2]

> a tube, at first of lead, in the ends of which I fitted two glass lenses, both plane on one side, but on the other side one spherically convex, and the other concave. Then applying my eye to the concave lens I saw objects satisfactorily large and near, for they appeared one-third of the distance off and nine times larger than when they are seen with the natural eye alone.

Galileo was so pleased with this result that he wrote immediately to his friends in Venice and, after mounting the lenses in a better tube, posted there himself. For over a month the instrument passed among the officials of Venice and never failed to excite their wonder and astonishment.

> Many noblemen and senators [Galileo writes][3] although of great age, mounted the steps of the highest church towers at Venice, in order to see sails and shipping that were so far off that it was two hours before they were seen steering full sail into the harbour without my spy-glass, for the effect of my instrument is such that it makes an object fifty miglia off appear as large and near as if it were only five.

The military importance of such a glass was not lost upon Galileo and he presented it to the Senate. In return, his professorship at Padua was conferred upon him for life, while his salary was increased to 1000 florins a year.

Galileo has often been referred to as the inventor of the telescope, but he himself never failed to point out that this credit is due to the 'Dutchman'. He remarks,

Fig. 14—Galileo Galilei

Portrait by Sustermans, 1635

(By courtesy of the Director of the Science Museum, London)

however, and with justice, that there is a big difference in value between an invention by ' hazard ' and ' chance ' and one which is the ' work of the reasoning and intelligent mind.' News of Lippershey's invention merely turned his thoughts in one particular direction—it did not assist in solving the problem.

> Thus, we are certain [he writes in *Il Saggiatore*][4] the first inventor of the telescope, was a simple spectacle-maker who, handling by chance different forms of glasses, looked, also by chance, through two of them, one convex and the other concave, held at different distances from the eye; saw and noted the unexpected result; and thus found the instrument. On the other hand, I, on the simple information of the effect obtained, discovered the same instrument, not by chance, but by the way of pure reasoning.

Upon his return to Padua, Galileo made a better telescope which magnified about 8 diameters and, a few days later, another of 20 magnification.[5] Through the latter he saw that Jupiter had a round disk while the moon appeared to have a

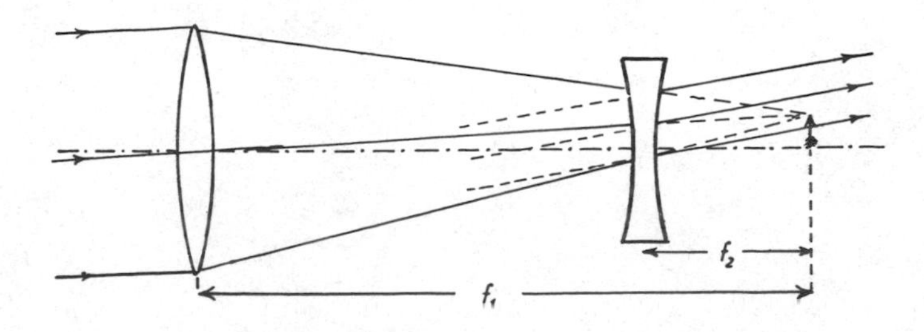

Fig. 15—The Galilean telescope

f_1 is the focal length of the object-glass, f_2 that of the eyepiece, usually taken as numerically negative. The magnification is given by $-f_1/f_2$, which comes out numerically positive, thereby indicating that images seen through the instrument are not inverted.

rough, mountainous surface.[6] On January 7, 1610, he observed Jupiter with a fourth and even better telescope of 30 magnification, having ground the lenses himself and spared ' neither labour nor expense.' This time he saw three small bright stars near the planet and took them to be fixed stars in the background. On the following night, to his amazement, they had changed their positions both relative to the planet and to one another. By January 13, he had seen a total of four stars near Jupiter, and observations up to March established beyond all doubt that they were not stars at all but four satellites.[7] The fourth telescope also confirmed his first impression that the moon's surface was not smooth like a mirror, as many ancient Greeks had thought, but rough with valleys and mountains. He noticed that these features accounted for the irregularities of the terminator and that some of the higher peaks, illumined by the rising sun, shone like stars from the dark side of the disk. He also remarked that the bright spots scattered over the full moon (certain craters and mountain masses), appeared in the darker parts (the *maria*) as well as in the brighter.

Galileo saw many more stars through his telescope than by the naked eye and found that they all appeared brighter. The brightest stars showed disks which were unaffected in size by changes in magnification—at least, not appreciably so. He thought that the telescope stripped stars of the false light which usually surrounds them when they are viewed with the naked eye, and that the spurious coronae were due to irradiation. Through his glass, the Milky Way presented an amazing spectacle, for 'upon whatever part of it you direct the telescope straightway a vast crowd of stars presents itself to view; many of them are tolerably large and extremely bright, but the number of small ones is quite beyond determination.'[8] Galileo found that these fainter stars were by no means confined to the Milky Way.

> Beyond the stars of the sixth magnitude [he continues] you will behold through the telescope a host of other stars, which escape the unassisted sight, so numerous as to be almost beyond belief, for you may see there more than six other differences of magnitude, and the largest of these, which I may call stars of the seventh magnitude, or of the first magnitude of invisible stars, appear with the aid of the telescope larger and brighter than stars of the second magnitude seen with the unassisted sight.

Galileo published the results of his observations in a small work called *Sidereus Nuncius*, dedicated to Grand Duke Cosmo II de Medici. Its publication in March, 1610, produced an extraordinary sensation among the learned. Galileo was content to state the bare facts of his observations, leaving them unrelated to Copernican ideas. Even so, their publication was met by strong opposition and adverse criticism. Rabid Aristotelians denied his observations *a priori.* Some refused to look through any telescope, others stated that it was impossible to see appearances which Aristotle had not mentioned. Even estimable men like Welser in Augsburg and Clavius at Rome could not credit Galileo's observations until they had made their own. Clavius laughed at the pretended satellites of Jupiter—'you must construct a telescope which would first make them and then show them.'[9] Kepler saw the satellites in August, 1610, and had a reprint of the *Sidereus Nuncius* issued at Prague. In the introduction he expressed his entire conviction of the truth of Galileo's statements and went to some pains to answer all objections. The animosity of the opposition found expression in Martin Horky's *Peregrinatio contra Nuncium Sidereum*, a libellous tract probably instigated by Magini, professor of astronomy at Bologna University.[10]

During the Easter recess of 1610, Galileo visited the Court of the Grand Duke Cosmo de Medici at Florence. He wished for a court position and the relief it offered from heavy lecturing duties. On July 12, he was appointed mathematician to the Grand Duke and first mathematician at Pisa. This move from republican protection and Padua University where, eighteen years before, he had received a warm welcome, was the beginning of his misfortunes.[11]

In July, 1610, Galileo observed the strange aspect of the planet Saturn. At first, he thought that the planet consisted of three bodies arranged in line and side by side. But the two lateral components diminished in size and brightness until, two years later, they disappeared altogether. Galileo was more than perplexed.

> What is to be said concerning so strange a metamorphosis? [he wrote to his friend Welser][12] Are the two lesser stars consumed after the manner of the solar spots? Have

they vanished and suddenly fled? Has Saturn perhaps devoured his own children? Or were the appearances indeed illusion or fraud, with which the glasses have so long deceived me, as well as many others, to whom I have shewn them?

After their disappearance, the 'globes' came again, when each observer gave them a different shape according to his eyesight and the power and definition of his telescope. Indeed, forty years elapsed before Huygens, with superior glasses, explained that all their 'oblongs', 'semicircles' and 'arcs' were due to the changes in position of a broad flat ring which encircled the planet.

Only a month after Galileo's arrival at Florence, he discovered the varying crescent form of the planet Venus. On December 30, 1610, and after nearly three months' observation, he informed by letter both Clavius and Castelli, his former pupil.[13] In another letter, dated January 1, 1611, to Julian de Medici at Prague, he concluded that none of the planets shines by its own light and that 'necessarily Venus and Mercury revolve round the sun.'[14]

> At first I saw Venus perfectly round [he wrote to Julian de Medici][15] neat and distinctly terminated, but very small; which figure she retained till she approached nearer to her greatest digression from the sun, increasing continually in apparent bulk. From that time her figure began to fail of its roundness on its eastern side which lay from the sun, and in a few days was reduced to a perfect semi-circle; and continued so without the least alteration, till she left the tangent to her orbit, and began to return towards the sun. At present the semi-circle becomes more and more hollow every day, its angles being changed into horns, which grow sharper and sharper, till they become so thin as to vanish with her passage into the beams of the sun.

At the end of March, 1611, Galileo went to Rome to induce the ecclesiastical authorities to look through his telescope.* On April 14, 1611, he was invited to an important banquet given in his honour by Prince Frederick Cesi, founder and president of the Accademia dei Lincei.[16] Sirturi, who wrote a book on the telescope in 1618, was present and says that the guests arrived before sunset and looked through the telescope at some writing over a doorway about a mile away. 'Taking my turn, I too saw it, and read the inscription to my heart's content. Then at nightfall, after dinner, we observed Jupiter and the motions [positions?] of its companion bodies.'[17] Later that night, Galileo dismantled the telescope so that the guests could examine the two lenses. From his sifting of a large mass of material, E. Rosen deduces that it was at this banquet that Cesi 'publicly unveiled' the term *telescope*. The new word, which replaced the *perspicillum* and *instrumentum* of Galileo and Kepler, was fostered if not invented by Johann Demisiani of Cephalonia, a guest at the banquet and 'a poet rather than a scientist'.[18] The first printed record of the word *telescope* appeared in the *Lunar Phenomena* published by Julius Caesar Lagalla in 1612 and in which he ascribed the introduction of the word to Demisiani.[19]

Galileo was received with the greatest honour in Rome. He met Cardinal Barberini, afterwards Pope Urban VIII, and exhibited his telescopes and the appearances through them to other cardinals and learned men. He had a long audience with Pope Paul V, who assured him of his unalterable good will. When Galileo

* His second visit, after an interval of almost a quarter of a century.

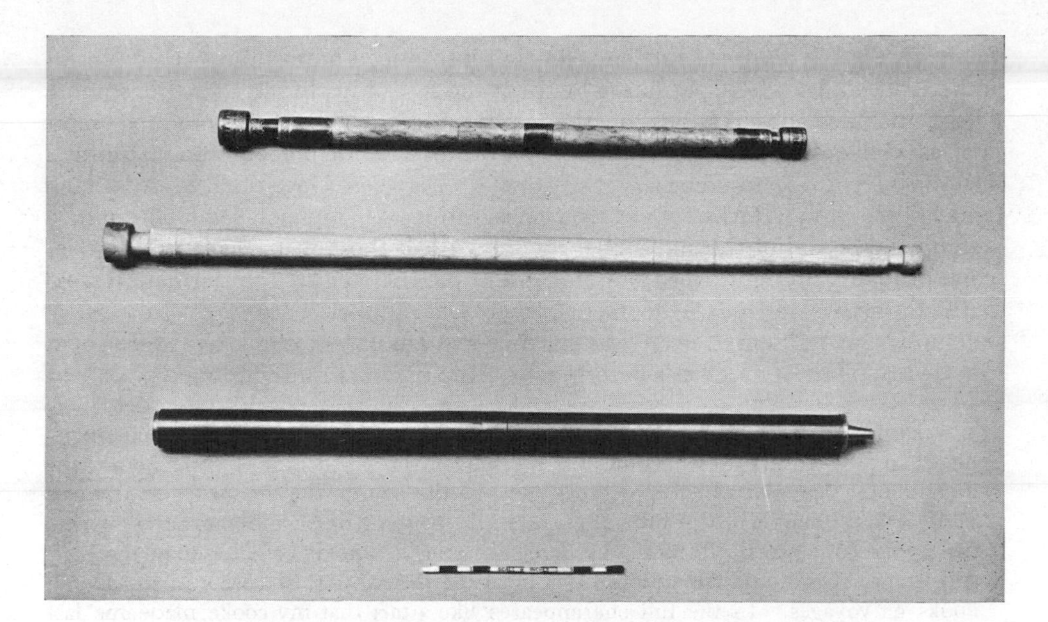

Fig. 16—Replicas of telescopes of Galileo and Toricelli

The scale is of inches with a total length of one foot

(*Science Museum, London. British Crown copyright*)

left Rome early in June, 1611, he had every reason to congratulate himself on his success.[20]

In June of the same year, Galileo observed spots on the sun. He soon found that they appeared to move slowly as they approached the edge of the disk and, at the same time, grew thinner, as seen in perspective, until they finally disappeared. They were, without doubt, integral parts of the sun's disk, being carried round by its rotation in about a month.[21] Yet they changed their form and relative positions so frequently that he thought they were dark clouds in the solar atmosphere.

As the news of Galileo's discoveries spread across Europe, so others claimed that they had already seen these wonders in the heavens. Simon Marius, with a telescope obtained from the Netherlands, saw the satellites of Jupiter in November, 1609, two months before Galileo, but he did not publish his observations until some years later.[22] Johann Fabricius apparently saw the solar spots from East Friesland in June, 1611, about the same time as had Galileo, and even preceded the latter in the publication of his observations.[23] Christopher Scheiner claimed that he saw the spots first from the palace of the Collegio Romano, but his observations date back only to November, 1611.[24]

Further evidence for the use of the telescope independent of Galileo's observations is found in the manuscript and letters of Thomas Harriot, mathematical tutor to Sir Walter Raleigh and an outstanding Elizabethan scientist. Harriot accompanied Sir Richard Grenville as surveyor to Virginia and, among other instruments, took ' a perspective glasse whereby was shewed many strange sights.'[25] On his return to England, Raleigh introduced him to the Earl of Northumberland, under whose patronage he spent the rest of his life in scientific studies. These included observations of the moon, made probably about the same time as Galileo's,[26] and observations on 'the newfound planets about Jupiter' (October 17, 1610, to February 26, 1612).[27] Between December 8, 1610, and January 18, 1613, Harriot made 199 observations of sunspots and determined thereby the period of the sun's axial rotation.[28] All these were made with a telescope probably brought over from Holland or constructed upon information from that country by Harriot's instrument-maker, Christopher Tooke. By July, 1609, his pupil, Sir William Lower, was making similar observations with telescopes supplied by Harriot. Lower's description of the moon is quite unique.

> According as you wished [he writes to Harriot][29] I have observed the moone in all his changes. In the new manifestlie I discover the earthshine a little before the dichotomie; that spot which represents unto me the man in the moone (but without a head) is first to be seene. A little after, neare the brimme of the gibbous parts towards the upper corner appeare luminous parts like starres; much brighter than the rest; and the whole brimme along looks like unto the description of coasts in the Dutch books of voyages. In the full she appeares like a tart that my cooke made me last weeke; here a vaine of bright stuffe, and there of darke, and so confusedlie all over. I must confesse I can see none of this without my cylinder.

Galileo, Fabricius, and Scheiner generally observed the sun by projecting its image on a white screen placed a convenient distance behind the eyepiece. In this way, sunspot positions could be more easily recorded and the eyes rendered free from the sun's glare and heat. As early as the fifteenth century, Apian had recommended coloured glass as a shield for the eyes, but both Galileo and Fabricius seem to have been unaware of his suggestion.[30] Galileo sometimes observed the sun when it was near the horizon or partially obscured by haze. Fabricius found that, by gradually introducing the sun's image into the field of view, his eye grew accustomed to its brightness[31]— a dangerous practice. Scheiner used the method of projection and also coloured screens which he placed between object-glass and eyepiece.[32] Harriot speaks of observing the sun 'thorough my coloured glasses',[33] a method of observation which he subsequently dropped, owing, no doubt, to the poor quality of the glass screens. He generally preferred to observe with unshaded eye when mist and thin cloud veiled the sun.[34] Sometimes he observed direct, and one observation, dated February 17, 1612, reads: 'all the sky being cleare, and the sonne, I saw the great spot with 10 and 20, but no more, my sight was after dim for an houre.'[35] John Greaves, later professor of astronomy at Oxford, took similar liberties and records that, in measuring the sun's diameter, he hurt his sight 'insomuch that for some days after, to that eye, with which I observed, there appeared, as it were, a company of crows flying together in the air at a good distance.'[36]

Until 1611, no theological scruples over the implications of Galileo's observations seem to have been felt in Rome. The first definite attack on theological grounds came early in that year from the monk Sizy who asserted that the moons of Jupiter were incompatible with the doctrines of Holy Scripture.[37] In 1612, Scheiner stressed the imperfection implied by a spotted sun. Its rotation on an axis, moreover, was contrary to Aristotelian and Ptolemaic teaching. Scheiner claimed that the spots were little planets which revolved about the earth and which sometimes appeared to cross the sun's disk. This interpretation, welcomed by the Aristotelians, was communicated to Welser at Augsburg in three letters and published there in 1612.[38] In three retaliatory letters,[39] Galileo challenged Scheiner on this point and questioned his priority as to the discovery of sunspots. He then published all six letters under the title *Istoria e Dimostrazioni intorno alle Macchie Solari*, and unreservedly took his stand with Copernicus.

In 1616, we again find Galileo of his own accord in Rome. This time he was warned, in view of the contents of his book on sunspots, neither to defend nor to hold the following two Copernican propositions:

1. The sun is the centre of the world, and immovable from its place.
2. The earth is not the centre of the world, and is not immovable, but moves, and also with a diurnal motion.

Both propositions were declared false and absurd philosophically and formally heretical. Galileo was advised to abandon his opinions. If he refused, then he would be commanded to abstain altogether from teaching or defending them.[40]

Galileo does not appear to have taken the admonition seriously for, in 1623, he published *Il Saggiatore*. In this work he refers to the Copernican hypothesis, talks of the idea of the motion of the earth as being false, but throughout veils his Copernicanism thinly. Barberini, then Urban VIII, was so pleased with the book that he had it read aloud to him at meals.[41] Galileo's next book, *Dialogo . . . Sopra I Due Massimi Sistemi Del Mondo*, published in 1632, was frankly Copernican. It relates an imaginary discussion between Salviati (Galileo), Sagredo (an ordinary man), and Simplicius (an Aristotelian). The book is a model of dialectic skill, and Salviati is made to argue the case for Copernicanism with great brilliance. The censors passed it after Galileo had made certain alterations, but regretted their action afterwards. Galileo's enemies soon penetrated its clever but thin disguises—it was a flagrant evasion of the 1616 prohibition.[42] Galileo was summoned to Rome, to be charged with deceiving the censor and disregarding the prohibition to the extent of openly teaching Copernicanism.[43] His recantation before the Inquisition followed and he was sentenced to virtual imprisonment in his villa at Arcetri, near Florence.[44] For nine years, until his death in 1642, Galileo was troubled by the surveillance of the officers of the Inquisition, failing health, and increasing blindness. Yet here he summed up and completed his discoveries in mechanics, publishing the *Dialoghi delle Nuove Scienze* (1636), after some delay due to restrictions, at Leyden in 1638. This book, in the form of a dialogue and distinctly modern in spirit, established mechanics as a science. Here we find the results of experiment developed by the accepted deductive methods of mathematical demonstration.

Galileo was more concerned with the optical qualities of his telescopes than with their mountings, which probably consisted of upright poles or tripod stands fitted with a small universal joint and cradle. Scheiner's *helioscope*[45] was a distinct improvement, in theory if not in use, on Galileo's stands. Scheiner made use of the idea of Tycho's polar axis, using it to carry a telescope in a frame fixed to the top of the axis. The lower end carried a circle graduated into twenty-four hour divisions. Once he had pointed the tube to the sun he had only to turn the polar axis at a slow and constant rate to follow the disk across the sky. The instrument constituted, in an elementary way, an equatorial telescope.

Before his total blindness in 1638, Galileo imparted his method of grinding and polishing lenses to Mariani of the Accademia dei Lincei.[46] Galileo seems to have used only the centre zones of his object-glasses, thereby improving image definition at the expense of image brightness. One of his telescopes of 5·1-cm full

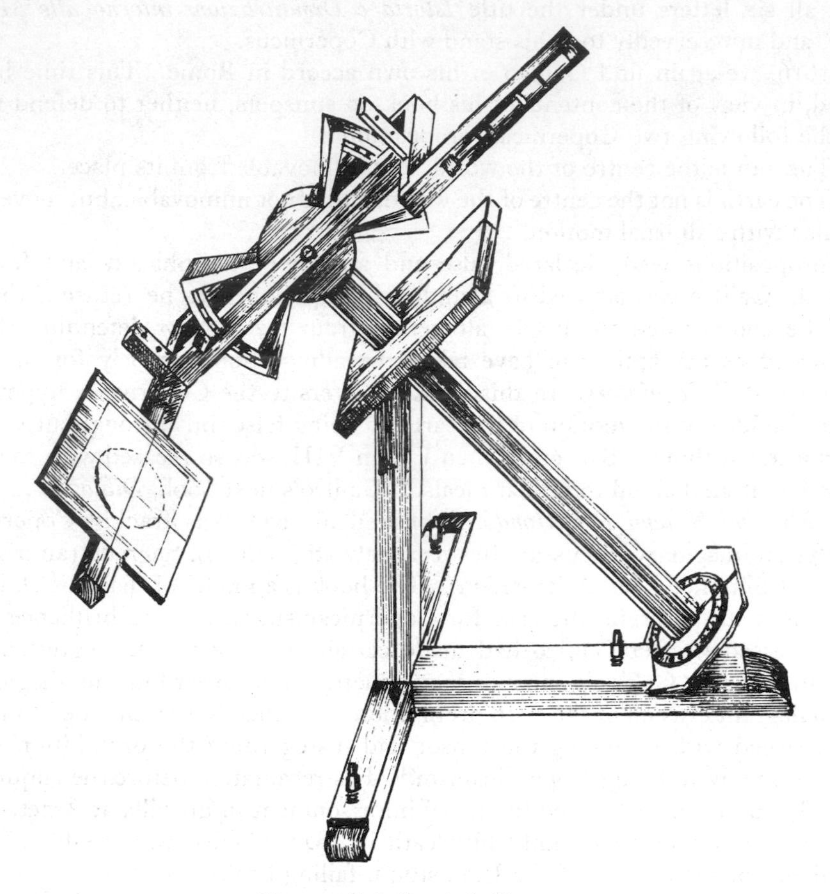

Fig. 17—Scheiner's helioscope

The first equatorially mounted telescope

aperture is stopped down to 2·6 cm, while the 2·6-cm diameter concave ocular is stopped down to 1·1 cm. This telescope, together with another of 3·7-cm clear aperture (stopped down to 1·6 cm) and a broken object-glass of 3·8-cm effective aperture, was tested by Abetti and Ronchi in the twenties of this century.[47] The first instrument, of magnification 14 diameters, showed the lunar craters Copernicus and Eratosthenes poorly defined and also four of Jupiter's satellites; the resolving power came out at 15 seconds of arc, the focal length 132·7 cm. The second telescope, of 1·6-cm clear aperture and about 20 magnification, gave fainter images than those shown by the first telescope. Saturn appeared olive-shaped, while the resolving power of 10 seconds caused more detail to be visible on the moon. The broken object-glass,* of 169-cm focal length, was easily seen to be better than the other two glasses. With one of Galileo's eyepieces it afforded views of a small group of five sunspots and lunar craters of the dimensions of Herschel (10″ to 15″).

The Galilean system gives a well-illuminated field and erect and bright images of distant objects. For these reasons it became popular, until the development of prism binoculars, as a naval and night glass. A disadvantage is its small field of view, even with low powers. The two telescopes just mentioned had fields of only 15 minutes, about a half of the moon's apparent diameter.[48] Two further defects, if the object-glass is a simple plano-convex or equi-biconvex lens, are spherical and chromatic aberration. The former is due to the fact that a lens with spherical surfaces fails to focus light from a point object to a point image. Rays incident

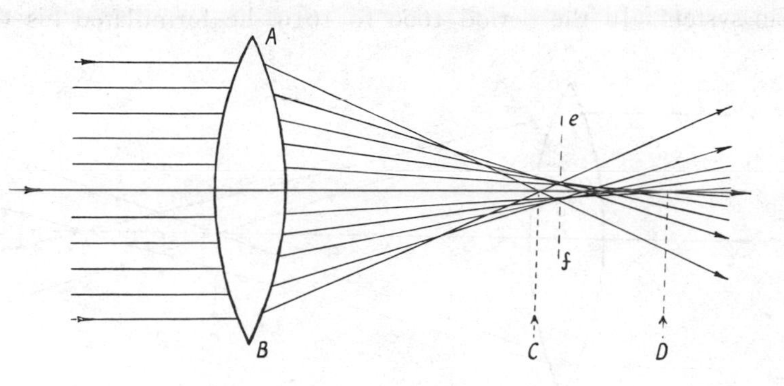

Fig. 18—Spherical aberration

AB	Biconvex lens
C	Focus for rays incident on marginal zone of lens
D	Focus for rays incident on axial zone of lens
CD	This distance is the longitudinal spherical aberration
ef	The position of the circle of least confusion

near the periphery, for example, focus at a point closer to the lens than those incident around the centre, with the result that the image, owing to this effect alone, is ill-defined and never comes into sharp focus (Fig. 18). A single lens also, because of the dispersive effect of glass, has a focal length longer for red rays than for violet rays. This means that the lens, when confronted by a distant white object, represents that object as a series of images arranged in the order of the colours of the spectrum and in increasing size from violet to red (Fig. 19). These images consequently subtend different angles when they are observed through an eyepiece. In practice, the image presented by an uncorrected telescope suffers from the joint effect of at least these two aberrations. When the object is both white and brilliant, like the planet Venus near maximum elongation, the image cannot be focused sharply and is marred by an encircling halo of an intense hue. The disk of the moon

* With which Galileo discovered the satellites of Jupiter. Viviani presented this broken glass to Prince Leopold of Medici who had it mounted in an ornamental frame. (Baxandall, D., *Trans. Opt. Soc.*, **24**, p. 309, 1923.)

under these conditions is likewise surrounded by a coloured blur, as are also the bright objects on it, all to the detriment of definition.

It was owing to these faults, no doubt, that Niccolo Zucchi, a Jesuit and professor of mathematics in the Collegio Romano, proposed in 1616 the use of a concave mirror in place of a glass objective.[49] He examined the image produced by a concave reflector with a negative eyepiece but, apparently, found so much aberration that he reverted to the lesser of two evils—the Galilean refractor.

Galileo corresponded with Johannes Kepler who, at that time, held the grossly underpaid but impressive-sounding post of 'Imperial Mathematician' to the Emperor Rudolph of Bavaria. Kepler worked as assistant to Tycho at Prague and, after the death of his teacher, undertook the task of reducing the Uraniborg observations. The study of these, together with the discussion of some of his own on the planet Mars, enabled him both to detect and to remedy weaknesses in the Copernican system. In the period 1609 to 1619, he formulated his three laws

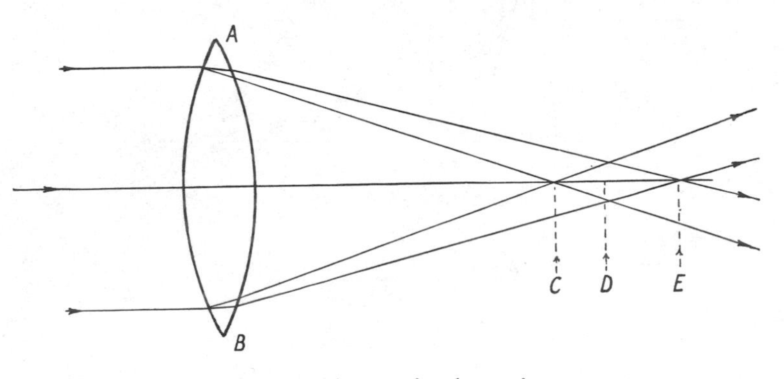

Fig. 19—Chromatic aberration

AB	Biconvex lens receiving parallel rays of white light
C	Focus for blue rays
D	Focus for yellow rays
E	Focus for red rays

of planetary motion, the basis of the *Rudolphine Tables* and a tribute to the ability of Tycho and the excellence of his instruments.

In reducing Tycho's data, Kepler had to eliminate errors due to atmospheric refraction. He was led to study optics and, in particular, the refraction of light. He established the principle that the amount of refraction depends on the amount of resistance offered by the medium, that is, on its density and not, as Tycho had supposed, on the nature of the object.[50] The density of the medium as well as its transparency caused, so he thought, the dispersion colours; he also gave light infinite velocity. He tried, unsuccessfully, to establish a law of refraction but, by giving glass a refractive index of 1·5 relative to air and by assuming that the angle of refraction is proportional to the angle of incidence, he was able to deal with simple lens problems.[51]

Kepler obtained a telescope in 1610, a gift from Ernest, Archbishop of Cologne, and, in his *Dioptrice* (1611), Kepler discussed its theory. In this work he enlarged upon his ideas on refraction and wrote about the anatomy of the eye. He described, for the first time, the defect of spherical aberration and stated that it could be overcome by giving optical surfaces hyperboloidal forms. This notion was suggested to him after he had studied the crystalline lens of the human eye, the anterior refracting surface of which is hyperboloidal. As far as he could judge, the eye was free from spherical aberration and he ascribed this to the form of the crystalline lens. He showed, also for the first time, that before an object can be seen distinctly, its image must be sharply formed on the retina. This theory of vision, to us obvious enough, was in Kepler's day in direct contrast to that held by some of his contemporaries, who still believed in the Euclidean idea that light proceeds from the eye to meet the object seen. Leonardo da Vinci had discussed retinal images towards

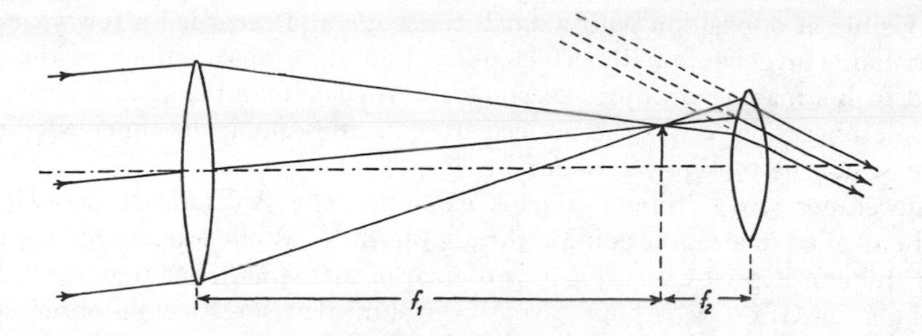

Fig. 20—The Keplerian telescope

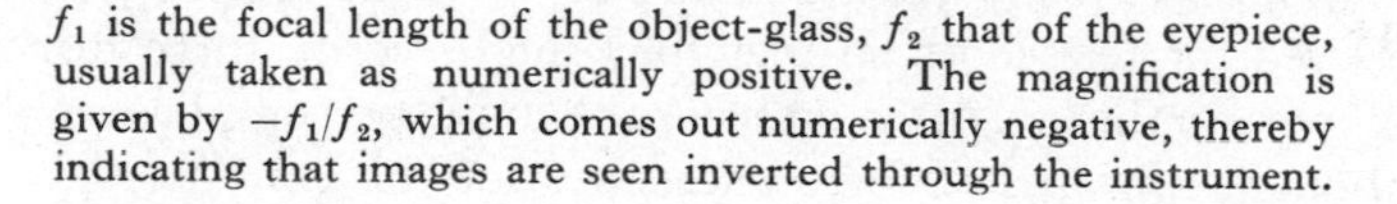

f_1 is the focal length of the object-glass, f_2 that of the eyepiece, usually taken as numerically positive. The magnification is given by $-f_1/f_2$, which comes out numerically negative, thereby indicating that images are seen inverted through the instrument.

the end of the fifteenth century, but Kepler was the first to establish their existence and the fact that they are inverted.

> Seeing [Kepler writes][52] amounts to feeling and stimulation of the retina, which is painted with the coloured rays of the visible world. The picture must then be transmitted to the brain by a mental current, and delivered at the seat of the visual faculty.

In the *Dioptrice* we find first mention of a refracting telescope with a positive eyepiece (Fig. 20). The latter offered a field of view larger than that of a Galilean system of the same power and proved invaluable, later in the seventeenth century, when reticules, cross-wires, and micrometers came into use. Kepler does not appear to have made a telescope with a positive eyepiece. He had an inaptitude for practical work and confessed that 'for observations his eye was dull and for mechanical operations, his hand was awkward.'[53] His eyepiece is not mentioned again until several years later, when Scheiner remarked in 1630 that he had made some observations in 1617, in the presence of Archduke Maximilian, with a telescope composed of two convex lenses.[54] Even after this date, so little was known of Kepler's telescope that when Anton Maria Schyrle of Rheita, Bohemia, mentioned it in 1645,[55] he was generally credited with its invention.

To Schyrle we owe the terrestrial eyepiece, for he introduced two further lenses, one the *field lens* to increase the field, the other the *erector* to erect the image.[56] In

1646, Francesco Fontana, a Neapolitan, claimed that he had conceived the use of a positive eyepiece before Kepler introduced his telescope but, for this as for his invention of the microscope, we have only his own testimony and that of two fellow Jesuits.[57] Fontana appears to have made a tube which contained eight convex lenses and which was bought by Cardinal Nepos for 800 crowns and presented by him to the Grand Duke of Florence. The latter complained that objects viewed through it appeared coloured ' so that the Tube could not be much used without injury to the Eyes.'[58]

Fontana was perhaps the first to use a Keplerian telescope for regular planetary observation. He saw the belts of Jupiter and a marking on Mars which he took to be a permanent feature of the planet.[59] This was probably the *Syrtis Major*, a region visible at opposition with a small telescope and recorded a few years later by Christian Huygens and Robert Hooke. Fontana's observations of the moon resulted in his making drawings to show the surface features at different phases and, while observing Venus, he noticed irregularities along the inner edge of the crescent which he took to be mountains.[60]

In December, 1612, Simon Marius examined the Andromeda nebula M31, which he likened to a candle shining through horn.[61] While looking for the comet of 1618, Johann Baptist Cysat, Jesuit professor of mathematics at Ingolstadt, found the nebula in Orion.[62] About 1643, Giovanni Battista Riccioli observed the shadows cast by Jupiter's satellites on the body of the planet, and he later noticed that ζ Ursae Majoris (Mizar) was a double star.[63] Little wonder, therefore, that others clamoured for telescopes so that they might verify these discoveries and, perhaps, add more to the list. It is on record that one person, the wealthy Nicholas de Peiresc, bought upwards of forty telescopes for this purpose.[64]

REFERENCES

1 Galileo Galilei, *Sidereus Nuncius*, translated by E. S. Carlos, 1880, p. 10.
2 *Ibid.*, pp. 10–11.
3 Albèri, E., *Opere di Galileo Galilei*, vi, pp. 75–77. Letter to Landucci, 29.8.1609, quoted by Gebler, Karl von, 1879, *Galileo Galilei and the Roman Curia*, p. 19.
4 *Il Saggiatore*, 1623; quoted by Fahie, J. J., 1903, *Galileo: his life and work*, p. 80.
5 Fahie, J. J., 1929, *Memorials of Galileo Galilei*, p. 84.
6 Gebler, *op. cit.*, p. 19.
7 Grant, R., 1852, *History of Physical Astronomy*, pp. 77–79.
8 Ref. 1, pp. 42–43.
9 Albèri, *op. cit.*, vi, p. 121, note 1; Gebler, *op. cit.*, p. 22.
10 Gebler, *op. cit.*, pp. 22–23.
11 *Ibid.*, p. 31.
12 Grant, *op. cit.*, p. 255, quotes *Opere di Galileo Galilei*, 1744, ii, p. 152.
13 Albèri, *op. cit.*, vi, pp. 130–133, 134–136, quoted by Gebler, *op. cit.*, p. 33.
14 Albèri, *op. cit.*, vi, pp. 137–138.
15 Letter to Julian de Medici. *Vide* Kepler, J., 1611, *Dioptrice*, Preface.
16 Rosen, E., 1947, *The Naming of the Telescope*, pp. 30–32.
17 Sirturi, 1618, *Telescopium sive ars perficiendi*, p. 27, quoted by Rosen, *op. cit.*, p. 53.
18 Rosen, *op. cit.*, pp. 66–67.
19 *Ibid.*, p. 54.
20 Gebler, *op. cit.*, pp. 35–36.
21 Grant, *op. cit.*, p. 216.
22 Marius, S., 1614, *Mundus Jovialis*. *Vide* Grant, *op. cit.*, p. 79. Johnson, J. H., *J.B.A.A.*, **41**, pp. 164–171, 1931. Pagnini, P., *J.B.A.A.*, **41**, pp. 415–422, 1931.
23 Fabricius, J., 1611, *Phrysii de Maculis in Sole Observatio*. *Vide* Grant, *op. cit.*, pp. 213, 215. Rigaud, S. P., 1833, Supplement to *Bradley's Miscellaneous Works*, pp. 35–36.
24 Grant, *op. cit.*, pp. 213–214, 216.
25 Harriot, T., 1588, *A Brief and True Report of . . . Virginia*, quoted by Baxandall, D., *Trans. Opt. Soc.*, **24**, p. 307, 1923.

[26] Rigaud, *op. cit.*, pp. 20–21.
[27] *Ibid.*, pp. 21–31.
[28] *Ibid.*, pp. 31–40.
[29] Quoted by Grant, *op. cit.*, p. 524.
[30] *Ibid.*, pp. 227–228.
[31] *Ibid.*, p. 227.
[32] Scheiner, C., 1630, *Rosa Ursina.*
[33] Rigaud, *op. cit.*, p. 33.
[34] *Ibid.*, pp 32–34.
[35] *Ibid.*, p. 34.
[36] *Ibid.*, p. 33, quotes Greaves, J., *Miscellaneous Works*, ii, p. 508.
[37] In the *Dianoja Astron.*, 1611. *Vide* Gebler, *op. cit.*, p. 39.
[38] Dated May 4, Aug. 14, Dec. 1, 1612. *Vide* Gebler, *op. cit.*, p. 43.
[39] Dated April 20, May 26, June 8, 1613. *Vide* Gebler, *op. cit.*, p. 44.
[40] Gebler, *op. cit.*, pp. 76–78.
[41] Albèri, *op. cit.*, ix, pp. 43–44. Gebler, *op. cit.*, p. 113.
[42] Gebler, *op. cit.*, pp. 171–174.
[43] *Ibid.*, pp. 201–229.
[44] *Ibid.*, pp. 230–248.
[45] Described and illustrated in Scheiner's *Rosa Ursina*, 1630.
[46] Baxandall, D., *Trans. Opt. Soc.*, **24**, p. 309, 1923.
[47] Abetti, G., *L'Universo*, **4**, Sept. 1923, No. 9. Ronchi, V., *L'Universo*, **4**, Oct. 1923, No. 10. Discussed by Baxandall, *op. cit.*, **25**, pp. 141–144, (1924).
[48] Baxandall, Ref. 47, p. 142.
[49] Zucchi, P., 1652, *Optica Philosophia*, i, cap. xiv, sect. V, p. 126.
[50] *J. Kepler* (1571-1630), *A Tercentenary Commemoration of his Life and Works*, 1932, pp. 16–17.
[51] *Ibid.*, p. 24. *Vide* Kepler, J., 1611, *Dioptrice.*
[52] Kepler, J., 1604, *Ad Vitellionem Patralipomena*, p. 168. *Vide* also Montucla, J. É., 1799–1802, *Histoire des Mathématiques*, ii, pp. 222 ff.
[53] Ref. 50, p. 34.
[54] Grant, *op. cit.*, p. 525.
[55] Schyrle, A. M., 1645, *Oculos Enoch et Eliae*, iv.
[56] Court, T. H., and Rohr, M. von, *Trans. Opt. Soc.*, **30**, pp. 210–212, 1929.
[57] Fontana, F., 1646, *Novae Coelestium Terrestriumque rerum observationes*, pp. 19–21.
[58] Gunther, R. T., 1923, *Early Science at Oxford*, ii, p. 334.
[59] Fontana, *op. cit.*, pp. 104–115.
[60] *Ibid.*, pp. 88–103.
[61] Marius, S., 1612, *Mundus Jovialis*, Preface. *Vide* Grant, *op. cit.*, p. 563.
[62] Wolf, R., 1853, *J. B. Cysat von Luzern.*
[63] Riccioli, G. B., 1651, *Almagestum novum*, i. p. 422. *Vide* Grant, *op. cit.*, p. 558.
[64] For an account of Peiresc's numerous activities with telescopes and microscopes *vide* Montucla, *op. cit.* ii, pp. 239–240. L'Abbé Rezzi, 1852, *Sulla Invenzione de Microscopio*, contains letters written by Peiresc between 1622 and 1624.

CHAPTER IV

Dans le temps qu'on s'occupait ainsi à perfectionner le travail des objectifs, on ne voyait point de bornes à la longueur des lunettes et aux découvertes du ciel. On pensait qu'il seroit possible de voir des animaux dans la lune.

A. Auzout

The beginning of the seventeenth century found the refracting telescope, so successfully applied to astronomy by Galileo, with two serious optical defects. The first determined effort to improve it was made by the French philosopher and mathematician, René Descartes. In his *Dioptrique*, published in 1637, Descartes described and explained the general optical properties of convex and concave lenses both singly and in combination. In possession of the sine law of refraction, originally formulated by Willebrord Snell in 1621, he showed that a lens with spherical surfaces cannot represent axial point objects as point images. His geometrical study of lenses with ellipsoidal and hyperboloidal surfaces revealed that a plano-convex lens, whose back, curved surface has been figured hyperboloidal, would meet this imagery criterion for point objects infinitely distant. He thought highly of this theorem and maintained that by this means it would be possible to increase the apertures and magnifications of telescopes.

While spherical aberration can be corrected in this way, Descartes failed to differentiate between this and chromatic aberration and overlooked the physical principles involved. He held, moreover, a theory of the nature of light and colour which, while interesting as a theory, was in the present instance of no practical value.* Neither Descartes nor any of his contemporaries was able to make a machine capable of polishing aspherical surfaces. Descartes did, in fact, design some machines and he commissioned a Paris optician named Ferrier to make aspherical lenses.[1] But the attempt failed, as did others that followed, despite the mechanical ingenuity displayed.

The same difficulty prevented the development of the reflecting telescope which Marin Mersenne, a Minorite friar, proposed at this time. Zucchi's proposed reflecting telescope had one serious defect—the image could not be viewed from in front of the mirror without obstructing the incident light. Mersenne suggested two concave paraboloidal mirrors, the larger perforated in the centre and faced by the smaller (Fig. 21).[2] As before, the main mirror or speculum took the place of the object-glass of the refracting telescope. The small secondary mirror conveniently reflected the light through the centre of the main mirror. Mersenne made no mention of an eyepiece and, apparently, intended the smaller mirror to act in that capacity. For this reason and the difficulty of grinding paraboloidal mirrors, Descartes thought the arrangement useless, and tried to dissuade Mersenne from

* *Vide* p. 67.

carrying it further. Mersenne does not appear to have put his plan into practice, although he mentions it in his *l'Harmonie Universelle* published in 1636.[3]

Astronomers were now guided by their experience that image quality is improved when the curves of an object-glass are made as shallow as possible. In taking this step, they decreased the aperture ratio until the effects of chromatic aberration

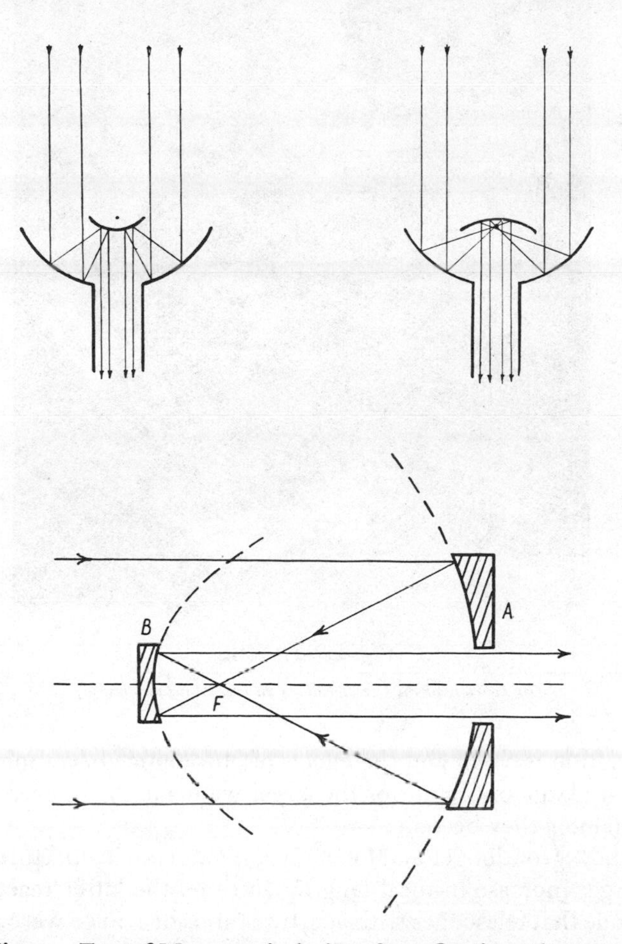

Fig. 21—Two of Mersenne's designs for reflecting telescopes

(*Adapted from 'L'Harmonie Universelle'*, 1636)

The lower diagram shows the optical system of one of the designs. *A* is the primary paraboloidal mirror centrally perforated, *B* is the secondary paraboloid, and *F* is the joint focus of the two mirrors.

became less noticeable. While, for any one convex lens, dispersive power remains constant with increased focal length, image size increases directly. The greater the focal length, the larger the image size and the less the *effect* of chromatic dispersion. Spherical aberration diminishes as the square of the focal length so that, for a lens of small aperture and large focal length, that is, a lens of small aperture ratio,

Fig. 22—Hevelius

(By courtesy of the Director of the Science Museum, London)

spherical aberration is small and the effects of chromatic aberration are less noticeable. Images are large but faint, for the larger we make them, keeping the aperture constant, the fainter they become.

Telescopes now grew longer and longer, every increase in aperture being followed by a much larger increase in focal length. In time, the latter reached such an extreme magnitude that telescopes with aperture ratios of 1 : 150 were not uncommon. Seventeenth-century prints convey the unconventional appearance of these instruments with their long, flimsy tubes, suspended from high masts by ropes and pulleys. It is not surprising that the lenses were seldom strictly in line, but the smallness of the aperture ratio greatly reduced extra-axial aberrations and allowed much latitude in focusing. The greatest advantage of long telescopes, apart from reduced chromatic effects, was undoubtedly the high magnification they provided with long-focus eyepieces.

One of the first to make long telescopes was Johannes Hevelius of Danzig, a brewer by profession and later councillor of his city (Fig. 22). Hevelius was introduced to astronomy by Peter Krüger, and built himself an observatory at his house in Danzig. Here he observed sun, moon, planets, and also occasional comets.[4]

In 1647 he published his first work, the *Selenographia.* This was the first complete lunar atlas. It embodied the results of four years' persistent observation with the telescope and, besides giving drawings of the moon in all her phases, included nearly 250 named lunar formations. It surpassed the inaccurate and less detailed maps of Scheiner and Langrenus and, although subject to the same errors of eye-observation, formed a sounder basis for further researches. It contained, besides, a short account of Hevelius' instruments, together with his remarks on telescopes. At this time none of Hevelius' telescopes apparently exceeded 12 feet focus and about 50 magnification.

In 1659, the publication of Christian Huygens' *Systema Saturnium* brought with it accounts of telescopes many times longer than those at Danzig. It appeared that the two brothers, Christian and Constantine Huygens, had grown dissatisfied with the small telescopes then obtainable and had made their own. Success rewarded their efforts and, in March, 1655, they directed a new telescope of just over 2 inches aperture and 12 feet focus towards Saturn, but nothing very definite was seen.[5] The aspect of the planet was such that the plane of the rings almost passed through the earth. On March 25, however, Christian Huygens discovered Titan, the brightest member of Saturn's large family of satellites.[6] Towards the end of 1655, the rings were still almost edge-on, but a new 23-foot telescope[7] made at this time revealed the tapering of the ansae at their extremities. Another feature, seen with both instruments, was a darkish line passing over the disk—the shadow of the rings upon the body of the planet. By January, 1656, Saturn appeared quite round and without an appendage, even when examined with a 123-foot telescope. By October, the rings had opened and, in the following year, showed appearances that left no doubt as to their circular nature. Huygens had already come to this conclusion when he hinted at their true nature in an anagram that he included in a small tract *De Saturni Luna Observatio Nova.* The solution came in 1659, when he divulged that the planet was accompanied by ' a ring, thin, plane, nowhere attached, and inclined to the ecliptic.'[8]

Another discovery due to Christian Huygens, and made without any knowledge of Cysat's nearly forty years previously, was that of the great Orion nebula.

> In the sword of Orion [Huygens writes][9] are three stars quite close together. In 1656 I chanced to be viewing the middle one of these with a telescope, instead of a single star twelve showed themselves (a not uncommon occurrence). Three of these almost touched each other, and with four others shone through the nebula, so that the space around them seemed far brighter than the rest of the heavens, which was entirely clear and appeared quite black, the effect being that of an opening in the sky through which a brighter region was visible.

Upon hearing of Huygens' discoveries, Hevelius planned several new telescopes. He had himself studied the problem of Saturn's appearance, and the Dutchman's work convinced him of the superiority of long-focus object-glasses. The outcome was that he made telescopes of 60 and 70 feet focus and, finally, one of 150 feet, all of which he described in his *Machinae Coelestis.*[10] The lenses for the 150-foot were made by a local glassworker ' expert in all kinds of mechanical as well as optical studies ',[11] and occasioned less trouble than the mounting. For this, Hevelius used

Fig. 23—Hevelius with a long telescope

(Science Museum, London. British Crown copyright)

wood. A paper tube, although light, would have been too flimsy and fragile, an iron tube too heavy and costly. The tube was sectional, each section consisting of two 40-foot wooden planks fixed at right angles to each other.[12] Three or four of these sections, joined end to end, made a two-sided trough; at the further end was the objective cell, at the other, the eyepiece. This arrangement, braced by wire stays, answered for night use but, during twilight or moonlight, the eyepiece had to be shielded from stray light. Hevelius, therefore, fixed wooden apertures or 'stops' at intervals along the tube.[13] These not only assisted in its re-alignment but added to the rigidity of each section. The entire apparatus was suspended from a mast 90 feet high[14] and was operated from below by means of ropes and pulleys.

Hevelius wrote convincingly of the ease and rapidity with which he could prepare the 150-foot telescope for observation,[15] but the opposite was more probably the truth. The tall mast was the only fixture, for the proximity of the site to the Baltic Sea precluded the permanent elevation of the tube. A band of assistants was required to work the different ropes that either raised or lowered the tube. Then

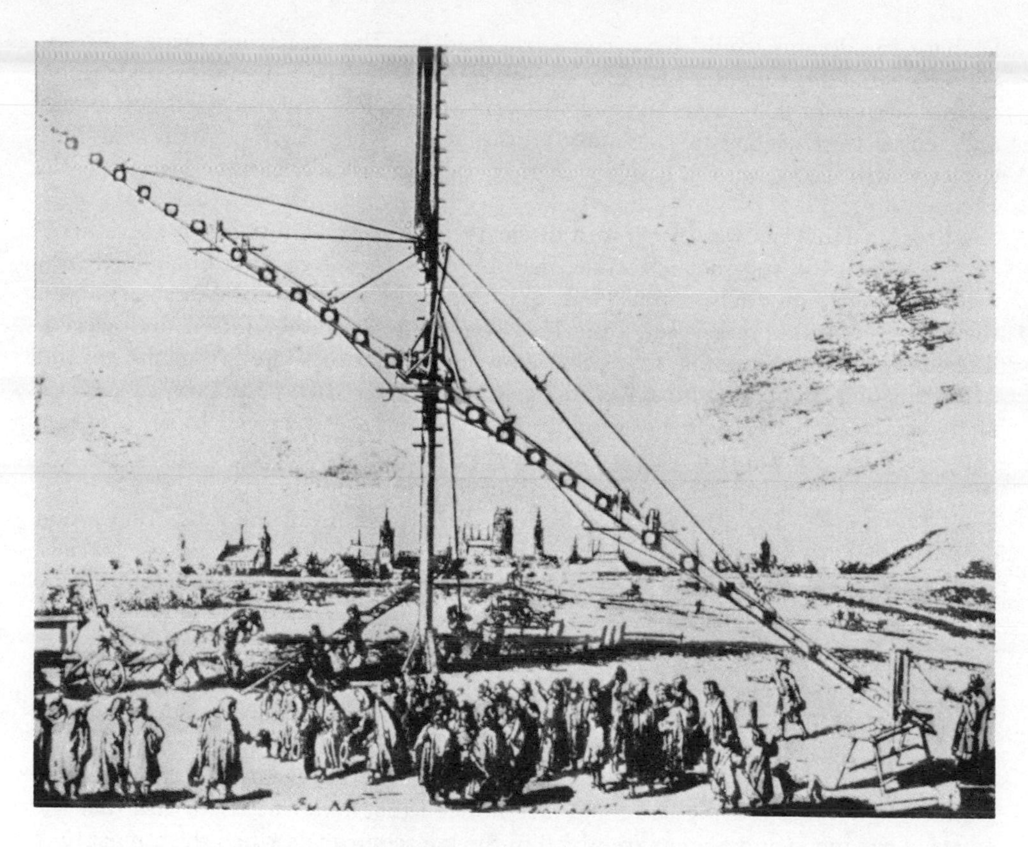

Fig. 24—Erecting the 150-foot telescope at Danzig

the object had to be located and ' followed ' with high magnifications. This is difficult enough with a small telescope, let alone with a long, flimsy tube moved by several pairs of rough hands. Only the finest nights were suitable for observing, for the slightest breeze would cause the framework to quiver like a reed. The warping of the wooden planks presented another difficulty, and ' give ' in the ropes due to changing humidity must have required the almost constant re-adjustment of the tube and guiding mechanism. Changes in the alignment of the tube were perhaps the greatest problem and Hevelius attempted to control them by running cords of adjustable tension from objective cell to eyepiece.[16] According to Halley, even these did not keep the lenses in a straight line, and he described Hevelius' telescope as ' useless ' because of this defect.[17] For these reasons, the 150-foot telescope was used only for occasional observations.

Hevelius spared neither time nor money in the construction of his telescopes. The *Machinae Coelestis*, with its fine plates and typography, is characteristic of the joy and labour which he derived from and gave to his work. The fame of Sternenburg, his Danzig observatory, spread throughout Europe and, if we judge his illustrations rightly, the erection of his giant 150-foot telescope was a big event in the city (Fig. 24). He conceived plans for an ' ideal ' tower-observatory, suitable for housing giant telescopes and with facilities for their immediate erection.

Perhaps his dreams would have come true had not the disastrous Danzig fire of 1679 consumed the greater part of his observatory.[18] He raised a new edifice upon the old, although then in his sixty-eighth year and, to this end, received new books and lenses from Halley in England. Observations were resumed with the same tireless enthusiasm, but the new observatory failed to rival the grandeur of the old Sternenburg.

Christian Huygens was the first to discard the long wooden tubes of Hevelius.[19] In the case of the 123-foot telescope, he mounted the object-glass in a short, iron tube and placed it on a high pole (Fig. 25). A groove running up the mast enabled the lens mount, counterpoised by a lead weight, to be either raised or lowered. The lens carrier was mounted on a ball-and-socket joint, operated from the ground by means of a connecting thread or string and reached, when necessary, by a series of triangular steps. The image produced by the lens was received by an eyepiece supported by two wooden feet and attached to the free end of the thread. Usually, the eyepiece was held in the hand, the observer steadying his arms on the wooden rest. The length of thread not only indicated the position of the focus but, when taut, aligned the object-glass with the eyepiece. Owing to its open or 'aerial' nature, great difficulty was at first experienced in pointing the glass to faint objects. Bright images were generally found by receiving them on a white, pasteboard ring fixed round the eyepiece or, more conveniently, upon an oiled and therefore translucent paper screen. On a dark night, the object-glass was illuminated by a lantern held by the observer.[20] The latter then searched for the lantern's reflection and, in so doing, brought the lenses into line.

Huygens was the first astronomer to appreciate the importance of atmospheric conditions upon telescopic 'seeing'. He noticed that stars twinkled, and that the edges of the moon and planets trembled in the telescope, even when the atmosphere appeared calm and serene. One could object to his aerial telescope on the grounds that the string would curl in windy weather, but he points out that a tube would shake far more and that observations made under such conditions would be of little value. So frequent are nights when the seeing is poor that Huygens warns the observer against too hastily blaming his telescope.[21]

Newton considered the aerial telescope a great improvement on the long tubes of the period:

> For very long tubes [he writes][22] are cumbersome, and scarce to be readily managed, and by reason of their length are very apt to bend, and shake by bending, so as to cause a continual trembling of the Objects, whereby it becomes difficult to see them distinctly, whereas by his [Huygens'] Contrivance the Glasses are readily manageable, and the Object-glass being fixed upon a strong upright Pole becomes more steady.

A further innovation due to Huygens is the compound negative eyepiece, which still retains its original form and the name of its inventor (Fig. 26). Two thin convex lenses of the same type of glass can be so separated as to give correction for transverse chromatic aberration—to yield an image of a white object which subtends the same angle for all colours. In its modern form the eyepiece consists of two plano-convex lenses of crown glass, one with a focal length of two to three times that of the other and separated by a distance equal to half the sum of their focal

Fig. 25—Huygens' aerial telescope

(*Science Museum, London. British Crown copyright*)

lengths. Huygens found that the combination gave better definition over a field of view larger than that given by a single lens—he was unaware of the achromatic properties of the combination. Today, we recognize that the system, in its basic form, gives rise to appreciable extra-axial aberrations and presumes an optically perfect eye. To work effectively, the eyepiece design should take into account the degree of correction of the objective and the aberrations and ocular refraction of the observer's eye.

Huygens appreciated the value of ' stops ' or circular apertures in lens-systems. Set along the main tube, these stops prevented light, reflected from the sides of the tube, from entering the eyepiece, while one in the eyepiece cut off extreme marginal rays and gave the apparent field a well-defined and circular edge. His eyepiece was certainly more useful than many of the compound eyepieces made by his predecessors. Schyrle of Rheita, for instance, made an eyepiece with five separated convex lenses; Fontana had one with eight lenses.[23] Eustachio of Naples broke all

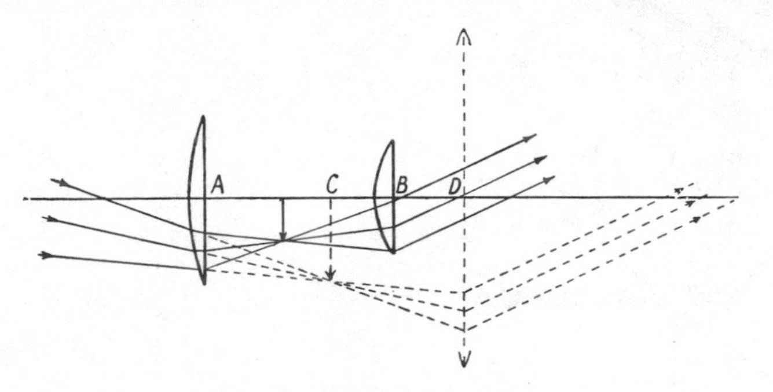

Fig. 26—Huygens' eyepiece

A Field-lens
B Eye-lens
D Position of single lens giving the same magnification
CD Focal length of equivalent lens

records with nineteen spaced convex lenses,[24] thereby producing a highly coloured, faint and distorted image.

Christian and Constantine Huygens made a large number of object-glasses and eyepieces, and their improved method of grinding and polishing aroused considerable interest among opticians and spectacle-makers. In 1661, Christian Huygens travelled to England and communicated his process to his friends at the Royal Society. His success in glassworking lay, not so much in the method he adopted, as in the skill of hand gained after many years of practical work in grinding and polishing. Some time after Huygens' visit, Samuel Molyneux gave an account of Huygens' technique in Robert Smith's *Opticks*.[25] His father, William Molyneux, visited Huygens in 1685 and was shown the various instruments in his garden at the Hague. Probably this personal contact with Huygens did much to kindle Molyneux's interest in the problems of the telescope.

Christian Huygens' method of making a long-focus objective, as told in Smith's *Opticks*, is briefly as follows. To simplify working, Huygens chose a biconvex form and, having determined the curves, cut a metal gauge to the required radius.

The grinding tool was of either copper or brass, and generally exceeded twice the proposed diameter of the lens. He cast the metal in a round, flat slab about $\frac{1}{2}$-inch thick and then ground it with emery and flat grinding-stones of increasing size until the surface fitted the gauge. Two grinding-stones usually sufficed, one about half the size of the tool, and the other equal to it. The tool was then polished by coating one of the stones with pitch and feeding emery between the pitch and metal surface as the two were rubbed together. To assist this abrasive action, Huygens worked the polisher under considerable pressure. Newton thought that too great a pressure tended to deform the shape of the tool and polisher, but Huygens went to the other extreme by placing a long and slightly bent pole between the polisher and the ceiling of the room. This acted like a strong spring and pressed down the polisher so hard that usually two persons were required to give it the necessary motions. When the brass tool had been brought to a good polish, it was ready to receive the glass blank. This was ground on the tool with moist emery, and once again the long pole controlled motions and pressure. Ideally, the apex of the blank's motion should have coincided with the centre of the tool's curvature but, with long-focus glasses of shallow curvature, this was not always practicable and strokes of shorter radius answered quite well. Huygens usually used one grade of emery throughout, washing off the coarser grains after periods of about fifteen minutes' continuous grinding. In this way, he was left with an ever finer emery residue, one that gave a progressively higher polish to the glass. Other workers, like the London spectacle-makers Cox and Scarlett, completed the polish with the grinding tool, but for this Huygens advocated a separate cloth polisher.

It will be seen how similar some of these early ideas are to modern lens-working techniques. Some opticians still use a one-grade abrasive and its ever finer residue, but the availability of selected grades of carborundum has now thrown the method into comparative disuse. Emery and sometimes a fine form of sandstone were used until late in the nineteenth century. The labour spent in grinding was considerable and a lens that took several hours to work can now be completed in as many minutes.

Huygens' cloth polisher consisted of a round block of slate, a little smaller than the glass and ground to the concavity of the brass tool. This slab was covered with a rough-textured cloth which in turn received a coating of wax and pitch. A central aperture in the slate enabled a hollow conical plate to penetrate almost to the cloth layer and the cone received the lower end of the guiding pole. The glass was polished with either putty powder (powdered copper sulphite and tin oxide) or tripoli.

Chérubin d'Orléans, a prolific microscope- and telescope-maker, both ground and polished his lenses with one and the same tool. He used fine sandstone for an abrasive and, after the grinding process, covered his tool with silk rubbed over with wet putty. On other occasions he paper-polished his lenses, but in both cases he departed from Huygens' use of a long pole. Instead, he stuck a short handle to the lens and ground it by hand on the brass tool.[26]

A less complicated but apparently reliable technique was published in 1694 by Nicolas Hartsoeker, a Dutch optician. Hartsoeker made several long-focus telescopes for the Paris Observatory. His process is similar to one used today. He took a plate of glass, about a third larger than the glass to be worked, and ground it with emery under a small glass disk. The glass plate eventually formed itself into a concave tool, and on this he worked the object-glass until it had acquired a spherical curve. As a check on the curvature of the lens surfaces he gave them both a rough polish, just sufficient to produce an image of the sun. If the focus was too short and the curves too steep, he would re-work one side using a different degree of pressure. The process would then be repeated until he obtained a focal length near the one required.[27]

Seventeenth-century opticians tested their object-glasses either in the workroom or upon some well-known celestial object. Light reflected from the lens surfaces gave some idea of the quality of the polish, and an examination of the refracted cone of light in the region of the focus decided the accuracy of the curves. Failing these tests, it was customary to view a distant object. A page of the *Philosophical Transactions* with its regular type was often used, and the ease with which this could be read with different magnifications gave a rough idea of the 'goodness' of the telescope. Another method, and one that decided the performance of the telescope under typical observing conditions, was to examine stellar and planetary images and small craters on the moon.

Hevelius often obtained large and measurable spurious diameters for bright stars when he diminished the aperture of his 150-foot telescope. These images, he thought, represented the actual disks of the stars, and he made laborious but useless measures of their diameters with a bar micrometer. The same appearances deceived J. Cassini as late as 1717,[28] but Huygens thought the disks insensible and due to an anomaly of vision. John Flamsteed, observing with a 14-foot refractor at Derby, found some disks to be of appreciable size. He thought that Huygens' inability to see them was due to his practice of slightly smoking the eyepiece lens[29] so as to minimize the effects of chromatic aberration.

By far the most prolific makers of long-focus object-glasses during the seventeenth century were Eustachio Divini of Bologna and Giuseppe Campani of Rome. Between them, these rival[30] craftsmen produced some of the finest object-glasses (and large compound microscopes[31]) of this period. The greater fame of Campani was due mainly to the remarkable discoveries made with his telescopes by Jean Dominique Cassini. In 1650, Cassini succeeded Cavalieri as professor of astronomy at the University of Bologna.[32] Here he became acquainted with the work of Campani and, under the clear Italian sky, made many planetary observations. Together with Campani, he studied the rotation periods of the planets and, in 1666, deduced that Mars rotated in 24 hours 40 minutes on an axis almost perpendicular to the plane of the planet's orbit.[33] From the motion of a bright spot on Venus he found a period of 23 hours 15 minutes, the axis in this case seeming almost in the plane of the planet's orbit.[34] He substantiated Hooke's suspicion that Jupiter revolved on an axis[35] and noticed a flattening of the disk which he estimated to be 14/15.[36] He gave attention to the eclipses of Jupiter's satellites, noticed their shadows on the body of the planet and drew up tables of their motions.[37] In 1669, Cassini came to Paris at the request of Louis XIV. Here he continued his observations with a 17-foot telescope, given to him by Campani, and extended them by determining more accurately the period of the sun's rotation.[38] The excellence of this instrument impressed the astronomers of the French Academy, particularly

J. B. Colbert, the King's influential minister. In consequence, a 34-foot telescope was ordered from Campani and presented by Louis XIV to the newly erected observatory of Paris.[39]

In September, 1671, Cassini received an invitation to superintend the work of the observatory. A month later, and with the 17-foot telescope, he discovered Iapetus, another of Saturn's satellites.[40] In December, 1672, he came across Rhea, Saturn's fifth satellite in order of distance, this time with the 34-foot telescope[41] and, in March, 1684, two more—a total of four satellites.[42] The last two, Tethys and Dione, were found with Campani telescopes of 100 and 136 feet focal length.

The novelty of Cassini's observations, made with telescopes of ever increasing length and power, raised the most extravagant hopes—indeed only such could have justified the absurd length of certain telescopes made about this time. The largest Campani glass at the Paris Observatory had a focal length of 136 feet and, because of its size, was used as an aerial telescope. Compared with some Dutch glasses it was quite small. Hartsoeker ground lenses of 155 and 220 feet focus,[43] Constantine Huygens possessed 150-, 170-, and 210-foot glasses, and A. Auzout, a French physicist, produced some with the completely unmanageable focal lengths of 300 and 600 feet.[44] Auzout even speculated on charging his longest glasses with magnifications of the order of 1000 in the hope of seeing animals on the moon.[45] The table of aperture ratios which he published in 1665 reveals the great importance which he attached to long telescopes.[46] Cassini found Campani's more modest glasses difficult enough to handle and usually mounted them in light, wooden ' tubes ' which were suspended from a high mast on the observatory terrace. For the giants, a more solid support was essential and, at one time, Cassini made use of the old wooden Marly water-tower. This was transported to the observatory, steps were provided to the top, and a balcony was fitted to prevent assistants from falling off on a dark night.[47] In this way a tower that once held water served the unique purpose of carrying both astronomers and their telescopes.

Cassini's work at Paris gave new life to the study of physical astronomy. The nature of his observations appealed to the public and, as his work met with the support of king and court, he did not lack good instruments. For many years he studied the surface of the moon, embodying his results in a map and in crayon drawings of particular regions.[48] In 1675, he discovered that Saturn's supposed ring consisted of two rings,[49] separated by a space now known as ' Cassini's division '. To several astronomers this work had little appeal; they regarded the Paris Observatory as an institution to be set apart for less spectacular although more valuable observations in fundamental astronomy. Probably they also blamed Claude Perrault for designing a building so unsuited for astronomical observations. But both Cassini and Perrault had higher authorities to contend with and, because of these, the observatory was never properly organized.

Cassini found Colbert a generous dispenser of royal favours. Many thousand francs were paid to Campani and Divini for object-glasses; the removal of the Marly tower alone cost nearly 11,000 francs. Louis even invited Campani and Divini to divulge the secret of their success in return for a large compensation, but they appear to have refused the offer.[50] Colbert thought their glasses the best in all France, much to the chagrin of Borel and other French opticians. Cassini writes of the many trials he made with the rival glasses, in company with Picard, Huygens, Perrault, and Römer. In his opinion, there was little to choose between those of Campani and those of Divini, but both were superior to French glasses. Colbert showed great foresight in his endeavours to secure large disks of optical glass. Correspondence was opened with Viviani of Venice and with Giovanni Borelli, both prolific telescope-makers and disciples of Galileo, and Pasquin and Hartsoeker were sent to make telescope lenses at Colbert's own glassworks near Cherbourg.[51]

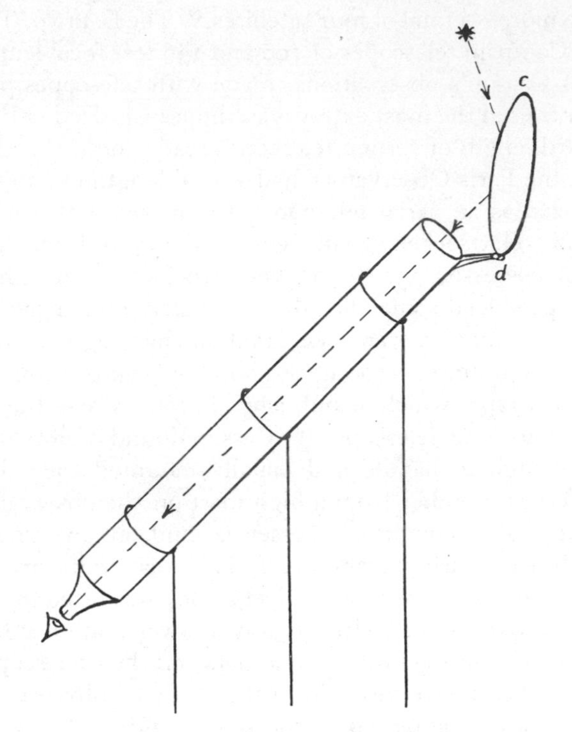

Fig. 27—R. Cotes' suggestion for a polar telescope

Tube and mirror rotate about the optical axis and so keep the star in the field of view. To bring another star into the field, mirror *cd* must be reset.

The unmanageability of long telescopes encouraged suggestions for keeping the tube in a fixed position while light was fed to the object-glass by a moving mirror. Such an arrangement, now called the *polar heliostat*, was proposed in 1682 by Boffat of Toulouse,[52] who fixed the tube in the direction of the celestial pole and fed it by means of a single movable mirror driven at a constant rate. A little later, Perrault brought out a similar arrangement.[53] He placed the tube horizontally and discussed devices whereby the mirror could be so operated from the eyepiece as to follow the apparent diurnal motion of the stars. Hooke, Halley, and Cotes all studied the problem,[54] but without practical success (Fig. 27). The first really useful solution did not appear until 1720 with s'Gravesande's *heliostat*.[55] But the polar heliostat never became a practical instrument owing to the difficulty of making flat mirrors and giving them a smooth and continuous motion. Another way out of the problem was to mount the object-glass upon an equatorial clock-drive and to follow the image by moving the eyepiece support. This apparatus was apparently first tried out by J. D. Cassini in 1685 and was briefly mentioned by W. Molyneux in his *Dioptrica Nova.*[56] Molyneux saw it whilst on a visit to Cassini and described it as ' a plain Piece of Clock-work, moved by a Spring and regulated by a Pendulum Vibrating half Seconds '.[57] Its weakness was the erratic motion of the clockwork, although Cassini thought so well of it that he attached circles for reading off the declination and hour angle of the object under observation.

The impossibility of polishing perfect optical flats defeated Hooke's ingenious plan for shortening telescopes. In 1668, he devised a system of mirrors which, by successive reflection of the light after it had passed through the object-glass, contracted a 60-foot telescope to a box only 12 feet long (Fig. 28).[58] Of greater practical

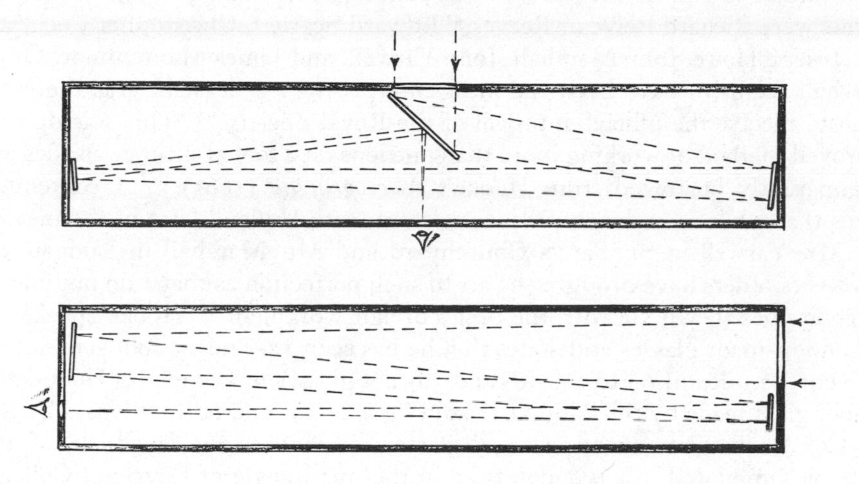

Fig. 28 Two of Hooke's proposals for shortening the length of long refracting telescopes

importance were the many mechanical devices put forward to facilitate the erection and operation of long telescope tubes and, in this field, France and Italy maintained the lead. In England, the Royal Society, represented by men like Hooke, Wren, Halley, and later, Bradley, encouraged research in the improvement of lenses and eyepiece combinations. Sprat in his *History of the Royal Society* tells of their early labours in this field and says that they constructed ' several excellent Telescopes of divers lengths of six, twelve, twenty-eight, thirty-six and sixty feet long, with a convenient Apparatus for the managing of them: and several contrivances in them for measuring the Diameters, and parts of the Planets, and for finding the true position, and distance of the small fix'd Stars, and Satellites.'[59] In 1661, the Royal Society appointed a committee ' to consider all sorts of tools and instruments for making glasses proper for perspectives.'[60] Hooke invented a machine for grinding lenses of long focal length[61] and Wren had some success with machine-ground aspherical surfaces.[62]

For ordinary terrestrial use, long telescopes were both heavy and clumsy, and there was a greater demand for the Dutch 'perspective glasses' as Galilean telescopes were then called. At first, the London opticians had neither the skill nor the glass to compete with Italian work and, for some years, 'perspectives' were imported in large numbers from the Continent. After 1628, many London spectaclė-makers turned their attention to optical instruments in whose manufacture they became more and more skilled. In that year, a group of craftsmen engaged in the 'art or mistry' of spectacle-making decided to follow the example of other bodies in the City of London and obtained a charter of incorporation.[63] The newly founded Worshipful Company of Spectacle Makers began to lay down rules for the general improvement of their trade. The new controls, enforced by the infliction of penalties, had a salutary effect on the trade, as they at once raised the spectacle-maker to a status above that of the haberdasher and unskilled retailer.

Towards the end of the seventeenth century, the leading London 'prospect' makers were Richard Reive or Reeves,[64] Edward Scarlett, Christopher Cock,[65] John Cox, Joseph Howe, John Marshall, John Yarwell, and James Mann junior. Of these, Marshall seems to have been the most enterprising—at least he was the first optician to receive the official approval of the Royal Society.[66] This was due to the improved method of working more than one lens on a tool at a time, an idea which he apparently borrowed from Hooke's *Micrographia* (1665).[67] A contemporary writes that while four dozen prospect glasses from Holland 'are of the mean sort, . . . Mr. Yarwell in St. Paul's Churchyard and Mr. Marshall in Ludgate Street and divers others have brought the art to such perfection as that I do not doubt but for good ones they'll vie with the best Foreign workmen.'[68] Hooke speaks highly of London-made glasses and states that he has seen 12- and 15-foot samples which were equal in definition to the 36- and 50-foot glasses of Campani. He mentions a 36-foot glass made by Sir Paul Neille and 'good' 50- and 60-foot lenses by Reeves and Cox.[69] The latter made several glasses for Hooke, one of which, of 36 feet focus, was mounted in a wooden tube in the quadrangle of Gresham College. It was with this telescope, fixed to the chimney stack in his chambers and pointed through a hole in the roof, that Hooke tried to detect displacements of zenith-culminating stars due to parallax.[70]

Examples of the first London-made telescopes are still in existence. Those up to 3 feet focus have only one tube, but longer glasses consist of several sliding or telescoping tubes. These tubes are made of either shagreen, leather, or parchment. Often they are arranged to take two object-glasses, one for day and the other (the larger) for night use. The lenses are fixed in wood or ivory cells while wooden rings or diaphragms prevent internal reflections.

Generally speaking, telescopes of extreme focal length never became popular in England. Observers like Halley and Flamsteed seemed content to leave discoveries in physical astronomy to the astronomers at Paris.

Halley landed at St Helena in February, 1677, and, with a $5\frac{1}{2}$-foot radius sextant, a 2-foot quadrant, two micrometers, a pendulum clock, and a 24-foot telescope, began a survey of stars near the south celestial pole.[71] After many difficulties and much disappointment due to fog and persistent cloud, he collected the positions of 341 stars in this then little-known part of the sky, and thus made the first southern star catalogue.[72]

Fig. 29—The Royal Observatory in Flamsteed's time

Observers are shown with quadrant and telescope

(By courtesy of the Director of the Science Museum, London)

Flamsteed, newly installed as first Astronomer Royal at Greenwich, no more comfortably pursued a similar plan. The observatory (Fig. 29) was founded in 1675 for the express purpose of assisting navigation by the provision of correct star positions and reliable tables of the sun, moon, and planets. No one was better fitted to see that this objective was effectively kept in view than was Flamsteed. He not only accumulated a large number of valuable observations, but himself directed their reduction. His instrumental equipment, contrary to that at the Paris Observatory, was particularly poor—it was, for the greater part, provided by himself and his friend, Sir Jonas Moore. It consisted of an iron sextant of 7 feet radius, two clocks, a 3-foot quadrant, and two telescopes of 7 and 15 feet focus.[73] Later, and at his own expense, Flamsteed added an object-glass of 90 feet focus,[74] the work of Borelli, and a mural arc by Abraham Sharp. Ill health, misunderstanding and the critical attitude of his would-be friends increased his difficulties but, despite all these, he became one of the most eminent practical astronomers of his age. His great star catalogue, the *Historia Coelestis*, formed a sound basis for precision astronomy for almost a century.

At least three of Huygens' object-glasses found their way into England. The 123-foot focus glass of $7\frac{1}{2}$ inches aperture, together with the supporting 'aerial' apparatus, was presented by Constantine Huygens to the Royal Society in 1692.

Its unmanageability was such that the Society considered erecting it for zenith observations on a high building, but none had the requisite height or stability.[75] Hooke and Halley at one time received orders to ' view the scaffolding of St Paul's Church, to see if that might not conveniently serve for the present, to erect the object-glass thereon, for viewing such of the celestial objects as now present themselves.'[76] In 1710, James Pound borrowed the glass and mounted it at Wanstead Park on a maypole just removed from the Strand. The fact that Sir Isaac Newton was instrumental in obtaining this pole, gave rise to the verse attached to it by a local wit:[77]

Once I adorned the Strand,
But now have found
My way to pound
In Baron Newton's land.

J. Crosthwaite, Flamsteed's assistant, was not impressed by the telescope's performance. Writing to Sharp on May 6, 1720, he says that Pound ' showed me Jupiter, which I could perceive very distinctly; so that I believe the glass is good: but then the motion of the air, the shaking of the pole, &c., renders it very difficult to trace the object, and makes me conclude that not many good observations can be made with a glass of 123 feet long in the open air.'[78]

Pound, however, seems to have turned the telescope to good account. He made micrometrical measures of the positions of the satellites of Jupiter and Saturn, the diameters of planets' disks and the rings of Saturn.[79] He furnished both Halley and Newton with data for their computations. At a later date, the object-glass came into the hands of Derham, but he failed to find a pole long enough to carry it.

> The chief inconvenience [he wrote in 1741][80] is the want of a long pole of 100 or more feet, to raise my long glass to such a height as to see the heavenly bodies above the thick vapours. . . . But, as I have been at considerable expenses already about these matters, and this I am informed would amount to £80 or £90, I thought it much too great a burden for the yearly income of my living.

Soon afterwards, Henry Cavendish mounted the glass and compared it with one of Peter Dollond's triple-lens achromatics, but he found its manipulation difficult.[81] About 1835, the astronomer W. H. Smyth thought of re-erecting it in order to re-assess its capabilities, but the project offered so many difficulties that it was laid aside.[82] Since then, the glass has rested in well-earned retirement at the Royal Society.

Newton possessed another Huygens' glass of 8 inches aperture and 170 feet focus which he presented to the Royal Society, while, in 1724, the Society received a third objective of $8\frac{3}{4}$-inch aperture and 210 feet focus from the Rev. G. Burnet.[83] All three glasses were made by Constantine Huygens and bear his signature with the date 1686.

After Campani's death, his workshop was purchased by Pope Benedict XIV and presented to the Academy at Bologna.[84] His instruments were used at Rome for many years by F. Bianchini, the Pope's domestic prelate. In 1726, Bianchini concluded, from observations made with 25- and 35-foot Campani aerial telescopes, that Venus rotates on an axis inclined at about 75° to the plane of its orbit in just over 24 days.[85] This result was at complete variance with the value of $23^h\ 15^m$

obtained by Cassini.[86] Cassini's son and Maraldi[87] later made a critical comparison between the two sets of observations but, with telescopes of 82 and 114 feet focus, they were unable to see any permanent markings on the disk. In this respect, it is interesting to note that the elder Cassini failed to see from Paris delicate markings on Venus, which were generally visible from Italy with smaller instruments.

Campani's telescopes remained at the Paris Observatory for a long time after his death. Some of his object-glasses are still preserved there, including the important 34-foot objective, the diameter of which is just over 5 inches. The glass is full of parallel streaks and gives round but strongly-coloured star images.[88] We know little of Campani's methods. His lenses were apparently paper-polished and, judging by the delicate nature of the observations made with them, were of fairly good quality judged by modern standards.

It will now be appreciated that the success of the long telescopes of the seventeenth century was due, very largely, to the painstaking and persistent efforts of men like Hevelius, Huygens, and J. D. Cassini. Indeed, after Cassini's death in 1712, his successors were unable to see what he had already discovered, let alone add to the list, and the aerial telescope gradually fell into disuse. But, in lieu of other means—the reflecting telescope was as yet little more than an interesting toy—long refractors held their own until well into the eighteenth century.

REFERENCES

1 Bailly, J. S., 1785, *Histoire de l'Astronomie Moderne*, ii, p. 199.

2 Mersenne, M., 1636, *L'Harmonie Universelle*, i. His diagrams appear in Danjon, A., and Couder, A., 1935, *Lunettes et Télescopes*, pp. 606–607.

3 Danjon and Couder, *op. cit.*, pp. 608–609.

4 MacPike, E. F., 1937, *Hevelius, Flamsteed and Halley*, pp. 1–9.

5 Huygens, C., 1659, *Systema Saturnium*, p. 11.

6 *Ibid.*, p. 9.

7 *Ibid.*, p. 4.

8 *Ibid.*, p. 47.

9 *Ibid.*, pp. 8–9, with drawing.

10 Hevelius, J., 1673, *Machinae Coelestis*, pp. 379–419.

11 *Ibid.* p. 404. Hevelius gives his name as Titus Livius Burattinus.

12 *Ibid.*, p. 407.

13 *Ibid.*, p. 406.

14 *Ibid.*, p. 415.

15 *Ibid.*, p. 410.

16 *Ibid.*, p. 413.

17 MacPike, *op. cit.*, p. 87. Gunther, R. T., 1923, *Early Science at Oxford*, ii, p. 298.

18 MacPike, *op. cit.*, pp. 9–11.

19 Huygens, C., 1684, *Astroscopia Compendiaria*; *Phil. Trans.*, **14**, p. 668, 1684.

20 *Miscellanea Beroliniensia*, **1**, p. 261, 1710.

21 *Œuvres Complètes de Christian Huygens*, 1944, xxi, p. 224.

22 Newton, I., *Opticks*, 1931 reprint of 4th edition, p. 102.

23 Gunther, *op. cit.*, ii, p. 334.

24 *Ibid.*

25 Smith, R., 1738, *A Compleat System of Opticks*, iii, pp. 281–301.

26 Chérubin d'Orléans, 1675, *Dioptrique Oculaire*, part 3. *Vide* also Twyman, F., 1952, *Prism and Lens Making*, 2nd edition, p. 12.

27 Montucla, J. É., 1799–1802, *Histoire des Mathématiques*, ii, pp. 509–510.

28 Grant, R., 1852, *History of Physical Astronomy*, p. 545.

29 Hiscock, W. G., 1937, *D. Gregory, I. Newton and their Circle*, p. 12.

30 *Phil. Trans.*, **1**, p. 209, 1665–1666.

31 Divini, E., *Phil. Trans.*, **3**, p. 842, 1668. Birch, T., 1756, *History of the Royal Society*, iv, p. 313.

32 Bailly, *op. cit.*, ii, p. 313.

33 Cassini, J. D., 1666, *Martis, circa axem proprium revolubilis, observationes*. Hooke comments on Cassini's observations of Mars in *Phil. Trans.*, **1**, pp. 242–245, 1665–1666. *Vide* also Cassini, J., 1740, *Éléments d'Astronomie*, i, p. 458. A similar rotation period had been discovered in 1659 by Christian Huygens but, as he was not sure, he did not publish.

[34] Cassini, *op. cit.*, p. 511; *Mém. Acad. Sc.*, p. 197, 1732; *Phil. Trans.*, **2**, p. 615, 1666–1667.
[35] Hooke, R., *Phil. Trans.*, **1**, pp. 3, 75, 173, 209, 245, 1665–1666. Cassini, J. D., *Phil. Trans.*, **1**, p. 143, 1665–1666; *Mém. Acad. Sc.*, p. 81, 1691.
[36] Grant, *op. cit.*, p. 245, cites *Anc. Mém. Acad. Sc.*, **2**, p. 130.
[37] Grant, *op. cit.*, p. 80.
[38] Cassini, J. D., *Mém. Acad. Sc.*, **10**, p. 727, 1666–1699.
[39] *Mém. Acad. Sc.*, p. 21, 1705.
[40] *Ibid.*, **10**, p. 584 ff, 1666–1699.
[41] Cassini, J. D., *Phil. Trans.*, **8**, p. 5178, 1673.
[42] Cassini, J. D., *ibid.*, **16**, p. 79, 1686.
[43] *Miscellanea Beroliniensia*, **1**, p. 261, 1710. Montucla, *op. cit.*, ii, p. 509.
[44] *Mém. Acad. Sc.*, **7**, i, p. 34 ff., 1666–1699. Hutton, C., 1795, *Mathematical and Philosophical Dictionary*, art. ' Telescope '.
[45] Hooke, R., *Phil. Trans.*, **1**, p. 63, 1665–1666.
[46] *Phil. Trans.*, **1**, p. 55, 1665–1666.
[47] Wolf, C., 1902, *Histoire de l'Observatoire de Paris*, p. 167; *Mém. Acad. Sc.*, **10**, p. 702, 1666–1699.
[48] Wolf, *op. cit.*, pp. 168–169.
[49] *Mém. Acad. Sc.*, **10**, p. 583, 1666–1699.
[50] *Lettres de Colbert*, v, p. 315, 1761.
[51] Wolf, *op. cit.*, p. 159, quotes *Compte des Bâtiments*, 1673, col. 1612; 1686, col. 1002.
[52] *Jour. des Sçavans*, **3**, pp. 410–412, 1682.
[53] Wolf, *op. cit.*, p. 163, cites *Machines et Inventions Approuvées par l'Académie*, **1**, p. 35.
[54] Edleston, *Correspondence*, p. 198; quoted by Plummer, H. C., *M.N.R.A.S.*, **101**, p. 167, 1941.
[55] s'Gravesande, G. J., 1748, *Physices elementa mathematica*, ii, pp. 715–725, Pl. 83, 84.
[56] Molyneux, W., 1692, *Dioptrica Nova.*
[57] *Ibid.*, p. 224.
[58] Hooke, R., 1676, *A Description of Helioscopes and some other Instruments made by Robert Hooke, F.R.S.*, p. 4.
[59] Sprat, T., 1667, *History of the Royal Society*, p. 250.
[60] Birch, T., 1756, *History of the Royal Society*, i, p. 20.
[61] Auzout, A., *Phil. Trans.*, **1**, p. 56, 1665–1666.
[62] Wren, C., *Phil. Trans.*, **4**, p. 1059, 1669.
[63] Champness, W. H., 1930, *The Worshipful Company of Spectacle Makers*. Court, T. H., and Rohr, M. von, *Trans. Opt. Soc.*, **31**, pp. 53–90, 1929–30.
[64] Court, T. H., and Rohr, M. von, *Trans. Opt. Soc.*, **32**, p. 121, 1930–31, give price list of Reeves' instruments.
[65] Baxandall, D., and Court, T. H., *Proc. Opt. Conv.*, **2**, pp. 529–537, 1926.
[66] Court, T. H., and Rohr, M. von, *Trans. Opt. Soc.*, **30**, p. 8, 1928–29, give copy of Marshall's shop print dated 1694.
[67] Walker, R., 1705, *The Posthumous Works of Robert Hooke*, p. x.
[68] Houghton, J., 1696, *A Collection of Papers for the Improvement of Trade* . . . No. 197, quoted by Court and von Rohr, *Trans. Opt. Soc.*, **31**, p. 75, 1929–30.
[69] Derham, W., 1726, *The Philosophical Experiments of Robert Hooke*, p. 261.
[70] Hooke, R., 1674, *An Attempt to prove the motion of the Earth*, p. 17 ff.
[71] MacPike, E. F., 1932, *Correspondence and Papers of Edmund Halley*, p. 2.
[72] Halley, E., 1679, *Catalogus Stellarum australium.*
[73] Grant, *op. cit.*, p. 468. Weld, C. R., 1848, *History of the Royal Society*, i, p. 255.
[74] Weld, *op. cit.*, the 90-ft o.g. was the one which Flamsteed hoped to use in his well (*vide* p. 111).
[75] *Phil. Trans.*, **30**, pp. 768–769, 1718.
[76] *Journal Book of the Royal Society*, **9**, p. 87, 1696–1702. Weld, *op. cit.*, i, p. 330.
[77] Gunther, Ref. 17, ii, p. 301.
[78] Baily, F., 1835, *An Account of the Revd. John Flamsteed*, p. 335.
[79] *Phil. Trans.*, **30**, p. 772, pp. 900–902, 1718.
[80] Derham, W., 1715, *Astro-Theology*, Preface, pp. 8–9.
[81] Kitchener, W., 1825, *The Economy of the Eyes*, ii, p. 22.
[82] Smyth, W. H., 1844, *A Cycle of Celestial Objects*, i, p. 370.
[83] Weld, *op. cit.*, p. 330.
[84] Priestley, J. 1772, *The History and Present State of Discoveries relating to Vision, Light and Colours*, p. 211.
[85] Bianchini, F., 1728, *Hesperi et Phosphori nova phenomena*, p. 22.
[86] *Mém. Acad. Sc.*, p. 197, 1732.
[87] Cassini, J., 1740, *Éléments d'Astronomie*, p. 526.
[88] Danjon and Couder, Ref. 2, p. 647.

CHAPTER V

Seeing therefore the Improvement of Telescopes of given length by Refractions is desperate, I contrived heretofore a Perspective by Reflexions, using instead of an Object-glass a concave Metal.

ISAAC NEWTON

LONG-FOCUS refracting telescopes were established observatory instruments when Isaac Newton made his first experiments in physical optics. Chromatic aberration remained a mystery, despite the obvious way in which it affected the definition by a telescope. It is strange that the separation of the red and violet images produced by lenses of 6 inches aperture and 20 or more feet focal length did not suggest a difference in the refractivity of glass for these two colours. The trouble lay in the theory of colour then current, a mixture of the ideas of Aristotle and Descartes. White light was regarded as fundamental, whilst colours were due to various admixtures of darkness. Descartes, in particular, regarded light as an instantaneously propagated statical pressure in a granulated continuum[1]—the *res extensa*. Colours, he said, are due to different rates of corpuscular rotation which are analogous to the ' cut ' of a tennis ball. Corpuscles which rotate most rapidly give the sensation of red, the slower ones yellow, and the slowest, green and blue.[2] This idea bears resemblance to the periodic aether vibrations inferred by Newton later in the seventeenth century, but it lacked experimental support.

Newton was the first to consider the alternative idea that colours are primary qualities and white light our perception of their combination. His interest in colour began in 1663 and was undoubtedly stimulated by the dispersive effects of object-glasses. He was then a student of twenty-one years, fresh from a study of the optical works of Kepler, Descartes, and Barrow. He made himself a grinding-machine and investigated the aberrations of lenses. In 1666, he repeated Grimaldi's experiment of passing sunlight through a glass prism. Where Grimaldi had assumed that the spectrum colours were due to a scattering effect of irregularities in the glass and on its polished surface,[3] Newton saw them as the primary constituents of white light.[4] Having established this step by experiments, Newton was in a position to differentiate between spherical and chromatic aberration. He concluded that whilst the former is due to the spherical surfaces of a lens, the latter has its origin in the dispersion which accompanies refraction. He was particularly interested in the extent of this dispersion in ordinary convex lenses, and his experiments with a $4\frac{1}{2}$-inch glass of 6 feet focus led to the following conclusions:[5]

. . . if they [rays of light] flow from a lucid Point, so very remote from the Lens, that before their Incidence they may be accounted parallel, the Focus of the most refrangible Rays shall be nearer to the Lens than the Focus of the least refrangible, by about the 27th or 28th Part of their whole distance from it. And the Diameter of the Circle in the middle Space between those two Foci [the circle of least confusion] . . . is about the

55th Part of the Diameter of the Aperture of the Glass. So that 'tis a wonder, that Telescopes represent Objects so distinctly as they do.

Conditions at the focus are, however, not quite so bad as this result suggested. This is due, Newton found, to the fact that the eye is particularly sensitive to yellow; the other colours appeared fainter by being spread over a much larger area.

> The sensible Image of a lucid Point [he writes in the *Opticks*][6] is therefore scarce broader than a Circle, whose Diameter is the 250th Part of the Diameter of the Aperture of the Object-glass of a good Telescope, or not much broader, if you except a faint and dark misty Light round about it, which a Spectator will scarce regard. And therefore in a Telescope, whose Aperture is four Inches, and Length an hundred Feet, it exceeds not 2″·75 or 3″ [of arc].

This 'sensible Image', or circle of least confusion, is greater than that due to the spherical figure of the lens. Newton found that spherical aberration alone gave

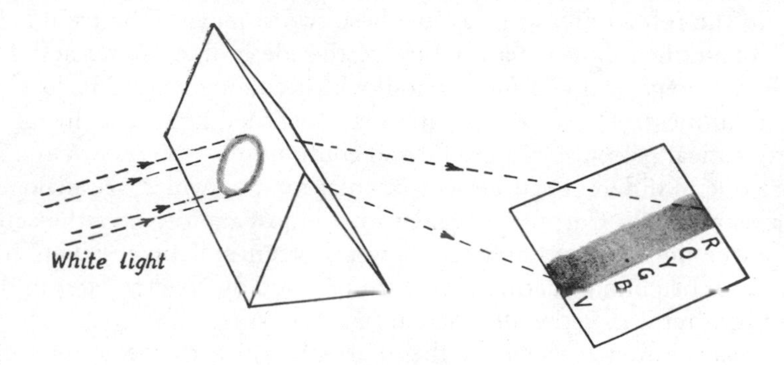

Fig. 30—The basis of Newton's prism experiments

The observation of the colours of the spectrum in this way led Newton to undertake many varied and ingenious prism experiments in light and colour

a circle of least confusion only 1/1000th the size of that due to chromatic aberration: 'And this sufficiently shows' he writes,[7] 'that it is not the spherical Figures of Glasses, but the different Refrangibility of the Rays which hinders the perfection of Telescopes.' That dispersion is the cause of the trouble was indicated by the fact that, to make telescopes of various lengths magnify 'with equal distinctness', their apertures and magnifications were usually in proportion to the square roots of their lengths. Thus a telescope of 64 feet focus and $2\frac{2}{3}$ inches aperture magnified 120 times with as much 'distinctness' as one of 12 inches focus, $\frac{1}{3}$ inch aperture, and 15 magnification.[8]

Newton's prism experiments (Fig. 30) led him to state that refraction is always accompanied by dispersion.[9] He studied both effects with different media, alone and in combination, but hasty measurements[10] led him to believe that they all had the same dispersive power. On this basis it was impossible to correct the dispersive effect of one lens by that of another, that is, to improve the refracting telescope; 'and the Refractions on the concave sides of the Glasses, will very much correct the

Fig. 31—Sir Isaac Newton

Fig. 32—James Gregory

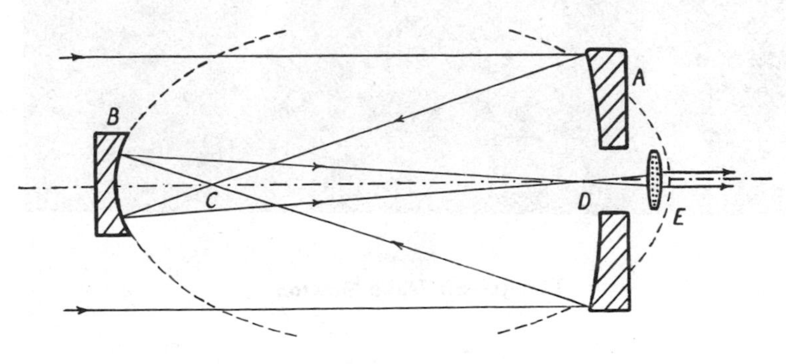

Fig. 33—The Gregorian reflector

A is the primary mirror, a concave paraboloid centrally perforated. *B* is the secondary concave ellipsoid. Light from a star is sent towards *C*, the joint focus of *A* and *B*, and reflected to *D*, the other focus of the ellipsoid. The image at *D* is then observed with an eyepiece, shown as a single equi-biconvex lens *E*.

Errors of the Refractions on the convex sides, so far as they rise from the sphericalness of the Figure. And by this means might Telescopes be brought to sufficient perfection, were it not for the different Refrangibility of several sorts of Rays.'[11]

This conclusion was unfortunate in that it delayed the invention of the achromatic lens by about half a century, but it gave us the first reflecting telescope. Zucchi,[12] Cavalieri,[13] Mersenne,[14] and Descartes[15] had all considered some form of telescope embodying concave mirrors and convex lenses, but they had not put their designs into practice. In 1663, James Gregory (Fig. 32) proposed in his *Optica Promota* a two-mirror combination (Fig. 33). A large paraboloidal concave collected the light, reflected it to a small ellipsoidal concave which, in turn, formed the final image at the centre of the main speculum. The latter was perforated in the centre to allow the introduction of an eyepiece.[16] This system, owing to the figures of the mirrors, is free from spherical aberration, and reasonably free from chromatic aberration,

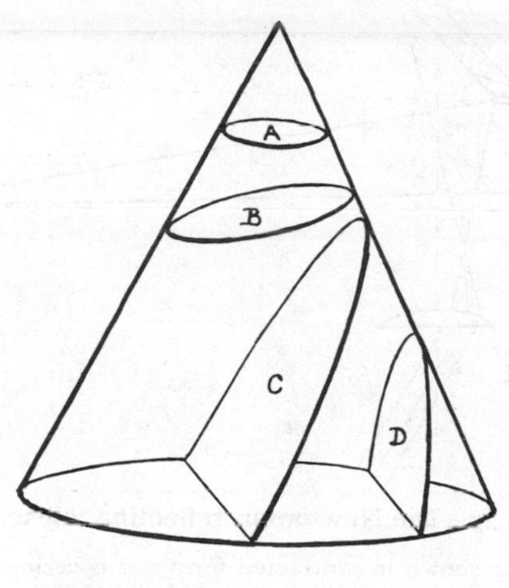

Fig. 34—Sections through a cone

A Circle — *C* Parabola
B Ellipse — *D* Hyperbola

the only refracting elements being the eyepiece and the eye of the observer. Gregory met with insuperable difficulties when he tried to make aspherical surfaces. On coming to London about 1664 or 1665, he ordered a 6-foot focus mirror from Richard Reeves and John Cox, but these opticians were also unable to produce satisfactory surfaces and their mirrors were worse than useless.[17] Gregory 'was much discouraged with the disappointment; and after a few imperfect trials made with an ill-polished spherical one, which did not succeed to his wish, he dropped the pursuit, and resolved to make a tour of Italy, then the mart of mathematical learning, that he might prosecute his favourite study with greater advantage.'[18]

Instead of Gregory's concave ellipsoidal secondary mirror, Newton used a small flat, inclined at 45° to the axis of the tube, which reflected the convergent beam to a hole in the side of the tube and so into the eyepiece (Fig. 35).[19] The first instrument on this plan was made in 1668, but it remained unknown, save to a few of his Cambridge friends, until about three years later. News of his work on colour had then come to the notice of the Royal Society, and he was invited to send an account of his telescope to London.

With the Telescope which I made [he wrote on March 16, 1671, to Henry Oldenburg][20] I have sometimes seen remote objects, and particularly the moon, very distinct, in those parts of it which were near the sides of the visible angle. . . . One of the fellows of the College is making such another Telescope, with which last night I looked on Jupiter, and he seemed as distinct and sharply defined as I have seen him in other Telescopes.

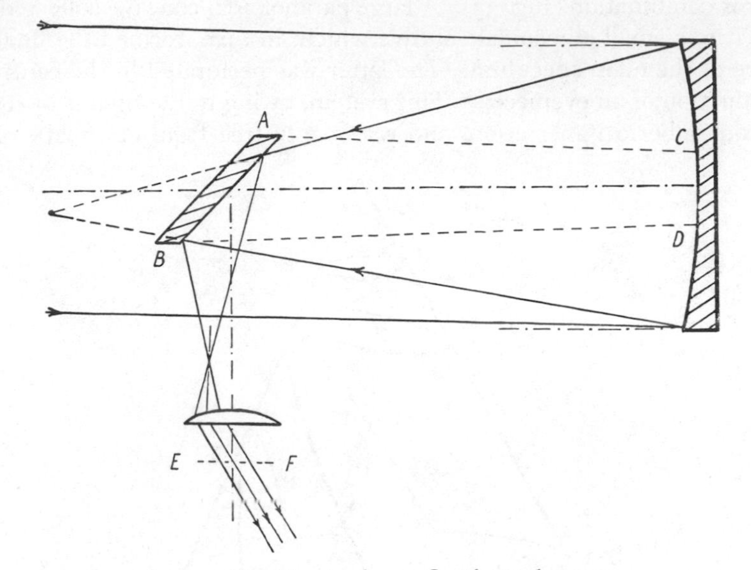

Fig. 35—The Newtonian reflecting telescope

The system is shown in contracted form and covering a true field of 5°. In practice a much smaller useful field is found, and the aperture ratio is usually $f/7$ to $f/10$. *AB* is the secondary flat; it must be large enough to reflect rays which are initially parallel and oblique. The region *CD* is entirely cut off from the incident light. *EF* is the plane of the exit pupil.

Newton's first telescope was little more than an interesting scientific toy. He writes:[21]

The diameter of the Sphere to which the Metal was ground concave was about 25 English Inches, and by consequence the length of the Instrument about six Inches and a quarter. The Eye-glass was Plano-convex, and the diameter of the Sphere to which the convex side was ground was about 1/5 of an Inch, or a little less, and by consequence it magnified between 30 and 40 times. By another way of measuring I found it magnified 35 times. The concave Metal bore an Aperture of an Inch and a third part, but the Aperture was limited not by an opake Circle, covering the Limb of the Metal round about, but by an opake Circle, placed between the Eyeglass and the Eye, and perforated in the middle with a little round hole for the Rays to pass through to the Eye. For this Circle being placed here, stopp'd much of the erroneous Light, which otherwise would have disturbed the Vision. By comparing it with a pretty good Perspective of four Feet in length, made with a concave Eye-glass, I could read at a

Fig. 36—Newton's reflecting telescope

The eyepiece and the mirror are original components of the instrument

(*The Royal Society*)

greater distance with my own Instrument than with the Glass. Yet objects appeared much darker in it than in the Glass, and that partly because more light was lost by Reflexion in the Metal, than by Refraction in the Glass, and partly because my instrument was overcharged. Had it magnified but 30 or 25 times, it would have made the Object appear more brisk and pleasant.

In the autumn of 1671, Newton made a second reflecting telescope, news of which so roused Oldenburg's curiosity that he asked Newton to send the instrument to the Royal Society for inspection by the Fellows (Fig. 36). Its appearance at the meeting of January 11, 1672, excited great interest. Among those who examined it were King Charles II, Sir Robert Moray, Robert Hooke, and Christopher Wren. At the same meeting, Newton was elected Fellow of the Society.[22]

The subsequent history of Newton's first telescope is obscure. Stone, writing in 1723, says that it was the worse of the two and that it was 'not long ago to be seen at Mr. Heath's, the Mathematical Instrument Maker in the Strand, having upon it, wrote with his own hand, Isaac Newton.'[23] Newton, after 1704, makes no further

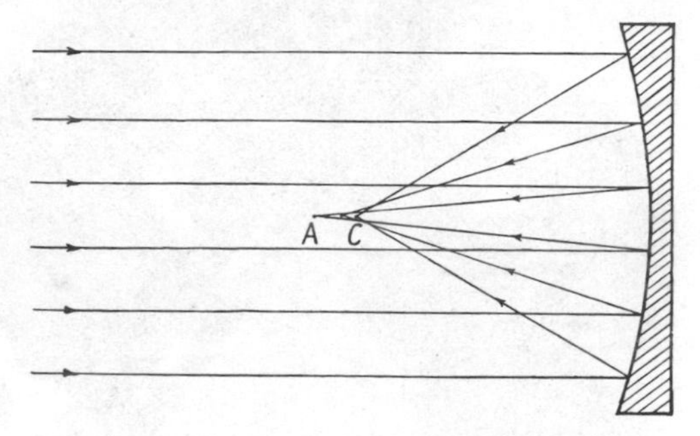

Fig. 37—Spherical aberration of a mirror

Rays incident on the marginal zone of the mirror are reflected to a focus at *C*. Rays incident near the axis are reflected to a focus at *A*.

reference to the instrument being in his possession. The second telescope remains in the possession of the Royal Society.

For the mirrors of his telescopes, Newton had to obtain a metal of suitable hardness, porosity, and reflectivity. He experimented with several alloys and, in so doing, raised problems which never left telescope-makers until Foucault made the first silver-on-glass mirror two centuries later. Newton's choice was an alloy known as bell-metal—six parts copper, two parts tin.[24] This alloy was well known to alchemists owing to its similarity in brightness to silver. Newton further added one part arsenic which, he thought, made it even whiter and capable of taking a better polish.[25] But the high copper content caused it to tarnish rapidly, and he had to repolish his mirrors from time to time in order to restore their original lustre.

Newton's account of the polishing process is important since he mentions, for the first time in published writing (1704), the use of a pitch polisher:[26]

The Polish I used was in this manner. I had two round Copper Plates, each six Inches in Diameter, the one convex, the other concave, ground very true to one another. On the convex I ground the Object-Metal or Concave which was to be polish'd, 'till it had taken the Figure of the Convex and was ready for a Polish. Then I pitched over the convex very thinly by dropping melted Pitch upon it. . . . Then I took Putty which I had made very fine by washing it from all its grosser Particles, and laying a little of this upon the Pitch, I ground it upon the Pitch with the concave Copper, till it had done making a Noise, and then upon the Pitch I ground the Object-Metal with a brisk motion, for about two or three Minutes of time, leaning hard upon it. Then I put fresh Putty upon the Pitch, and ground it again till it had done making a noise, and afterwards ground the Object-Metal upon it as before. And this Work I repeated till the Metal was polished, grinding it the last time with all my strength for a good while together, and frequently breathing upon the Pitch, to keep it moist without laying on any more fresh Putty.

Owing to this method of working, Newton gave his concave mirrors what is now called turned-down edge—a curvature for the marginal parts less than that for the centre zones. This defect he suppressed by placing a tiny hole between the eyepiece and his eye;[27] this aperture also reduced the effects of spherical aberration (Fig. 37), for he made no attempt to parabolize his mirrors.

In 1672, news came from France that a Frenchman named Cassegrain had invented another new reflecting telescope. Some say that Cassegrain was professor at the College of Chartres;[28] Bell [29] identifies him with Sieur Guillaume Cassegrain, a sculptor and, therefore, one familiar with foundry methods and metalwork. Whatever Cassegrain's identity, de Bercé's representation of his invention was both misleading and inaccurate (Fig. 38). De Bercé described it to the French Academy and insinuated, on untenable grounds, that it was superior to the Newtonian. He claimed priority of discovery for Cassegrain, dating his countryman's invention several weeks before the time when Newton sent his second paper to the Royal Society.[30]

Instead of Gregory's small concave mirror, Cassegrain suggested a convex mirror, so placed as to intercept the rays from the centrally perforated primary mirror before they came to the focus (Fig. 39). His arrangement has advantages over both Gregorian and Newtonian, but these advantages were overlooked by his critical English contemporaries. In 1779, Jesse Ramsden showed[31] that the combination of a concave and convex mirror tends to correct the aberrations of each surface whereas, in the Gregorian, they are additive. Cassegrain's telescope is also shorter than a Gregorian of equal power by twice the focal length of the secondary mirror. In 1672, however, Newton was more concerned with the disadvantages of the Cassegrainian which he did not delay in enumerating to the Royal Society. His annoyance with de Bercé is apparent in his May paper '. . . concerning the Catadioptrical Telescope, pretended to be improved and refined by M. Cassegrain.'[32] Newton says that the idea of the French telescope was contained in James Gregory's *Optica Promota* and remarks:

I had thence an occasion of considering that sort of constructions, and found their disadvantages so great, that I saw it necessary, before I attempted any thing in the Practique, to alter the design of them, and place the Eye-glass at the side of the Tube rather than at the middle.

In his list of disadvantages Newton says that, unless the secondary convex mirror had a true hyperboloidal figure, it would add to the aberrations of the large mirror and so make the instrument almost useless. He was also misled in thinking that the small mirror bore too great a share in magnifying the image in relation to its effective diameter:

By reason that the little convex conduces very much to the magnifying virtue of the instrument, which the Oval plane [flat] doth not, it will magnify much more in proportion to the Sphere, on which the great concave is ground, than in the other design; And so magnifying Objects much more than it ought to do in proportion to its aperture, it must represent them very obscure and dark; and not only so, but also confused by reason of its being overcharged. Nor is there any convenient remedy for this. For, if the Little Convex be made of a larger Sphere, that will cause a greater inconvenience by intercepting too many of the best rayes; or if the Charge of the Eye-glass be made so much shallower as is necessary, the angle of vision will thereby become so little, that it will be very difficult and troublesome to find an object, and of that object, when found, there will be but a very small part seen at once.

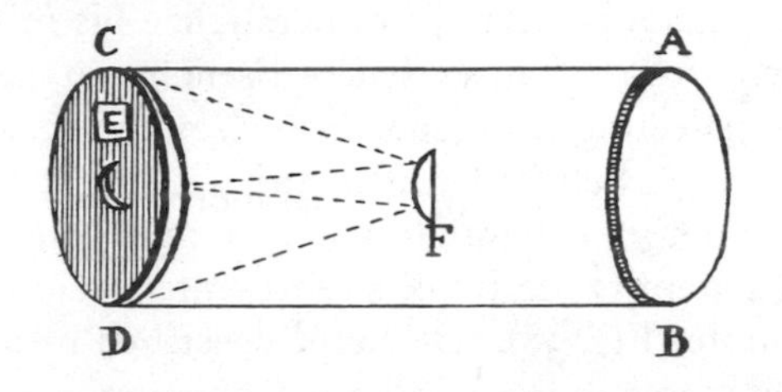

Fig. 38—Cassegrain's telescope as drawn by de Bercé

'*ABCD* is a strong tube, in the bottom of which there is a great concave *Speculum*, *CD*, pierced in the middle, *E*. *F* is a convex *Speculum*, so disposed, as to its convexity, that it reflects the *Species*, which it receives from the great *Speculum*, towards the hole *E*, where there is an Eye-glass, which one looketh through.'

Newton concludes with the remark:

The advantages of this design are none, but the disadvantages so great and unavoidable, that I fear it will never be put in practise with good effect. . . . I could wish, therefore, Mr. Cassegrain had tryed his design before he divulged it. But if, for further satisfaction, he please hereafter to try it, I believe the success will inform him, that such projects are of little moment till they be put in practise.

Newton's scathing criticism passed unchallenged. De Bercé was apparently unwilling to enter the lists against so formidable an opponent. Cassegrain returned to obscurity feeling, no doubt, that perhaps his idea was unworthy of consideration.

To ensure Newton's priority in the matter of making a reflecting telescope, the Royal Society sent a description of it to Huygens, who in turn acquainted the French Academy. The invention soon became generally known on the Continent, and we find Newton referred to as the ' telescope-maker of England '.[33] Huygens was impressed by the novelty and latent possibilities of Newton's telescope. He stated that far less light was lost by reflection than by refraction through an object-glass, a statement hardly true in this case.[34] Newton's mirrors were of comparatively low reflectivity owing to their high copper content and to the readiness with which they tarnished. Compared with later standards, their polish was poor; between them, they reflected only about 16 per cent of the incident light. But it was the difficulty of making satisfactory mirrors that retarded the further progress of the reflecting telescope.

Robert Hooke had some success with specula and, in 1674, made a small Gregorian which he showed to the Royal Society.[35] During his tenure of the professorship of geometry at Gresham College and his participation in the rebuilding of London, Hooke found time in which to grind and polish mirrors, some as large as 7 inches diameter and 9 feet focus. He experimented with both glass and metal mirrors, but the results never justified the many hours which he spent at this work.[36] At this time also, the Royal Society ordered a 4-foot long reflector from Cox, but he was unable to bring the mirrors to a good polish.[37] There the matter rested for another thirty years until John Hadley produced a Newtonian reflector which was

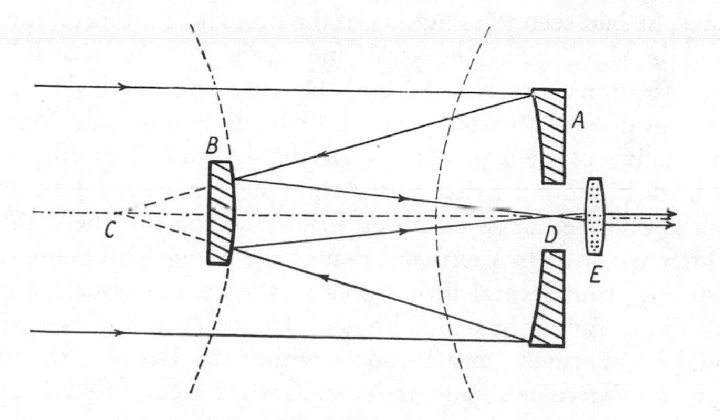

Fig. 39—The Cassegrain reflecting telescope

A is the primary mirror, a concave paraboloid centrally perforated. *B* is the secondary convex hyperboloid. Light from a star is sent towards *C*, the joint focus of *A* and *B*, and reflected to *D*, the other focus of the hyperboloid. The image at *D* is then observed with an eyepiece, shown as a single equi-biconvex lens *E*.

equal, if not superior, in performance to the 123-foot aerial telescope of Christian Huygens (Fig. 40).

Hadley was born in 1682. As a young man he seems to have shown an inventive bent. Desaguliers, in lectures published in 1744, says that 'the contrivance for raising and falling the water-wheel' at some new waterworks at London Bridge 'was the invention of Mr. Hadley, who put up the first of that kind at Worcester.'[38] At the age of thirty-five, Hadley became a Fellow of the Royal Society and proved an active and valuable member. About this time John Hadley tried his hand at speculum-grinding and, with the assistance of his two brothers, George and Henry, succeeded in making a 6-inch Newtonian reflector of 62 inches focus, mounted on an original and effective stand.[39] This instrument made its first public appearance at the January 12 meeting of the Royal Society in 1721.

> Mr. Hadley [the *Journal Book* records] was pleased to show the society his reflecting telescope, made according to our President's [Newton] directions in his Optics, but curiously executed by his own hand, the force of which was such, as to enlarge an

object near two hundred times, though the length thereof scarce exceeds six feet, and, having shewn it, he made a present thereof to the Society, who ordered their hearty thanks to be recorded for so valuable a gift.

In the following week, Hadley explained the motions of the stand and was ' entreated to lend Dr. Halley this apparatus, to have one like it made at the Society's charge, to be used with his noble present.'

At Wanstead, Bradley and Pound made trials with the new telescope ' on the bodies and satellites of the superior planets ' and compared its performance with that of Huygens' 123-foot refractor.[40] The definition of both instruments was about the same, although the Huygenian gave the brighter image. Both observers were impressed by the ease with which the reflector could be directed and focused. Meanwhile, Hadley had given the Society an account [41] of his own observations on the satellites of Jupiter and Saturn, and stated that, ' Mr. Folkes and Dr. Jurin being present ', he had seen the shadows of the first and third satellites on the body of Jupiter and, in May, 1722, ' the dark Line on the Ring of Saturn.'

Pound's observations corroborated both Halley's and Hadley's. Together with Bradley, then professor of astronomy at Oxford, Pound saw four of Saturn's satellites, the shadow of the largest in transit, the division of the ring, and the ring's shadow on the disk. Pound writes [42] that the reflector ' represents Objects as distinct, though not altogether so clear and bright; which may be occasioned partly from the Difference of their Apertures (that of the Huygenian being somewhat the larger) and partly from several little Spots in the concave Surface of the Object Metal, which did not admit of a good Polish.' He speaks of the ease with which the reflector could be managed ' at a Window within the House ', though it is to be hoped that the tests were not made under such adverse conditions. The advantage of the reflector's enclosed tube became marked during twilight, when the aerial telescope was almost useless.

Hadley himself left no written account of the grinding and polishing processes, but he gave assistance to others. Both S. Molyneux and Bradley caught his enthusiasm, and they, in turn, instructed two London opticians, Scarlett and Hearne. It was Hearne who supplied the metal castings for Molyneux's mirrors, and Molyneux refers to him as ' that excellent workman, Mr. Hearne, of Dogwell Court, Whitefriars.'[43] More complete information was handed to Robert Smith and published by him in his *A Compleat System of Opticks* (1738), a comprehensive but diffuse work which was translated into both German and French. Molyneux and Smith wrote the greater part of the account, Hadley adding just his method of parabolizing. The references are not so complete as if they had come direct from Hadley, but they reveal the advance he made despite innumerable difficulties.

The surface of an *f*/10* mirror such as Hadley's need not be paraboloidal to work effectively. The tolerance on the paraboloidal figure is, in this case, larger than the difference between the sphere and the paraboloid, and small reflectors with

* An *f*-number provides a convenient way of expressing the aperture ratio of an optical system; *f*/10 means that the focal length is ten times the effective aperture. Systems with numerically low *f*-numbers yield bright images and are said to be of high aperture ratio. Since, in photography, bright images require short exposure times, a high aperture ratio system is said to be ' fast '. An *f*/10 system, by comparison with an *f*/2 would have a smaller aperture ratio and would be classed as ' slow '.

Fig. 40—Hadley's reflector

(*Science Museum, London. British Crown copyright*)

Fig. 41—John Hadley

By courtesy of the Director of the Science Museum, London)

aperture ratios of $f/10$ and under can be given spherical surfaces. Even so, the difficulty of polishing a spherical mirror is evidenced by the mediocre results achieved by Bradley and Molyneux. Bradley's assistance, interrupted by a change of residence and by many demands on his time, was of necessity spasmodic, but Molyneux was considered expert in optics. The only positive evidence of their united work was a few small telescopes and a number of mirrors, but their experiments smoothed the way for further workers.

Molyneux tried 150 different alloys for specula.[44] Only one of these contained arsenic, a constituent added by Newton, who thought it made the metal whiter and more solid. The remainder were just different proportions of tin, copper, and brass. The copper, by reason of its high melting point, was melted first, and then the other constituents were added. The molten alloy was either poured off directly into the mould or allowed to cool for subsequent casting. The mould was an iron chamber or ' flask ', lined with clay or sand so that the inside space took the form of the speculum required. To ensure this, a pattern of pewter was turned on a lathe to the intended size. When cast, one surface of the metal was ground ' quite bright upon a common grindstone ' to remove the larger irregularities. Further grinding to the gauge limit was done on a convex stone of finer texture. For abrasives, Hadley used sand and emery,[45] controlling the curvature by varying the stroke; circular strokes increased the curvature, diagonal strokes reduced it.

A polisher was now required and, for making it, a brass tool. The latter was turned on a lathe, roughly ground concave on a convex marble slab, and fine-ground on a pavement of hone squares. With this, a glass disk was ground to the corresponding convex curvature with emery. In the next operation, this glass, covered with pitch-impregnated silk, formed the polisher.[46] Brass tool and speculum were alternately ground on the hones until each had taken a ' good and even brightness ' and the speculum was then polished on the pitch-polisher with fine putty and water. After these laborious and lengthy steps, the worker, with luck, had a speculum something near the intended figure. Both the grinding and polishing tools took their form from the brass tool, and the accuracy of this decided the curvature of the speculum. In every case, surfaces were altered to suit the polisher, and if the speculum took an unequal polish, it was figured again on the hones with a different motion. By this means, Hadley produced a tolerably good spherical mirror near the focal length required. This he attempted to parabolize by ' bruisers ', or what we now call ' local polishers ', and found that ' the oftener the mud is washed away the more truly spherical the figure of the speculum will generally be; but the leaving a little more of this mud on the stones has sometimes seemed to give the metal a parabolick figure.'[47] A spiral motion of the speculum on the hones for about 30 seconds produced a similar effect.

Before the final polishing, the speculum received a fair polish and was examined for figure. For this, Hadley devised the first laboratory method and one which contains the germ of Foucault's test, perfected some 150 years later. He placed an illuminated pin-hole at the centre of curvature of the mirror and examined, with an eyepiece, the reflected cone of light in the vicinity of the image. He gives the following interpretations to the focal pictures so obtained:[48]

> If the area of the light, just as it comes to or parts from the point, appears not round but oval, squarish, or triangular etc., it is a sign that the sections of the specular surface, through several diameters of it, have not the same curvature. If the light, just before it comes to a point, have a brighter circle round the circumference, and a greater darkness near the centre, than after it has crossed and is parting again; the surface is more curve[d] towards the circumference and flatter about the centre, like that of a prolate spheroid round the extremities of its axis; and the ill effects of this figure will be more sensible when it comes to be used in the telescope. But if the light appears more hazy and undefined near the edges, and brighter in the middle before its meeting than afterwards, the metal is then more curve[d] at its centre and less towards the circumference; and if it be in a proper degree, may probably come near the true parabolick figure. But the skill to judge well of this, must be acquired by observation.

While Hadley did so much to make the reflecting telescope a practical proposition, he is perhaps better known as the inventor of a nautical octant, precursor of the modern sextant. The forerunners of the sextant were the cross-staff and astrolabe, instruments which enabled early navigators to measure the sun's altitude and thence to determine their latitude. To use either of these instruments, the observer had to expose his eyes to the full glare of the sun and, at the same time, to keep his hands steady. To remedy the first drawback, John Davis invented the back-staff or Davis quadrant (1590).[49] Hooke, in 1664, and Grandjean, in 1732, tried to improve the back-staff but failed to make it fulfil the ' constant coincidence ' condition.[50] Newton solved this difficulty in his octant,[51] the principle of which is that of the modern nautical sextant (Fig. 42). Newton told Halley about his invention, but they do not appear to have put the design to practical test.

In 1731, Thomas Godfrey, a young American glazier, independently both designed and used a sea quadrant.[52] In the following year, Hadley described his own instrument (Fig. 43) to the Royal Society.[53] As soon as Bradley, from a yacht, obtained readings with Hadley's octant to 1 minute of arc, the instrument was universally adopted. It went through many stages of development before it

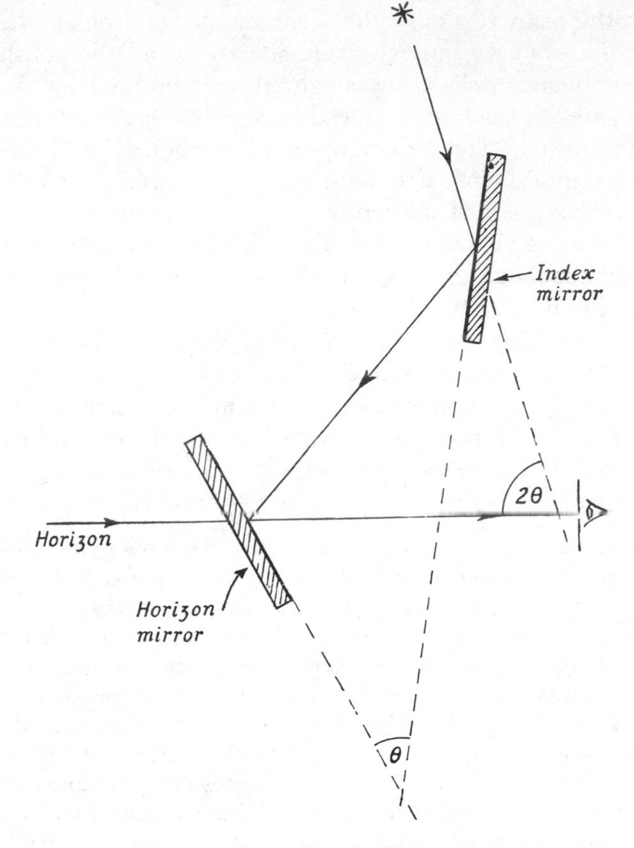

Fig. 42—The principle of the sextant

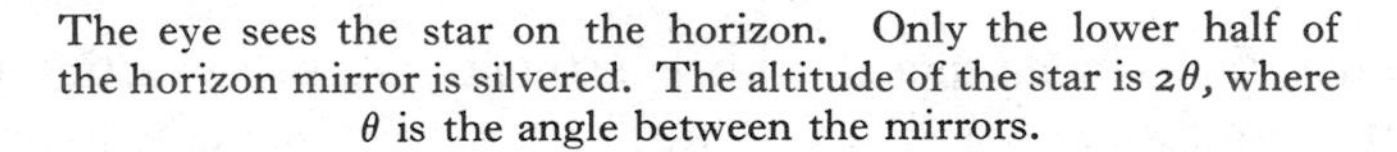

The eye sees the star on the horizon. Only the lower half of the horizon mirror is silvered. The altitude of the star is 2θ, where θ is the angle between the mirrors.

reached anything like the accuracy of the modern sextant. Hadley suggested extending the arc to 220° by the addition of a back sight, but the means he proposed were complicated and many years elapsed before his idea became a practical possibility in the repeating circles of Mayer and Borda.

Besides his Newtonian telescope and octants in wood, Hadley made several small Cassegrain and Gregorian reflectors, devising new stands and drawing up rules for determining the sizes of the mirrors. One of his Gregorians, of 2 inches

Fig. 43—Hadley's octant

(*Science Museum, London. British Crown copyright*)

aperture and 8 inches focus, and probably made in 1726, is still in existence.[54] Nothing is known of Hadley's personal life—we know only of his scientific activities. In 1726, he was one of the committee appointed to examine and to report upon Halley's new instruments at the Royal Observatory, Greenwich. He was elected vice-president of the Royal Society two years later, having been an active Fellow and member of the Council.[55] His interest in astronomical instruments continued until his death.

By the middle of the eighteenth century, the optical trade was following the example set by Scarlett and Hearne, and small reflectors were in regular production by James Mann the third, Chaplain and others. Francis Hawksbee, an optician in Fleet Street, made an instrument of 40 inches focus and 4½ inches aperture which, according to Smith,[56] was 'scarce inferior to Mr. Hadley's of 5 feet 2½ inch focal distance; since with the same eye-glass that gave it this power, it showed not only the minute parts of the new moon exceedingly distinct, but also the belts of Jupiter and the black line or division of Saturn's ring.' Hawksbee is also said to have worked at other mirrors of 6 feet and 12 feet focus.[57]

In 1740, these small but promising telescopes were outclassed by the reflectors of James Short. Short had taken his M.A. degree and was a qualified preacher of the Church of Scotland when lectures by Colin Maclaurin, at the University of Edinburgh, led him to consider telescope-making. Short first tried his hand at figuring glass surfaces, afterwards backing them with quicksilver.[58] Veins in the glass and difficulty in obtaining films of uniform reflectivity defeated this plan. He found that he could work metals with equal facility and with better results. Maclaurin, no doubt, assisted in these experiments at least, he placed his rooms at Short's disposal and made the latter's activities known to the London opticians.

A letter from Maclaurin to Robert Smith mentions the results obtained by Short up to 1734:[59]

> Mr. Short, an ingenious person well versed in the theory and practice of making telescopes, has improved the reflecting ones so much, that I am fully satisfied he has far outdone what has yet been executed in this kind. He has not only succeeded in giving so true a figure to his speculums of glass quicksilvered behind, as to make the image from them perfectly distinct, but has made telescopes with metal speculums which far surpass those I have seen of any other workman. . . .

According to Maclaurin, Short's metal specula had larger aperture ratios than usual and were carefully mounted in the finished instrument. One of 15 inches focus showed the five satellites of Saturn, 'which very much surprised me till I found that Mr. Cassini had sometimes seen them all with a seventeen-foot refracting telescope.'[60] London-made specula proved inferior to Short's. Maclaurin continues:[61]

> I have compared some of these with such as have been brought from London; and find one of Mr. Short's of six inches focal distance, compared with one of the best I have seen from London of nine inches and three tenths focal distance, to exceed it in brightness, distinctness and magnifying power; and when I called an indifferent person, who knew not who had made the instruments, to give his opinion, he very

readily preferr'd that of six inches focal distance. It also manifestly exceeded another I had from London of eleven inches and a half focal distance. The same was the result of some other comparisons.

As a result of this letter, Robert Smith arranged for some samples of Short's work to be sent to London. We read[62] of comparisons made with different instruments by G. Graham, E. Scarlett, Caleb Smith, G. Bevis and others, in the presence of Martin Folkes, president of the Royal Society. At one of these gatherings, the observers tried a Gregorian of $5\frac{1}{2}$ inches focus by Short—' the first reflector he ever made '—and, at another, one of 9 inches focus and 2·3 inches aperture. Comparison instruments were supplied by Chaplain, S. Molyneux, J. Mann, and E. Scarlett.

To add to his slender income, Short sold a few telescopes and, upon the invitation of Queen Caroline, tutored the young Duke of Cumberland in mathematics. Further monetary help came when he assisted the Earl of Morton in a survey of the Orkney Islands. By 1740, he was able to open a business of his own.[63] His workshop was situated in Surrey Street, off the Strand, and, with the growing demands of his business, he seldom left this vicinity. In 1737, a year after his arrival in London, Short was elected Fellow of the Royal Society, and his interest and ability in astronomy and optics, together with the influence of friends like Hadley, Maclaurin, George Graham, and Dr George Bevis, facilitated his entry into cultured society. This introduction had important financial aspects. In those days, a good telescope was a rarity and, as a gem of art and industry, commanded a high price. Short's largest telescopes were few and costly, reserved for aristocratic dilettantes who used them for anything save serious research. For this reason, Short soon found himself in an enviable financial position; for the same reason, his instruments led to few, if any, discoveries of importance.

The majority of Short's telescopes were Gregorians, and ranged in size from small hand-perspectives to giants (for those times) of 18 inches aperture and 12 feet focus. Short, no doubt, chose the Gregorian owing to the comparative ease with which a secondary concave mirror could be made compared with a convex, and because it adapted itself to terrestrial observations. With each instrument he supplied a terrestrial eyepiece and several high-power astronomical eyepieces. The public, apparently, favoured high magnifications and Short felt no disinclination to provide them, even if it led to some exaggeration of their power. He often overrated the highest powers of his telescopes, an unfortunate procedure, for it led his successors to follow his example in order to keep up high prices. An example of this practice was revealed some years later when William Herschel examined Short's 2-, 2·6-, 10-, and 12-foot reflectors at the Edinburgh Observatory:[64]

I viewed several land objects thro' them. The large one which has the reputation of being a very bad instrument appeared to me not to be very defective, but the presumption seems to be in favour of its being good. I could only see the top of a steeple, and even there great part of the light was lost by being confined in the opening of the roof. . . . I saw again the large reflector at the observatory and measured its power, which was said to be 800; I found it 130, the aperture is 12 inches.

The following is a copy of one of Short's price lists—it shows that high prices accompanied high magnifications.[65]

No.	*Focal length Inches*	*Aperture Inches*	*Magnifications*		*Price Guineas*
1	3	1·1	1 Power 18	times	3
2	4½	1·3	1 ,, 25	,,	4
3	7	1·9	1 ,, 40	,,	6
4	9½	2·5	2 ,, 40, 60	,,	8
{5	12	3·0	2 ,, 55, 85	,,	10
{6	12	3·0	4 ,, 35, 55, 85, 110 ..	,,	14
7	18	3·8	4 ,, 55, 95, 130, 200 ..	,,	20
8	24	4·5	4 ,, 90, 150, 230, 300	,,	35
9	36	6·3	4 ,, 100, 200, 300, 400	,,	75
10	48	7·6	4 ,, 120, 260, 380, 500	,,	100
11	72	12·2	4 ,, 200, 400, 600, 800	,,	300
12	144	18·0	4 ,, 300, 600, 900, 1200	,,	800

Short no doubt justified these prices on the basis that nearly every instrument was his own work or constructed under his supervision. But labour and materials were cheap in those days and he had few, if any, serious competitors. His price for a 3-foot reflector was 75 guineas. Fifty years later, the same instrument could be purchased for 38 guineas, complete with rackwork and accessories.

William Kitchener, a wealthy collector of telescopes, possessed a Short Gregorian. It was of 15 inches focus, 4·3 inches aperture and 170 magnification—an aperture ratio of 1 : 3·5 and an excessive power of 40 to 1 inch of aperture. A Cassegrain, known as ' Short's Dumpy ', had an aperture of 6 inches and worked at *f*/4, a remarkably high aperture ratio for the period. This instrument eventually came into the possession of Alexander Aubert, an amateur astronomer, who assessed its maximum magnification as 231.[66] With 355, the power given by the eyepiece provided, it ' broke down ' completely. Among Short's larger instruments was a 12-inch aperture Gregorian made for Lord Thomas Spencer at a cost of £630 complete with accessories. This is apparently the instrument mentioned by J. Bernoulli[67] and which was kept as a curiosity at Marlborough House, London, for some twenty years. Another Gregorian, of 18 inches aperture and made in 1752 for the King of Spain, cost £1200 and remained, for many years, the largest instrument of its kind. Short also made a few Newtonians; one at the Royal Observatory, Greenwich, had an aperture of 9¼ inches.

Short was reticent about his method of speculum-making and left no account of his work. He is said[68] to have destroyed all his tools before his death, through professional jealousy, but, had they been preserved, it is doubtful whether his optical colleagues would have had the ability to copy, still less to improve upon, his technique. After Hadley, he was the first to produce paraboloidal specula. Perhaps he discovered the ' overhang ' method of concaving a metal or glass disk by grinding it on another of the same diameter. It is more likely that his method was similar to that practised by John Mudge (page 88), who discussed these matters with him. At one time, false hopes were raised when Short deposited with the Royal Society a sealed paper which purported to deal with his technique. This was opened after his death and was found to contain *A Method of Grinding Object-Glasses of Refracting Telescopes truly Spherical*,[69] with little information above that already known. The simplicity of Short's method, however, contrasts with the elaborate instructions given by Huygens and Molyneux. Two brass tools were turned to gauge size and ground together with fine emery. A further concave brass tool, covered with pitch, served as a polisher. The glass disk was first fine-ground on the concave tool and then pitch-polished on the other with putty. To ensure that the pitch had requisite hardness, Short indented it with his thumb-nail, inferring that if it took an impression, the pitch was too soft. This simple test is familiar to modern amateur telescope-makers and is often their only test. It covers an important point, for if too hard, pitch can give rise to scratches over the surface. If too soft, the pitch is too easily deformed by the heat and pressure of the polishing operation.

Short's telescopes were of brass and were nearly all mounted on altazimuth stands, although some of his large tubes were fitted to a type of fork equatorial. In the latter, the tube could be rotated on trunnions mounted at the top of a U-shaped fitment. The latter was attached to a conical polar axis which could be set to any particular latitude. About 1749, Short invented a ' universal ' portable equatorial mount for small telescopes.[70] The tube was attached to a movable declination circle, and other circles, each divided into degrees and thence to every 20 minutes of arc, represented the horizon, meridian, and equator. The circles were only 6 inches in diameter for a 2-foot tube, and the absence of proper counterpoises made the instrument unstable. Short fixed a high price to this mounting and this did not encourage its adoption by astronomers. Indeed, it met with little success until Ramsden and later opticians improved its stability. Another feature of all Short's telescopes was the numbers which he engraved on their tubes, for example, 211/1172–12, and 37/1037–24. After examining a series of Short's telescopes, including some dated ones, D. Baxandall found[71] that the numerator referred to the serial number of the instrument of that particular size, the denominator referred to the number of telescopes made to that date, and the third number gave the focal length in inches. Both Thomas Short and Edward Nairne used a similar system.

The transits of Mercury in 1753 and of Venus in 1769 offered opportunities for determining afresh the solar parallax. Short observed the transit of Venus from Savile House ' by the Command of the Duke of York ' and surrounded by royalty[72] —an atmosphere not conducive to accurate observations. The results were disappointing. Observers differed over the times of ingress and egress; consequently the two chief computers at that time, Short in London and Pingré in Paris, obtained different results. The transit was also observed by several American astronomers, among whom David Rittenhouse and Dr Smith obtained a value for the solar parallax near that accepted today. The transit method cannot give results of a high order of accuracy and has now been superseded by more trustworthy methods.

Short died in 1768, taking with him the accumulated experience of thirty-five years of telescope-making. Kitchener called him ' one of our best practical opticians ' and ' that truly excellent artist, Mr. James Short.'[73] He attributed Short's success to two factors. First, to the pains he took to ensure that the optical elements were all correctly aligned in the tube. Second, to the ' patient industry with which he worked his large metals, and the very great care he bestowed in adapting the small speculum to the large one; he made a great many small specula of the same focus, and tried them one after the other, till he made a good match.'[74] This matching was part of a trial-and-error process which Short termed ' marrying ' the specula.

To his brother, Alexander, Short left his business and some £20,000. Sixty-four unmounted mirrors and a 12-foot Gregorian reflector passed to Thomas, another brother and optician at Leith. Thomas Short arranged for the telescope to be erected on Calton Hill but, after his death, the observatory (Edinburgh's first) fell into complete disuse and the instrument appears to have been demolished. In his finished instruments, Short left a valuable legacy to contemporary opticians, but the latter seem to have been unable to improve upon them. ' Opticians have spent the greater part of their days in similar pursuits ' Tulley wrote in 1817, ' and the only improvement since Short's time is in the composition of the speculums.'[75]

Short corresponded with John Mudge, brother to Thomas Mudge the horologist. Mudge was a physician whose hobby was mirror-making and whose researches into suitable alloys for specula earned him the 1777 Copley Medal of the Royal Society. Mudge considered that Short's method of figuring was little different from his own, and there is no doubt that, had Short published the results of his experiments he, and not Mudge, would have received the award. For the composition of specula, Mudge rejected small additions of silver, brass, and arsenic and recommended a mixture of copper and tin in the proportion 2·2 to 1 by weight. Of the grinding process he spoke as follows:[76]

> Four tools are all that are necessary: viz. the rough grinder to work off the rough face of the metal; a brass convex grinder, on which the metal is to receive its spherical figure; a bed of hones which is to perfect that figure and to give the metal a fine smooth face; and a concave tool or bruiser, with which both the brass grinder, and the hones are to be formed. A polisher may be considered as an additional tool; but as the brass grinder is used for this purpose, and its pitchy surface is expeditiously, and without difficulty formed by the bruiser, the apparatus is therefore not enlarged.

When ready for polishing, ' the most difficult part of the whole process ', the metal was moved over the polisher with a circular motion. This drove the pitch inwards and caused the polish to begin at the centre. When the polish almost extended to the edges, the circular motion was changed to short straight strokes in all directions across the centre. This process extended the polish over the entire surface which it made to assume the spherical form, further circular strokes being required for parabolizing. For testing his mirrors, Mudge covered them successively with two diaphragms which occluded first the centre, then the edge. He then observed the image of a distant object with a high-power eyepiece. This was the germ of modern zone tests.[77] We know little about the quality and size of his instruments, but it is recorded that he made two large ones, one of which passed to Count Brühl and then to the Gotha Observatory.[78]

Mudge's ideas were readily applied by Joseph Jackson, a maker of mathematical instruments at Angel Court, Strand, whom Mudge calls ' a most excellent workman '. Jackson made his mirrors with 33 per cent tin as Mudge had advised, and found them so hard that he had to travel 200 miles to find a stone sufficiently hard to cut them. 'I have seen several of his finished metals' Mudge writes; 'they were indeed perfectly hard and white, but the kind of stone with which he ground them, he kept a secret.'[79]

Another experimenter, contemporaneous with Mudge, was the Rev. John Edwards of Ludlow, Shropshire, who wrote a short treatise on the casting and composition of metal specula.[80] In his endeavours to find a good reflecting alloy, he tried over a hundred different combinations and experimented with ingredients like iron, lead, bismuth, and zinc. Of all these, one of 32 parts copper, 15 tin, 1 brass, 1 silver, and 1 arsenic, gave a metal that was by far 'the whitest, hardest and most reflective.' This was because the quantity of tin was just sufficient to 'saturate' the copper; more tin reduced the metal's brilliancy and made the surface 'of a grey blue and dull colour.' The small quantity of arsenic made the alloy 'more compact and solid' and, as Newton had observed, much whiter. The brass made it 'more tough and not so excessively brittle', while silver increased its lustre.[81]

Both Mudge and Edwards appreciated the importance of using pure metals and of first casting the metal into ingots. Copper has a melting point much higher than that of tin (1083°C and 232°C respectively), and Newton had found that the longer the latter remained in the fluid state, the more porous became the final casting. Tin was therefore added after the copper had sufficiently melted and at a temperature just sufficient to fuse the two metals. Edwards melted the copper first, added brass and silver and, finally, tin. He then chilled the metal suddenly by plunging it into cold water. The ingot remelted at a comparatively low temperature, and he then dropped in arsenic which, had it been added to the first melt, would have volatilized and escaped from the crucible. The second melt was now poured into a sand-and-clay mould, formed previously by means of a wooden pattern. While still red hot, the disk was annealed by being put into a 'pot with hot Ashes or Coals'. 'Let the speculum remain in the Ashes' Edwards concludes, 'till the whole is become quite cold.'[82]

The resulting disk had two disadvantages—it was both hard and brittle. Edwards found that an ordinary file could not mark it and that it was difficult to abrade with the slow-cutting emery of those days. Grinding was done by hand and was a slow and laborious task. The extreme brittleness, due to a highly crystalline structure, meant that the heat of the hand or a slight blow often shivered quite a large mass. Edwards also drew attention to the fact that atmospheric acidity attacked the copper and turned a bright metal 'into a dirty or dingy coloured speculum.' This propensity to tarnish meant that, from the moment the metal received the final polish, its reflectivity gradually decreased and could be restored only by repolishing. Speculum metal, however, owing to its high initial reflectivity and to the absence of a better alternative, was used for reflecting telescopes until well into the nineteenth century.

When he came to grind the metal, Edwards dispensed with Hadley's 'bruiser' or local grinder. He used instead two convex tools and a common grindstone, which was finished to gauge curvature, and on which he rough-ground the speculum. He then transferred it to a slightly elliptical convex tool, made of tin and lead, and then fine-ground it on a circular convex tool covered with hone pieces.[83]

For the polishing operation, Edwards covered the elliptical rough grinder with hard pitch—hard, because it did not alter its shape during the polishing proc s and imparted a brighter polish to the metal. The mirror was pressed on a covering some $\frac{3}{4}$ inch thick and thereby imparted its curvature to the pitchy surface. The latter was then divided by grooves into squares. ' These squares ' Edwards wrote, ' by receiving the small portion of the metal that works off it in polishing, will cause the figures of the speculum to be more correct than if no such squares had been made.'[84]

Edwards preferred to polish with ' colcothar of vitriol ' (rouge) instead of putty, since the former gave a ' very fine and high black lustre, so as to give the metal finished with it the complexion of polished steel.' ' In regard to the parabolic figure to be given to the metal ' he continues, ' no particular caution is required in the polishing; the elliptical tool will always cause the speculum to work into an accurate parabolical figure, supposing the transverse and conjugate diameters bear the true proportion to each other (10 to 9), and the metal is not too thick to prevent it always from adhering firmly and uniformly to the polisher.'[85] Undue softness of the pitch sometimes caused it to ' give way ' and so alter the figure, but ' by a little perseverance the correct figure is very easily acquired.' As long as the two axes of the polisher were in the ratio 10 : 9 and the strokes were cross-strokes regularly changed in direction, the figure became a paraboloid. If, through some mischance, the minor diameter of the ellipse was decreased, the mirror became over-corrected or hyperboloidal.[86]

Edwards' elliptical polisher never became popular with opticians. Apart from the difficulty of making it true to shape, it was less accommodating than other methods. Opticians found that various systems of strokes with circular polishers produced a variety of aspherical surfaces—the greatest problem was to detect rather than to acquire the desirable paraboloidal figure.

Like Mudge, Edwards was unable to offer a test suitable for controlling a mirror's figure during the final polishing stages. When the two mirrors of a Gregorian, say, had received their final polish, Edwards mounted them in a tube and viewed a black circle of about $\frac{1}{2}$-inch diameter placed some 50 to 100 yards from the instrument. He then noted the appearance of the extra-focal image of the ring by moving the secondary mirror slightly nearer and further from the eyepiece. If the circle blurred to the same extent either way, the large mirror was considered paraboloidal; if the blurring was more pronounced on one side than the other, the large mirror was either over- or under-corrected.[87] These conclusions take for granted a well-corrected secondary mirror, the importance of which Edwards seems to have overlooked.

Edwards made several small telescopes, mainly Gregorians and Newtonians. We know little about them but for Maskelyne's testimony on the brightness of their specula:

> Mr. Edwards' telescopes [he writes][88] shew a white object perfectly white, and all objects of their natural colours; very different from common reflecting telescopes which give a dingy copperish appearance to objects. I found, by a careful experiment that they shew objects as bright as a treble object-glass achromatic telescope, both being put under equal circumstances of areas of the aperture of the object-metal and

object-glass, and equal magnifying powers; whereas the diameter of the aperture of a common reflecting telescope must be to that of an achromatic telescope as 8 to 5, to produce an equal effect.

Following the example set by Short, two workers made large reflectors—so large, in fact, that they were useless. At Paris, Father Noël, a Bernardin priest and custodian of King Louis XV's private collection of scientific instruments, made a Gregorian 22 feet long with a primary speculum $23\frac{1}{2}$ inches in diameter. This telescope was finished in 1761, but gave such poor results that Noël asked for a further sum (it had already cost over 60,000 francs) to put it right before the 1769 transit of Venus. He refigured the mirrors and, in 1765, the instrument was housed in its own observatory in the Muette. We hear no more of its performance and can safely assume that it was of low standard. At any rate Carroché, in 1796, made a new primary mirror and converted the instrument into a Newtonian. It was then transferred to the Paris Observatory, to become an ornament until dismantled in 1841.[89]

A less pretentious, but no more useful, telescope was the large Gregorian equatorial made about 1780 by the Rev. J. Michell of Thornhill, Yorkshire. The primary speculum had an aperture of $29\frac{1}{2}$ inches and a focal length of 10 feet—an unusually large aperture ratio of $f/4 \cdot 1$.[90] It was reported to perform well on 'day-objects' but astronomy was none the richer for its existence. After Michell's death it was, although broken, bought by Sir William Herschel, together with the polisher, a 12-foot iron tube, and several concave secondary mirrors.[91]

REFERENCES

1 Descartes, R., 1650, *Principia Philosophiae*, iii, p. 64.
2 Descartes, R., 1637, *Météores*, Disc. 8.
3 Grimaldi, F. M., 1665, *Physico-Mathesis de lumine, coloribus, et iride*, i.
4 Newton, I., *Phil. Trans.*, **6**, pp. 3075–3087, 1671.
5 Newton, I., 1718, *Opticks*, p. 83.
6 *Ibid.*, p. 87.
7 *Ibid.*, p. 89.
8 *Ibid.*, pp. 89–90.
9 *Ibid.*, pp. 16–22, Prop. I–IV. *Phil. Trans.*, **6**, p. 3081, 1671.
10 Brewster, D., 1855, *Memoirs of the Life, Writings, and Discoveries of Sir Isaac Newton*, i, p. 110. Herschel, J. F. W., 1848, *Encyclopaedia Metropolitana*, iv, p. 411. These authors suggest that Newton mixed sugar of lead with the water to increase its refractive index. Newton sometimes resorted to this expedient (*Opticks*, 2nd edition, p. 62), but there is no evidence that he used impregnated water in this experiment.
11 Newton, Ref. 5, p. 90.
12 Zucchi, P., 1652–56, *Optica Philosophia*, i, p. 126.
13 Cavalieri, B., 1650, *Specchio Ustorio*.
14 Mersenne, M., 1636, *L'Harmonie Universelle*, i, pp. 61–62.
15 Descartes, R, *Œuvres de Descartes*, 1898, ii, pp. 559, 589. *Lettres de Descartes*, 1659, ii, Nos. 3, 29, 32.
16 Gregory, J., 1663, *Optica Promota*, p. 94.
17 Rigaud, S. J., 1841, *Correspondence of Scientific Men of the 17th Century*, pp. 244, 247, 357.
18 Hutton, C., 1796, *Mathematical and Philosophical Dictionary*, p. 553.
19 Newton, I., Ref. 5, pp. 91–92.
20 Weld, C. R., 1848, *History of the Royal Society*, p. 235.
21 Newton, I., *Opticks*, 1931 edition of 4th edition, p. 103. For a further description vide *Phil. Trans.*, **7**, pp. 4004–4010, 1672.
22 Birch, T., 1756, *History of the Royal Society*, iii, pp. 2, 4.
23 Stone, E., 1758, Supplement to Bion's *The Construction and principal uses of Mathematical Instruments*, pp. 283, 298.
24 Newton, I., *Phil. Trans.*, **7**, p. 4006, 1672.
25 *Ibid.*, pp. 4006–4007.
26 Newton, I., *Opticks*, 1931 reprint of 4th edition, p. 104.

[27] Newton, I., *Opticks*, 1931 reprint of 4th edition, p. 108.
[28] Danjon, A., and Couder, A., 1935, *Lunettes et Télescopes*, p. 613.
[29] Bell, L., 1922, *The Telescope*, pp. 22–23.
[30] *Jour. des Sçavans*, **3**, pp. 81–84, 1672. *Phil. Trans.*, **7**, p. 4056, 1672.
[31] Ramsden, J., *Phil. Trans.*, **69**, pp. 419–425, 1779.
[32] *Phil. Trans.*, **7**, pp. 4057–4059, 1672.
[33] Weld, *op. cit.*, i, p. 236.
[34] He changed his mind when he tried to make specula—*vide* Bell, A. E., 1947, *Christian Huygens*, pp. 70–71.
[35] Birch, *op. cit.*, iii, p. 122.
[36] Robinson, H. W., and Adams, W., 1935, *Diary of Robert Hooke.* For mirror-making *vide* dates Aug., Sept., 1672; Jan., Feb., 1673; Sept., 1675; Jan., 1678.
[37] Birch, *op. cit.*, iii, pp. 19, 43, 49, 57. Rigaud, *op. cit.*, pp. 290–291.
[38] Desaguliers, J. T., 1744, *A Course of Experimental Philosophy*, ii, p. 528.
[39] Hadley, J., *Phil. Trans.*, **32**, p. 303, 1723.
[40] Pound, J., *ibid.*, p. 382.
[41] Hadley, J., *ibid.*, p. 384.
[42] Pound, J., *ibid.*, p. 383.
[43] Smith, R., 1738, *A Compleat System of Opticks*, ii, p. 304.
[44] *Ibid.*, p. 304.
[45] *Ibid.*, p. 305.
[46] *Ibid.*, pp. 307–308.
[47] *Ibid.*, p. 311.
[48] *Ibid.*, p. 310.
[49] Martin, L. C., *Trans. Opt. Soc.*, **24**, p. 293, 1923.
[50] *Ibid.*, *p.* 297.
[51] *Phil. Trans.*, **42**, p. 155, 1742.
[52] Dreyer, J. L. E., *Astr. Nach.*, **115**, pp. 34 ff., 1886.
[53] *Phil. Trans.*, **37**, p. 147, 1731.
[54] *Biographical Account of John Hadley*, pp. 27–28, an unsigned tract in the Library of the Royal Astronomical Society. Rigaud, S. P., 1832, *Miscellaneous Works and Correspondence of James Bradley*, p. xi, says that Bradley possessed a 5-ft Hadley Newtonian.
[55] *Biographical Account of John Hadley*, p. 11, Ref. 54.
[56] Smith, *op. cit.*, ii, p. 79.
[57] *Ibid.*, p. 303.
[58] *Ibid.*, Remarks, p. 80.
[59] *Ibid.*, Remarks, p. 80.
[60] *Ibid.*, p. 81.
[61] *Ibid.*
[62] Court, T. H., and Rohr, M. von, *Trans. Opt. Soc.*, **30**, p. 220, 1929.
[63] *Dict. Nat. Biog.*, 1893, art. ‘James Short’. *Gentleman's Magazine*, p. 303, 1768.
[64] Dreyer, J. L. E., 1912, Introduction to *Scientific Papers W. Herschel*, i, p. lx.
[65] *Nautical Almanac*, 1787, Appendix, p. 39.
[66] Kitchener, W., 1825, *The Economy of the Eyes*, ii, Of Telescopes . . ., pp. 70–71.
[67] Bernoulli, Jean, junior, 1771, *Lettres Astronomiques*, p. 111.
[68] *Dict. Nat. Biog.*, 1893, art. ‘James Short’.
[69] *Phil. Trans.*, **59**, pp. 507–511, 1769.
[70] *Phil. Trans.*, **46**, pp. 241–246, 1741.
[71] *Trans. Opt. Soc.*, **24**, p. 316, 1923; **30**, p. 224, 1929.
[72] *Phil. Trans.*, **48**, p. 192, 1754; **52**, p. 178, 1761.
[73] Kitchener, *op. cit.*, pp. 76, 92.
[74] *Ibid.*, pp. 114–115.
[75] *Ibid.*, p. 163.
[76] Mudge, J., *Phil. Trans.*, **67**, p. 304, 1777.
[77] *Ibid.*, pp. 336–338.
[78] *Dict. Nat. Biog.*, 1893, art. ‘Mudge’.
[79] Mudge, *op. cit.*, p. 298.
[80] *Nautical Almanac*, 1787, Appendix, pp. 3–22, 23–48.
[81] *Ibid.*, pp. 4–7.
[82] *Ibid.*, p. 14.
[83] *Ibid.*, pp. 15–16.
[84] *Ibid.*, p. 19.
[85] *Ibid.*, p. 21.
[86] *Ibid.*, p. 22.
[87] *Ibid.*, pp. 31–32.
[88] *Ibid.*, p. 3.
[89] Wolf, C., 1902, *Histoire de l'Observatoire de Paris*, pp. 334–336.
[90] Dreyer, Ref. 64, i, p. xxxii.
[91] *Ibid.*

CHAPTER VI

The application of the telescope to divided instruments, so as to serve in ascertaining the apparent direction of a celestial body, was another of those great improvements which distinguished the progress of practical astronomy in the seventeenth century.

R. GRANT

NEXT in importance to the invention of the telescope was its application to divided instruments, which did not occur until over fifty years after Galileo made his discoveries. In consequence, Tycho's observations represented the best record of the state of the heavens at that time, and his instruments, the *ne plus ultra* of the instrument-maker's art. The first to improve upon Kepler's *Rudolphine Tables* was Jeremiah Horrocks, a young clergyman of Hoole, near Liverpool. From his tables, Kepler predicted that Mercury and Venus would both cross the sun's disk in 1631 and announced that the latter would not transit again until 1761. The transit of Mercury was seen by Pierre Gassendi from Paris on November 7, the day predicted, but early by several hours.[1] The transit of Venus, unknown to Gassendi, happened at night, and he made fruitless observations both before and after the predicted time.[2] From a study of Kepler's and other tables, corrected by his own observations made with a half-crown telescope,[3] Horrocks predicted another transit of Venus at 3 p.m. on November 24 (O.S.), 1639, and forewarned his friend, William Crabtree. We can imagine his excitement when, adopting Gassendi's method, he projected the sun's image on a circle divided round into degrees and, at 3.17 p.m. 'beheld a most agreeable spectacle . . . a spot of unusual magnitude and of perfectly circular shape which had wholly entered upon the Sun's disk to the left.'[4] Crabtree, at Broughton, near Manchester, was less fortunate, for the sky remained overcast all day, and only when the sun was nearly on the horizon did the clouds break. Then, to his delight, he saw Venus silhouetted against the sun and, according to Horrocks, was so taken aback by the spectacle that, by the time he had recovered himself, the clouds had again obscured the sun.[5]

Besides making this important observation, Horrocks recorded the positions of the moon and planets. Before he was eighteen, he had discovered discrepancies which, at first, he attributed to his own malobservation. On comparing notes with Crabtree, however, he found that the errors were in the tables. He was a great admirer of Tycho and Kepler and defended rather than disparaged their work, for he realized that their tables were far more accurate than those of Lansberg and others.[6] In seeking to improve the *Rudolphine Tables*, therefore, he did much to promote the general acceptance of Kepler's laws. He was the first to show that the moon travelled round the earth in an ellipse and to detect the long inequalities of the motions of Jupiter and Saturn.[7] Astronomy suffered a double loss when

Horrocks died in 1641, at the age of twenty-two, and when Crabtree perished soon afterwards in the Civil War.

Pierre Vernier, of Ornans in Burgundy, was the first to show how astronomers might improve upon Tycho's method of reading arcs and circles. In 1631, he described a quadrant[8] on which the divisions were read off against a small scale instead of against a pointer—and to an accuracy of 1 minute of arc. The circle itself was divided to every 30 minutes, while the length of the separate scale was obtained by taking thirty-one of the 30-minute divisions and dividing them into thirty equal parts. Consequently, when the two scales were placed edge to edge, every division of the vernier advanced one-thirtieth of 30 minutes, or 1 minute of arc. This was Vernier's original arrangement; the same result is obtained if twenty-nine divisions of the main scale are divided into thirty equal parts, for then every division of the main scale advances 1 minute of arc as before. The procedure is the same in both cases—the eye looks for the line without a break and reads direct from the vernier scale the number of minutes which must be added to the last $\frac{1}{2}^{\circ}$ division on the main scale. The success of the arrangement depends upon the observer's ability to detect the slightest break in an otherwise straight line, so that the line of coincidence is readily picked out from its neighbours.

The replacement of plain sights by a telescope was a step first taken by Jean Baptiste Morin in 1634.[9] Morin was a French astronomer, notorious for his opposition to the Copernican theory and for his devotion to astrology. He found that he could see stars in daytime through a Galilean telescope and thought that, by fitting one to a divided circle, he could materially assist in the problem of finding longitude at sea.

The invention of the telescopic sight rightly belongs to William Gascoigne. Gascoigne, an amateur astronomer of Middleton, near Leeds, was the youngest member of an enthusiastic little band of Midland astronomers which comprised Horrocks, Crabtree, and Oughtred. He was only twenty when he received the following communication from Crabtree, the contents of which indicate that astronomy had long been a subject for study.

> You told me, as I remember [Crabtree writes][10] you doubted not in time to be able to make observations to seconds. I cannot but admire it, and yet, by what I saw, believe it; but long to have some farther hints of your conceit for that purpose. One means, I think, you told me was, by a single glass in a cane, upon the index of your sextant, by which, as I remember, you find the exact point of the sun's rays. . . . Could I purchase it with travel, or procure it for gold, I would not be without a telescope for observing small angles in the heavens; nor want the use of your other device of a glass in a cane upon the movable ruler of your sextant, as I remember, for helping to the exact point of the sun's rays.

The contents of this and other letters reveal the intense interest these amateurs took in any invention likely to make their instruments and observations more accurate. We can imagine with what interest they heard that an obliging spider had presented Gascoigne with a telescopic sight (Fig. 44).

> This is that admirable secret [Gascoigne writes to Oughtred][11] which, as all other things, appeared when it pleased the All Disposer, at whose directions a spider's line

drawn in an open case could first give me by its perfect apparition, when I was with two convexes trying experiments about the sun, the unexpected knowledge.

The importance of this occurrence lay in the fact that the thread appeared in the plane of the primary image. The eyepiece magnified both alike; by arranging crossed threads at the centre of the field, the telescope could be pointed precisely to a star.

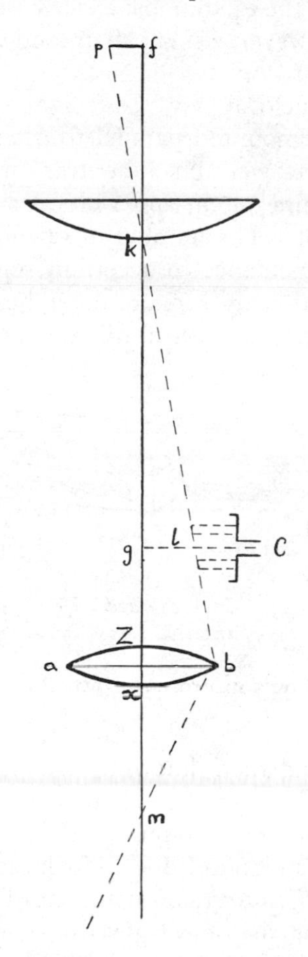

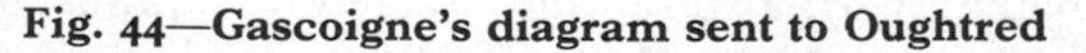

Fig. 44—Gascoigne's diagram sent to Oughtred

The diagram shows where he placed the spider's thread. He is using the lens at *k* as an object-glass. The web is *glC*.

Gascoigne seems to have appreciated the advantages offered by spiders' webs but he did not apply them to his telescope owing, no doubt, to difficulty in mounting the delicate strands. In another letter, dated January 25, 1641, this time to Crabtree, Gascoigne tells how he managed to illuminate an ordinary hair in the telescope:[12]

If the night be so dark that the hair, or the pointers of the scale not to be seen, I place a candle in a lanthorn, so as to cast light sufficient into the glass, which I find very helpful when the moon appeareth not, or it is not otherwise light enough.

So far as we know, Gascoigne invented the eyepiece micrometer soon after he applied hairs to the field of his telescope, but he may have been using it as early as 1639. In a letter to Horrocks in that year,[13] Crabtree says he has just visited Gascoigne and seen ' a large telescope, amplified and adorned with new inventions ' whereby the owner could ' take the diameters of the sun or moon, or any small angle in the heavens or upon the earth, most exactly through the glass to a second.' The first micrometer was, however, a small affair, and was used with quite a modest Keplerian telescope (Fig. 45).

For reference lines in the field of view, Gascoigne used the finely-ground edges of two thin pieces of metal, mounted parallel to each other in the focal plane of the object-glass. These edges were given contrary motions by means of a right- and a left-handed screw of fine pitch, and were at all times equidistant from the optical axis of the instrument. The number of revolutions of the screw necessary to move the edges from coincidence to a certain separation was registered by a scale, and parts of a revolution by a micrometer head, divided into 100 parts.[14] In practice, Gascoigne turned this head until each edge became tangential to the

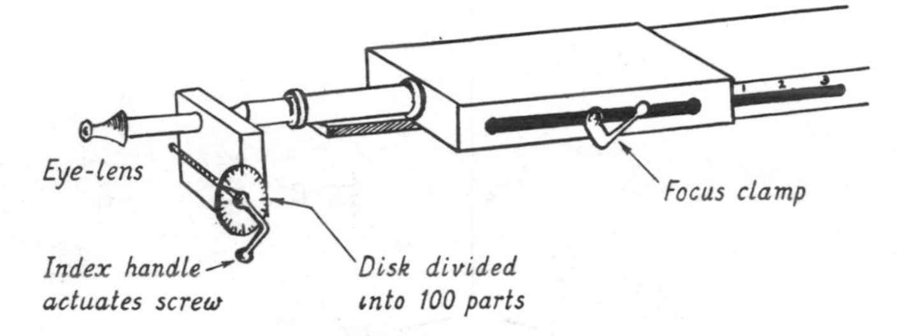

Fig. 45—Gascoigne's micrometer attached to the telescope

body under measurement and then, knowing the focal length of the object-glass and the reading of the micrometer, he calculated the object's angular diameter.

Gascoigne says little about the mounting of his telescope, but it apparently differed little from the one handed to Robert Hooke some years later and described by him in the *Philosophical Transactions* of the Royal Society.[15] There were two tubes. The inner containing the object-glass was probably of cardboard, paper-covered and graduated ' with a spiral line divided by equidistants ' for measuring the distance of the eye-lens from the object-glass. The outer tube, or ' case ' as Gascoigne called it, was oblong in section and contained the micrometer fittings. The eye-lens was a single equi-biconvex lens, and Gascoigne pointed out how much larger the field appeared in the Keplerian form compared with that of the Galilean.

Townley states[16] that Gascoigne had a treatise on optics just ready for the press, but that he was unable to find any trace of the manuscript. From the diagrams with which Gascoigne illustrated his letters it is clear that he was an expert ray-tracer, but his methods remained unknown until 1692. In that year, William Molyneux published his *Dioptrica Nova* and mentioned[17] that ' the Geometrical Method of calculating a Ray's Progress, which in many particulars is so amply delivered hereafter, is wholly new, and never before publish'd. And for the first Intimation thereof, I must acknowledge my self obliged to my worthy Friend Mr. Flamstead Astron. Reg. who had it from some unpublished Papers of Mr. Gascoigne's.'

In his letters to his friends, Gascoigne deplores the absence of a ' skilful workman in iron or brass ' in his district, and complains of the bad workmanship of the local joiner.[18] Yet Gascoigne managed to construct several graduated instruments, to fit up a Galilean from ' London best sale glasses ' and to apply telescopic sights to his quadrant and sextant.[19] From 1639, until his premature death at the age of twenty-four in the Battle of Marston Moor, he made several measures of the diameters of the moon and planets, together with other observations now lost. Unfortunately, the knowledge of all this work, together with that connected with the micrometer, was limited to the small circle of his friends, all of whom died before or during the Civil War. Had not his friend Townley and then Flamsteed recognized the value of his letters, it is probable that even these would have been lost.

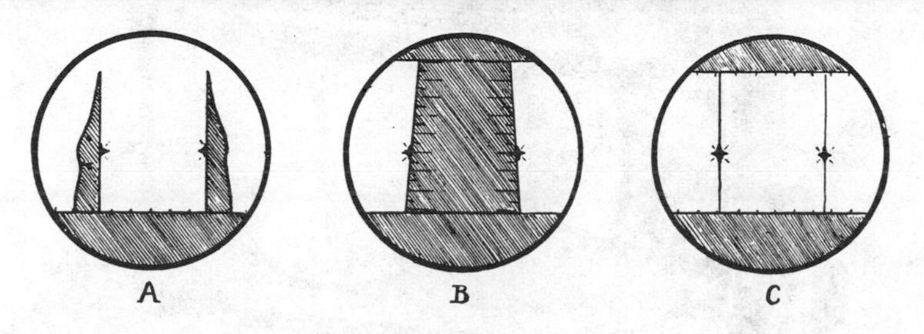

Fig. 46—The apparent distance between two stars measured in three ways

A Gascoigne's knife-edge micrometer
B Huygens' 'virgula'
C Hooke's hair micrometer

The first published account of an eyepiece micrometer appeared in 1659, in Huygens' *Systema Saturnium*, when Gascoigne's previous work remained unknown. Huygens placed a *virgula* or tapered metal bar in the focal plane of the ocular of his telescope and moved the wedge until it exactly covered the object to be measured (Fig. 46). The breadth of the bar at this point, compared with the total extent of the field of view (ascertained by noting with a pendulum clock the time taken by a star near the equator to cross its diameter), enabled him to deduce the apparent diameter of the object.[20] In this manner, Huygens measured the diameters of the planets, the positions of their satellites and the apparent size of Saturn's ring. Irradiation, and the defects of his telescopes, account for nearly all of the excessive values he obtained. Before the invention of the moving-wire micrometer, astronomers had only the known breadth of the field of view or the apparent separation of two proximate stars to guide them. Until Huygens introduced the compound eyepiece, the former was of indefinite extent; the latter was subject to the errors of previous observers working under similar or worse conditions. Then, apart from peculiarities of vision such as irradiation, there was the effect due to chromatic aberration. The latter, as Newton pointed out, caused the spurious enlargement of bright disks like those of the moon and planets.

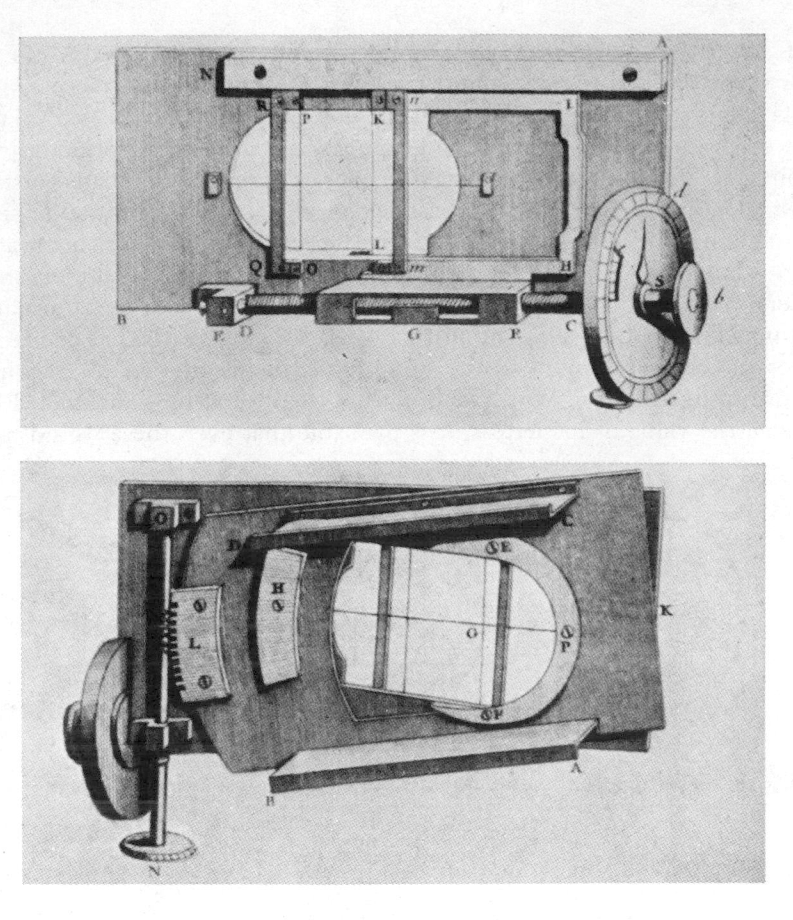

Fig. 47—An early wire micrometer, *circa* 1700

(From Lalande's 'Astronomie')

Huygens' micrometer was adopted by Hevelius who used it to measure the apparent diameters of Venus and Jupiter. These he compared with the apparent sizes of certain spots on the moon, but his measures were again too large.[21] A less clumsy micrometer was the reticule devised in 1662 by the Marquis Cornelio Malvasia of Bologna. It consisted of a network of fine silver wires which divided the field into a number of equal squares of known size.[22] Malvasia used it for plotting the positions of craters on the moon, measuring the distance between any two craters by counting the number of whole squares between them. Seldom did two craters or two stars fall exactly on the wires and the observation was thus subject to the errors of eye-estimation. With the intention of removing this disadvantage, Auzout, quite independently of Gascoigne, made a moving-wire or filar micrometer.[23] He dispensed entirely with Malvasia's reticule, and mounted two silk threads parallel to each other, one fixed and the other movable. In practice, the movable thread traversed the diameter of the object, the number of revolutions and part of a revolution of the dial necessary to effect this being a measure of the object's angular diameter. In measuring the diameter of the moon, for instance, the fixed thread would be placed to form a tangent to the disk, and the other thread moved until it too formed a tangent on the opposite side. A little later, Auzout elaborated on this simple device by using nine threads—six fixed at regular intervals across the field and three mounted in a movable frame. He sent an account of his first micrometer to the Royal Society in 1666,[24] where it was roundly criticized by Hooke and Wren. Both these Fellows had experimented with micrometers in the form of diaphragms and reticules, Hooke for the purpose of determining the parallax of stars and planets, Wren with reference to the more accurate determination of longitude. Both Hooke and Wren claimed priority over Auzout, but Auzout's claim had been printed only a few months when all three claimants had to give first place to Gascoigne.

In March, 1667, the following letter from Townley was received by the council of the Royal Society:[25]

> Finding in one of the last *Philosophical Transactions* how much M. Auzout esteems his invention of dividing a foot into near 30,000 parts, and taking thereby angles to a very great exactness, I am told I shall be looked upon as a great wronger of our nation, should I not let the world know that I have, out of some scattered papers and letters that formerly came to my hands of a gentleman of these parts, one Mr. Gascoigne, found out that, before our late civil wars, he had not only devised an instrument of as great a power as M. Auzout's but had also for some years made use of it, not only for taking the diameters of the planets and distances upon land; but had further endeavoured out of its preciseness to gather many certainties in the heavens. . . . I shall only say of it that it is small, not exceeding in weight, nor much in bigness, an ordinary Pocket-watch, exactly marking above 40,000 divisions in a foot, by the help of two indexes; the one showing hundreds of divisions, the other divisions of the hundred.

In concluding his letter, Townley said that he was willing to loan the instrument to the Royal Society for its inspection. His offer was accepted, the micrometer was on show at the July, 1667, meeting of the Society and Gascoigne was acclaimed the first inventor. The micrometer was not exactly as Gascoigne had left it, for Townley had made a few improvements. In this condition it was described and illustrated by Hooke in a subsequent number of the *Philosophical Transactions*.[26] To Hooke also, must be given credit for replacing Gascoigne's solid edges by hairs.

Owing to the delay of twenty-six years in making known Gascoigne's activities, French astronomers received full credit for the invention of telescopic sights. In 1667, the Abbé Picard made observations of the sun with a 9½-foot quadrant and 6-foot sextant, both fitted with telescopic sights.[27] In his geodetical work, Picard used a 38-inch quadrant with a telescope attached to the alidade and, when he published his first results in 1670, left no doubt but that he was sole inventor of telescopic sights.[28] Probably Auzout,[29] and perhaps Römer, had a share in the design of Picard's instruments. In any case, the work of these astronomers was instrumental in establishing the new school of precision astronomy which developed rapidly in the early eighteenth century both in England and on the Continent.

Picard applied his telescopic sight to several instruments, including an optical level, the use of which he described in his *Traité du Nivellement*.[30] About this time, several large aqueducts were under construction in France and there grew an increasing demand for the more precise determination of relative altitudes and the surveying of large areas. In Picard's level, the direction of the vertical was indicated by a long plummet contained in a glass case to exclude draughts.[31] The telescope itself formed part of the weight and so earned it the name 'pendulum level'. The first bubble level was described in 1661 by Melchisedech Thévènot, a French scholar and traveller but, owing to the difficulty of manufacturing accurate bubble tubes, his suggestion was shelved for some years.[32] In 1702, A. M. Mallet described several spirit-levels, none of them very accurate. In these, the tubes were selected from a tube of uniform bore and with smooth surfaces, to be hermetically sealed after the introduction of the liquid.[33] The modern process of internal grinding was introduced by Chezy some fifty years later.[34]

There remained one astronomer who stoutly denied the advantages of telescopic sights. Hevelius was an observer of the Tychonic school—he always used plain sights and pinnules on his large sextants and quadrants (Fig. 48). When he published his *Cometographia* in 1668, Hevelius sent copies to his friends at the Royal Society, among them Hooke. In return, Hooke sent a description of a refracting telescope and recommended its use, instead of plain sights, on graduated instruments. Many letters followed in consequence, the main point of debate being whether it was possible to measure with pinnules to any limit smaller than 1 minute of arc. Hooke thought not. He argued that, while he could measure with a telescope an angle of 1 second, the minimum angular separation given by a normal eye was, at the most, about 30 seconds and, more generally, only 1 minute. Unless two stars were separated further than this amount, they would appear as one. Hevelius did not deny this. He claimed, however, that good eyesight and long practice had enabled him to exceed these limits. In proof, he sent a set of measures to Hooke, requesting him to check them with his own instruments. With the publication of the *Machinae Coelestis* in 1673, Hevelius further urged the accuracy of his instruments, and stated his reasons for not adopting the telescopic sight. He writes:[35]

> Add to this that if at any time the Observator chances not to look directly and precisely through the midst of the Glasses (as believe me it may often happen) some Varieties may easily intermingle with the Observations, which in time may egregiously deceive the Astronomer. Moreover, seeing the Needle or cross Threads, do stand so close to the Eye-Glass, and near the Eye of the Observator; I question whether these Sights, so near the Eye, can discover the smallest Stars much more accurately, than our plain Sights, which are distant from each other Six or Nine Feet. For, though by these Telescopick Sights, one may see the Object more distinctly; yet because they are so nigh to the Eye, one may err, more than 'tis possible by our plain Sights, that are so far asunder; so that I shall take no further notice of another Inconvenience, which is, that the Intersection of the Threads shall cover the smallest Stars from your Sight.

These remarks, and others like them, received the strong and tactless censure of Hooke in his retaliatory *Animadversions upon the Machinae Coelestis of Hevelius*. Hooke passed over Hevelius' optical difficulties but ridiculed his instruments and slighted the value of his observations. 'This' Molyneux writes,[36] 'did but

Fig. 48—Hevelius' sextant in an observatory

(*Science Museum, London. British Crown copyright*)

exasperate the Noble old Man, and made him adhere more obstinately to his former Practice.' Hevelius was never sure of his position with Hooke; like Newton, he was subjected to a tone as patronizing as it was sometimes accusing. Eventually, the dispute became so heated that the Royal Society intervened and, in 1679, sent young Halley to Danzig to ease the tension.[37] Hevelius received him warmly, and the two astronomers, one at the beginning and the other near the end of his career, made and compared observations.[38] Halley found that Hevelius' claims were justified, but he could not convince his host of the greater accuracy of the telescopic sight. Even after this, a further passage of arms ensued between Hevelius and Hooke in 1685, but to the gain of neither. Hooke had to moderate his scathing tone, but he remained firm to his original arguments. The justice of these was amply confirmed by subsequent developments in practical astronomy.

With the increasing application of the refracting telescope to graduated instruments, astronomers found themselves becoming more dependent on the skill of the instrument-maker. The scientific instrument-maker, in the modern sense, did not appear until about 1740. Until then, his place was taken by craftsmen skilled in general metalwork or in horology—indeed by anyone who possessed the requisite mechanical ability. Consequently, astronomers were often hard put to it to find craftsmen skilled enough to undertake accurate work. The situation was particularly difficult in France, where there had been little demand for graduated instruments, but a large one for object-glasses. Auzout did not exaggerate when he told Louis XIV in 1664 that he did not know any instrument in the kingdom accurate enough for obtaining a reliable and consistent value of the altitude of the Pole Star.[39] In England, things were little better; the London opticians were all busy supplying the public demand for spectacles, microscopes, and telescopes. For this reason, Hooke constructed his own instruments, Flamsteed divided his own quadrant, and two clockmakers, Tompion and Graham, made many important astronomical instruments.

Fortunately for the progress of precision astronomy, several skilled instrument-makers appeared in the eighteenth century, all ' artists ' in the broad sense, all ready to study the astronomer's needs and to work under his guidance. The eighteenth century saw many fine specimens of the instrument-maker's art. It was a century dominated, in the first half, by large sextants and quadrants and, in the second, by the achromatic telescope and transit instrument. At first, the possibilities of the telescopic sight seemed endless, limited only by the power of the telescope and the size and accuracy of the divided arc. But the use of a telescope for angular measurements meant that its optical system magnified defects in the instrument as a whole. In the case of a quadrant, errors of graduation of the limb, defects due to the wearing of movable parts, possible warping of the frame due to unrelieved strains or unequal expansion—these defects and others influenced the readings and had to be allowed for when reducing the observations. Difficulties in graduation were partly overcome by making instruments of long radius; a large arc was better to work on and could be graduated to small divisions. These divisions were more easily seen than on arcs of small radius, and the greater length and aperture of the telescope increased its magnification and improved the definition.

Fig. 49—Römer's transit instrument, 1684

(*Science Museum, London. British Crown copyright*)

Disadvantages appeared, however, when the mass of a large quadrant was suspended vertically in the plane of the meridian. Its weight alone was sufficient to cause appreciable flexure, especially if the frame was unequally supported by wall brackets. In the early quadrants, also, little attempt was made to counterpoise the weight of the telescope; in consequence, the central pivot upon which it rotated became eccentric after a few years' use.

In later quadrants, the telescope was counterpoised and the weight of the frame was shared equally by carefully spaced bearing surfaces embedded in a wall. These devices were also found inadequate—alternatives were either to make a complete mural circle or, better still, to mount the circle on a horizontal axis supported at each end by two walls. Yet, with all its drawbacks, the quadrant, as in Tycho's day, continued the main instrument in observatories. Once fixed in the meridian, it enabled astronomers to determine fundamental reference points in the sky, the altitudes and hence declinations of stars and, with the aid of a reliable clock, their right ascensions.

We have seen how the poor timepieces of the sixteenth century prevented Tycho from measuring right ascensions in this way but, since then, Huygens had invented a much more accurate clock. In 1656, he adapted to clock movements[40] the principle of isochronism of the pendulum (a principle discovered by Galileo), with the result that his new pendulum clocks indicated minutes with greater accuracy than the old foliot clocks could indicate hours.

Picard was one of the first to appreciate the value of the pendulum clock when used in conjunction with a quadrant. One of these clocks in his possession had an error of only one minute over a period of two months, whilst its audible tick enabled him to time transits to every second and part of a second.[41] For many years, therefore, he tried to procure a mural quadrant for the Paris Observatory, but to no avail, and he made shift with a small telescope fixed in the meridian.[42] Thus necessity drove him to devise the first transit instrument.

The invention of the transit is generally ascribed to Olaus Römer, Picard's assistant, who apparently obtained the idea from Picard, but who was the first to use it extensively and to describe it in detail. Römer's transit was mounted in a window-frame of his house in Copenhagen (Fig. 49). It consisted of a telescope fixed at one end of a 5-foot horizontal axis and at right angles to it. The axis rotated on brackets attached to two opposite walls and was counterpoised so as to prevent flexure and undue pressure on the brackets. The tube was composed of two hollow cones, joined together at their bases, a form more rigid than a plain cylinder. A pointer attached to one end of the axis moved over a graduated arc as the telescope rotated in altitude. The scale read direct to every 10 minutes of arc and by microscope to 1 minute.[43] This is the first time that we find the use of hollow cones for telescope tubes and of a microscope for assisting the eye in taking pointer readings.

In 1704, Römer moved to a ' rural ' observatory on the outskirts of Copenhagen, where he erected another valuable instrument—the first meridian circle. The adoption of a complete circle in place of a quadrant or sextant was a great advance. ' I would place more confidence in a circle of four feet, than I would in a quadrant of ten feet ' Römer once wrote.[44] In its general design, the new instrument was

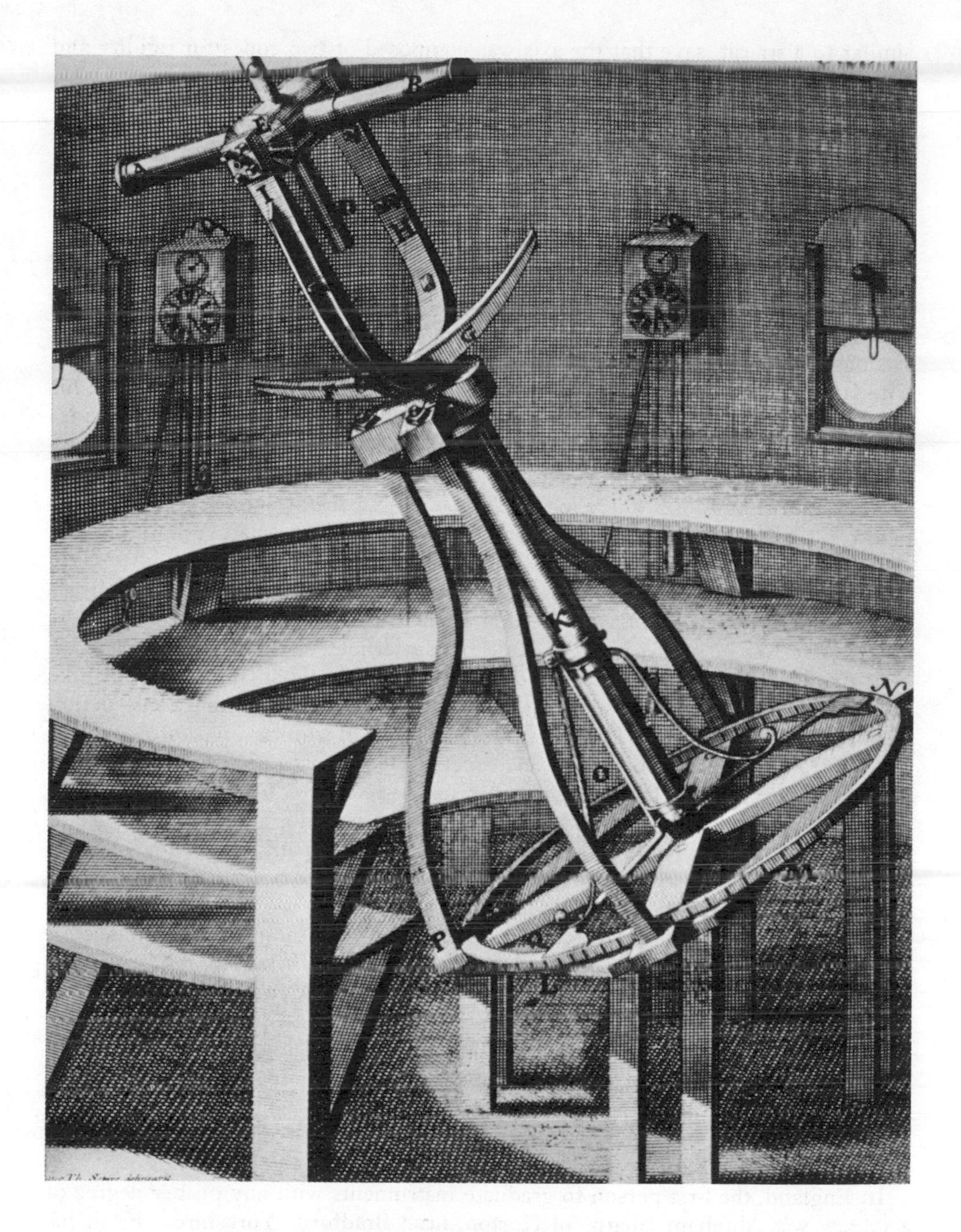

Fig. 50—Römer's equatorial instrument, 1690

(*Science Museum, London. British Crown copyright*)

similar to a transit, save that the axis was composed of two cones for rigidity and that the telescope had a $5\frac{1}{2}$-foot circle attached to it.

Römer was also the first to construct, or at least to use, an altazimuth circle, an instrument which, with its $3\frac{1}{2}$-foot circles, enabled him to take observations for checking his clocks and meridian instruments. He could also determine the positions of objects without having to wait until they crossed the meridian.[45] Similar service was given by an equatorial which he also used for determining the parallaxes of planets by measuring their apparent distances from nearby stars at different times[46] (Fig. 50).

Römer took precautions to ensure that his instruments remained in adjustment and, for this purpose, instituted a set of observations for checking such important settings as horizontality of axis, collimation, and azimuth. His first residence, with its restricted horizon, was by no means an ideal observatory although, in his estimation, it was as efficient as the palatial Paris Observatory. ' I differ very widely ' he once wrote to Leibniz,[47] ' from those who have hitherto decked out observatories more for show than for use, accommodating the instruments to the buildings rather than the buildings to the instruments.' This attitude is evident in the comparative simplicity of his instruments; they lacked the ornate decorations so characteristic of the instruments of Tycho and Hevelius.

The first accurate quadrant in the Paris Observatory was constructed by J. Langlois in 1732. Langlois was the first French instrument-maker of any consequence. He was introduced to the trade by Butterfield,[48] an Englishman who, at the ' Armes de l'Angleterre ' at St Germain, made and sold small instruments like pocket dials and graduated rings. Langlois' predecessors in quadrant construction were Sevin, Gosselin, and Lagny, workmen skilled in gunmaking, horology, and general ironwork. Langlois, however, concentrated on astronomical instruments. His mural arc for the Paris Observatory had a radius of 6 feet.[49] Frame and telescope were without counterpoises and the brass limb, pinned to the solid cast iron frame, was liable to the effects of unequal expansion. The limb was divided into minutes by transversals and a tangent micrometer gave readings to seconds. In 1738, Langlois made another quadrant for use to the north of the zenith,[50] this time only $3\frac{1}{2}$ feet in radius, while in the same year he finished a 6-foot sextant. The latter was used by Cassini II and Maraldi for geodetical observations. In 1751, Lacaille took the sextant to the Cape of Good Hope for use in his survey of the brighter stars in the southern hemisphere.[51]

Langlois made several instruments for the French Academy. Among them may be mentioned a 6-foot movable iron quadrant, covered by a rotatable conical ' dome '.[52] He also made smaller instruments for geodetical surveys in Peru and Lapland. His last instrument appears to have been a 3-foot portable quadrant, constructed in 1756 for the Paris Observatory.[53]

In England, the first person to graduate instruments with any proper degree of accuracy was Abraham Sharp, of Horton, near Bradford, Yorkshire. From his boyhood, Sharp devoted his spare time to astronomy and mathematics and, in 1683, his abilities brought him to the notice of Flamsteed. He was engaged to assist in the work of the Royal Observatory, Greenwich, and, during 1688–1689, strengthened, re-aligned, and re-graduated Flamsteed's meridian arc.[54] Flamsteed

Fig. 51—Flamsteed's equatorial sextant, 1677

(*From Lalande's 'Astronomie'*)

had always wanted a meridian instrument, for by no other means could he determine his latitude with the requisite degree of accuracy, the position of the equinoxes and other fundamental reference points. In vain he petitioned the Government for the necessary funds. Only his generous patron, Sir Jonas Moore, Surveyor of the Ordnance, realized the extremity of his need. At Moore's request, Hooke made a 10-foot mural quadrant, but it proved useless.[55] The death of Moore in 1679 terminated further work in this direction and Flamsteed was once more left to his own devices. He had to rely on extremely modest equipment and, failing better means, used a 7-foot sextant as a meridian instrument.

There could be no better proof of the superiority of telescopic over plain sights than the work which Flamsteed accomplished with his 7-foot sextant. This instrument consisted of an iron frame mounted on a stout polar axis and two subsidiary half-circles that allowed the limb's movement into any plane. It was made by Thomas Tompion, a leading horologist of the day, and the only person to whom Flamsteed could then apply for careful workmanship. Flamsteed divided the limb himself to every 5 minutes of arc and then, by transversals, to single seconds read off directly by means of a movable index arm. There were two telescopes, both movable by tangent screws which engaged in the racked edge of the limb.[56] The instrument is shown in Fig. 51.

In 1681, Flamsteed decided to have a meridian instrument, even if he had to make one himself and, two years later, we find him observing with a 140° meridian mural arc of nearly 7 feet radius. It was a poor instrument and gave false readings owing to its weak construction.[57]* Flamsteed patiently determined these errors and, for some years, used it as a meridian instrument. In 1688, his father died, and he found himself successor to the family property. He thereupon resolved upon a stronger arc and, with the assistance of Sharp, reconstructed his own 7-foot arc. The 3000 star positions in the *Historia Coelestis* testify to the accuracy that Sharp gave to this instrument. Flamsteed determined his latitude afresh, checked his sextant measures, and then devoted his attention to the re-determination of right ascensions. Certain bright stars were selected as fundamental reference points, and their co-ordinates determined with the aid of the mural arc and a pendulum clock. These stars formed a framework in which to position other stars, the sextant being used to determine their relative angular separations.

Flamsteed found that two persistent errors crept into his observations, on both of which he kept a close watch, allowing for them when reducing his measures. He thought they were caused by a slight sinking of the wall which supported the quadrant and to its being out of the plane of the meridian. No doubt they were, in part, due to these causes, but it has since been shown that they were due more to the effects of aberration and nutation, then unknown.

Flamsteed said that Sharp made the parts of the arc so well that it ' was a source of admiration to every experienced workman who beheld it.'[58] These words came to be applicable to all Sharp's instruments. When he ' retired ' to Horton, in 1694, Sharp made several sextants, quadrants, and dials of unsurpassed accuracy. In connection with the telescope, he made his own tools and both ground and polished his own lenses. He had the habit of asking for only the finest materials, a point not always appreciated by his London suppliers. The optician Yarwell found him a difficult customer to please.

> You may be sure [Yarwell wrote to Sharp in April, 1705][59] I should be glad to please both you and all men, but I must confess I have had more complaint from you than from all the rest of mankind . . . for after all you can find no fault with the work of the glass but the glass itself, which you may be sure I should get as good as possible I can, tho' not soe good sometimes as I could wish.

Yarwell's complaint against the quality of the glass he was able to buy was to become a common one with opticians for the next hundred and fifty years.

After Sharp's death in 1742, his work was continued by an artist no less skilled. Born in Cumberland, George Graham was originally apprenticed to Henry Aske, a London watchmaker.[60] Later, he joined Thomas Tompion in his shop in Fleet Street. Graham married Tompion's niece[61] and, at the death of his partner in 1713, found himself in sole possession of a flourishing business. This naturally demanded a great deal of his time, but not enough to prevent him from following his own inclinations. Like his predecessor, he gave particular attention to the irregularities of clocks and their correction. To this end, he invented a ' dead-beat ' form of anchor escapement[62] and, in order to minimize the effects of temperature changes on pendulum length, developed the compensating mercury and bi-metallic grid pendulums.[63] With his many scientific attainments and skill in

* Flamsteed writes that he was ' forced to make use of an ill workman, who respected nothing but the getting of wages by his work.' (Baily, F., 1835, *An Account of the Revd. John Flamsteed*, p. 51.)

Fig. 52—Large mural quadrant by Graham

(*From Lalande's 'Astronomie'*)

metalwork, he became a valuable ally to Halley and Bradley, both of whom were anxious to possess more accurate instruments. Indeed, his skill in this direction earned him the title of 'first general mechanician of his day', while one writer was of the opinion that 'Mr. Graham was, without competition, the most eminent of his profession.' He continues with the remarks:[64]

> He was the best mechanician of his time, and had a complete knowledge of practical astronomy; so that he not only gave to various movements for the mensuration of time, a degree of perfection which had never before been attained, but he invented several astronomical instruments by which considerable advances have been made in that science; he made great improvements in those which had before been in use, and by wonderful manual dexterity constructed them with greater precision and accuracy than any other person in the world.

By making better sidereal clocks, Graham materially assisted practical astronomy. During Halley's first years at Greenwich, the clocks there were most unsatisfactory. Pendulums were not compensated for changes in temperature and the mechanism stopped in the act of being wound. Irregularities in the rates (sometimes due to the pendulum bob striking against the sides of the clock-case) were their greatest failing. On one occasion, it would appear that a clock actually stopped during an important transit observation.

In 1725, Graham made his first important astronomical instrument—the Greenwich 8-foot mural quadrant (Fig. 52). The relatives of Flamsteed had so effectively cleared the observatory of instruments, that Halley found the building almost empty. More fortunate than his predecessor, Halley obtained a grant of £500 from the Board of Ordnance. With £73 of this he obtained from Graham a meridian or transit telescope of 5 feet focus and 2 inches aperture. To obtain more accurate measures in both right ascension and declination of stars, he ordered an 8-foot mural quadrant from Graham, intending to use it like a meridian circle. This instrument, fixed to a massive pier where Flamsteed's arc had stood, was used until Halley's death for the complete observation of the positions of the moon, planets, and zodiacal stars. Using both instruments, Halley never once lost a meridian transit of the moon, either by day or by night (when the weather permitted) over a period of seventeen years. By the method of repeated bisections, Graham divided the quadrant into two sets of graduations. Each scale served as a check on the other and, by eye-estimates, the readings could be taken down to about 5 seconds of arc.[65]

Of equal historical importance with the mural arc were two zenith sectors which Graham made for Samuel Molyneux and Bradley and which led to the discovery of two new apparent stellar motions—those due to the effects of nutation and the aberration of light. These instruments were limited to stars near the zenith. The tube of each could be moved pendulum-like on a horizontal axis near the objective. The long focal length of the latter made readily perceptible the smallest deviation in a star's zenith distance as it culminated. To assist the eye in detecting these small displacements, the field of view was furnished with cross-wires.[66]

The idea of a fixed vertical telescope was by no means new for, as early as 1669, Hooke made some observations at Gresham College with a fixed 36-foot zenith telescope in an endeavour to measure the parallax of γ Draconis.[67] His value of

27″ for this star was treated with reserve at the time, the more so when Picard failed to find an appreciable parallax for the bright star Vega,[68] yet both Picard and Flamsteed reported annual irregularities in the declination of the Pole Star to the extent of 40″, and this stimulated others to continue the investigation. Flamsteed toyed with the idea of using a deep well for a telescope tube, fixing the object-glass at the top and the eyepiece at the bottom. There are no records to show that Flamsteed made or used such a well, and the site is not known with accuracy.

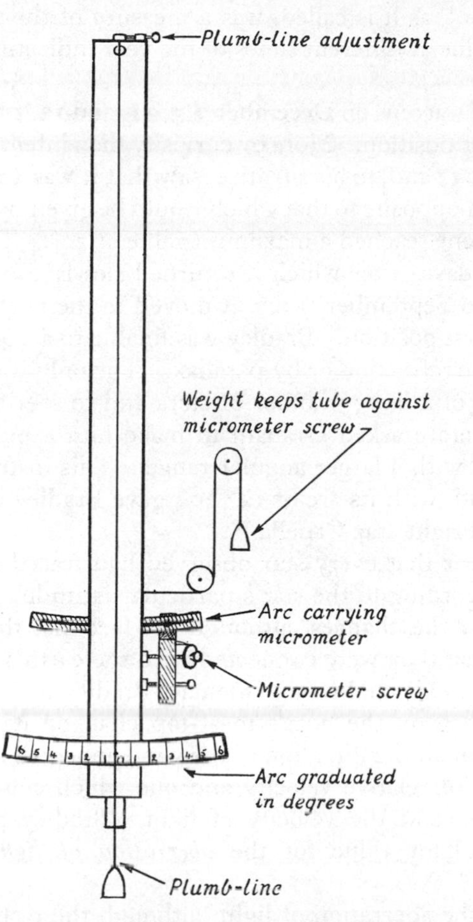

Fig. 53—Molyneux's zenith sector

The instrument, here shown diagrammatically, was made by G. Graham.

A wood-engraving at Greenwich,[69] however, shows the well in section, with a spiral staircase leading to the bottom where there is a couch for the observer. The overall depth is 100 feet and the opening is shown covered by a small brick building. If Flamsteed ever used a shaft like this, he must have found it not at all helpful to the headaches and neuralgia about which he complained so frequently.

Molyneux's 24-foot sector differed from Hooke's inasmuch as the tube was not fixed but could be moved slightly in the meridional plane. It was built entirely within Molyneux's residence at Kew and, on account of its length, passed through holes in the ceilings and roof. The tin-plate tube was suspended at its upper end by ironwork embedded in a chimney stack. The tube's inclination was changed by means of a long screw, kept pressed against the lower end by a weight that passed over a pulley. A large, graduated wheel attached to the screw indicated the inclination and, in default of a spirit-level, a plumb-line suspended from the top axis enabled the tube to be set vertical. The number of revolutions indicated by this ' external micrometer ', as it is called, was a measure of the star's zenith distance, differences in this value at different times of the year indicating changes in declination[70] (Fig 53).

Observations of γ Draconis on December 3, 5, 11, and 12, 1725, showed no change in the star's apparent position. More by curiosity than intention, Bradley observed the star on December 17 and, to his surprise, saw that it was transitting more southerly and in a direction opposite to that which would be given by parallax. By March, 1726, this displacement reached a maximum value of 20″. The star then remained stationary for a few days, after which it returned slowly to its original zenith distance. From June to September, 1726, it moved to the north and, by December, had returned to its first position. Bradley was unable to account for these changes, either by atmospheric refraction or by parallax. The only way open to him was to observe the motions of a large number of stars and to see if they all acted in the same way. He therefore asked Graham to make him a more versatile sector of $12\frac{1}{2}$ feet radius—one with a larger angular range. This instrument was erected at Wanstead in 1727 and, with its arc of 12° 30′, gave Bradley a choice of nearly 200 stars, including the bright star Capella.[71]

By 1728 it was clear that every star observed had traced out an apparent path whose size varied according to the star's particular latitude. But Bradley was still unable to account for the changes, although the fact that they occurred annually led him to surmise that they were connected with the earth's annual motion. The solution came in an unexpected way. One day, Bradley took a trip on the Thames and noted that the flag on the boat's mast-top changed its direction with every turn of the boat, although the direction of the wind appeared to be unaltered. This was clearly an effect of relative velocity and one which could occur with respect to the earth's motion and the velocity of light. Bradley pursued the idea and found that the resultant value for the *aberration of light* accounted for the observed star shifts.[72]

The discovery of the aberration of light, although the result of failure to detect stellar parallax, was one of the greatest if not the greatest achievement in positional astronomy in the eighteenth century. Based on the results of direct observation of star displacements and Römer's value for the velocity of light it had, in consequence, two important issues. The observations provided further evidence of the orbital motion of the earth and also indicated that stars are very remote. Bradley himself stated that, had the parallax of γ Draconis been as much as 1″·0, he would have detected it.[73] Naturally, all other observations of star positions, whether in early catalogues or in Bradley's own work, had to be corrected for the aberrational constant. Yet Bradley's explanation still failed to account for all the observed changes, and he was led to seek the cause of a subsidiary and superimposed motion.

Further observations disclosed the effect of nutation or the periodical 'nodding' of the earth's axis. Sixty years previously, Newton had stated the grounds for a nutation effect, but he supposed that its amount was inappreciable. Bradley, however, with Graham's zenith sector, found variations in the declination of γ Draconis and 35 Camelopardalis which had a period of about nineteen years.[74] This period enabled Bradley to correlate nutation with the inclination of the moon's orbit to the plane of the earth's equator, for the line joining the moon's nodes describes a complete circle in about nineteen years.

Bradley succeeded Halley as Astronomer Royal in 1742 and, like his two predecessors, came up against the usual instrumental difficulties. In his report,[75] he tells how the counterpoise of the 8-foot quadrant rubbed on the roof, whilst screws holding the telescope were broken for want of a little oil. Graham made the necessary repairs and, no doubt, assisted his friend in planning the illumination of the wires of both quadrant and transit instrument. Halley had been satisfied with candle-light reflected from one of the roof shutters for this purpose, but the arrangement was hopeless in windy weather. Bradley enclosed the candle in a box and sent its light down the telescope by means of an elliptical aperture, placed at 45° to the axis of the tube.

Whereas Flamsteed and Halley were content with an apparent instrumental limit of 5 seconds of arc in their observations, Bradley succeeded in reading-off intervals of 1 second of arc. The 8-foot quadrant gave him this degree of accuracy after Graham had added a micrometer screw to the vernier,[76] although he found later that the centre suspension axis had become eccentric. This defect, together with others, he allowed for in his reductions, for he made the peculiarities of each instrument his especial study. Both Graham's quadrant and the transit instrument remained at the observatory until after Bradley's death, together with the 12½-foot sector, bought for the observatory by a Government grant in 1749.

Another of Graham's zenith sectors contributed to the success of a party of French astronomers led by Maupertuis when they set off, in 1736, to measure the length of a degree of the meridian at Torneo in Lapland. Their observations, compared with some taken at the same time in Peru and studied in conjunction with Picard's 1669 measures, led to the first reliable value of the earth's ellipticity.[77]

Graham possessed several instruments, with which he observed eclipses, occultations, the comet of 1723, and variations in the earth's magnetism. He assisted both Halley and Bradley in their observations and, in this way, learnt best how to meet their needs. Together with Bradley, he took an active part in investigations on the shape of the earth by means of seconds-pendulum observations. He was noted for his modesty and generosity, attributes which earned for him the nickname 'Honest George Graham'. He was a Fellow of the Royal Society and seems to have spent a life of constant scientific activity. Bradley admitted that Graham had made his own work possible. Writing a few years after Graham's death in 1751, he says:[78]

> I am sensible that, if my own endeavours have, in any respect, been effectual to the advancement of astronomy, it has principally been owing to the advice and assistance given me by our worthy member, Mr. George Graham, whose great skill and

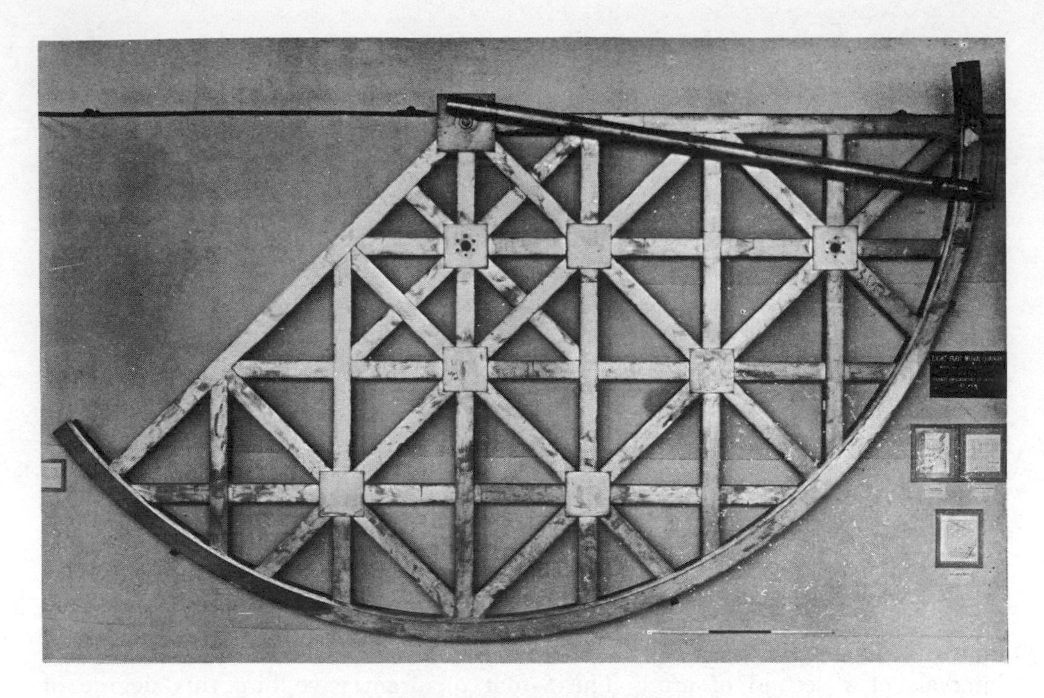

Fig. 54—A Sisson 8-foot mural quadrant

(*Science Museum, London. British Crown copyright*)

judgement in mechanics, joined with a complete and practical knowledge of the uses of astronomical instruments, enabled him to contrive and execute them in the most perfect manner.

Graham's assistant was Jonathan Sisson, who later opened a business of his own in the Strand. Sisson became well known for the accuracy of his arcs and circles and for his improved altazimuth theodolites.

The principle of the theodolite originated with the *dioptra* made by Hero of Alexandria about A.D. 100. This was one of the first surveying instruments to be concerned with the measurement of angles rather than distances.[79] A sighted alidade moved over a circular plate which could be rotated into any desired plane. Screws and cog-wheels provided adjustment for verticality of the altitude scale, which was set by means of a water-level. In 1571, Thomas Digges described a type of theodolite designed and made by his father, Leonard Digges, in which the sighted rule was mounted on a graduated azimuth circle.[80]

While Sisson's theodolites are reminiscent of these earlier devices, they have many points in common with modern instruments. The circles are usually from 4 to 6 inches in diameter, read to about 5 minutes of arc by vernier, and are actuated by pinion wheels. Levelling-screws, in conjunction with spirit-levels, enable the base to be adjusted for horizontality, and a compass fixed to the base plate indicates the direction of magnetic north. For stability, the vertical axis is conical and rotates in a conical bearing.

To Sisson we owe also the development of the ' English ' equatorial, a telescope mounting originally proposed by Henry Hindley of York in 1741.[81] From the illustration in Vince's *A Treatise on Practical Astronomy* of Sisson's equatorial for Maskelyne,[82] we notice that the telescope was attached to one side and towards the upper end of a square polar axis.* Modern telescopes mounted in a similar manner are the Victoria, Toronto, Perkins, and Pretoria reflectors. In every case, the telescope is counterpoised by a weight on the other side of the polar axis. A no less efficient alternative, but one not used by Sisson, is to mount the telescope centrally within a polar axis constructed so as to form a cradle. Early examples of this form were Ramsden's Shuckburgh equatorial of 1793, Troughton's 5-foot stand for Huddart in 1797, and his not too successful mounting for South's 11¾-inch objective (1821). The 100-inch Hooker and 200-inch Hale telescopes are mounted in this way.

Sisson was further notable for his large mural quadrants of 6 and 8 feet radius, made of brass for lightness and ease of manufacture and well braced for rigidity[83] (Fig. 54).

Mention of large brass quadrants brings us to consider the work of John Bird. He was born in 1709 and, like his contemporary, John Dollond (*vide* Chapter VIII), started out in life as a weaver. In his spare time, he engraved divisions on clock faces, for he was struck by the careless way a certain clockmaker had finished his dials.[84] In 1740, he came from Durham to London and entered Sisson's workshop, where he assisted in cutting the divisions on sextants and quadrants. He then decided to open his own workshop and, with Graham's help, started a small business in the Strand at 'The Sea Quadrant'. These premises soon became the centre of the astronomical instrument trade, especially after Bradley, with £1000 available for new instruments, gave Bird his first large order.

Bradley invited Bird to examine Graham's quadrant with a view to making another for observations to the north of the zenith. Bird records his visit as follows:[85]

> In the year 1748, I was informed by the late Dr. Bradley, that application had been made for a new Mural Quadrant, to be fixed to the west side of the pier in the Royal Observatory, in order to take observations to the North; that he had great hopes of success, and therefore desired that I would consider how to prevent a fault which he had found in the old Quadrant; which was, that it had altered its figure by its own weight, so as to render the whole Arc 16″ less than a Quadrant or 90°. Accordingly, I made myself fully acquainted with the general construction of the old Quadrant, which was executed under the direction of the late Mr. Graham, and found the general plan, though little taken notice of at that time, to be such, as, I think, will be a lasting testimony of his great skill in mechanics. The reason of the alteration by its own weight, seems to be a defect in the manner of fastening the several parts together, probably owing to the cocks, and plates for that purpose, being of iron, which could not be forged in that advantageous shape, which I afterwards contrived to give those that were cast of brass, for the new Quadrant; an order for the making of which, I received in February 1749.

Eventually, the old quadrant was redivided by Bird and turned towards the north, where it remained a meridian instrument throughout the period of office of

* Sisson made another of 4-inch aperture for the Brera Observatory, Milan. The instrument is illustrated in L. Ambronn, 1889, *Handbuch der astronomische Instrumentenkunde*, ii, p. 1067.

Bradley, Maskelyne, and Pond. The new 8-foot quadrant was ready for use in June, 1750, and cost £300. Whilst it embodied many of the ideas of Graham and Sisson, the instrument contained much that was new.

Following Sisson, Bird made the entire frame of brass in order to reduce its weight, but even then it weighed some 8 cwt. The assembly of all the component pieces involved a great deal of soldering and required as many as two thousand screws, all of which had to be cut, turned, and threaded in the workshop. Bird took every possible precaution to ensure that temperature changes did not alter the shape of the limb whilst it was being divided. The temperature of the workroom was kept as nearly constant as possible and only two persons were allowed there at a time. The actual engraving was done only during warm weather, and Bird avoided placing his hands in contact with the metal. Both frame and telescope were well braced against possible flexure, and the telescope was supported at the eyepiece end by two little rollers. Bird found that the quadrant required only two supports when it was erected in the meridian. He made one of them adjustable, so that the instrument could easily be re-set in a vertical plane with the aid of a plumb-line. The line was centred over the 0° division, its vibrations being damped in a bowl of mercury. The pains which Bird took over this quadrant were such that, in 1753, the arc was found to be only 0″·5 greater than 90°. Six years later, the whole arc was exactly 90°.[86] Bird himself was highly pleased with its performance and wrote that, although it had been 'upwards of seventeen years in one, and the same position, without any sensible alteration of its figure' it would, with care and attention, 'last for many ages, without any diminution of its value.'

The success of the Greenwich quadrant caused other astronomers to inquire for a similar instrument, and duplicates were sent to the observatories of St Petersburg, Cadiz, and Paris. Smaller instruments of the same design were scattered over Europe and surpassed in accuracy those of French and German manufacture.

With a Bird 6-foot quadrant, Tobias Mayer undertook an important series of observations of the moon's position. George II presented the instrument to the University Observatory of Göttingen in 1751 and Mayer, after submitting it to a rigorous examination, began observations which resulted in the publication of new solar and lunar tables.[87]

In 1752, Le Monnier added an 8-foot quadrant by Bird to Sisson's 5-foot quadrant, already installed in the Observatoire des Capucins. This same instrument later found its way into the Paris Observatory, where it was used until as late as 1823. A similar quadrant by Bird was sent in 1778 to Bergeret, Receiver-General of the Finances, who loaned it to the observatory of the École Militaire.[88] The quadrant was entrusted to Lepaute d'Agelet, a pupil of J. J. Lalande, and at the latter's suggestion, d'Agelet began to use it to form a new star catalogue. In 1785, d'Agelet left Paris to act as astronomer on a long voyage of discovery. He did not return. In the same year Bergeret died and work began on the demolition of the then old observatory. Baron von Zach, a German astronomer, endeavoured to purchase the Bird quadrant with the intention of transporting it to an observatory that he was building near Gotha. Happily for France, however, the precious instrument was purchased by the council of the École Militaire who housed it in a new observatory.[89] It was then used by J. J. Lalande, assisted by his nephew and niece, to determine the positions of about 50,000 stars, the content of the *Histoire Céleste Française* published in 1801.

With the foundation of the Radcliffe Observatory at Oxford in 1771, Bird was considered the person most competent to construct the required instruments. His health at this time was far from good, and the Radcliffe trustees lost no time in securing his services. Among the instruments ordered were two 8-foot mural quadrants of pattern similar to the 8-foot Greenwich quadrant, one for northern and the other for southern transits. They were both of brass and the telescopes, at the suggestion of Dr Hornsby, the first Radcliffe observer, were fitted with achromatic object-glasses (*vide* page 144 ff.).[90] The telescope tubes were counterpoised and attached to brass frames to avoid flexure. The object-glasses were of 3 inches diameter and, with their eyepieces, magnified 60 diameters.

The first quadrant was ready for shipment by river in September, 1772; the second followed a year later. Of the first, Bird wrote that, 'as it is made with greater solidity and with more improvements than any hitherto made, I scruple not to call it by far the best instrument of the kind in the world.'[91] Bird also made a transit instrument and a 12-foot zenith sector[92] about this time, both instruments being sent by Thames barge to Oxford in 1773. The objective of the transit was an achromatic of 4 inches aperture and 8 feet focus by Dollond. A similar instrument had been made previously for Bradley, and Hornsby followed his example by using his transit in conjunction with a good clock for right ascension determinations. Bradley's transit was used until 1812, when it was replaced by Troughton's circle. Observations with the Oxford instruments continued until 1838.

Bird was concerned more with the mechanical construction and division of instruments than with the manufacture of lenses, although a few examples of his optical work have been traced. He made a few reflecting telescopes, besides barometers, thermometers, and drawing instruments. Astronomy was his favourite study and he frequently observed from the top of his high premises in the Strand. In 1761, he observed the transit of Venus from Greenwich with one of his own reflectors and, on April 1, 1764, an annular eclipse of the sun.[93] In horology, he watched the patient labours of Harrison and others to produce a reliable timepiece and, together with Maskelyne, reported on the mechanism of Harrison's historic watch. The standard yards of 1758 and 1760 were of his making;[94] they were destroyed in the 1834 fire of the Houses of Parliament.[95]

In May, 1769, Jean Bernoulli junior called on Bird. 'It is unfortunate' he writes,[96] 'that this excellent artist is not younger. He is even obliged to use spectacles; but he is, however, still vigorous and he has a firm hand.'

Bird died on March 31, 1776, after a long period of ill health.[97] The Oxford instruments kept him busy until the end and, although he received about £1300 for them, he had little left for himself. At no time could he be said to have been a rich man; to meet initial expenses for the Oxford instruments, he had to ask for money in advance. Materials and labour were difficult to procure—whenever possible, he executed the work himself.

Bird made no secret of his methods. His first treatise, *The Method of Dividing Astronomical Instruments* (1767), was published at the request of the Commissioners of Longitude. For this, Bird received £500 with an additional £60 to defray the cost of the plates. A further condition was that he took an apprentice for seven years, 'instructing him in his art and method of making Astronomical Instruments; and also in instructing in like manner, such other persons as the said Commissioners shall from time to time direct.' In the following year, 1768, the Commissioners published Bird's *Method of Constructing Mural Quadrants: Exemplified by a Description of the Brass Mural Quadrant in the Royal Observatory at Greenwich.* Both works had a preface by Nevil Maskelyne, the Astronomer Royal, and they were re-issued as one volume in 1785.

REFERENCES

1 Grant, R., 1852, *History of Physical Astronomy*, pp. 415–416.
2 *Ibid.*, p. 419.
3 Stratton, F. J. M., 'Horrox and the Transit of Venus', *Occasional Notes of the R.A.S.*, **1**, No. 7, p. 90, 1939.
4 Stratton, *op. cit.*, p. 93, quoting Whatton's translation of Horrocks's *Venus in Sole Visa.*
5 Grant, *op. cit.*, p. 422.
6 Stratton, *op. cit.*, p. 90.
7 Grant, *op. cit.*, pp. 426–427.
8 Vernier, P., 1631, *Construction, l'Usage, et les Propriétés du Quadrant Nouveau.*
9 Grant, *op. cit.*, pp. 453–454.
10 *Ibid.*, pp. 454–455.
11 Rigaud, S., 1841, *Correspondence of Scientific Men of the 17th Century*, letter xx, p. 46.
12 Grant, *op. cit.*, p. 454.
13 *Ibid.*, p. 452, citing Sherburne's translation of *The Sphere of Manilius*, 1675.
14 *Phil. Trans.*, **1**, pp. 457, 541, 1665–1666.
15 *Ibid.*, p. 541.
16 *Ibid.*, p. 457.
17 Molyneux, W., 1692, *Dioptrica Nova*, section 'Admonition to the Reader'.
18 Rigaud, *op. cit.*, i, p. 48; *Phil. Trans.*, **30**, p. 605, 1718.
19 Rigaud, *op. cit.*, i, pp. 35–39.
20 Huygens, C., 1659, *Systema Saturnium*, p. 91.
21 Bailly, J. S., 1785, *Histoire de l'Astronomie Moderne*, ii, p. 266.
22 *Ibid.*, p. 267.
23 Bailly, *op. cit.*, pp. 268–269. Auzout, A., *Phil. Trans.*, **1**, p. 373, 1665–1666. *Mém. Acad. Sc.*, **4**, pp. 97–112, 1666–1669.
24 *Phil. Trans.*, **1**, p. 373, 1666.
25 *Ibid.*, **1**, p. 457, 1667.
26 *Ibid.*, **1**, p. 541, 1667.
27 Wolf, C., 1902, *Histoire de l'Observatoire de Paris*, pp. 11–12, p. 136. Le Monnier, P. C., 1741, *Histoire Céleste*, p. 11. Bailly, *op. cit.*, p. 273. Grant, *op. cit.*, p. 456.
28 *Hist. Acad. Sc.*, **4**, pp. 1–59, 1736.
29 Bailly, *op. cit.*, ii, p. 273.
30 Picard, J., *Mém. Acad. Sc.*, **4**, pp. 233–283.
31 *Ibid.*, p. 249.
32 Martin, L. C., *Trans. Opt. Soc.*, **24**, p. 297, 1923.
33 Mallet, A. M., 1702, *Géométrie Pratique.*
34 Martin, *op. cit.*, p. 297.
35 Molyneux, W., 1692, *Dioptrica Nova*, gives translation, p. 230.
36 *Ibid.*, p. 231.
37 MacPike, E. F., 1937, *Hevelius, Flamsteed and Halley*, pp. 85–87.
38 *Phil. Trans.*, **15**, pp. 1164, 1167, 1168, 1685.
39 Bailly, *op. cit.*, ii, p. 278.
40 Grant, *op. cit.*, p. 448.
41 Danjon, A., and Couder, A., 1935, *Lunettes et Télescopes*, p. 631.
42 Wolf, *op. cit.*, pp. 139–140; Grant, *op. cit.*, p. 458.
43 Horrebow, P., 1735, *Basis Astronomiae*, p. 48.
44 Grant, *op. cit.*, p. 463.
45 *Ibid.*, p. 465.
46 *Ibid.*, p. 466.
47 *Miscellanea Beroliniensia*, **3**, p. 277, quoted by Rigaud, S., 1832, *Miscellaneous Works and Correspondence of J. Bradley*, p. lxxvii.
48 Clay, R. S., and Court, T. H., 1932, *History of the Microscope*, arts. 'Butterfield' and 'Langlois' in appendix. For references to Butterfield's instruments *vide* Gunther, R. T., 1923, *Early Science at Oxford*, ii, pp. 134 and 153; Martin, *op. cit.*, p. 292.
49 Wolf, *op. cit.*, p. 176.
50 *Ibid.*, p. 177.
51 *Ibid.*, pp. 177–178.
52 *Ibid.*, pp. 178–179.
53 *Ibid.*, p. 185.
54 Cudworth, W., 1889, *Life and Correspondence of Abraham Sharp*, pp. 11–14.
55 Grant, *op. cit.*, p. 468.
56 *Historia Coelestis*, Prolegomena; Pearson, W., 1829, *Introduction to Practical Astronomy*, ii, pp. 297–298.

[57] Grant, *op. cit.*, p. 469. Baily, F., 1835, *An Account of the Rev. John Flamsteed*, pp. 52, 54.
[58] Cudworth, *op. cit.*, p. 14.
[59] *Ibid.*, p. 31.
[60] Milham, W. I., 1923, *Time and Timekeepers*, p. 544.
[61] Atkins, S. E., 1881, *Account of the Worshipful Company of Clockmakers*, p. 166.
[62] Milham, *op. cit.*, p. 146.
[63] *Phil. Trans.*, **34**, p. 40, 1726; **47**, pp. 9, 11, 1752.
[64] *Gentleman's Magazine*, **21**, pp. 523–524, 1751.
[65] Grant, *op. cit.*, p. 478. Smith, R., 1738, *Compleat System of Opticks*, ii, pp. 332–341.
[66] Rigaud, S. P., 1832, *Miscellaneous Works and Correspondence of James Bradley*, pp. xiii–xiv.
[67] Hooke, R., 1674, *Attempt to prove the Motion of the Earth . . .*, p. 7 ff. Rigaud, *op. cit.*, p. xiii.
[68] Le Monnier, *op. cit.*, p. 252.
[69] *Observatory*, **26**, pp. 102, 138, 179, 1903.
[70] *Vide* Ref. 66, pp. xii–xiv, pp. 93–115. The idea of the external micrometer originated with Louville, 1712; *vide* Danjon and Couder, *op. cit.*, p. 629.
[71] *Vide* Ref. 66, pp. xxiii, xxvi–xxviii, pp. 201–202.
[72] *Ibid.*, pp. xxx–xxxi.
[73] *Ibid.*, p. 15.
[74] *Ibid.*, pp. lxii–lxix.
[75] *Ibid.*, p. 381.
[76] Grant, *op. cit.*, pp. 483–484.
[77] Wolf, *op. cit.*, pp. 223 ff., Maupertuis, P.L.M. de, *Mem. Acad., Sc.*, p. 408, 1737.
[78] Rigaud, Ref. 66, p. 20.
[79] Cajori, F., 1924, *History of Mathematics*, pp. 43–44. Heath, T. L., 1921, *Greek Mathematics*, ii, p. 345. Gunther, *op. cit.*, ii, pp. 9–10. Great uncertainty exists as to both the period and works of Hero. *Vide* also Martin, L. C., *Trans. Opt. Soc.*, **24**, p. 289, 1923.
[80] Martin, *op. cit.*, p. 293, cites *Pantometria*, 1571 and 1591.
[81] Pearson, *op. cit.*, pp. 517–518. Hindley was also responsible for the introduction of a dividing engine for marking circles. *Vide* Smeaton, *Phil. Trans.*, **76**, p. 1, 1786.
[82] Vince, S., 1790, *A Treatise on Practical Astronomy*, pp. 141 ff.
[83] In the Science Museum, London, there is preserved an 8-foot Sisson mural quadrant made for George III's private observatory at Kew in 1770.
[84] *Dict. Nat. Biog.*, 1893, art. ' John Bird '.
[85] Bird, J., 1768, *The Method of Constructing Mural Quadrants: Exemplified by a Description of the Brass Mural Quadrant in the Royal Observatory at Greenwich*, pp. 7–8.
[86] *Ibid.*, p. 24.
[87] Grant, *op. cit.*, p. 487.
[88] Wolf, *op. cit.*, p. 246.
[89] Lalande, J. J., *Histoire Abrégée de l'Astronomie*, p. 669. Tisserand, M., *Astron. Register*, **19**, p. 218, 1881.
[90] Gunther, *op. cit.*, pp. 319–321.
[91] *Ibid.*, p. 323.
[92] *Ibid.*, p. 324.
[93] *Phil. Trans.*, **52**, i, p. 175, 1761; **54**, p. 142, 1764.
[94] Weld, C., 1848, *History of the Royal Society*, ii, p. 256.
[95] *Mem. R.A.S.*, **9**, pp. 80–81, 1836.
[96] Bernoulli, Jean, junior, 1771, *Lettres Astronomiques*, p. 128.
[97] Gunther, *op. cit.*, p. 88.

CHAPTER VII

It was a mighty bewilderment of slanted masts, spars and ladders and ropes from the midst of which a vast tube, looking as if it might be a piece of ordnance such as the revolted angels battered the walls of heaven with, according to Milton, lifted its mighty muzzle defiantly to the sky.

OLIVER WENDELL HOLMES

WILLIAM HERSCHEL'S boyhood interest in astronomy did not take a practical turn until 1773, when he was thirty-five years of age. That year found him living in Rivers Street, Bath, with his sister Caroline and brother Alexander. Most of his time was spent with music pupils or in his duties as Director of Music in the City of Bath. He had to engage artists for private and public concerts, compose 'glees and catches' for different singers, and anthems and chants for the choir of the Octagon Chapel. The pursuit of his new hobby had, therefore, to be confined to odd moments during the day and continued by stealing hours from sleep. But in June, 1773, the end of the winter season, many of Herschel's pupils left Bath and he made preparations for viewing the heavens, as he put it, with his own eyes.

> I procured some short object-glasses [he writes][1] and had tubes made for them, beginning with a 4 feet one of the Huyghenian construction. . . . It magnified about 40 times. In the next place I attempted a 12 feet one and contrived a stand for it. . . . After this I made a 15 feet and also a 30 feet refractor and observed with them. The great trouble occasioned by such long tubes, which I found it almost impossible to manage, induced me to turn my thoughts to reflectors, and about the 8th September [1773] I hired a 2 feet Gregorian one.

It is interesting to see that Herschel's first efforts were directed to long refractors. No doubt he was attracted by the comparative cheapness and procurability of the lenses for, although there was no manufacturing optician in Bath, the London opticians could then supply suitable glasses. The borrowed Gregorian was so much more convenient to operate than the long pasteboard and tin tubes that Herschel wrote to London to inquire the price of a 5- or 6-foot reflector. He was informed that none of this size was made, although one person offered to supply one at a price above what he intended to give. This decided him to make one himself, with the help of the instructions in Smith's *Opticks*, and in this project he was assisted by a Quaker resident in Bath. This person used to amuse himself with mirror-making, and he sold Herschel all his tools and apparatus,[2] at the same time instructing him in their use.[3] Before the end of 1773, Herschel had cast and polished several mirrors, among them one for a $5\frac{1}{2}$-foot Gregorian. This he mounted in a square wooden tube, but the alignment of the mirrors proved so troublesome that he turned to the Newtonian design. Here he was more successful and, in the same year, records that with a magnification of 40 he saw the rings of Saturn and the Great Nebula in Orion. Encouraged by this result

Fig. 55 Sir William Herschel

A pastel portrait by J. Russell, 1794

(By courtesy of the Director of the Science Museum, London)

Fig. 56—Caroline Lucretia Herschel

Of this drawing, which shows Caroline Herschel in 1847 at the age of ninety-seven, her friend Miss Beckedorff wrote that it does not 'do justice to her intelligent countenance; the features are too strong, not feminine enough, and the expression too fierce.'

he made a 7-foot telescope ' with many different object mirrors, keeping always the best of them for use and working on the rest at leisure.'[4]

To her sorrow, Caroline found that every room in the house began to take on the appearance of a workshop. A cabinet-maker made tubes and various stands in the drawing-room and Alexander erected a lathe in a bedroom for turning patterns and eyepieces, and grinding glasses. ' At the same time ' she wrote, ' music durst not lie entirely dormant during the summer.'[5] A further cause for concern was ' the uncommon precipitancy ' which accompanied her brother's actions. His energy was as boundless as his enthusiasm, and his disregard for his health and all temporal restrictions gave her endless anxiety.[6]

In 1774, Herschel moved to a larger house near Walcot turnpike, Bath, where there was a convenient grass plot for the erection of telescopes. Here he prepared several 9-inch mirrors for a 10-foot telescope, also one with a central perforation as he intended to try the Gregorian form again. He sought a satisfactory method of figuring and, like Newton, worked his mirrors on the tool without pressure, preventing them from moving ' loosely and unevenly about ' by filing a double cross-channel in the brass tool. He first ground the convex tool on the speculum with emery and then, with a separate pitch-covered polisher, used rouge to produce the final figure. To keep the pitch fairly soft, he frequently immersed the polisher in warm water, but the change in temperature tended to alter its figure and he found a little grease answered just as well.[7]

This work, performed with slow abrasive on hard, stubborn metal and without machinery was a slow and laborious task. Caroline gives many instances of how her brother spent hours of unremitting toil over his mirrors and how, on one occasion, ' in order to finish a seven foot mirror, he had not taken his hands from it for sixteen hours together.'[8] Progress, if slow, was encouraging. In May, 1776, the 7-foot showed Herschel ' Saturn's ring and two belts in great perfection ',[9] whilst the 10-foot took a power of 240 on minute parts of the moon. At this time Herschel was also at work on a 20-foot Newtonian for which he made three mirrors.

In 1777, the Herschels moved to 19 New King Street near the centre of Bath.[10] Here Herschel resumed his experiments on mirrors and, in the long, narrow

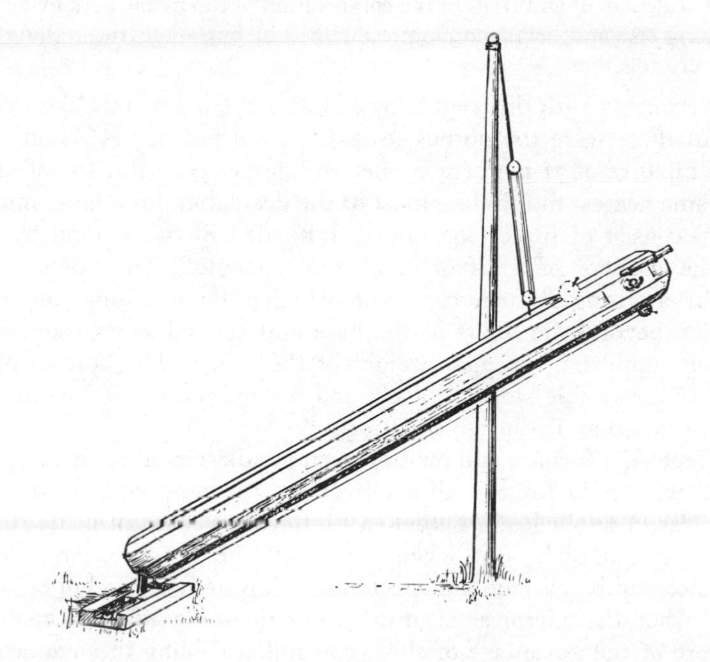

Fig. 57—Herschel's method of mounting the 10-foot reflector at Bath

garden at the back of his premises, tried out various mountings for the 20-foot. He found that the direction, length, and type of stroke greatly influenced a mirror's ultimate figure. Instead of passing through the circular strokes advocated by Mudge to the direct production of a paraboloid, he usually found it better to produce a spherical figure first, give it a fine polish, and then complete the ' so much wished for parabola ' with ' a few round strokes'. Experience told him that it was easier to restore an under-corrected rather than an over-corrected mirror. As he became more skilled, he found that every speculum seemed to require a different treatment and, in this way, he tried every possible kind of stroke and combination of strokes.[11]

In July, 1778, Herschel polished a 10-foot Newtonian mirror, using straight strokes regularly changed in direction, and a polisher on which the pitch was cut into squares. In November of the same year he polished ' a most capital speculum ' of 6·2 inches aperture, which he mounted in a 7-foot tube.[12] With this instrument and a magnification of 227, he made a ' review ' of the heavens which resulted in his first catalogue of double stars and, about two years later, the discovery of Uranus.*

Meanwhile, the work of mirror-making continued without a break. Caroline tells us:[13]

> Many attempts were made by way of experiment against a mirror before an intended thirty-foot telescope could be completed, for which, between whiles (not interrupting the observations with seven, ten and twenty-foot, and writing papers for both the Royal and Bath Philosophical Societies) gauges, shapes, weight &c., of the mirror were calculated, and trials of the composition of the metal were made. In short, I saw nothing else and heard nothing else talked of but about these things when my brothers were together.

In his experiments with different alloys, Herschel tried metals like wrought and cast iron, but these were too porous to take a good polish. At length, he chose Molyneux's mixture of 71 per cent copper and 29 per cent tin, for of all possible alloys this came nearest to fulfilling most of the desiderata for a large mirror. The stand was to consist of three long poles, embedded in the ground by brickwork placed at the corners of a 4-foot equilateral triangle. The poles were joined at the top by a strong circular cap upon which rotated a long horizontal arm. This extended beyond the limits of the base and carried at its free end a set of pulleys which supported the main weight of the tube. The bottom of the tube was pivoted to a movable wooden block, and the observer stood within the three poles on an enclosed and adjustable platform.[14]

Herschel erected a furnace and melting oven in a basement room and, on August 11, 1781, all was ready for casting a mirror for the proposed 30-foot telescope. The mould was of sifted loam, hardened by burning charcoal in it; the metal at first ran in ' very quietly '. But, when nearly full, the mould began to leak, which caused a deficiency in part of the speculum. Any remaining hopes for success were dashed when the mirror cracked in two or three places while cooling. Herschel, unaware of the advantage of slow, controlled cooling in an annealing oven, believed that his mixture was too brittle. He remelted the pieces, adding more copper to reduce the tin content to 27 per cent. The total weight of metal was 538 lb. Hopes fell again when some of the fluid metal began to drop through the bottom of the melting oven into the fire. The crack increased, the metal flowed over the ash-hole and, when it came to the stone floor, ' the flags began to crack and some of them to blow up.' The workmen ran for their lives while Herschel, exhausted by heat and exertion, fell on a heap of bricks.[15] After these setbacks Herschel abandoned the idea of a 30-foot telescope.

William Watson, later Sir William Watson, is often mentioned by Caroline during these busy years at Bath. Some six years younger than Herschel, Watson enjoyed, partly through the medical and electrical discoveries of his father, the

* On the night of March 13, 1781.

Fig. 58—A 7-foot Herschel telescope

(*Science Museum, London. British Crown copyright*)

friendship and regard of many leading men of science. A Fellow of the Royal Society, Watson invited Herschel to communicate his astronomical findings to this body; Watson also introduced him to the short-lived Bath Philosophical Society.[16] Watson was one of the first to appreciate the growing impossibility of Herschel's position as musician by day and astronomer by night. He proved a valuable intermediary between his friend and George III and exerted himself to obtain a sum of money sufficient to release Herschel from financial worry.

Herschel first met Maskelyne, the Astronomer Royal, at Walcot in 1777.[17] About the same time he was introduced to Hornsby, Savilian professor of astronomy at Oxford[18] and, through Hornsby, to the amateur astronomer, Alexander Aubert.[19] Aubert's observatory at Loampit Hill, near Deptford, was furnished with the best instruments that Short, Bird, Dollond, and Ramsden could provide. Herschel profited by Maskelyne's mature knowledge of physical astronomy, and Aubert's opinion was welcome when some doubtful observation required confirmation.

Through the efforts of Watson and Banks, president of the Royal Society, steps were taken to relieve Herschel of his duties as a musician. Both men were anxious to see him appointed successor to Demainbray, superintendent of the 'King's Observatory' in Richmond Park and, to further this object, they arranged for him to have an audience with George III. In June, 1782, Herschel accordingly left Bath with his 7-foot telescope and 'with everything necessary for viewing double stars'. First he visited Maskelyne at Greenwich.

> These two last nights [he wrote home to his sister][20] I have been star-gazing at Greenwich with Dr. Maskelyne and Mr. Aubert. We have compared our telescopes together, and mine was found very superior to any of the Royal Observatory. Double stars which they could not see with their instruments I had the pleasure to show them very plainly, and my mechanism is so much approved of that Dr. Maskelyne has already ordered a model to be taken from mine and a stand to be made by it to his reflector. He is, however, now so much out of love with his instrument that he begins to doubt whether it *deserves* a new stand.

A similar trial at Aubert's observatory led Herschel to write that his telescope performed better than any of Aubert's. 'I can now say' he concluded, 'that I absolutely have the best telescopes that were ever made.'[21]

The result of Herschel's audience with the King was that he became 'Royal Astronomer' with an annual salary of £200. The post at Kew went to Demainbray's son, already in royal favour as tutor to the King's children. This decision was a fortunate one for Herschel. The allowance was small, yet sufficient to release him from teaching and other duties at Bath. He now had time in which to make and use large telescopes, in which to develop original lines of research.

Herschel and his sister moved to Datchet in the summer of 1782, taking over a disused and dilapidated house not far from the river Thames.[22] The house was large, commanded a good view of the sky, and was reasonably near Windsor Castle where the King often spent his weekends. Herschel erected the 20-foot with the minimum delay and, before he left Bath, cast another 12-inch speculum for it. Observations continued throughout the following winter and scarcely a single hour of starlight weather was lost. 'I used either to watch myself' Herschel wrote, 'or to keep up somebody to watch; and my leisure hours in the day-time were spent in preparing and improving telescopes.'[23] Caroline gives a truer picture of the work accomplished during these so-called 'leisure' hours:[24]

> For the assiduity with which the measurements on the diameter of the Georgium Sidus, and observations of other planets, double stars, etc., etc., were made, was incredible, as may be seen by the various papers that were given to the Royal Society in 1783, which papers were written in the day-time, or when cloudy nights interfered. Besides this, the twelve-inch speculum was perfected before the spring, and many hours were spent at the turning bench, as not a night clear enough for observing ever passed but that some improvements were planned for perfecting the mounting and motions of the various instruments then in use, or some trials were made of new constructed eye-pieces, which were mostly executed by my brother's own hands.

Herschel had not been long at Datchet before he found that his salary failed to meet all the expenses of his trips to Bath, London, and Greenwich and, in response to many pressing inquiries, he made and sold telescopes privately. This extra work occupied much of his time, as did his periodical visits to Windsor to show the King objects through the 7-foot reflector. Had he been able to delegate the making of tubes and stands to local workmen, the demands on his time and energy would have been far less, but outside work seldom answered expectation and only the very rough parts were entrusted to the smiths and carpenters of Datchet. His own list[25] of instruments sold embraces about sixty telescopes, most of them of 7 and 10 feet focus and constructed during his three years at Datchet.

The Datchet observations were the continuation of his 'third review' of the heavens, begun at Bath with 7- and 20-foot reflectors. Observations continued from dusk to dawn and for many hours at a time. In reviewing all Flamsteed's stars Herschel often examined 400 stars a night.[26] This was the first time that these stars had been systematically examined in themselves, as distinct from their use as reference points for observing the moon and planets. Nothing was known of their distribution and motions, of the nature and number of double stars, star clusters, and nebulae. Herschel had an untrodden field before him.

The first extensive catalogue of nebulae was due to Messier of Paris[27] who discovered so many comets that he was known as the 'ferret of comets'. While observing a comet in Taurus, Messier found the Crab nebula (M1) and decided to form a catalogue of nebulae and clusters to facilitate comet-searching. By 1784, Messier had drawn up a list of 103 objects, all visible through 3½- and 5-foot achromatics. On seeing Messier's catalogue, Herschel decided to examine nebulae with a new telescope he had just completed. He saw with pleasure that they 'yielded' to the power of this instrument; many objects were resolved into stars. This decided him to 'sweep' the heavens, or to subject certain areas of the sky to an exhaustive search, first to find more nebulae, and second to 'gage' (Herschel's spelling) the depth of the sidereal universe by taking star counts.

The new telescope set aside for this work was a 20-foot reflector of 18·8 inches aperture, called the 'large 20-foot' to distinguish it from the 12-inch. It became his most useful instrument and, in later years, he preferred it to the 40-foot. The mounting was convenient to operate and the mirrors performed well. The tube, supported by a triangular system of poles and ladders, was moved by ropes and pulleys, while the whole apparatus could be turned in azimuth. 'The machinery of my twenty-feet telescope' Herschel wrote, 'is so complete that I have been able to take up the planet [Saturn] at an early hour in the evening and to continue the observations of its own motion, together with that of its satellites, for seven, eight, or nine hours successively.'[28] The telescope had two mirrors, both brittle, but of good reflectivity. Herschel first used it as a Newtonian, mounting the eyepiece so that, when the tube was horizontal, it made an angle of 45° with the vertical. This position, resembling that of a reading desk, was 'preferable to the perpendicular one commonly used in the Newtonian construction.'[29] Herschel then dispensed with the diagonal mirror altogether and used the instrument as a 'front-view'. 'I repeated some former experiments' he wrote, 'by looking into the telescope at the front without the small reflecting mirror, and found the image as good as at the side; the light is incomparably more brilliant, and I thought sometimes that the stars were, if not better, at least full as well defined as in the Newtonian way, so that it seems I have heretofore too hastily laid it aside. In high sweeps, the position of looking is a very convenient one; and in no other situation can it be a very bad one.'[30]

Herschel's method of 'sweeping'[31] consisted of first rotating the stand until the tube pointed in the right direction. A lateral motion of the speculum box then allowed a change of 15 degrees in azimuth while Herschel, standing in the gallery, could draw the tube along another 15 degrees. He thus obtained a range of 30 degrees in azimuth without moving the stand. In sweeping, he drew the tube along in slow oscillations of 12 or 14 degrees, made notes of the objects seen, and then raised or lowered the tube 8 or 10 minutes of arc to make another oscillation. Ten to thirty of these traverses constituted a sweep. The main disadvantage of this method was that, whenever he saw an interesting nebula or cluster, he plotted the neighbouring stars visible in the finder and then made notes by lamplight, thereby impairing the sensitivity of his eye. The process was also fatiguing without assistance and, after forty-one sweeps, he changed it for a series of vertical sweeps, engaging a workman to raise and lower the tube, and Caroline to record the observations at his dictation. In this way, he observed and described over 2000 nebulae and star clusters.

These observations were not made without considerable personal risk.

> My brother began his series of sweeps [Caroline wrote][32] when the instrument was yet in a very unfinished state, and my feelings were not very comfortable when every moment I was alarmed by a crack or fall, knowing him to be elevated fifteen feet or more on a temporary cross-beam instead of a safe gallery. The ladders had not even their braces at the bottom; and one night, in a very high wind, he had hardly touched the ground before the whole apparatus came down.

Caroline once had a serious accident when she fell on a large hook, hidden under a blanket of snow; Piazzi of Palermo Observatory, on a visit to Datchet, fell over a protruding rack bar. 'I could give a pretty long list of accidents' Caroline concluded,[33] 'which were near proving fatal to my brother as well as myself.'

In 1786, Herschel moved to new and better premises at Slough, where he made immediate preparations for constructing a 40-foot reflector of 48 inches aperture. Through the influence of Watson, the King granted £2000 towards its cost and promised a further £2000 when it was almost complete.* Herschel designed the entire telescope himself. He supervised as many as thirty to forty workpeople as they prepared the ground, laid the foundations, and gradually erected the different parts. He made trips to London to arrange for casting the speculum, and directed the manufacture of the tools, gauges, and metal parts.

> The garden and workrooms [wrote Caroline][34] were swarming with labourers and workmen, smiths and carpenters going to and fro between the forge and the forty-foot machinery, and I ought not to forget that there is not one screw-bolt about the whole apparatus but what was fixed under the immediate eye of my brother. I have seen him lie stretched many an hour in a burning sun, across the top beam whilst the iron-work for the various motions was being fixed.

The first mirror was cast in London in 1785 but, ' by a mismanagement of the person who cast it ',[35] came out thinner in the centre than was intended. This was considered no great drawback, however, and the disk was mounted in an iron ring ready for grinding and polishing. The ring's circumference was provided with handles so that twelve men could direct it over the convex iron tool which, under this motion and the abrasive action of coarse emery, imparted its curvature to the casting. The tool was then covered with pitch and the mirror, suspended by a crane, was dipped in a shallow tub of hot water and then laid on the polisher. This was repeated until the gauge indicated that the pitch had assumed the right curvature. A deep cross-gutter was then cut in the pitch and the work of polishing began.[36] The size and weight of the overlying mirror made progress difficult and Herschel had to admit that the number of men performing the work did not allow of ' those delicate attentions which are required in polishing mirrors.' Nevertheless, the mirror was brought to a good polish and, in February, 1787, eyepiece in hand, Herschel crawled into the tube and searched for the focus. ' The object I viewed ' he wrote,[37] ' was the nebula in the belt of Orion, and I found the figure of the mirror, though far from perfect, better than I had expected. It showed the four small stars in the Nebula and many more. The nebula was extremely bright.'

Early the following year, Herschel had a second mirror cast, but the brittle metal cracked in the cooling. A third casting, containing a greater proportion of copper, turned out well and was polished, as before, with twenty-four men (in groups of twelve) at the handles. Herschel found the work wearisome. When, in October, 1788, he tried the speculum on Saturn, the figure was not so perfect as he had hoped. He now thought of making a polishing machine, and for some months he experimented with machines for 7- and 40-foot focus mirrors. By August 23, 1789, he had so perfected a machine[38] for 48-inch disks that he began to polish the second mirror with it. The speculum was held by flanges attached to a flat, ratched ' polishing ring ', which moved freely in an outer frame. The latter was connected to a lever which, as it moved to and fro, imparted the same motion to the mirror on the tool. Further levers prevented the mirror from wandering and gave it the necessary side motion. At the same time, the levers caused two claws to engage the ratchet on the ring and so rotated the mirror. As Herschel arranged it, the mirror made eight strokes every minute, the length of each stroke being 36 inches

* Herschel received both sums, also £200 a year for the instrument's upkeep.

Fig. 59—The 40-foot telescope

and the amplitude of the side motion never more than 10 inches. After eight hours on the polisher, the surface, still far from perfect, was good enough to afford a view of Windsor Castle. Herschel thereupon decided to try the mirror on the stars.

By this time the massive wooden structure behind Herschel's house was becoming visible from coaches travelling along the London–Bath road nearby. Travellers must have gazed with interest at the rising mass of poles and spars, and it probably formed a well-worn topic of conversation for many miles before and after Slough. In its general scheme, the great telescope was not unlike the 'large' 20-foot altazimuth. A triangular latticework of poles and ladders,[39] some 50 feet high, supported the observing galleries and the main weight of the tube. This framework, reared on an octagonal platform, rotated by sets of rollers on two concentric brickwork circles. The tube could thus be turned in azimuth to any position, an operation effected by running a rope from a windlass around a pole (set firm in the ground) to one corner of the platform.

To accommodate the mirror, the tube had to be over 40 feet long and nearly 5 feet in diameter. Herschel had it made in a large barn near Upton Church, about a quarter of a mile from his house. The tube was constructed of large pieces of sheet-iron seamed together and strengthened by iron hoops and longitudinal bars. At its lower end, at the speculum housing, it was reinforced by

Fig. 60—The remains of the 40-foot telescope

The only surviving part of the tube resting in the garden of Observatory House, Slough

(Photograph by Dr W. H. Steavenson, 1924)

further iron bars and supported by the axle of two rollers which ran on short lengths of rail. The upper end was suspended between the two main sets of ladders by a system of ropes and pulleys attached to a high transverse beam some 40 feet above ground. Further ropes passed down from the beam to a winch which raised and lowered the tube and which, through a system of gears, indicated polar distances on a graduated dial. Besides this vertical motion, the tube could be moved laterally by the observer as he stood on his little observing platform. A further platform or gallery ran up the ladders to just beneath the open end of the tube and was large enough to accommodate several people. Herschel also considered introducing a running chair which could carry him from the ground to the gallery, but his agility in ascending the ladders, even at the age of seventy, made this an unnecessary luxury. The telescope was used as a front view and was provided with an observing seat so placed that Herschel could get his head just inside the mouth of the tube. From this lofty station he could give the tube ' micrometer

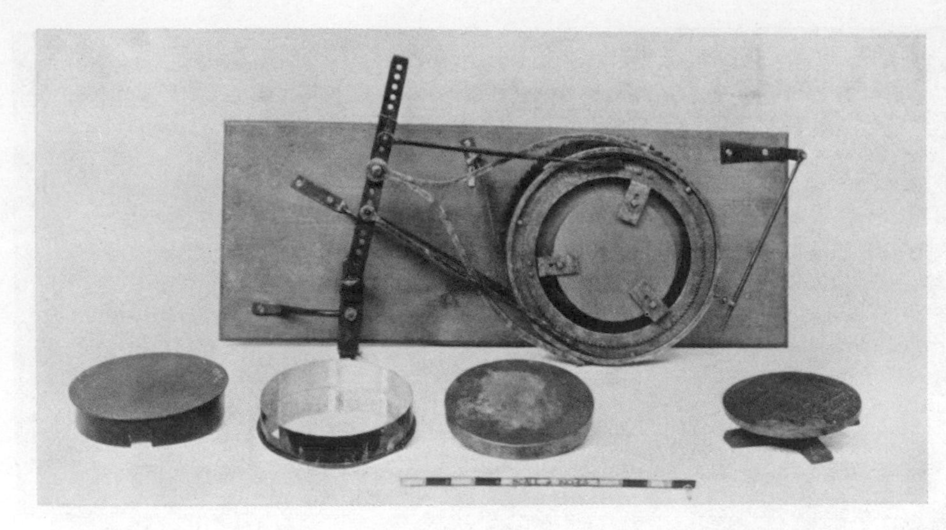

Fig. 61—Herschel's machine for polishing small mirrors

(Science Museum, London. British Crown copyright)

motion ' up and down or from side to side but, for conducting a ' sweep ', a workman had to work the winch below. This was housed in a small hut or ' observatory ' on the ground platform, which contained also the right ascension and polar distance apparatus, a polar distance ' clock ', and a sidereal clock. Here also Caroline sat with her candle and zoned catalogue of stars. The range in altitude was limited by the striking of a bell on the polar distance ' clock ' and it was Caroline's duty to note the polar distances and the time indicated by the sidereal clock. Her register consisted of sheets of paper, ruled into squares representing quarters of a degree, a cross in any square showing that the corresponding portion of the sky had been observed.

On August 27, 1789, the second mirror was drawn by trolley to the telescope and lifted by a crane into the speculum box. Tried on the stars, the mirror gave ' pretty sharp ' images accompanied by a certain amount of ' flare ' due to scratches on the mirror's surface. That its light-grasping power excelled that of any other telescope, despite the imperfect polish, was revealed on the following night when Herschel discovered Saturn's sixth satellite (Enceladus)* and saw some spots on the planet ' much stronger than I have ever seen them before '.[40]

Regular work with the 40-foot began towards the end of 1789 with Saturn the chief object of study. Enceladus was observed during September whenever weather permitted and, on the 17th of that month, Herschel detected a seventh satellite, Mimas. In 1790, he appears to have made only one observation of the planet, concentrating instead on a search for new nebulae. In this field, the light-grasp brought many new objects into view, but it revealed no fresh details of structure in the old. Globular clusters appeared brighter and were more readily resolved into stars. Of the satellites of Uranus there are only three recorded observations—the rest were made with the 7-, 10-, and large 20-foot reflectors. No

* Seen August 19, 1787, with the 20-foot but not followed up.

observations of Saturn are recorded between 1791 and 1798. In April, 1801, there appears the entry: 'The speculum is much injured by time, I see the phenomena, however, of the ring, the belts and satellites &c. very well.'[41] The mirror was repolished, but observations of Saturn continued few and far between. On July 29, 1813, there is the entry: 'I viewed Saturn, it was very bright and I saw many of the satellites but the stars of the milky way being scattered over the neighbourhood the satellites could not be identified unless their situation had been calculated before. The mirror is so much tarnished that the image of Saturn was very imperfect. At all events the light collected is much more than is required for viewing this planet even with a high power, for which reason the delicacy of a 10-foot mirror in perfection is fully adequate to critical observations of Saturn's phenomena.'[42] In later years, Saturn was always Herschel's favourite object of study, and it was fitting, therefore, that the planet should have been the last thing seen through the 40-foot. The last observation was made in August, 1815. 'Saturn was very bright and considerably well-defined. The 4th satellite appears like a star of the first magnitude to the naked eye. The mirror is extremely tarnished.'[43]

The paucity and irregularity of Herschel's observations with the 40-foot leave little doubt that the great telescope failed to meet its maker's expectations. In the first place, the weather was seldom good enough to allow full use of its aperture and, when conditions were favourable, Herschel preferred the smaller and more manageable 20-foot. He found there were few objects visible in the 40-foot which he could not see in its smaller counterpart. Both Enceladus and Mimas were well within the light-grasp of the 20-foot and were both seen previously with this instrument, but the observations were not followed up. Much valuable time was lost in uncovering the large mirror and in preparing the instrument for a night's work. At least two assistants were required to operate the motions and to take down the observations. With only a hundred hours in the year when seeing conditions were really good, Herschel could not afford to waste those occasions in setting up and working a clumsy telescope. 'A 40-foot telescope' he once wrote, 'should be used only for examining objects that other instruments will not reach. To look through one larger than required is loss of time, which in a fine night, an astronomer has not to spare.'[44]

Time wasted preparing the 40-foot for observation was insignificant compared with the valuable hours lost in explaining its construction and motions to sightseers who had little real interest in astronomy. As the telescope neared completion, so the number of visitors increased, and seldom a day passed but some high personage called to inspect the work and to look for an invitation to walk through the tube as it lay on the ground. Of the callers, the King and his suite were among the most frequent, and Caroline records that, on one occasion, the King conducted the Archbishop of Canterbury through the tube with the words: 'Come, my Lord Bishop, I will show you the way to Heaven!'[45] Princes, dukes, lords, and bishops, both English and foreign, came to see the new wonder of the age. The Prince of Orange, Archduke Michael of Russia, Prince Galitzin, Prince of Hesse Homburg, the Duke of Sussex, Princess Sophia of Gloucester, Lord Darnley, Lord Kirkwall, and Admiral Boston are a few of the names in the *Visitors' Book* at Slough. In the same book appear the signatures of Lalande, Méchain, Legendre, Cassini, and Carroché from Paris, Pictet from Geneva, Oriani from Milan, Piazzi from Palermo, and Sniadecki from Cracow. We find also the names of Sir William Watson, Alexander Wilson, Mr and Mrs Maskelyne, Mr and Mrs James Watt, and Professor Vince. No doubt many of these visitors contrived to call on a fine night. If Herschel deplored the loss of many observing hours, Caroline grieved to see how exhausted her brother became in answering questions when he should have been 'renewing his strength by going to rest.'[46]

The 48-inch mirror was susceptible to temperature changes by reason of its size and the rough way in which it was mounted. It cooled at a rate slower than that of the surrounding air and, in changeable weather, the focal length varied appreciably. Sometimes, when the air was humid, the surface misted over and put a temporary stop to the night's work. On frosty nights, mist froze to the surface and there it had to remain until thaw set in.[47] Had the disk been thicker, effects due to changing temperature would have been even worse, for the larger mass of metal would have taken longer to adjust itself to the temperature of the outside air. While its very thinness in this respect was an advantage, it was a drawback from the point of view of rigidity. As the tube was lowered from the zenith position, so the mirror flexed under its own weight, to the detriment of image definition. Supports at the back of the disk, arranged to act as levers, would have remedied this fault to some extent but, as Herschel arranged it, the entire weight of the speculum in low altitudes was borne by the lower parts of the retaining ring. The mirror's greatest defect was its propensity to tarnish, owing to the high copper content. The work of frequent polishings was a great strain on Herschel, then approaching his sixtieth year. This factor, more than any other, probably prevented him from troubling to keep the telescope in proper working order.

Besides the large 20-foot and 40-foot reflectors, Herschel kept smaller instruments ready for action should the weather or time available prevent the use of large apertures. One of these reserve telescopes was a 'large' 10-foot Newtonian of 24 inches aperture, called the 'X feet' to distinguish it from ordinary 10-foot telescopes of 9 inches aperture. Herschel kept it fitted in a 'light common wood' stand for his own use, its short focal length making it, as Dreyer has remarked,[48] a convenient instrument for an old man. It was eventually sold to Lucien Bonaparte in 1814. Another favourite instrument was a 7-foot Newtonian of $6\frac{1}{4}$ inches aperture, fitted 'commodiously' for his own special use. With this instrument he made most of his Saturn observations for, while the strong light of the 20-foot fatigued his eye, that given by the 7-foot was sufficient, with a power of 287, to show the belts and rings of Saturn 'completely well'. Such was his preference for this telescope that he made ten object-specula for it and fourteen diagonal flats.

Herschel mounted his object-specula loosely in a brass or iron box, fitted with a narrow retaining ring and secured at the back by two adjusting screws. For disks above 5 inches diameter and with the tube near the zenith, he found that springs yielded to their weight; when made stiffer, the springs distorted the figure. Each box was provided with a close-fitting lid to protect the polished surface. The space allowed for the expansion of the speculum was, however, too generous.

Tulley and subsequent opticians remark[49] that Herschel's mirrors got out of adjustment when the tubes were in any position other than the vertical. The small elliptical flats were more securely mounted in cylinders of thin sheet-brass and were supported by a radial arm at right angles to the tube. Herschel's favourite 7-foot Newtonian was mounted in a mahogany stand, in appearance something like a narrow chair. The tube was octagonal in section and had a hinged door at the lower end to allow the removal of the mirror's cover. Motion from the horizontal nearly to the vertical was effected by cords, pulleys and rackwork, while to set the tube in azimuth the whole stand was shifted on its casters. Small adjustments in azimuth were given by rack-motion attached to the main tube support. Nearly all Herschel's smaller telescopes were mounted in this way, including the one with which he discovered Uranus. Two notable exceptions were a 27-inch focus and a 5-foot focus Newtonian comet sweeper. The first consisted of a low, round table, well braced for rigidity, in the centre of which rotated a vertical wooden pillar. To this was attached an upright wooden board for carrying the telescope and its operating ropes and pulleys. The tube moved in altitude on a pivot fixed opposite the eyepiece, so that the arrangement constituted something new in telescope design—a fixed-eyepiece reflecting telescope. Caroline had sole use of this instrument for her searches for comets, of which objects she discovered eight. The field of view was 2° 12′, the power 20.[50] Presumably the 5-foot Newtonian of 9¼-inch aperture was mounted in the same way, but of this we have no direct proof. A further instrument belonging to Herschel's private collection was a 7-foot skeleton Newtonian, the tube and stand of which consisted of metal bars which could be screwed together and taken apart so as to occupy very little space.[51]

Herschel corresponded with many eminent astronomers in France and Germany, sending them his papers in full or in abstract, mirrors ready for mounting, or sometimes complete instruments. To one correspondent, Johann H. Schröter, falls a rare distinction—he used a Herschel telescope for systematic observation, built and perfected his own instruments, and emulated Herschel in enthusiasm and perseverance. Schröter was bailiff of Lilienthal, a small village about 12 miles from Hanover. He knew the Herschel family and at one time received from Alexander Herschel a small Dollond telescope.[52] In 1784, Schröter purchased from William Herschel a complete 4-foot telescope and, two years later, a 7-foot and mirrors for larger instruments. Schröter mounted these mirrors, assisted by two officials infected by his enthusiasm. In 1792, he purchased another 7-foot reflector from Schräder of Kiel and later, a 13-foot Newtonian of 9½ inches aperture.[53] Furthermore, he began to cast and figure his own mirrors, a gardener named Gefken being trained to perform much of the work. In November, 1793, Schröter wrote to Herschel that he was engaged on ' a very bold undertaking '—the construction of a 27-foot reflector of 18½ inches aperture. ' The enterprise ' he continued, ' has been successful from first to last and with all its apparatus it is now finished.'[54] The instrument was used in the open and bore the proud inscription: *Telescopum Newtonianum XXVII pedum constructum Lilienthalii 1793*.[55] Despite the demands of his official duties and frequent ill health, Schröter became a most persistent lunar and planetary observer. From his observations of apparent irregularities near the cusps of Venus he concluded that Venus has a highly mountainous surface. Herschel could not agree with this conclusion. 'You will find' he wrote in 1793, 'that I have never been able to see any mountains in Venus; that to me the horns were always equally sharp.'[56] Schröter's book on the moon, the *Selenotopographischen Fragmente*, printed in 1791 and 1802, is one of the classics of selenography.

Activities so unusual and so earnestly conducted could not fail to attract others. In 1800, a 'Societas Liliatalica' was founded at Schröter's residence. Its members decided on a detailed survey of zodiacal stars in an attempt to discover the 'missing' planet between the orbits of Mars and Jupiter predicted by Bode's law. Schröter was elected president, and Baron von Zach's journal, the *Monatliche Correspondenz*, spread news of the society's work across Europe. In January, 1801, G. Piazzi at Palermo discovered the minor planet Ceres only to lose it soon afterwards. Later that year, Baron von Zach and Olbers at Lilienthal rediscovered the planet. In March, 1802, Olbers discovered Pallas and in 1804 and 1807 M. Harding, Schröter's assistant, discovered Juno and Vesta respectively. Gauss, a frequent visitor to Schröter's residence, reduced the observations and computed orbits which, as was expected, lie between those of Mars and Jupiter. Herschel was kept informed of these discoveries and suggested the name *asteroids* for the new bodies. He invented a micrometer which he called a 'lucid disk micrometer' and spent many hours in endeavours to measure their apparent diameters.

Schröter's elation at the direct and indirect results of his work was shortlived. In 1813, the French army under Vandamme sacked Bremen, and Lilienthal was practically burnt to the ground. Schröter lost home, library, and papers, and saw his observatory pillaged. Harding's successor, F. W. Bessel, had gone to Königsberg to assist in the new observatory there, and the aged Schröter had neither strength nor inclination to start again. The usable remains of his instruments were sent to Göttingen University. After his death in 1816, the observatory fell into disrepair and was finally demolished; the site is now marked by a small tablet.

Always interested in the performance of his telescopes, Herschel attempted to assess numerically what he called their 'power of penetrating into space.' This 'space-penetrating power' was the ability of an observer to detect faint objects regardless of magnifying power.

> Objects are viewed in their greatest perfection [Herschel writes][57] when, in penetrating space, the magnifying power is so low as only to be sufficient to show the object well; and when, in magnifying objects, by way of examining them minutely, the space penetrating power is no higher than what will suffice for the purpose; for, in the use of either power, the injudicious overcharge of the other will prove hurtful to perfect vision.

The amount of light reaching an observer's retina is governed by the area of his pupil, an aperture of 0·2 inch for a dark-adapted eye, if we adopt Herschel's figure. A telescope admits light in an amount varying with the square of the diameter of the object-glass. Some of this light is lost by reflection and absorption at the lenses; the remainder, if the pupillary aperture is large enough, passes into the observer's eye. To the effect of increasing telescope aperture, Herschel added that produced by increasing distance of a point object. The amount of light received from a star varies inversely as the square of its distance. If a star is just visible in a 6-inch telescope, another star of the same intrinsic brightness but at twice the distance will be just visible in a 12-inch—assuming the same percentage loss in both instruments. In other words, a 12-inch telescope has twice the 'space-penetrating power' of a 6-inch. To bring in all the contributory factors, Herschel used the expression $\sqrt{\{x(A^2 - b^2)/a\}}$ for 'space-penetrating power', where A is the aperture of the large mirror, b the minor axis of the elliptical flat, a the diameter of the observer's pupil, and x the percentage light transmission within the instrument.[58] Herschel measured his diameters in inches and, to assess x, experimented with a plane speculum. For normal incidence, he found that the mirror absorbed 33 per cent of the incident light and thus 55 per cent after two normal reflections. Eyepieces of the single-lens type transmitted 95 per cent of the incident light so that, for a front-view telescope, the eye received 63 per cent of the incident light and, for a single-lens-eyepiece Newtonian, only 43 per cent.[59] On the assumption that these losses were constant, Herschel obtained the following 'space-penetrating powers' for his telescopes.[60]

7-foot reflector,	aperture	6·0 inch,	space-penetrating power	20·25
10 ,,	,,	8·8 ,,	,, ,,	28·67
20 ,,	,,	12·0 ,,	,, ,,	61·18
20 ,,	,,	18·0 ,,	,, ,,	75·08
40 ,,	,,	48·0 ,,	,, ,,	191·69

While the choice of a suitable magnifying power depended on atmospheric conditions, resolving power of the eye, figure of the mirrors, and the nature of the object, 'space-penetrating power' was limited by brightness of the night sky. Herschel writes, quite rightly, 'that the natural limit seems to be an equation between the faintest star that can be made visible, by any means, and the united brilliancy of star-light.'[61] It is this background luminosity, often supplemented by town and city light, which sets limits to the photography of very faint objects in modern astronomy, however large the aperture ratio of the telescope.

As early as 1774, Herschel found that he could not distinguish a square from a circle if they subtended an angle of less than 2′ 17″.[62] 'If we would distinctly perceive and measure or estimate extremely small quantities, such as a tenth of a second' he wrote in 1781, 'it appears, that when we use . . . a power of 1500 [this] will be but 2′ 30″, which is a quantity not much more than sufficient to judge well of objects and distinguish them from each other, such as a circle from a square, triangle or polygon.'[63] In other words, he estimated the visual efficiency or contour acuity of his eye at 2′ 17″ and realized that, to see very fine details, he had to magnify them so that they exceeded this limit.

Herschel appreciated that, even during best seeing conditions, only the most perfect mirror would bear high powers, while the brightness of the object imposed a further restriction. On one occasion, when he was anxious to ascertain the diameter of Juno, he tried to estimate the resolving power of his 10-foot telescope.[64] He placed a row of little globules of sealing-wax some 200 feet from the mirror and, with his eye at the eyepiece, selected the smallest globule which presented 'a visible disk, a round, well defined appearance.'[65] Other experiments with globules of silver, pitch, beeswax and brimstone led to the conclusion that, with the full aperture of his telescope, 8·8 inches, he could just recognize the disk of one which subtended an angle of $0''\cdot25$. As he reduced the aperture, the spurious diameters increased. The brighter the object, the larger the spurious disk, the central parts of the mirror being more responsible for the size of the disk than the outer. When a globule subtended an angle less than the minimum for resolution, no magnification, however great, could make it 'round and well-defined.'

In one of his first papers in which he discussed his double-star method of determining stellar parallax,[66] Herschel had made frequent reference to the use of magnifications as great as 2000, 3000, and 6000. These unprecedented powers aroused both considerable interest and some suspicion and, at Watson's advice, Herschel wrote a paper 'to obviate some doubts concerning the great Magnifying Powers used.'[67] He told his critics that he had measured the focal lengths of his eyepieces and that measured powers agreed very well with calculated powers. Yet it would be wrong to suppose that he was addicted to high powers. No one knew better than he their serious limitations—how they restricted the field of view and magnified the imperfections of the mirror, vibrations of the stand, and turbulence in the atmosphere.

> By enlarging the aperture of the telescope [he once wrote][68] we increase the evil that attends magnifying . . . the object without magnifying the medium. . . . From my long experience in these matters, I am led to apprehend, that the highest power of magnifying may possibly not exceed the reach of a 20 or 25 feet telescope, or may even lie in a less compass than either. However, in beautiful nights, when the outside of our telescopes is dripping with moisture discharged from the atmosphere, there are now and then favourable hours, in which it is hardly possible to put a limit to magnifying power. But such valuable opportunities are extremely scarce; and, with large instruments, it will always be lost labour to observe at other times.

Proof of this is seen in the fact that he mentions only two occasions when he used a power of 6450—once on α Lyrae and once on γ Leonis. Herschel used high powers when the state of the atmosphere and performance of his mirrors allowed, and then usually on close double stars. With the 20- and 40-foot telescopes, he seldom employed more than 200, for this gave a wide field and rendered faint nebulae, his special study, more easily seen. On the other hand, he was of the opinion that if the object had sufficient light 'as the stars undoubtedly have' there was no reason 'why we should limit the powers of our instruments by any theory.'[69] This view he expressed in 1781, twenty years before Thomas Young, by his interpretation of interference phenomena, established the wave theory of light and, indirectly, set an unavoidable limit to the usefulness of very high magnifications.

Herschel seldom used Huygenian eyepieces and his high powers were all achieved with single convex lenses. The favour in which he held convexes is clearly shown in the following extract from his 1781 paper 'On the Parallax of the Fixed Stars'.[70]

> I have tried both the single and double eye-glass of equal powers, and always found that the single eye-glass had much the superiority in point of light and distinctness. With the double eye-glass I could not see the belts on Saturn, which I very plainly saw with one single one. I would, however, except all those cases where a large field is absolutely necessary, and where power joined to distinctness is not the sole object of our view.

As early as 1776, he experimented with biconcave and plano-concave lenses, comparing them with convexes of equal focal length.[71] At first, it seemed that the concaves were superior. They gave bright and more distinct images and their small field of view was no great inconvenience when it came to observing double stars or isolated satellites. But concave lenses cannot be used with a micrometer and this, added to the apparent ease with which Herschel made his convexes, turned him against concaves.

Herschel's favourite power for Saturn with the 7-foot reflector was 287. This necessitated an eyepiece of only 0·3 inch focus, but for the third review he used '18 higher magnifiers' which gave a variety of powers from 460 to 6000. Taking the focal length of the 7-foot as 84 to 88 inches, these smaller lenses must have had focal lengths of from 0·2 inch to just over 0·01 inch. In 1924, W. H. Steavenson examined forty-eight of Herschel's eyepieces and found that this actually was the case.[72] He measured the focal lengths of the smaller lenses with a micro-focometer and found values from 0·064 inch down to 0·0111 inch. With the 7-foot telescope these would yield powers from 1331 to 7676. All were well-formed lenses with the exception of one of 0·024 inch focus which was spherical. Tested on celestial objects with a 6-inch Wray refractor, Steavenson found that even the smallest lens of 0·0111 inch focus 'showed the spurious disk of Vega, with portions of the first diffraction ring, and also exhibited the general outlines of the planet Saturn. The field of view, however, was only some 20 seconds of arc in diameter and the image would hardly have been examined, or even held in view, without the help of a clock-drive.'[73] Yet Herschel, in his experiments with high powers, had nothing better than an altazimuth stand with hand-operated slow motions. Just how he contrived to make such small lenses remains a mystery. The smallest is but 1/45 inch in diameter and we assume only that it was made from a small glass globule. The lenses were mounted by making the eyepiece in two parts; these screwed together and confined the lens between them. The mount was turned in brass or cocos wood and screwed into the eyepiece tube either direct or through an adaptor.

Herschel found that, when his eyes were fully dark-adapted, he could see objects which were invisible when he first looked into the telescope. To preserve this condition he sometimes covered his head with a black hood. 'This increased sensibility' he writes, 'was such, that if a star of the 3rd magnitude came towards the field of view, I found it necessary to withdraw the eye before its entrance, in order not to injure the delicacy of vision acquired by long continuance in the dark.'[74] On another occasion, after a long sweep with the 40-foot, 'the appearance of Sirius announced itself, . . . and came on by degrees, increasing in brightness, till this brilliant star at last entered the field of view of the telescope, with all the splendour of the rising sun, and forced me to take the eye from that beautiful sight.'[75]

During his residence at Datchet, Herschel encountered very varied weather conditions owing to the severity of the winters and the proximity of the river Thames with its dampness and mists. He treated inclement weather as an opportunity for studying the reactions of mirrors. Humid, still air and even fog proved no detriment to definition, although the telescope might run with water and mist settle on the eyepieces. 'The water condensing on my tube' he writes on one occasion, 'keeps running down; yet I have seen very well all night. I was obliged to wipe the object-glass of my finder almost continually. The specula, however, are not in the least affected with the damp. The ground was so wet that, in the morning, several people believed there had been much rain in the night, and were surprised when I told them there had not been a drop.'[76] Mirror performance during frost was unreliable. He made a number of 'delicate observations' with the thermometer at 11°F; the ink froze indoors and a 20-foot speculum eventually broke into two pieces. On another night he records: 'My telescope will not act well, even at an altitude of 70 or 80 degrees. There is a strong frost.' Dry air and wind were detrimental, causing stars to appear 'tremulous and confused', whilst a night that promised well often proved a failure. 'The stars appear fine to the naked eye, so that I can see ϵ Lyrae very distinctly to be two stars; yet my telescope will show nothing well,'[77] is a typical observation on such a night. Every owner of a reflecting telescope knows that his instrument never performs well when it is first brought out into the open. Herschel investigated the cause of this. 'To all appearance, the morning was very fine, but still the telescope, when first brought out, would not act well. After half an hour's exposure, it performed better.'[78] He realized that a mirror, in cooling to air temperature, departed slightly from its true figure—or rather from the figure given to it during the final polishing stage. Experiments soon revealed that changes of temperature led to changes in focal length so that the latter was longer in a falling temperature, shorter in a rising temperature. On these grounds he accounted for his failure to use high powers while observing the sun, the only cure for which, in his opinion, was 'an elliptical speculum . . . to counteract the assumed hyperbolical form.'[79]

Mention of the sun's heating action invites us to consider Herschel's experiments with absorbing media. In order to examine the details of sunspots he often used one of the 20-foot reflectors. If coloured glasses placed behind the eyepiece did not crack, some of them reduced the glare but did not prevent an intolerable sensation of heat. Since, through different combinations of coloured media, the sun's image appeared differently coloured, it occurred to Herschel 'that the prismatic rays might have the power of heating bodies very unequally distributed among them. . . . If certain colours should be more apt to occasion heat, others might, on the contrary, be more fit for vision, by possessing superior illuminating power.'[80]

In one of his experiments made at the close of 1799 to test this hypothesis Herschel cast in turn the various colours of a solar spectrum on a guinea and observed it through a microscope. As the colours progressed from red to green, so details on the coin became more distinct, reaching a maximum clearness between yellow and green. With blue light, illumination was poor, and with violet, dropped almost to darkness. He then placed thermometers at different parts of a bright solar spectrum and found that those placed in the red section indicated a greater temperature rise than when they were placed in the green. Maximum heating intensity was registered by a thermometer placed just beyond the red, and Herschel concluded that 'radiant heat will at least partly, if not chiefly, consist, if I may be permitted the expression, of invisible light.'[81]

In view of these results, when he came to re-observe the sun Herschel adopted dark green or green filters used in conjunction with smoked green. These

Fig. 62—Apparatus used by Herschel to demonstrate the refraction of 'invisible heat rays'

The three thermometers are shielded from the direct light and heat of the fire

glasses gave comfortable and clear vision but invariably cracked, one of them 'with a very disagreeable explosion, that endangered the eye.' Nearly thirty different combinations were tried but variations of dense green answered the best.

In April, 1800, Herschel read his second paper[82] on the discovery of these 'invisible rays', the near infra-red, to the Royal Society. In May and November of the same year, he read a third paper[83] on this subject—the result of over two hundred experiments. He found that the infra-red rays, like visible radiation, could be reflected and refracted. They obeyed the sine law and, when refracted by a large convex lens, came to a focus different from that of visible light. Yet they seemed independent of the visible rays, for while deep-red glass transmitted nearly all the heat rays, it absorbed a large proportion of red light. From this and other considerations, Herschel concluded that radiant heat and visible light are fundamentally independent of each other. In so doing he unwittingly retarded progress in radiant-heat studies more, indeed, than did the preconceived views of others about caloric. He thus passed to Thomas Young the important task of initiating the classical radiation concept.

In 1801, Herschel viewed the sun through a mixture of ink diluted with water and filtered through paper. It gave an image of the sun 'as white as snow'. An eyepiece was then filled with a solution of ferrous sulphate with 'tincture of galls'. This gave a dark blue solar image which changed, on adding more ferrous sulphate to the solution, to an image 'of a deep red colour.' Herschel expressed no surprise at this interesting case of dichromatism and failed to appreciate its significance.

Herschel died on August 25, 1822, after some years of ill health and increasing feebleness of body. Summer visits to Edinburgh and Glasgow, and changes of air at Bath and Brighton, failed to bring back the vigour and strength which his work for so long had exacted. He lies buried in the little church of St Lawrence, Upton, in which he was married and near which he assembled the 40-foot tube.

In less than two months after her brother's death, Caroline Herschel left England for Hanover, where she lived alone, but not forgotten, for another twenty-six years. Apart from such news as came from England concerning the activities of her nephew, John Herschel (later Sir John Herschel), and the recognition of her work by learned bodies, she found life a dull business.

> You will see what a solitary and useless life I have led these 17 years [she writes in her last days][84] all owing to not finding Hanover, nor anyone in it, like what I left, when the best of brothers took me with him to England in August, 1772.

REFERENCES

1 Dreyer, J. L. E., 1912, *Scientific Papers of Sir William Herschel*, Introduction, p. xxiv.
2 Herschel, Mrs J., 1876, *Memoir and Correspondence of Caroline Herschel*, p. 35.
3 Dreyer, *op. cit.*, p. xxiv.
4 Dreyer, *op. cit.*, p. xxv.
5 Hersch el, *op. cit.*, p. 36.
6 *Ibid.*, p. 37.
7 Dreyer, *op. cit.*, p. xxv.
8 Herschel, *op. cit.*, p. 37.
9 Dreyer, *op. cit.*, p. xxv.
10 Lubbock, C. A., 1933, *The Herschel Chronicle*, p. 72.
11 Dreyer, *op. cit.*, p. xxvi.
12 *Ibid.*, p. xxvi.
13 Herschel, *op. cit.*, p. 41.
14 Dreyer, *op. cit.*, p. xxvii.
15 *Ibid.*, pp. xxvii–xxviii; Herschel, *op. cit.*, p. 44.
16 Lubbock, *op. cit.*, p. 73.
17 Herschel, *op. cit.*, p. 41. Lubbock, *op. cit.*, p. 75.
18 Lubbock, *op. cit.*, p. 75.
19 *Ibid.*
20 Herschel, *op. cit.*, p. 47.
21 Lubbock, *op. cit.*, pp. 116–117.
22 Herschel, *op. cit.*, p. 50.
23 Dreyer, *op. cit.*, p. xxxvii.
24 Herschel, *op. cit.*, p. 53.
25 Dreyer, *op. cit.*, pp. l–li.
26 *Ibid.*, p. xxxix.
27 *Connaissance des Temps*, 1784.
28 Lubbock, *op. cit.*, p. 167.
29 Herschel, W., in Dreyer, *op. cit.*, i, p. 260.
30 Dreyer, *op. cit.*, p. xlii. The idea of a tilted mirror was put forward by Le Maire in 1728.
31 *Ibid.*, p. xxxix.
32 Herschel, Ref. 2, p. 54.
33 *Ibid.*, p. 55.
34 *Ibid.*, p. 73.
35 Herschel, Ref. 29, i, p. 486.
36 Dreyer, *op. cit.*, p. xlviii.
37 *Ibid.*, p. xlvii.
38 *Ibid.*, pp. xlviii, xlvix, l.
39 For full description *vide Phil. Trans.*, **85**, pp. 347–409, 1795. Herschel, Ref. 29, i, pp. 485–527.
40 Dreyer, *op. cit.*, lii.
41 *Ibid.*, liii.
42 *Ibid.*
43 *Ibid.*
44 Herschel, Ref. 29, ii, p. 536.
45 Herschel, Ref. 2, p. 309.
46 *Ibid.*, p. 74.
47 Herschel, Ref. 29, ii, p. 536.
48 Dreyer, *op. cit.*, p. lv.
49 Kitchener, W., 1825, *The Economy of the Eyes*, ii, pp. 126, 129.
50 Smyth, W. H., 1881, *A Cycle of Celestial Objects*, p. 675.
51 Dreyer, *op. cit.*, p. lv.
52 *J.B.A.A.*, **61**, pp. 192 ff., 1951, quoting V. v. Marnitz, *Sternenwelt*, Sept. 1950.
53 Neison, E., 1876, *The Moon*, Preface.
54 Lubbock, *op. cit.*, pp. 213–214.
55 *Vide* Ref. 52.
56 Lubbock, *op. cit.*, p. 212.
57 Herschel, Ref. 29, ii, p. 50.
58 *Ibid.*, p. 41.
59 *Ibid.*, p. 40.
60 *Ibid.*, pp. 44–46.
61 *Ibid.*, p. 49.
62 *Ibid.*, p. 297.
63 *Ibid.*, i, p. 47.
64 *Phil. Trans.*, **95**, i, pp. 31–64, 1805; Herschel, Ref. 29, ii, pp. 297–316.
65 Herschel, Ref. 29, ii, p. 297.

[66] *Phil. Trans.*, **77**, pp. 82–111, 1782; Herschel, Ref. 29, i, pp. 39–57.

[67] *Phil. Trans.*, **77**, pp. 173–178, 1782; Herschel, Ref. 29, i, pp. 97–99.

[68] Herschel, Ref. 29, ii, p. 49.

[69] *Ibid.*, i, p. 47.

[70] *Ibid.*, i, p. 46.

[71] *Ibid.*, ii, pp. 544–5.

[72] *Trans. Opt. Soc.*, **26**, pp. 210–220, 1925.

[73] *Ibid.*

[74] Herschel, Ref. 29, ii, p. 34.

[75] *Ibid.*

[76] Herschel, Ref. 29, ii, p. 240.

[77] *Ibid.*, p. 243.

[78] *Ibid.*

[79] *Ibid.*, p. 249.

[80] *Ibid.*, p. 53.

[81] *Ibid.*, p. 63.

[82] *Phil. Trans.*, **90**, pp. 284–292, 1800; Herschel, Ref. 29, ii, pp. 70–76.

[83] *Phil. Trans.*, **90**, pp .293–326, 437–538, 1800; Herschel, Ref. 29, ii, pp. 77–146.

[84] Sime, J., 1900, *William Herschel and his Work*, p. 254.

CHAPTER VIII

It were wished that astronomers, concerned in observations, might be accommodated with achromatic telescopes of the most perfect construction; as such are the only instruments whereby a great knowledge of the celestial bodies can be acquired, for the improvement and perfection of astronomy.

MESSIER

WHILE the reflecting telescope was receiving so much attention at the hands of Short, opticians did not entirely neglect the refractor. Its development, however, was hindered by the conclusive appearance of Newton's researches on the behaviour of light. His generalization that refraction is always accompanied by dispersion went unquestioned for nearly a century. One exception to this belief, although it was not a direct refutation, was due to David Gregory, nephew of James Gregory and Savilian professor of astronomy at Oxford.

> Perhaps it would be of Service [he wrote in 1695][1] to make the Object-Lens of a different Medium, as we see done in the Fabric of the Eye, where the crystalline Humour (whose Power of refracting the Rays of Light differs very little from that of Glass) is by Nature, who never does anything in vain, joined with the aqueous and vitreous Humours (not differing from Water as to their Power of Refraction) in order that the Image may be painted as distinct as possible upon the Bottom of the Eye.

True, we are not aware of any want of achromaticity when we look at bright points of light, but the optical system of the eye is now known to be far from achromatic.

In 1729, Chester Moor Hall, reasoning along the same lines as Gregory, obtained achromatic lenses by combining together two glasses of opposite powers, the concave of flint and the convex of crown glass. Hall was a gentleman-barrister who amused himself in his spare time by making optical experiments.[2] Besides designing the first achromatic lens, he is known to have invented a new type of sea-quadrant.[3] He does not appear to have ground his own lenses for, in 1733, he asked Edward Scarlett of Soho to grind one component and James Mann senior of Ludgate Street to grind the other.[4] These opticians both sub-contracted the order to George Bass, a jobbing optician in Bridewell Precinct who, finding that the two glasses were for the same person, put them together and so disclosed Hall's secret.[5] This compound glass, the first of its type, was 2½ inches in diameter and had a focal length of 20 inches. Having satisfied himself that this glass and others (also made by Bass), were a great improvement on the single convex lens, Hall gave written instructions for making his lens to both Bird and Ayscough.[6] But the former was too busy with his quadrants, and the latter too occupied by business worries, to consider the project. There is evidence, however, that both Bird and Ayscough had an achromatic lens in their possession several years before similar lenses were marketed by Dollond.[7]

Hall was too modest and too fond of privacy to urge any claims for his invention. News of his lens spread slowly among the London opticians, for none of them grasped its full significance. This apathetic attitude was changed by John Dollond (Fig. 63).

Accounts differ as to the manner in which Dollond heard of Hall's invention. At the 1764 trial case, Peter Dollond versus the London opticians, it was quite clearly stated that Dollond heard of the invention in 1750 through Robert Rew, but Ramsden says that he heard from Dollond himself that his informer was Bass. Jesse Ramsden was, as we shall see later, an instrument-maker of repute, known also for his honesty and fairness. The letter which he wrote to the Royal Society about the invention of the achromatic telescope is, therefore, of considerable importance.

> Sometime after this [Ramsden writes][8] a reading Glass was bespoke of Mr. Dollond for the late Duke of York. He applied to Bass the private working optician before mention'd to supply him with one. He shew'd him several and Mr. Dollond seem'd to give the preference to one of flint Glass from its clearness and transparency; but Bass told him that the fault of that Glass was that letters seen through it towards the edges were much more ting'd with Colours than in one made of Crown glass adding at the same time that he work'd the Concaves for Mr. Hall's object-Glasses of that glass, that is, the flint glass.

Like many of his contemporaries in optics, Dollond was born in humble circumstances. Following the revocation of the Edict of Nantes in 1685 his father, Jean Dollond, fled to England and joined the silk-weaving community then growing in Spitalfields, London. As a boy, John Dollond developed strong scientific leanings and, in his hours away from the loom, studied mathematics, optics, and astronomy. By 1750, he was known in scientific circles in London for his wide optical knowledge.[9] Three years previously, the distinguished mathematician Leonhard Euler (Fig. 64), knowing nothing of Hall's work, had communicated a memoir[10] to the Berlin Academy of Sciences in which he endeavoured to show that it was possible to correct both spherical and chromatic aberration in object-glasses. Like Gregory, he was led to this idea by a consideration of the supposed achromatism of the eye. He evolved an object-glass which consisted of two meniscus lenses with water between them, and calculated that colour effects would be nullified by the combination. Dollond received a copy of this paper through his friend Short and hastened to inform Euler that he was in error since Newton had expressly stated that refraction was impossible without dispersion. ' It is therefore, Sir ' Dollond wrote to Short, ' somewhat strange that anybody now-a-days should attempt to do that, which so long ago has been demonstrated impossible.'[11] An interchange of letters followed, in consequence of which Dollond ' made experiments on the different dispersive powers of Water and of Glass and from the result of them had attempted to make object-glasses that would represent objects free from colour.'[12] These experiments were unsuccessful and Dollond gave up hopes of making workable lenses by this method. Euler, on the other hand, admitted that his theory was based on a law of refractivity different from Newton's but pointed out that this difference was very small. He stated that his new glasses were better than single lenses for freedom from colour but that they had marked

Fig. 63—John Dollond

spherical aberration. 'Our eminent Lieberkuhn' he wrote, 'is applying himself to the working of lenses of which the curvature of the surfaces decreases from the middle towards the edge, and he foresees great improvements. For these reasons, I think, my theory suffers nothing on this account.'[13]

In 1755, S. Klingenstierna, professor of mathematics at Upsala, wrote to Dollond through M. Mallet to say that he thought Newton had been misled in his dispersion experiments. He showed, by a geometrical method, that Newton's results applied only for prisms of small apical angle and were far from correct for larger prisms. 'The whole affair' he concludes, 'deserves to be more accurately examined by experiments.'[14] Klingenstierna's letter, coupled with the information about Hall's work which he received a little earlier from Bass or Rew, completely changed Dollond's former attitude. He put the question to experimental investigation with the result that, three years later, there appeared his important 'Account of some Experiments concerning the different Refrangibility of Light'.[15] His experiments completely disproved Newton's conclusions:[16]

> I cemented together two plates of parallel glass at their edges, so as to form a prismatic or wedge-like vessel, when stopped at the ends or bases; and its edge being turned downwards, I placed therein a glass prism with one of its edges upwards, and filled up the vacancy with clear water. . . . As I found the water to refract more or less than the glass prism, I diminished or increased the angle between the glass plates, till I

Fig. 64—Leonhard Euler

found the two contrary refractions to be equal; which I discovered by viewing an object through this double prism. . . . Now according to the prevailing opinion the object should have appeared through this double prism quite of its Natural colour . . . but the experiment fully proved the fallacy of this received opinion, by showing the divergency of the light by the prism to be almost double of that by the water; for the object, though not at all refracted, was yet as much infected with prismatic colours, as if it had been seen through a glass wedge only, whose refracting angle was near 30 degrees.

To obtain refraction without dispersion, Dollond saw that he would have to increase the angle of the water prism to an impracticable extent. He therefore chose smaller-angled prisms of about 5 to 10 degrees and, on adjusting them so that together they acted like a parallel plate, found there was residual colour. He correctly inferred that the dispersion of glass was greater than that of water and, by increasing the angle of the water prism to compensate for it, obtained refraction without dispersion.[17]*

The next step was to compute, grind, and polish object-glasses compounded of two spherical lenses with water between them. These glasses were ' free from errors arising from the different refrangibility of light, . . . [but] the images formed at the foci were still very far from being so distinct as might have been expected from the

* Or so it first appeared. *Vide*, however, p. 156.

Fig. 65—Peter Dollond

removal of so great a disturbance.' Dollond realized that much of the trouble was due to the large amount of spherical aberration occasioned by the steep curvature of the spherical surfaces. He looked, therefore, for a suitable medium to replace water and found one in English flint glass, which possessed a higher dispersive power than ordinary 'Venice' glass and English crown glass. We have already seen, on Ramsden's testimony, that it was in the workshop of Bass that Dollond stumbled upon the choice of flint glass. In 1757, Dollond found that a 25-degree flint prism and a 29-degree crown prism gave little resultant deviation but strong dispersion when combined.[18] Further crown prisms were made, and by trial and error one was found which, when combined with the other two, rendered the triple combination achromatic. From this arrangement he deduced that the focal length of the concave flint to that of the convex crown must be as the ratio 6 : 4, that is, in the ratio of their dispersive powers. To reduce spherical aberration he 'plainly saw the possibility of making the aberrations of any two glasses equal; and as in this case the refractions of the two glasses were contrary to each other, their aberrations being equal would entirely vanish.'[19] On these principles, 'after numerous trials, and a resolute perseverance', Dollond made his first achromatic lenses.

Nowhere in his paper does Dollond refer to the practical work of Hall or to Klingenstierna's theorem*—even Euler's name is omitted. He merely states that, while the notion of refraction accompanied by dispersion has been 'generally adopted as an incontestible truth', he is the first to test it by 'evident experiments.'[20]

* There is, preserved at Upsala, Dollond's own acknowledgement of the receipt of Klingenstierna's paper sent to him through Mallet.

Fig. 66—Antoine Darquier

In this print, dated 1774, he is seen observing with a quadrant. The telescope was certainly by Dollond, but it is not known if the arc and mounting were also.

In 1750, Peter Dollond (Fig. 65), the eldest son of John Dollond, ' having received some information upon philosophical subjects from his father, and observing the great value which was set upon his father's knowledge in the theory of optics by professional men ',[21] opened a small optical business in Vine Street, Spitalfields.

Two years later, he was joined by his father and moved to a house, ' The Sign of the Golden Spectacles and Sea Quadrant ', near Exeter Change in the Strand.[22] Here John Dollond performed his experiments and made his first achromatic lenses. His paper previously quoted was sent to Short, who passed it to the Royal Society. Although this was not Dollond's first contribution to the Society (he sent Short a letter[23] about some improvements he had made in terrestrial eyepieces in March, 1753), the practical importance of its contents was so great that he was awarded the Copley Medal in 1761 and elected a Fellow of the Society.[24]

Peter Dollond was an astute business man and he urged his father to apply for a patent for the new telescope. In this he was successful, the patent covering the ' new invented method of refracting telescopes, by corresponding mediums of different refractive qualities, whereby the errors arising from the different refrangibility of light, as well as those which are produced by the spherical surfaces of the glasses, are perfectly corrected.'[25]

At the same time, for financial reasons, Dollond and Son received a partner into the business, but the alliance was of short duration.[26] The patent was small comfort to the optical fraternity in London. They saw the new house flourishing and, no doubt, envied its success. Benjamin Martin straightway advertised telescopes by ' Doland ', George Adams sold telescopes ' with 5 or 6 glasses ' at a guinea a foot,[27] and Henry Pyefinch listed both ' ordinary ' and ' achromatic ' telescopes.[28] All these early telescopes, termed ' achromatic ' by the amateur astronomer John Bevis,[29] had small objectives owing to the scarcity of good flint glass. Their compact size and superior optical qualities made them suitable for divided instruments, and Maskelyne introduced them at Greenwich at an early date. French astronomers followed suit and, as early as 1761, M. Darquier, an astronomer of Toulouse, was observing with a 14-inch Dollond telescope. ' This telescope is particularly fine ' he wrote in 1777. ' It carries at its right extremity an alidade fitted with an English-type micrometer which indicates zenith distances on a small eighteen-inch mural arc.'[30] (Fig. 66.)

John Dollond died on November 30, 1761, and did not live to witness the full development of his object-glass. In addition to his work on dispersion, Dollond gave much thought to the improvement and construction of the divided-glass micrometer (Fig. 67) and, in 1754, read to the Royal Society his paper on this instrument.[31] It consisted of a telescope with its object-glass divided into two equal parts, each semi-lens being set in a separate brass frame. These frames slid laterally on each diameter and rotated round a common optical axis; when they were together the observer saw only one image. By moving the semi-lenses along their edges, the superimposed images separated, and the amount of their separation could be read from a scale and vernier to single seconds of angular displacement. This was not an entirely new idea, as Dollond himself pointed out. Römer had suggested[32] something of the kind as early as 1675 for observing eclipses, that is, for dividing the diameters of the sun and moon into twelve equal parts. Many years

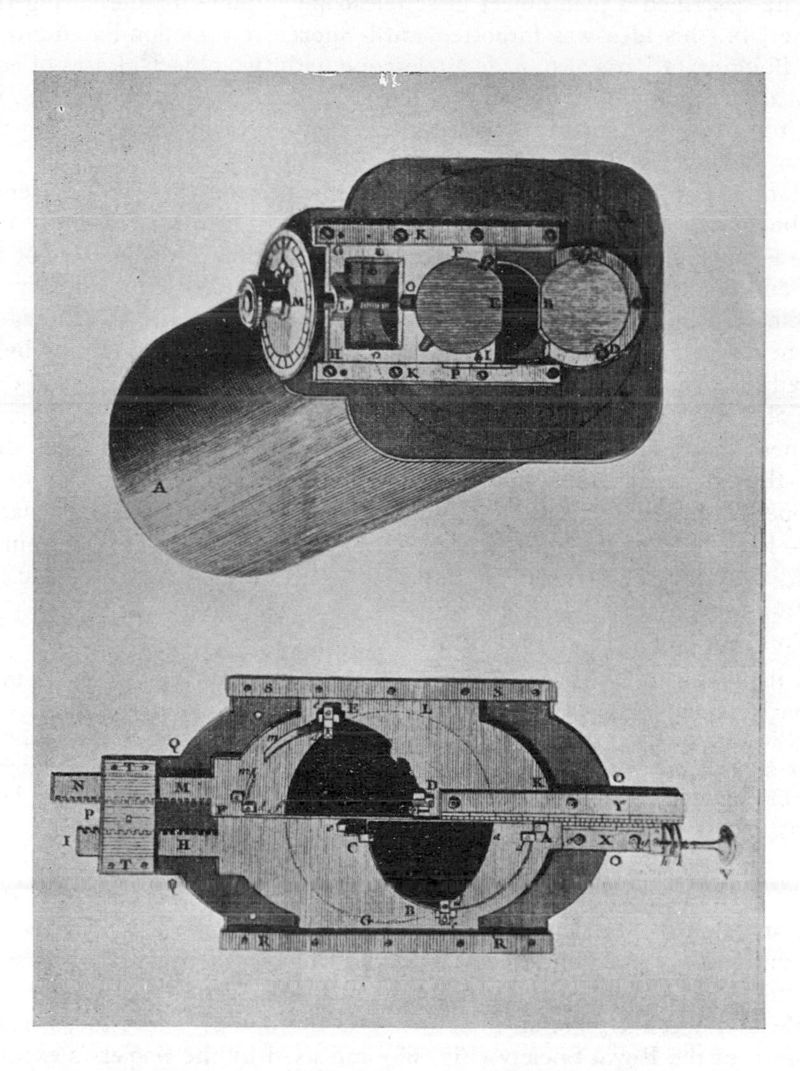

Fig. 67—(*Top*) Bouguer object-glass micrometer
(*Bottom*) Dollond divided object-glass micrometer

(*From Lalande's 'Astronomie'*)

later (1743), Servington Savery of Oxford sent a paper[33] to the Royal Society in which he described a twin object-glass telescope suitable for measuring the sun's diameter, but his idea was forgotten until Short drew Dollond's attention to it. Pierre Bouguer of Paris also made a telescope with two object-glasses of equal size and focal length and so arranged that one eyepiece served for both (Fig. 67). He called it a *heliometer* or *astrometer* and, like Savery, used it for measuring the sun's apparent diameter at different times of the year.[34]

Dollond's plan of dividing a single object-glass meant that the two semi-lenses could be applied to reflecting telescopes, which Short immediately did. The principle was, moreover, beautifully adapted for measuring the diameters of the sun, moon, and planets and the separations of double stars. With an ordinary eyepiece micrometer, a high power made it impossible to see the entire solar image, whilst in the heliometer only the contacting limbs had to be in the field. The heliometer required no special illumination system, and the accuracy of results depended on the observer's estimate of contact, which was found to be more reliable than his judgement of transits over wires. There were, however, three serious disadvantages—the two images were only half as bright as the image given by an unmodified telescope of the same size, the resolving power was halved, and the separation of the two lenses caused the images to get out of focus. The first two, of course, were irremediable, but the German instrument-maker Repsold corrected the third by mounting the lenses in cylindrical slides.

Dollond proposed three types of divided object-glass micrometer:[35]

> In the first place it may be fixed at the end of a tube, of suitable length to its focal distance, as an object-glass; the other end of the tube having an eye-glass fitted as usual in astronomical telescopes. Secondly, it may be applied to the end of a tube much shorter than its focal distance, by having another convex glass within the tube, to shorten the focal distance of that, which is cut in two. Lastly, it may be applied to the open end of a reflecting telescope.

Only the first arrangement became of any use to astronomy. The second was scarcely ever used, whilst Short applied the third to Gregorians. Unfortunately, of all instruments, the last were the most ill-adapted for micrometrical work owing to the difficulty in maintaining the true centring of the mirrors. It is significant that, after Short's death, we hear no more of this particular application. A typical example of Short's Gregorian micrometer is the one of 2 feet focus now in the possession of the Royal Society (Fig. 68) and used by the Society's expedition to observe the 1769 transit of Venus.

Maskelyne showed particular interest in Dollond's micrometer, which he made more accurate by placing rotatable cross-wires at the focus of the objective.[36] He then changed it for an ordinary telescope inside which two Dollond achromatic prisms slid along the length of the tube and so duplicated the image.[37] The length of the tube thus became a scale for measuring angular separations, and quite a small movement of the prisms produced a deviation of 1 minute of arc. Peter Dollond made the achromatic prisms which were the first used for astronomical work. Maskelyne used his modified Dollond micrometer until about 1776 when Ramsden introduced A. M. de Rochon's superior rock-crystal double achromatic wedge.[38]

Fig. 68—Gregorian telescope by Short

Also shown is a divided object-glass micrometer

(*Science Museum, London. By permission of the Royal Society*)

Left on his own, Peter Dollond soon found himself in trouble with the other partner. The latter's unsatisfactory conduct led to his dismissal after Dollond had paid him £200.[39] In defiance of the patent, this person made and sold achromatic telescopes on his own, whereupon Dollond, finding that he had paid £200 to no purpose, prosecuted him. He was, apparently, successful, for he proceeded further and warned other opticians that if they sold achromatic telescopes without first paying him royalty, he would sue them for damages. At this ultimatum, thirty-five London opticians rose as one hostile body and, in 1764, addressed a petition to the Privy Council urging it to revoke the patent.[40] The signatories to this petition included Addison Smith, James Champneys (alias Champness), Benjamin Martin, John Cuff, John Troughton, George Bast (alias Bass), Robert Rew, James Burton, and John Bird. The Worshipful Company of Spectacle Makers, of which both the Dollonds were members, contributed £20 towards the cost of the petition. The petitioners complained that John Dollond did not invent the achromatic telescope, but that Hall was the inventor, his glasses being sold in London before 1758. John Dollond had 'permitted them to Enjoy the benefit thereof in Common with himself rather than Risque a Contest with them. . . . But since the death of the said John Dollond his Son and Administrator (under Colour of the said Patent) hath Threatened to bring Actions against your Petitioners and any others of the Trade who shall make and sell the said Glasses. . . . And the said Peter Dollond is now attempting to Establish a Monopoly of the said glasses for his sole Benefit by Virtue or Colour of the said Exclusive Grant.'[41]

In support of their statements, the petitioners produced Bass, 'maker of the aforesaid Glass in the year 1733', and Robert Rew, 'who in the year 1755 Inform'd Mr. John Dollond of this Compound Object-glass.' Nothing came of the petition. In 1766, Peter Dollond sued James Champneys, an optician of Cornhill and one of the subscribers to the petition. Champneys stated that the patent was both unjust and invalid, and Hall showed the Court of Common Pleas some of his achromatic lenses, but the infringement cost Champneys £204 and resulted in his bankruptcy. The Court, however, agreed that Hall was the inventor, but as he summed up the case Lord Camden observed that 'it was not the person who locked his invention in his scritoire that ought to profit by a patent for such invention, but he who brought it forth for the benefit of the public.'[42] Before judgement was given in this case, Peter Dollond had also commenced an action against Francis Watkins and Addison Smith, opticians of St Martin's Lane, and once again the plaintiff was successful. The last action was in 1768, when proceedings were taken against Henry Pyefinch of Cornhill for making telescopes regardless of the patent.[43]

Peter Dollond took great care to urge the claims of his father but, in so doing, he never once recognized the earlier work of Hall. On one occasion, in 1789, his refutation of certain remarks which, he thought, belittled his father's work was read to the Royal Society. He tried to persuade this body to publish it in the *Philosophical Transactions*, but the council probably thought it too controversial. Dollond, perforce, had it printed privately.[44] The derogatory remarks in question were made by Lalande in his *Astronomie*, by Fuss in his eulogy of Euler, and by Cassini in his *Paris Observations* for 1787. In the several editions of Kelly's *Life of John*

Dollond which Peter Dollond financed, Hall is never once mentioned. Indeed, the inventor was forgotten until a letter by an anonymous author who signed himself 'Veritas' appeared in the *Gentleman's Magazine* for 1790, sixty years after Hall's invention. 'Veritas' was obviously in full possession of the facts relating to the work of Hall.

> Mr. Hall [he wrote][45] used to employ the working opticians to grind his lenses; at the same time he furnished them with the radii of the surfaces, not only to correct the different refrangibility of the rays, but also the aberration arising from the spherical figures of the lenses. Old Mr. Bass, who at that time lived in Bridewell precinct, was one of these working opticians, from whom Mr. Hall's invention seems to have been obtained.

This article was well received, but Dollond made no reply. It is not unlikely that 'Veritas' was Ramsden,[46] who married Peter Dollond's sister[47] and from whose letter to the Royal Society we have already quoted. Some years later, de Rochon wrote two inaccurate narratives[48] which purported to stress the claims of Hall over those of Dollond. As late as 1802, another French version appeared in J. É. Montucla's *History of Mathematics*[49] in which Hall is given absolute priority. This account is ascribed to Lalande but, by this time, the patent had become void and Dollond, now head of a flourishing business, had little time for lawsuits and controversy.

Lalande expressed the general feeling on the Continent when he wrote that Dollond first heard of the invention from one of Hall's workmen, and that both Dollond and Hall had obtained their ideas from Euler. For his part, Euler remained suspicious of Dollond's claims. In the numerous papers which he contributed to the Berlin Academy of Sciences on the improvement of optical instruments, he offered explanations of the 'goodness' of Dollond's glasses. His main argument was that, in choosing a fairly small aperture ratio and in correcting for spherical aberration, Dollond had considerably reduced the effects of chromatic aberration.[50] In addition, residual colours were perhaps less significant by the greenish tint of the crown glass which absorbed some red and orange rays.[51] Alternatively, the particular design of Dollond's eyepieces corrected or partly corrected the aberrations of the object-glass.[52] Short's declaration that Dollond's object-glasses were quite achromatic served only to strengthen Euler's belief that, in reducing 'the confusion which is caused by the aperture of glasses . . . the diverse refrangibility of rays scarcely affected the appearance of objects.'[53] It was his firm belief that a single lens, carefully designed and so polished as to remove all microscopic pits, would produce images as fine as those seen through Dollond's glasses.[54]

Euler was reluctant to admit that Dollond's achievement necessitated the modification of his own particular theory of refraction. It was his firm belief that no doublet could be made to correct for chromatic aberration. His experiments with glass-and-water combinations met with no success whatever, but as late as 1761 he was persistently experimenting with fluid lenses.[55] He was convinced that the dispersion by these two media was 'the greatest that Nature offers us' and was sceptical of the superior dispersion by flint glass.[56] It is abundantly clear that he dreaded the overthrow of his theory which was based entirely on mathematical reasoning, 'and which' he wrote in 1761, 'I flattered myself to have brought to a very high degree of accuracy.'[57] It was not until 1764, when Zicher of Petersburg informed him that he had obtained, by the addition of lead, pieces of glass with a dispersive power four times as great as that of other kinds of glass, that Euler renounced all his former opinions.[58]

In the year following John Dollond's death, and when his telescopes were little known in northern Europe, the Petersburg Academy made the correction of the defects of microscopes and telescopes the subject for a prize.[59] The prize essay was contributed by Klingenstierna who, in 1760, published in Swedish a pioneer mathematical theory of achromatic and aplanatic lenses* after he had learned of Dollond's success. The Latin translation of this[60] was published in the *Philosophical Transactions* for 1761,[61] much against Dollond's wishes, while Clairaut translated it for the French Academy. Clairaut repeated Dollond's experiments and noticed that objects seen through a supposedly achromatic lens or prism were surrounded by green and wine-coloured fringes, an effect which he correctly

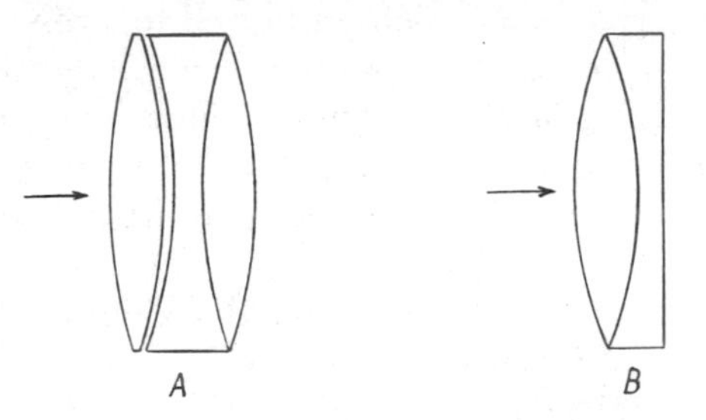

Fig. 69—*A* Peter Dollond's triple object-glass
***B* One of Clairaut's object-glasses**

B is generally called the 'Clairaut objective'. The concave lens is of flint glass in both cases. The curves in the diagrams are exaggerated for clearness.

attributed to the fact that the dispersive powers of the two glasses were not correctly matched. He realized also that crown and flint glass had different partial dispersions† and he showed that the image given by any two-lens combination must inevitably be accompanied by secondary spectrum. To reduce this defect, he made determinations of the refractive indices and dispersive powers of several glasses‡ and deduced theorems for pairing them together. His results appeared in three papers to the French Academy.[62] In the first, he introduced the famous 'Clairaut' objective with its equal interior curves, the radii of which were approximately one-fifth the radius of the front crown surface. Lenses of this type, joined together by Canada Balsam, appear in many modern instruments. In Clairaut's day the two lenses were uncemented and light was reflected from the back and second surfaces to be focused in the focal plane together with the principal image. The second

* 'aplanatic' in the sense of being free from spherical aberration, i.e. not 'aplanatic' in the full Abbe sense of being corrected for spherical aberration and coma. *Vide* also p. 180 and footnote.

† E g. $(n_D - n_A)$ $(n_F - n_D)$, $(n_G - n_C)$.

‡ But with no great accuracy owing to the absence of suitable monochromatic sources and observing instruments (e.g. spectrometers). This was remedied by Fraunhofer about thirty years later (*vide* pp. 180, 185–186).

image or ' ghost ' was quite faint, but unless the lenses were carefully centred, single stars appeared double. Spherical aberration was small; Clairaut could annul it for only one colour and it existed in a small degree for all the others. His second memoir contained several suggestions for other combinations, some with the flint lens in front, others similar to types in use today. In the third memoir, he dealt with the important problem of extra-axial aberrations in double and triple object-glasses. He found that, if he corrected a lens for chromatic and spherical aberration, parallel rays incident on the objective obliquely to the optical axis were not refracted to the focal plane but fell outside it. By tracing many rays of different degrees of obliquity he saw that they focused on a surface of complicated form—a form due to a mixture of radial astigmatism, curvature, and coma, three extra-axial aberrations. Clairaut managed to reduce astigmatism and curvature to within reasonable limits, but he could do nothing for coma and considered it an irremediable evil of two-lens combinations. Fortunately, the restricted actual field of view

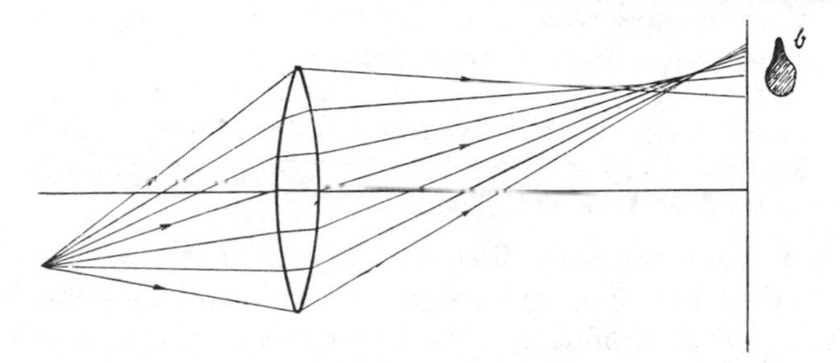

Fig. 70—Coma in a single biconvex lens

The image of a point object placed off the axis is a pear-shaped or comet-shaped patch of light with the greatest concentration of light at *b*. The flare may be directed either towards the axis, as in the figure, or away from it, depending on the form of the lens.

imposed by uncorrected coma is no great drawback in visual telescopic work but, when an extensive field is required, a coma-free *aplanat* is necessary. This is the case when the aperture is relatively large, that is, $f/9$ or $f/10$, as in the case of meridian and photographic telescopes. The formation of a comatic image is illustrated in Fig. 70.

D'Alembert, another French analyst, made similar investigations.[63] He applied his great mathematical skill to the design of triple object-glasses and looked slightingly on Peter Dollond's arrangement ' which it appears he [Dollond] came by only by a process of trial and error.'[64] He prescribed lens forms which, in his opinion, produced better imagery for the same aperture. He distinguished between the ' longitudinal ' and ' transverse ' features of the two axial aberrations (spherical and chromatic) and investigated extra-axial aberrations.[65] He showed, moreover, how to correct a system for spherical aberration for two colours instead of for just yellow. It was left to Gauss many years later to design a lens which would give good definition for more than two colours.

The first achromatic objective to be made in France was probably the one of 3 feet focus which Passement presented to Louis XV in May, 1763.[66] Clairaut had his glasses made by the amateur opticians Anthéaulme, de l'Étang, and Bouriot. 'M. de l'Étang' he writes, 'to whom I owe my first encouragement in optics, by the success my researches have met in his able hands, has also given me the satisfaction of knowing that the objective that I thought the most perfect in theory was the best in practice.'[67] This was a crown-flint doublet 38·3 mm in diameter and 73·8 cm in focal length. Two of de l'Étang's telescopes of $2\frac{1}{2}$ feet focus were fitted to a half-circle invented by the Duc de Chaulnes—an early application of the achromatic telescope to a divided instrument.[68] The Duc de Chaulnes found that his telescopes gave the same magnification as an 8-foot quadrant telescope, and surpassed the latter in definition and size of field. Jean Bernoulli junior thought Chaulnes' instrument rather complicated but conceded that de l'Étang and his colleagues made excellent glasses.

> There are three amateurs in Paris [Jean Bernoulli junior wrote][69] who occupy themselves in making achromatic telescopes and for the reason that they are amateurs and not professional artists, work with infinite care and leave nothing to be desired in their glasses; I speak of l'abbé Bouriot, M. Anthéaulme and de l'Étang; perfect success witnesses to the patience of these gentlemen, and as they are so diligent, friends to whom they have left their glasses have already profited by their efforts; I have heard mention, of telescopes of 10 and 12 feet long made by them and have seen some myself that are quite as good as those of Dollond.

John Dollond was apparently the first to try to correct secondary spectrum by the introduction of a third lens in his object-glasses, but he did not publish his results.[70] R. J. Boscovitch, a Jesuit professor at the University of Padua, entertained similar ideas and devised an instrument which he called the *vitrometrum*[71] for investigating residual dispersions. This consisted of a pair of prisms 'achromatized' after Dollond's plan (*vide* p. 147). On passing sunlight through the vitrometrum, Boscovitch was able to cast a short spectrum on a screen. He then altered the angle of the water prism until the best colour correction was obtained, that is, until the secondary spectrum had the shortest extension. By slightly altering the water angle he was able to over-correct or under-correct a little for colour. He could not, however, calculate the radii for the surfaces of a telescope objective free from secondary spectrum, since it was then impossible to give the refractive index of a medium for a specified colour.

Peter Dollond made the first triple objectives about 1763, and, in the following year, offered them for sale complete with tube and stand. One of his first triplets had an aperture of $3\frac{3}{4}$ inches and a focal length of 42 inches—an aperture ratio of $f/11$. According to Short, it gave an image 'distinctly bright and free from colours' when charged with a magnification of 150.[72] Dollond made many objectives of this size, for the demand was great and he appears to have come across a pot of very fine glass from which he procured a number of disks free from striae and major defects.[73] The Scottish astronomer, J. Ferguson, says that this is a 'vulgar error', the proprietor of the glass-house having assured his friend that the original methods of manufacture of optical glass were still being followed.[74] At any rate, the number of disks obtained from such a pot would not have been great, and Dollond, in common with other opticians, experienced considerable difficulty in obtaining what were then large disks. Had he been able to secure a steady flow of blanks, his triple objectives would have been even more popular, despite the high price he charged for them.

Glassmakers were lucky if they could manufacture disks larger than 4 inches diameter, and they saw to it that such prizes went only to their best customers. The only crown-glassworks of any importance were at Lambeth and Ratcliffe,[75] and the latter were presumably destroyed at the beginning of the nineteenth century during the construction of the London Dock. These factories made flint glass in smaller quantity and were surpassed in dense-glass output by the Russel glassworks in West Street, East London. Jean Bernoulli junior writes[76] that the premises were 'very considerable' although he deplores the fact that only poor-quality glass leaves this and other factories for foreign export. 'Only the opticians which the glassmaker wishes to favour can easily choose the better glass . . . .' The rest was cast into flat but rough-ground blanks into which nobody could see unless he first polished one face and, when the ordinary buyer had made his choice, the remaining blanks were marketed abroad. Dollond's monopoly of the telescope trade did not encourage attempts to remedy this deplorable situation. According to Ferguson, 'the principal opticians'—and he may well have had the Dollonds in mind—'always complain of the bad quality of the glass, but never fail to take the whole quantity he makes at their request; and that, when they renew their orders, they always desire it may be exactly the same as the last.'[77] One consolation, if such it can be called, was that, before the beginning of the nineteenth century, conditions abroad were even worse, despite the efforts of Colbert and J. D. Cassini junior to encourage research on optical glass and to establish a permanent and national glassworks.

Maskelyne was among the first to buy one of the new $3\frac{3}{4}$-inch telescopes, and he was so pleased with it that he mounted it in a small room made especially for the purpose. Another triple achromat, of 46 inches focus, was purchased by the Hon. T. Beauclerc from whom it passed to Aubert. Ramsden made all the mechanical parts; Dollond made the objective, which he told Kitchener was 'one of the things which is to make me immortal.'[78] Kitchener bought this instrument at the sale of Aubert's equipment, in 1806, and spoke of it as 'one of those perfect Instruments which are rarely produced, and only attainable, by a happy concurrence of the various circumstances which combine to form these compound Object-glasses.'[79]

> Since Mr. Beauclerc's 46-inch telescope has been in my possession [he wrote in 1818][80] I have had opportunities of carefully and attentively comparing it with nine achromatics of 5 feet focus, with double object-glasses of $3\frac{3}{4}$ inch in the clear aperture, with three 7 feet of 4 inches aperture, and a 10 feet of 4 inches aperture; and when the test was a star, the 46-inch has always been acknowledged to be the more perfect instrument, as it showed every thing the others did; and with it, some delicately minute objects could be easily discerned, which some of its competitors were not perfect enough to define at all. The superiority was most manifest when the instruments were turned to double and coloured stars.

Such was Aubert's ample fortune that he possessed four telescopes of this size and type, and his observatory contained, besides, thirty-three other instruments. His favourite telescope was not the Beauclerc refractor but another triple achromat of the same size, which also passed to Kitchener. Darquier of Toulouse had another triple objective of $3\frac{3}{4}$ inches aperture and 42 inches focus. 'This telescope' he wrote in 1777, 'is entirely of brass with a small star-finder fixed on the top; . . . it is of admirable clearness and definition; I have often seen four of Saturn's satellites with a high magnification, and sometimes five, but seldom.'[81]

From these and other statements it is apparent that Dollond's triple objectives were of higher quality than anything previously produced. The third lens not only reduced secondary spectrum, but it gave better correction for spherical aberration. The latter, however, can be completely corrected by only two lenses—as Fraunhofer showed in the next century—but Dollond was not aware of this.* Much was expected of his glasses, both triplets and doublets. Astronomers wished to see Sir William Herschel's double stars for themselves and they began to assess the performance of their telescopes by the ease with which they resolved these objects. This meant that the corrections had to be more rigorous and the lenses more carefully matched, mounted, and centred than hitherto. This Dollond seems to have achieved, for his triple achromats easily resolved double stars like ϵ and ι Bootis (2″·6 and 2″·9 separation†), γ Leonis (2″·5), and η Coronae Borealis (0″·9), and took powers of 350, or 90 to the inch of aperture, without 'breaking down'. One was compared with Huygens' 123-foot refractor, to the great advantage of the smaller instrument, and Maskelyne found that his instrument performed as well as a double achromat of over twice the length. Kitchener possessed a 5-foot double achromatic of 3·8 inches aperture which was 'one of the chef d'œuvres of the late Mr. Peter Dollond, . . . it shews the disc of the Moon and of Jupiter as white and as free from colour as a Reflector;—to its perfect Achromatism I attribute in great measure its power of very distinctly shewing the division in the ring of Saturn.'[82] Dollond once said of this instrument: 'It is one of the best I ever made, and such as I cannot expect to be able to equal.'[83] His largest telescope was a 5-inch double achromat of 10 feet focus.[84] It was, for a short time, the largest instrument of its type and was erected at the Royal Observatory, Greenwich, for observations of the eclipses of Jupiter's satellites and occultations of faint stars by the moon. A more important glass was the $4\frac{1}{2}$-inch diameter ($3\frac{3}{4}$ inches clear) doublet which Peter Dollond made in 1797 for Captain Huddart.[85] This was elaborately mounted in brasswork by Troughton and was used by John Herschel and James South in their classic double-star investigations.‡

We know little of Peter Dollond's method of working his glasses. This information, being a trade secret, he kept to himself and his chosen employees. Jean Bernoulli junior, who visited Dollond's workshop in 1769, received the impression that Peter Dollond did not possess his father's theoretical knowledge.[86] He stated that Dollond worked his objectives, even those with three lenses, on the most elementary principles, and that he depended for his success on methods of trial and error. He apparently made several lenses of crown and flint glass at a time, and then used those combinations which yielded the best results. This procedure, with large disks so expensive and difficult to procure, would have hardly recompensed him for his labours or have produced such excellent results, and we must accord him a large

* Or of the fact that the secondary spectrum can be kept to a minimum if the condition (Minimum focal length) = 5 (Aperture)^2 is observed.

† All for epoch 1830.

‡ This instrument is now in the Science Museum, London.

measure of practical skill. Kitchener, who knew Peter Dollond well, called him the 'Father of Practical Optics' and stated that, for practical work, Dollond had 'as good and as well-educated an Eye as perhaps ever Man had.'[87]

In 1766, Peter Dollond moved to more commodious premises at 59 St Paul's Churchyard, where he was joined by his brother John. In 1783, the Dollonds began to provide their telescopes with brass draw-tubes.[88] They had been the first to use brass-bound mahogany tubes, a great improvement on the old paper-covered vellum tubes. Joshua L. Martin was the first to use plated brass draw-tubes, which he patented,[89] and Peter Dollond probably bought the patent from him. The process was first to place the soldered or brazed tube on a mandril of diameter the same as that required for the interior of the tube; the latter was then drawn on the mandril through holes of smaller and smaller diameter until the right thickness was obtained. With these tubes, the Dollonds offered a wide range of sturdy stands. Small glasses were mounted on table stands of the folding pillar and claw type, large ones up to 8 feet focus on folding mahogany tripods with slow motions in altitude and azimuth or, if preferred, in declination and right ascension. Prices ranged from £2 for a 1-foot telescope to over £70 for a 5-foot.

Dollond created and fostered the public demand for portable telescopes, being helped by the requirements of the army and navy during the Napoleonic Wars. About 1780, he introduced an 'Army telescope' with mahogany brass-bound body and brass collapsible tubes. The length of these 'Improved Achromatic Telescopes', as they were advertised, ranged when open from 14 to 52 inches, the objectives from 1 to 2¾ inches aperture, and the prices from 2½ to 12 guineas. Dollond also advertised 'Achromatic Pocket Perspective Glasses', night glasses, portable equatorials, object-glass micrometers, small reflecting telescopes, compound and 'solar' microscopes, and glass prisms 'arranged to demonstrate the principle of the achromatic objective.' In addition, sextants, octants, cameras obscura, reflecting circles, theodolites, dynameters, and spectacles, were being made by the house of Dollond at the close of the eighteenth century. Dollond took advantage of W. H. Wollaston's periscopic spectacle lens, deliberately altering the curves so as to circumvent the patent.[90] He also gave considerable attention to microscopes and made several useful additions to the Cuff instrument, one of the best of the period. Many examples of his work, including one of his solar microscopes, can be seen in the Court Collection in the Science Museum, London.

The patent for achromatic telescopes became void in 1772, and at this time most of the London opticians lost their animosity towards Dollond—at least he became the centre of a large circle of scientific friends. He was, for many years, a member of the American Philosophical Society and, in 1774, became Master of the Worshipful Company of Spectacle Makers, a post which he held in 1778 and again in 1779.[91] He took a great interest in the French Hospital which was founded early in the eighteenth century and of which, in 1794, he was elected director. He left his workshop at 59 St Paul's Churchyard, during his eighty-seventh year, retiring to a stately residence on Richmond Hill. His retirement after many years of business activity was destined to be of short duration for, on July 2, 1820, he died, only a few days after he had moved to a new house at Kennington. The business thereupon passed to his nephew, George Huggins, later known as George Dollond.

Bernoulli was not particularly impressed by Dollond's larger astronomical instruments.

> You make a great mistake [he wrote][92] when you imagine that an astronomical instrument bearing the name Dollond must be excellent in every feature. If you receive one which has this quality, it is a sign that it has not been finished by one of Dollond's workmen: often to maintain his reputation he has the mountings and divisions made for him by his brother-in-law, M. Ramsden, who passes for one of the best artists for this work in London.

Of the instrument-makers of this period, Jesse Ramsden, a great-nephew of Abraham Sharp, was undoubtedly the most capable. He started life as a clerk in a cloth warehouse and then became apprenticed to Burton, an instrument-maker in Denmark Street, Strand.[93] In 1762, he opened on his own in the Haymarket and, three years later, married John Dollond's daughter, Sarah, receiving as part of the marriage settlement a share in the patent for manufacturing achromatic lenses.[94]

Ramsden first turned his attention to the improvement of Short's portable equatorial mounting. This he provided, for the first time, with circles and counterpoises of correct size and weight[95] and, by 1773, had made at least four improved instruments. One was bought by the Earl of Bute, one by Sir Joseph Banks, a third by S. McKenzie and the fourth by Sir George Shuckburgh, who used it for surveys in France and Italy.[96] The telescopes were triple achromatics of 2 inches aperture and 15 inches focal length, and were probably made by Peter Dollond. The stands were characteristic of all subsequent work executed by Ramsden and his workmen—original, well designed, compact, and graduated in the most perfect manner. In 1775, Ramsden moved to larger workshops in Piccadilly, where he concentrated his efforts on a problem which had occupied his attention for a number of years. We have seen how Bird divided arcs with considerable accuracy. Where Bird worked to limits of 3 seconds of arc, Ramsden with his engines went to lower than $0''\cdot5$. He made his first dividing machine in 1766, but this, while it gave more accurate divisions than the ordinary plate method, was not delicate enough for the graduation of astronomical circles.[97] A superior instrument was made in 1775,[98] and a sextant graduated by it earned the warm approval of Bird. The circle or arc to be graduated was first clamped to a 45-inch diameter horizontal wheel. One downward stroke of a treadle turned this wheel through 10 minutes of arc and a division line was then cut by hand, the cutting point being carried in a radial frame. A brass plate on the screw arbor was further divided into 60 parts, so that Ramsden could carry the graduations down to every 10 seconds of arc. For this invention he received £615 from the Commissioners of Longitude, on condition that he published an account of it and divided sextants and octants at the rate of six shillings per sextant and three shillings per octant. In consequence, the Commissioners published a *Description of an Engine for Dividing Mathematical Instruments* in 1777, and Ramsden issued his own *Description of an Engine for Dividing Straight Lines on Mathematical Instruments* two years later. His second engine was used continuously until his death in 1800, and then by his successors to as late as 1890.[99]

In 1779, also, Ramsden sent a paper to the Royal Society through Sir Joseph Banks, the President, entitled *A Description of two New Micrometers*.[100] A study of the Short–Dollond Gregorian micrometer led him to a simpler and, theoretically,

Fig. 71 Astronomical equatorial instrument by Ramsden

(*Science Museum, London. British Crown copyright*)

more accurate double-image device—a special Cassegrain telescope. His proposal was to duplicate the image by bisecting the small secondary mirror. Rotation of the two halves on a common axis perpendicular to the optical axis of the main mirror would then give the images the necessary angular separation, the amount of rotation being a measure of this effect in the focal plane. The arrangement thus had the obvious advantage of requiring no additional optical elements. Ramsden called it a *catoptric* micrometer to distinguish it from his second and generally more popular device. This was designed for use with refracting telescopes and consisted of two semi-lenses placed in the focal plane of the eyepiece. The semi-lenses slid laterally against each other and so duplicated the image produced by the object-glass. ' It has been before observed ' Ramsden writes, ' that if a micrometer is applied at the object glass, the imperfections of its glass are magnified by the whole power of the telescope; but in *this* position [i.e. just in front of the eyepiece], the image being considerably magnified before it comes to the micrometer, any imperfection in its glass will be magnified only by the remaining eye glasses which in any telescope seldom exceeds five or six times.'[101]

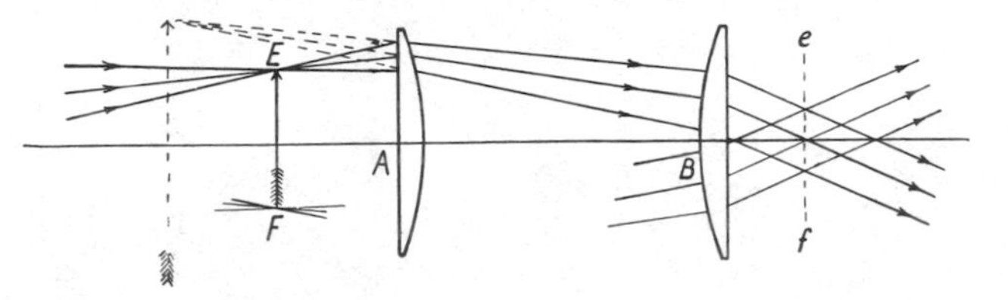

Fig. 72—The Ramsden eyepiece

The telescope is pointed to clear sky and the object-glass forms an image *EF* in its focal plane. *A* is the field-lens, and *B* the eye-lens, of the eyepiece combination. The Ramsden disk occurs in the plane *ef*.

Ramsden's preference for the Cassegrain over the Gregorian was due to the fact that the aberration of the large mirror is partly corrected by the aberration of the small one. In the Gregorian, the small mirror increases the aberration and, in Ramsden's opinion, the residual aberrations of the two instruments were in magnitude as 3 to 5. Yet, with this apparent advantage, Ramsden's reflecting micrometer met with little success owing, mainly, to difficulties of construction and the ease with which different parts got out of adjustment.

A second paper by Ramsden concerned *A Description of a new Construction of Eye-glasses for such Telescopes as may be applied to Mathematical Instruments.*[102] His eyepiece was an attempt to produce a flat achromatic field* for viewing micrometer wires and, as such, is characteristic of his attention to detail. The term ' Ramsden

* The Ramsden eyepiece consists of two separated plano-convex lenses of equal power arranged with their plano surfaces on the outside. For transverse achromatism the lenses should be separated by a distance equal to half the sum of the focal lengths; but this would mean that the field-lens, and every speck of dust on it, would be in the anterior focal plane of the eye-lens. In practice, therefore, and at the expense of full chromatic correction, the lenses are separated by about two-thirds of this distance. The eyepiece is termed positive because the anterior focal plane of the system is in front of the field-lens; the usable field is about 40 degrees in angular extent. Kellner and Steinheil subsequently improved this basic form (*vide* p. 185 footnote).

disk ' does not appear to have originated with Ramsden, although he was the first to realize the importance of the size and position of this area in telescopic vision.

> It has been usual [he writes][103] to consider that form and position of the eye-glasses best that would make the pencils from every part of the field intersect each other in the axis of the telescope at the place of the eye; but this will be found of little consequence, seeing the diameter of a pencil here is generally much less than the pupil, nothing more is requisite than that the eye may take in the pencils from different parts of the field at the same time: but the field of a telescope will be most perfect when the construction of the eye-glasses is such, that the focus of an extreme and of a central pencil are at the same distance from the eye.

The Ramsden disk, or exit-pupil as it is generally called, is the image of the objective formed by the eyepiece. Its size is therefore obtained by dividing the diameter of the objective by the apparent magnification. Alternatively, knowing the

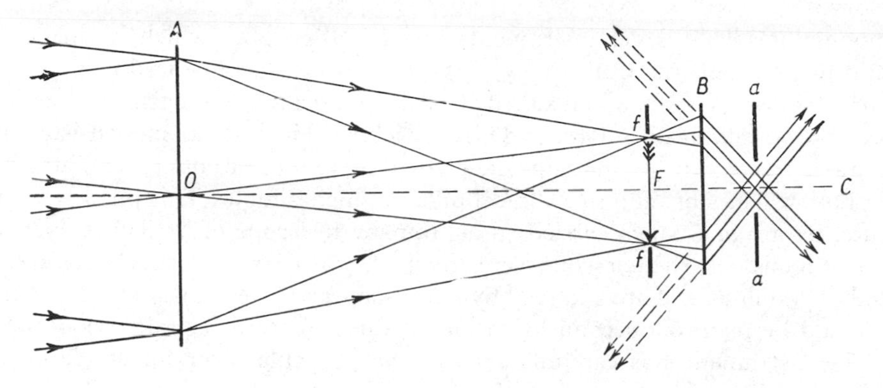

Fig. 73—The Ramsden disk

Groups of parallel rays leaving the eyepiece *B* of an astronomical telescope all meet in the plane *aa*. The Ramsden disk is this area of greatest ray constriction. *A* is the objective, *ff* the field stop, and *OC* the optical axis. The pupil of the observer's eye must be placed at *aa* to embrace all rays leaving the instrument.

diameter of the objective and exit-pupil, one can, by simple division, obtain the apparent magnification. Ramsden recognized this and, about 1775, invented a dynameter or special eyepiece micrometer for measuring with accuracy the diameters of the exit-pupils of telescopes.[104]

Ramsden was one of the first to apply his positive compound eyepiece to the microscope as a means of reading circle graduations. The Duc de Chaulnes had applied microscopes to the dividing engine which he described in 1768,[105] and Römer, J. L. Martin, and Louville[106] used them on the graduated circles of their instruments. According to Francis Wollaston,[107] Ramsden mentioned the idea of reading-off divisions by a micrometer microscope ' one evening, at a meeting of our Society in the beginning of 1787.' At any rate, after this date this application of the new positive eyepiece appears on all Ramsden's theodolites.

Ramsden made many large instruments for observatories in this country and abroad and, at one period, the demand was so great that he had fifty workmen in his employ.[108] The demand for graduated instruments continued to grow, and Ramsden, in common with Bird, the Dollonds, and Short, was never idle for want of orders. Many new public observatories were being founded at this time, while many individuals, inspired by Sir William Herschel's discoveries, built their own private observatories. In existing institutions like the Greenwich, Oxford, and Paris Observatories, the advantages of achromatic telescopes over long, single-lens refractors called for a complete change in instrumental equipment. Add to this activity the urgent need for achromatics for military and naval uses during the Napoleonic Wars, and it becomes difficult to imagine how these opticians were able even to satisfy the demand. Yet all through these busy years, English instruments underwent great improvements and, in accuracy, generally surpassed Continental makes. Indeed, it would be no exaggeration to say that instruments by Bird, Short, the Dollonds, or Ramsden, were to be found in every up-to-date European observatory.

Among Ramsden's larger instruments was the theodolite which he made for General Roy's trigonometrical survey of the British Isles commenced in 1791, and the earlier survey of 1787, which aimed at establishing a trigonometrical connection between the observatories of Paris and Greenwich.[109] The instrument consisted of a large brass circle 36 inches in diameter, divided to 10 minutes of arc, and attached by ten hollow cones or radii to a conical pillar 24 inches high. This pillar was surmounted by a frame which carried an achromatic telescope of 2½ inches aperture and 3 feet focus. To the telescope were attached two 10½-inch vertical circles, each divided to 30 minutes of arc and read by microscopes to 3 seconds. The horizontal circle could be read to 1 second by means of three micrometer microscopes and, when the instrument was carefully set up, the probable error for single observations was as little as 2 seconds of arc for distances up to 70 miles. The entire instrument consisted of sixty-six main parts and could be accommodated in four cases for ease of transport.[110]

Roy commenced his operations in 1784 with the measurement of a base line on Hounslow Heath with standard glass rods made by Ramsden.[111] In 1795, the distance was re-measured with a Ramsden steel chain and found to be only 2¾ inches greater than before.[112] Ramsden received the order for the theodolite in 1784 and, to the consternation of Roy, took three years over its execution.

> In the spring of 1787 [Roy writes][113] there were indeed appearances, that Mr. Ramsden would have enabled us to embrace the early part of the season, by proceeding with the execution of the main design. . . . For several months of the spring and summer of 1787, Mr. Ramsden had been seriously at work in endeavouring to finish the instrument. Not having employed a sufficient number of workmen upon it at the outset, it was now evident that he had even deceived himself by leaving too much to be done at the latter end.

Later, Roy writes about ' necessary improvements, which might have been executed in a short space of time. With these alterations the instrument was at last returned, but so late, that it could not be placed on Goudhurst Steeple until the 9th of August, 1788.'[114] A prism tube for zenith observations, promised for 1787, was not obtained until the following year and then ' with great difficulty '.

Another large theodolite-type instrument was the 5-foot vertical circle[115] which

Fig. 74—Ramsden's Palermo altitude and azimuth circle

(*From Pearson's 'Introduction to Practical Astronomy'*, 1829)

Fig. 75—Jesse Ramsden

He is seated beside his dividing engine. The Palermo circle is in the background.

Ramsden made for Piazzi's new observatory at Palermo. Both Ramsden and Piazzi worked at this instrument, twice abandoned before its completion in August, 1789, The horizontal circle gave readings in azimuth read by a micrometer microscope, and the vertical, readings in altitude by two diametrically opposed microscopes. The divisions on the circles were illuminated by an inclined silver mirror fixed to each microscope, and the wires in the telescope eyepiece by transmitting light from a small lamp through the hollow tube-axis—two Ramsden innovations. This instrument, the finest complete circle hitherto made, formed the basis of Piazzi's astronomical work. With it, he catalogued nearly 8000 stars, compiled a valuable table of refractions, and attempted to measure the parallax of several bright stars. The great labour involved in the catalogue is the more remarkable when we consider that each star was observed several times before its position was decided.

Encouraged by the success of the Palermo circle, Ramsden now planned grander instruments and was in this mood when he met Dr Ussher. Ussher was one of the original founders of the Dublin Observatory and, as such, was anxious to provide the new institution with a telescope and circle of the largest possible dimensions. Ramsden entered wholeheartedly into the project and designed a circle 10 feet in diameter which met with the full approval of the observatory board. But as the instrument progressed, so difficulties arose and Ramsden had to reduce the circle to 9 feet and then to 8 feet.[116] These changes caused a delay of seven years. Ussher died in 1790, and, despite the assurances and promises of Ramsden, yet another seven years passed and Dublin, or more precisely the Dunsink Observatory, was still without its instrument. By 1799, Ramsden had fallen into ill health and the board threatened proceedings; but all its inquiries, pleadings and threats were to no avail, for Ramsden died in the following year.[117] The board had now apparently lost both instrument and money and, as a last resource, the provost appealed to Maskelyne for advice. Maskelyne replied that Ramsden had not died intestate and that the business was in the hands of Matthew Berge who was finishing the instrument. But Berge was as deliberate as his master had been and, although the instrument was promised for 1807, it was not erected at Dunsink until the next year, about twenty-three years after the date of the original order.[118]

Ramsden was notorious for his disregard of time. Although a brilliant and thorough workman when the occasion suited, he had the inconsistencies usually associated with genius. Only Piazzi's constant supervision of the 5-foot circle, and his presence in the Piccadilly workshop, spurred Ramsden to complete the instrument on the very day for which he had promised it, but in the wrong year! On another occasion Ramsden attended, so the story goes, at Buckingham Palace precisely, he supposed, at the time stated in a royal mandate. But he was, as the King remarked, punctual to the day and hour while late a whole year. Ramsden had, apparently, mislaid and forgotten the King's memorandum until it accidentally turned up a year later.[119]

Ramsden constructed several large portable zenith sectors for various trigonometrical surveys but they were, owing to their great size, difficult both to transport and to adjust. His largest sector was made for a survey of England, directed by the Duke of Richmond, Master General of the Ordnance. But Ramsden did not live to complete the instrument and it was finished by Berge in 1802. In the official account of the survey, it is stated that 'Mr. Ramsden has obviated the inconveniences attendant on the use of former sectors; and has also diminished in a very considerable degree the errors unavoidably arising from their imperfect construction.'[120] After Colby's survey of Scotland, the instrument was placed for safety in the Tower of London, where it was destroyed in the fire of 1841.[121] Sir George Everest used a similar Ramsden instrument for his India survey.

Ramsden was elected Fellow of the Royal Society on January 12, 1786, and, in 1795, received the award of the Copley Medal for his 'various inventions and improvements in philosophical instruments.'[122] He made new instruments of the sextant and theodolite, devising new methods for their mounting, counterpoising, illumination, and testing. Besides single telescopes fitted with achromatic object-glasses of his own manufacture, he made improved barometers, pyrometers, standard lengths, surveying chains, balances, electrical machines, levels, manometers, and drawing instruments. He seldom enjoyed good health, which his constant application to work did much to weaken. His habits were simple; he studied far more than he ate and slept, and when engaged on particular work, disregarded all temporal and material demands.[123] His profits, what they were, went to further the perfection of his instruments and the comfort of his workmen. He studied, not how cheap or how dear to make an instrument, but rather how best to improve it.[124] In this way, and with a politeness, modesty, and reserve not always found in business men, he earned the goodwill of all who had contact with him.

Kitchener, in his book on telescopes, gives the following intimate and often quoted anecdote of Ramsden's home life:[125]

> It was his custom to retire in the evening to what he considered the most comfortable corner in the house, and take his seat close to the kitchen fire-side, in order to draw some plan for the forming of a new instrument, or scheme for the improvement of one already made. There, with his drawing implements on a table before him, a cat sitting on one side, and a certain portion of bread, butter, and a small mug of porter placed on the other, while four or five apprentices commonly made up the circle, he amused himself with whistling the favourite air or sometimes singing . . . and appeared, in this domestic group, contentedly happy. When he occasionally sent for a workman, to give him necessary directions concerning what he wished to have done, he first showed the recent finished plan, then explained the different parts of it, and generally concluded by saying, with the greatest good humour, ' Now, see man, let us try to find fault with it! '

In person Ramsden was ' above the middle size, slender, extremely well made, and to a late period of life, possessed great activity. His countenance was a faithful index of his mind, full of intelligence and sweetness. His forehead was open and high, with a very projecting and expansive brow. His eyes were dark hazel, sparkling with animation.'[126] The truth of this description is in part shown in Robert Home's portrait of Ramsden, engraved by Jones in 1791 (Fig. 75). Ramsden is shown seated before the Palermo circle, clad in a fur coat, introduced to commemorate an order that he executed for the Emperor of Russia. Ramsden was never very pleased with this representation for, although he was elected member of the Imperial Academy of St Petersburg in 1794, he never wore a Russian fur in his life.

In England, Ramsden's only serious rivals, if such they can be called, were Edward Nairne and William Cary. Nairne turned to a great variety of ' philosophical apparatus '—barometers, air-pumps, levels, manometers, thermometers, electrical machines, and small portable equatorials based on Short's instrument. He was born in 1726, apprenticed to one Matthew Loft at the age of fifteen, made F.R.S. in 1776 and, in 1797, was Master of the Spectacle Makers Company.[127] He apprenticed Thomas Blunt and later took him into partnership at 20 Cornhill. In 1794, Blunt opened his own workshop at 22 Cornhill and, like Nairne at No. 20, traded under his own name. Nairne retired in 1801; Blunt died on March 16, 1823, and his business was taken over by Thomas Harris.[128]

William Cary served his apprenticeship under Ramsden and, besides larger instruments, made globes, rules, microscopes, clinometers, compasses, and transit theodolites. He appears to have first set up in business at 272 Strand, in 1790, but later directories refer to J. and W. Cary of 181 Strand. It was to William Cary that David Brewster, in 1805, first communicated the idea of a pancratic or variable-power eyepiece, a device which Kitchener brought out later as a new invention.[129]

Fig. 76—Ramsden's Shuckburgh equatorial

(*By courtesy of the Director of the Science Museum, London*)

Among Cary's larger instruments was an impressive altazimuth circle made for Colonel Beaufoy,[130] and a transit instrument[131] made in 1805 for the Moscow Observatory. Of greater importance, however, was the transit circle made in 1793 to Francis Wollaston's design[132]—the first instrument of its kind to be made in England. Wollaston already possessed a transit instrument and conceived the idea of attaching a graduated circle to it so that he could measure altitudes. He laid the idea before Ramsden, but the latter was too busy with other work at the time. Troughton was similarly placed, so he had recourse to Cary who, with minimum delay, made and set up the instrument at Wollaston's private observatory at Chislehurst, Kent.

With the intention of improving his health, Ramsden moved to Brighton, but died soon afterwards on November 5, 1800.[133] His only son, John, was a captain in the East India Mercantile Marine and the business passed to Matthew Berge, Ramsden's head workman. It seems likely that Berge was either son or brother to one John Berge who was apprenticed to the Dollonds and who remained in their employ for thirty-five years until 1791.[134] After that date, John Berge appears to have opened his own business in Johnson's Court, Fleet Street.[135] Matthew Berge's first responsibility was to complete Ramsden's unfinished instruments—a task which, as we have seen, he undertook in his master's leisurely way.[136] His work was accurate, however, and when he completed a large 4·2-inch aperture English-type equatorial for Sir George Shuckburgh, the latter wrote:[137] 'I think I am entitled to believe that the accuracy of these divisions under consideration is hardly to be equalled, and still less to be excelled, by that of any astronomical instrument in Europe; and, from the unexampled diligence and care, with which the skilful artist Mr. Matthew Berge, workman to Mr. Ramsden, has executed them, I feel myself bound to bear this testimony to his merit.'

Following Berge's demise, the famous Piccadilly workshops were bought by a Mr Worthington[138] and then passed into obscurity. One would have thought that either Blunt or Cary would have taken advantage of this respite to consolidate his position, but the lead was snatched from them, first by Thomas Jones, and then by Edward Troughton.

Little is known about Jones save that he worked for Ramsden from 1789 (when only fourteen years old), was a Fellow of the Royal Society and a member of the Institute of Civil Engineers. He assisted Pearson, Troughton, Smyth and others in the formation of the Astronomical Society [139] and, at his workshop at 62 Charing Cross and later in Rupert Street, constructed large transits, mural circles and the usual run of surveying instruments. A transit instrument made by him for W. H. Smyth is shown in Fig. 77. Of his larger work should be mentioned 6-foot mural circles for Greenwich, Oxford, and Cambridge, a 5-foot equatorial for Cambridge, transit circles for Paramatta and Oxford, and a 10-foot transit, 6-foot mural circle, and 5-foot equatorial for Cadiz. Jones died July 29, 1852.[140]

Fig. 77—The transit room at W. H. Smyth's observatory, Hartwell, Buckinghamshire, in 1832

The transit instrument was made by Thomas Jones of Charing Cross. The clock, with Graham dead-beat escapement, was by B. L. Vulliamy of Pall Mall.

(*From W. H. Smyth's 'Speculum Hartwellianum'*, 1860)

REFERENCES

[1] Gregory, D., 1735, *Elements of Catoptricks and Dioptricks*, 2nd edition, pp. 110–111.
[2] *Gentleman's Magazine*, **60**, ii, p. 890, 1790.
[3] *The Description and Use of a New Instrument or Sea Quadrant*, London, about 1735, 40 pp. Quoted by Court, T. H., and Rohr, M. von, *Trans. Opt. Soc.*, **30**, p. 228, 1929.
[4] Ramsden, J., *Some Observations on the Invention of Achromatic Telescopes*, MS. Letter No. 138, Royal Society's *Record Book.*
[5] *Ibid.*
[6] *Ibid.*
[7] *Ibid.*
[8] *Ibid.*
[9] Kelly, J., 1808, *The Life of John Dollond*... 3rd edition, p. 7.
[10] Euler, L., *Mem. Acad. Berlin*, pp. 274–296, 1747.
[11] *Phil. Trans.*, **48**, i, pp. 289–90, 1754.
[12] *Ibid.*, p. 294.
[13] *Ibid.*
[14] Kelly, *op. cit.*, p. 70. Kelly, pp. 68–70, gives a translation of this paper. Another copy is in the *Record Book* of the Royal Society, London.
[15] *Phil. Trans.*, **50**, ii, pp. 733–743, 1758. This paper was read June 8, 1758.
[16] *Ibid.*, pp. 735–736.
[17] *Ibid.*, p. 738.
[18] *Ibid.*, p. 740.
[19] *Ibid.*, p. 742.
[20] *Ibid.*, p. 735.
[21] *European Magazine*, **78**, ii, p. 99, 1820.
[22] Kelly, *op. cit.*, p. 8. Short, J. Cuff, Burton and J. Smeaton all had workshops in this district.
[23] *Phil. Trans.*, **48**, i, pp. 103–107, 1754.
[24] Kelly, *op. cit.*, pp. 11, 18.
[25] Prosser, R. B., *The Optician*, **53**, p. 203, 1917, quotes from Privy Council Papers at Public Record Office, London (P.C. 1/7, Bundle 37).
[26] *Vide* Ref. 21, p. 100.
[27] Court, T. H., and Rohr, M. von, quote in *Trans. Opt. Soc.*, **30**, p. 229, 1929. Price List of G. Adams for 1766 (p. 231) and 1771 (p. 231). For Price Lists Benjamin Martin for 1757 and 1758, *vide* Court Collection, Science Museum, London.
[28] Bernoulli, Jean, junior, 1771, *Lettres Astronomiques*, p. 71, quotes from the List of H. Pyefinch.
[29] Smyth, W. H., 1844, *A Cycle of Celestial Objects*, i, p. 370, footnote.
[30] Darquier, A., 1777, *Observations Astronomiques faites à Toulouse*, Preface, p.v.
[31] *Phil. Trans.*, **48**, ii, pp. 551–564, 1754.
[32] Horrebow, P., 1735, *Basis Astronomiae* . . ., p. 88, p. 97.
[33] *Phil. Trans.*, **48**, i, p. 165, 1754.
[34] *Mém. Acad. Sc.*, p. 15, 1748.
[35] *Vide* Ref. 31, p. 554.
[36] *Phil. Trans.*, **67**, ii, p. 799, 1777.
[37] Pearson, W., 1829, *Introduction to Practical Astronomy*, ii, p. 98.
[38] Bailly, J. S., 1785, *Histoire de l'Astronomie Moderne*, iii, p. 125.
[39] *Vide* Ref. 21, p. 100.
[40] *Vide* Ref. 25.
[41] *Ibid.*
[42] *Vide* Ref. 2, p. 891.
[43] Ranyard, C., *M.N.R.A.S.*, **46**, p. 460, 1886.
[44] Dollond, P., *Some Account of the Discovery* . . ., read to the Royal Society, May 21, 1789, afterwards published by J. Johnson, 4to Kelly, *op. cit.*, quotes it in full, pp. 61–77.
[45] *Vide* Ref. 2, pp. 890–891.
[46] Tulloch, *Phil. Mag.*, **2**, p. 177, 1798, who states ' which, we have been informed by Mr. Ramsden, contains a true statement of the facts '—in referring to this article by ' Veritas '.
[47] *Gentleman's Magazine*, **70**, ii, p. 1116, 1800.
[48] *Phil. Mag.*, **2**, pp. 19–27, 170–177, 1798. *Nicholson's Journal*, **4**, pp. 110 119, 1803.
[49] Montucla, J. É., 1802, *Histoire des Mathématiques*, Nouv. edition, iii, p. 448.
[50] Euler, L., *Mem. Acad. Berlin*, pp. 245, 260, 1762.
[51] *Ibid.*, pp. 245, 234.
[52] *Ibid.*, p. 245.
[53] *Ibid.*, p. 260.
[54] *Ibid.*, pp. 120–121.
[55] *Ibid.*, pp. 231–40, 1761.
[56] *Ibid.*, p. 227, 1762.
[57] *Ibid.*
[58] Littrow, J. J., *Mem. R.A.S.*, **4**, pp. 491–492, 1830.
[59] Montucla, *op. cit.*, pp. 455–456.
[60] *Phil. Trans.*, **51**, p. 944 ff., 1761.
[61] *Ibid.*
[62] *Mém. Acad. Sc.*, 1761, 1762, 1764. Clairaut's papers are dated 1756, p. 127; 1757, p. 153; 1762, p. 160.
[63] Alembert, J. le Rond D', *Opuscules Mathématiques*, iii, 1764; iv, 1768. *Vide* also *Mém. Acad. Sc.*, 1764, 1765, 1767.
[64] *Mém. Acad. Sc.*, p. 102, 1764.
[65] *Ibid.*, p. 165, 1762.
[66] Passement, Cl. S., 1763, *Description et usage des téléscopes, microscopes, ouvrages et inventions.*
[67] *Vide* Ref. 65.
[68] Montucla, *op. cit.*, iii, p. 471.

69 Bernoulli, Jean, junior, 1771, *Lettres Astronomiques*, letter 14, pp. 172–173.
70 *Phil. Trans.*, **55**, p. 55, 1765.
71 Boscovitch, R. J., 1785, *Opera pertinentia ad opticam et astronomiam ex parte nova . . .*, i, pp. 137–172.
72 *Vide* Ref. 70, p. 54.
73 *A Companion to the Telescope*, anon., 1811. *Vide* also Kitchener, W., 1825, *The Economy of the Eyes*, ii, p. 16.
74 Ferguson, J., 1837, *Lectures on Select Subjects*, p. 262.
75 Lysons, S., 1795, *Environs of London*, iii, p. 473. Rees, A., 1819, *Cyclopaedia*, Art. Glass.
76 Bernoulli, *op. cit.*, p. 67.
77 Ferguson, *op. cit.*, p. 262.
78 Kitchener, *op. cit.*, p. 26.
79 *Ibid.*, p. 27.
80 Kitchener, W., 1818, *Practical Observations on Telescopes*, p. 109 ff.
81 Darquier, *op. cit.*, p. 5.
82 Kitchener, Ref. 73, i, pp. 68–69.
83 *Ibid.*, ii, p. 48.
84 *Ibid.*, p. 17.
85 *Ibid.*, p. 375; Baxandall, D., *Trans. Opt. Soc.*, **24**, p. 317, 1923.
86 Bernoulli, *op. cit.*, p. 65.
87 Kitchener, Ref. 73, ii, p. 17, p. 427.
88 Court Collection, Price List of P. Dollond. Also mentioned by Court, T. H., and Rohr, M. von, *Trans. Opt. Soc.*, **30**, p. 242, 1929.
89 Martin, J. L., *New-invented art of drawing tubes . . .* Brit. Patent No. 1316, May, 1782.
90 *Trans. Opt. Soc.*, **30**, p. 247, 1929.
91 Court, T. H., and Rohr, M. von, *Trans. Opt. Soc.*, **31**, p. 82, 1930.
92 Bernoulli, *op. cit.*, pp. 68–69.
93 Hutton, C., 1815, *Mathematical and Philosophical Dictionary*, p. 283; *Dict. Nat. Biog.*, art. ' Ramsden '.
94 *Dict. Nat. Biog.*, art. ' Ramsden '.
95 *Phil. Trans.*, **83**, pp. 71–73, 1793.
96 *Ibid.*, p. 72.
97 *Vide* note attached to the dividing engine, said to have been made by Ramsden, in the Science Museum, London. This exhibit was once in the possession of Messrs Dollond and was used until 1828.
98 Ramsden, J., 1777, *Description of an Engine for Dividing Mathematical Instruments.* Watkins, J. E., *The Ramsden Dividing Engine*, Annual Report Smithsonian Institution, Washington, pp. 721–739, 1890.
99 Since 1890 this engine has been preserved in the U.S. National Museum, Washington.
100 *Phil. Trans.*, **69**, pp. 419–431, 1779.
101 *Ibid.*, p. 429.
102 *Ibid.*, **73**, p. 94, 1782.
103 *Ibid.*, p. 97.
104 Kitchener, Ref. 73, ii, pp. 238–243.
105 Chaulnes, le Duc de, 1768, *Nouvelle méthode pour diviser les instruments.*
106 Martin, L. C., *Trans. Opt. Soc.*, **24**, p. 300, 1923.
107 *Phil. Trans.*, **83**, p. 133, 1793.
108 Wolf, C., *Histoire de l'Observatoire de Paris*, 1902, p. 290, gives between 40 and 50. Dutens, L., 1813, *Aikins General Biography*, viii, p. 454, says ' near 60 '.
109 Weld, C. R., 1848, *History of the Royal Society*, ii, p. 186.
110 *Phil. Trans.*, **80**, p. 134 ff., 1790.
111 *Ibid.*, **75**, p. 385., 1785.
112 *Ibid.*, **111**, p. 76, 1821.
113 *Ibid.*, **80**, p. 112, 1790.
114 *Ibid.*, p. 117.
115 Piazzi, G., 1792, *Della Specola Astronomica . . .*; Pearson, W., 1829, *Introduction to Practical Astronomy*, ii, pp. 413–417.
116 Ball, R., 1907, *Great Astronomers*, p. 241.
117 *Ibid.*, p. 242.
118 *Ibid.*, pp. 242–243.
119 Weld, *op. cit.*, pp. 187–188.
120 Pearson, *op. cit.*, p. 534.
121 Grant, R., 1852, *History of Physical Astronomy*, p. 149.
122 Weld, *op. cit.*, p. 569.
123 Kitchener, Ref. 73, ii, p. 239.
124 *Ibid.*
125 *Ibid.*, pp. 239–240.
126 Dutens, *op. cit.*, viii, p. 454.
127 Court, T. H., and Rohr, M. von, *Trans. Opt. Soc.*, **31**, p. 82, 1930.
128 *Ibid.*, pp. 82, 84, 85. Baxandall, D., *J.B.A.A.*, **41**, p. 133, 1931.
129 Brewster, D., 1853, *A Treatise on Optics*, p. 509.
130 Smyth, W. H., 1844, *A Cycle of Celestial Objects*, i, p. 335 and illustration.
131 Pearson, *op. cit.*, pp. 362–365.
132 *Ibid.*, p. 402.
133 *Vide* Ref. 47.
134 Court and Rohr, *op. cit.*, p. 84.
135 *Ibid.*
136 Ball, *op. cit.*, pp. 242–243.
137 *Phil. Trans.*, **83**, p. 102, 1793.
138 *Ibid.*, **111**, p. 79, 1821, where Worthington is described as Berge's ' successor '.
139 *Mem. R.A.S.*, **22**, p. 221, 1854.
140 *Ibid.*, pp. 220–221.

CHAPTER IX

One of the greatest obstructions to the construction of large achromatic telescopes is, the difficulty of procuring large disks of flint glass of a uniform density—of good colour, and free from veins.

THOMAS DICK

FOR many years after the invention of the achromatic telescope, the London opticians still used English flint glass. Glassmakers clung to their old methods and produced small crown and flint disks fairly free from major defects, but glasses above 4 or 5 inches were seldom worth the labour of working, still less the expense of a suitable mounting. These larger disks contained streaks and veins prejudicial to good definition, imperfections which multiplied with fatal rapidity with every increase in size. With the glass industry in this backward state, it seemed that the refractor would never compete with, let alone rival, the light-grasping power of Herschel's reflectors. This unfortunate position would probably have continued but for the discoveries of Pierre Louis Guinand, a clock-case- and cabinet-maker of les Brenets, Switzerland. After many years of patient experiment, Guinand, by new stirring methods, produced flawless disks of flint glass.

Guinand first saw an English reflecting telescope when he was about twenty to twenty-three years of age (*circa* 1770). The instrument, probably by Short, was an ornament in the home of Jaquet Droz, a clockmaker of les Brenets. Guinand obtained permission to borrow the instrument. He took it to pieces, measured the curvatures of the mirrors and examined the lenses. At this time he made metal clock-cases and had received some instruction in casting processes from a neighbouring buckle-maker. He was able, therefore, to cast his own metal mirrors and eventually to make a duplicate instrument.

Guinand turned to flint-glass manufacture about the year 1783, by which time Dollond achromatics had percolated into central Europe. Jaquet Droz procured one of them and allowed Guinand to dismantle it. Guinand sought means to duplicate the instrument but found that suitable flint glass was unobtainable. He procured the necessary blanks some years later when a neighbour, returning from a visit to London, brought back some flint and crown pieces. The specimens were full of striae as were others obtained later, whereupon Guinand decided to cast his own glass. That he was already thirty-five years old, wholly dependent on his trade for a living and without systematic scientific training, had no effect on his decision.[1]

Between the years 1784 and 1790, Guinand taught himself the rudiments of chemistry and experimented with small melts of 3 or 4 lb at a time. He soon discovered that lead was a constituent of flint glass since this metal came out of solution when samples were melted over a small furnace. News that various scientific bodies in France were offering prizes for homogeneous flint glass stimulated our investigator. Meanwhile, he needed money for his experiments and, at the age of forty (1787), turned to the more lucrative trade of making bells for repeaters. He purchased ground on the banks of the river Doubs, near les Brenets, and there built a furnace capable of melting 2 cwt of raw material at a time.[2]

The investigations did not proceed smoothly. Once the furnace nearly burst and had to be rebuilt. On another occasion fire caught the roof of the building and, when the flames were extinguished, it was discovered that water had found its way into the annealing oven and had spoilt the melt. Large quantities of wood were needed to heat and to maintain the heat of pot and melt. Often, when the pot was charged, a flaw would widen into a crack and the precious contents would pour into the fire. Above all was the growing cost of wood, furnace fitments, and raw materials of the required purity. The amount of heavy manual labour involved was considerable. Guinand, alternately excited and depressed by the vicissitudes of the work, shared his life between his furnace and his trade. During this busy period, to add to his troubles, he lost his first wife.[3]

At length, Guinand obtained a solid block of glass of some 2 cwt. This he sawed into two and polished on the cut faces. On looking within the mass he saw globules, bubbles and threads, mostly near the bottom and top of the block. There were many specks which ended in a tail or train of substance less dense than the rest of the glass—these he called ' comets '. He explained these defects as follows. The ' drops ' at the top of the block were air bubbles and globules due to lead which separated from the glass and which eventually appeared on the surface in its metallic state. This lead then oxidized and re-entered the solution. Owing to its specific gravity it sank to the bottom of the melt, according to the temperature, and left in its passage a train of small threads. Having reached the bottom it went into solution with impurities in the clay of the pot and formed a vitreous compound of lesser density.

Guinand exerted himself to the full to remove these defects. For several days, assisted at times by his second wife, Rosalie, and son, Aimé, he would alternately toil and brood over his furnace. He retained his first procedure of allowing the red-hot mix to cool fairly quickly and so to split into irregular lumps. The largest and best of these were then cut and polished on one side. Guinand used to examine the glass through the polished face and then cut out the largest possible blank.[4]

Disks of 4 and 6 inches diameter began to accumulate and, in 1798, Guinand visited Lalande at Paris and, no doubt, met several of the opticians there. Lalande advised Guinand to work his blanks into object-glasses and this he eventually did, constructing 4- and 5-inch doublets ground and polished by a machine of his own contriving. He made no attempt to qualify for the prizes offered by the French Academy, fearing, no doubt, that he would have to divulge the secrets of his process.

Visitors now began to appear at les Brenets, among them a Captain Grouner of Berne, a mining expert, who later spoke of Guinand's work to friends in Munich. In 1804, Grouner, on behalf of Utzschneider, a wealthy Munich lawyer, asked Guinand for samples of his glass. In the spring of 1805, Utzschneider met Guinand at Aarau.

Utzschneider was at this time financing a company known as the 'Mathematical-Mechanical Institute Reichenbach, Utzschneider and Liebherr.' The company started in 1802 when G. Reichenbach, a young Bavarian artillery officer, joined forces with J. Liebherr, a watchmaker, who already possessed a small workshop.[5] They aimed at producing high-quality surveying instruments, mainly for military purposes. Utzschneider realized that a good supply of optical glass was essential for the success of their project and, on seeing Guinand's samples, did not hesitate to finance his work.[6] In return, and after some trial melts, Guinand prepared a memorandum on his results.

Utzschneider met Guinand at an important period for, in July, 1805, Guinand used a fireclay stirrer for the first time. Hitherto he had used a long wooden pole which caused the lead oxide to mix unequally with the other ingredients and to form streaks. The motion of a porous fireclay rod, however, brought bubbles to the surface and kept the mass thoroughly mixed until, very slowly cooling, it became too viscous to stir longer.[7] The fireclay stirrer, introduced by Guinand, continues to be important in the modern stirring process of optical glass.

Samples from the centre of the resulting blocks were sent to Munich and their excellence caused Utzschneider to travel post-haste to les Brenets. Guinand and his family were induced to leave Switzerland and, in 1806, we find them housed at the Institute's glassworks at Benediktbeuern.

The old Benedictine monastery at Benediktbeuern struck Utzschneider as an ideal location for his glassworks. Firewood abounded for the furnaces, there was a good water supply and a mill-stream could be harnessed to drive machinery. Under the terms of an agreement, Guinand was to be assisted by his wife, Rosalie, and was to instruct one of his sons in his processes. Comparative secrecy was thereby assured. In addition to his fixed salary of 500 florins a year he received a further 500 florins a year for his discovery and 20 per cent of the profits of the establishment. Labourers were available for cutting and carrying fuel and for dealing with other rough work. Utzschneider treated Guinand quite well, providing living accommodation and unlimited fuel. At the same time he stipulated that all the glass made should remain the property of the Institute.[8]

A year later, 1807, a second agreement was signed. Guinand was apparently unsatisfied with his remuneration and a new sum, 1600 florins per annum, was fixed. He was engaged for a term of ten years, after which period he became entitled to an annual pension of 800 florins. The same secrecy was demanded, but Guinand was now required to instruct a person, nominated by Utzschneider, in his methods.

The newcomer proved to be Joseph Fraunhofer, then a young man of twenty years, who was destined to become one of the greatest figures in the history of German optics.

Fraunhofer was born at Staubing, Bavaria, on March 6, 1787. He was the eleventh and last child of Xaver Fraunhofer, a master-glazier and worker in decorative glass. Joseph lost his mother when he was eleven and his father a year later. His guardians then apprenticed him to P. A. Weichselberger, a mirror-maker and ornamental-glass cutter at Munich. Weichselberger gave little instruction and no pay. Fraunhofer, perforce, had to attend school on Sundays at the so-called Holiday Schools for apprentices, but before long Weichselberger refused him even this slender freedom. After two years, however, the apprenticeship ceased dramatically when Weichselberger's house collapsed, killing his wife and burying Fraunhofer in the ruins.[9] The boy, miraculously protected by a cross-beam, was rescued injured but alive. News of his escape came to the notice of the Elector Maximilian who gave him money whilst Utzschneider began to take an active interest in his welfare.

Fraunhofer continued to work for Weichselberger, but now attended the Holiday Schools and, with part of the Elector's gift, bought his own glass-cutting machine. With the remainder he eventually bought himself out of his contract. Anxious to be his own master, he imprudently set up in business as an engraver of copper plates for visiting cards.[10] After six months' comparative freedom he found that, to be sure of his meals, he must return to Weichselberger. The date of his return, the end of 1804, more or less coincided with the date of Utzschneider's entry into partnership with Reichenbach and Liebherr. In 1806, Utzschneider offered Fraunhofer a junior post in the Munich Institute.

In the generous atmosphere of the Munich Institute, Fraunhofer's ability and enthusiasm for his work grew beyond usual bounds. Within a year he was grinding lenses by himself. A little later, at Benediktbeuern, he had sole charge of workshop and apprentices. He and Guinand did not get on well together—Guinand was ambitious and sanguine, Fraunhofer was critical and patient and much younger. Fraunhofer, as grinder and polisher, often had occasion to criticize Guinand's work. Unpleasantness arose and Utzschneider had to intervene.[11] In 1809, Fraunhofer was given sole charge of Benediktbeuern and made junior partner in the firm. Guinand was asked to instruct him in his methods; we are not surprised to find Guinand and his family returning to les Brenets in May, 1814. Utzschneider appears to have welcomed Guinand's return; he promised him an annual pension of 800 florins if he and his wife ceased to interest themselves in glassmaking.[12] Safe back in les Brenets, Guinand attempted to start his own works again, attempts which did not escape the notice of Utzschneider. When, therefore, Guinand wrote to Munich to persuade Utzschneider to take him back, he received a firm refusal and lost his pension.

Guinand spent his remaining years in the construction of refracting telescopes, making stands and metal parts himself, even to the soldering of the tubes and the preparation of varnishes.[13] He discovered the modern method of making large optical blanks—that if a piece of homogeneous glass was softened, pressed into a circular mould and slowly cooled, it provided a disk fit for working into a lens. This new method saved the labour of sawing out the required portions from the blocks, entailed far less waste, and provided disks of 12 inches and, in one instance, 18 inches diameter.[14] Such sizes were made possible in the first place, by the homogeneity of the first melts, a condition which Guinand achieved by his use of fireclay stirring rods.

During his last years, Guinand received a visit from J. N. Lerebours, a French optician, who purchased all the flint glass in his possession. From the largest blanks, Lerebours made objectives for the Paris Observatory which were reputed to be better than those of the Dollonds and, in 1819, when he received a gold medal for his optical work, he had made achromatic telescopes of 4, 6, and 7½ inches aperture.[15] One of his largest glasses, finished in 1823, had a diameter of 9½ inches and an aperture ratio of $f/13$. At a later date, Lerebours furnished Paris with even larger telescopes—a 12·8-inch and a 15-inch refractor—but of such mediocre quality that, in common with the 9½-inch, the outer zones had to be masked out by a diaphragm.[16] Lerebours' rival, Cauchoix, also heard of Guinand's work and he too posted 300 miles to les Brenets, where he bought all the glass he could find and ordered several much larger disks. The English astronomer, James South, purchased one of Cauchoix's object-glasses, an 11¾-inch of 19 feet focus, for his private observatory at Camden Hill, Kensington, and the Duke of Northumberland paid 15,000 francs for another of the same size for the Cambridge Observatory.[17]* Cooper bought a still larger lens of 14 inches diameter for his new observatory at Markree Castle, Sligo, where he used it in compiling an extensive and valuable catalogue of stars (Fig. 78).[18]

Meanwhile, Fraunhofer at Benediktbeuern had computed a set of lens forms for achromatic doublets and was giving serious attention to the removal of surface imperfections and the remaining defects in optical glass. He improved the furnaces and annealing ovens and, in 1812, produced a 7-inch object-glass and had high hopes of proceeding to a 10-inch. These and others he ground and polished on machines designed by Liebherr. He used Newton's coloured rings for the first time for testing lens surfaces and, as a further check on their accuracy, designed special spherometers and micrometers.[19] When he examined some of Dollond's achromatic lenses he could not help noticing their pronounced over-corrected spherical aberration and under-corrected chromatic aberration. He at once questioned Dollond's particular method of computation. As he saw the problem, residual aberrations could be effaced only by measuring the partial dispersions of optical glass more accurately, and these varied for different pieces of the same block and with the slightest change in chemical composition. At first, he measured the refractive indices of different glasses for differently coloured lights. The results were not particularly consistent, but they led him to the discovery that the spectrum of a sodium lamp consists of two close and bright yellow lines. With low dispersions, these lines appeared as one and constituted a reliable and useful source of monochromatic light for measuring refractive indices. Following his discovery of the dark spectrum lines, Fraunhofer and his friend Soldner compiled a list of the refractive indices of different glasses for different colours.[20]

This advance enabled Fraunhofer to design and construct some of the finest object-glasses of the early nineteenth century. His 9½-inch Dorpat equatorial, with its well-corrected and aplanatic† object-glass and original equatorial mount, was for many years the largest refractor in the world. This instrument, installed at

* *Vide* p. 238.

† *Aplanatic.* A term introduced by J. Herschel (1827, *Light*, p. 386) for lens systems which, for a given object distance, are free from spherical aberration. The axial object point and the conjugate image point are then known as *aplanatic* points. The system is often loosely referred to as an *aplanat*, but it must be remembered that it is aplanatic only for a certain pair or pairs of conjugates. For telescope objectives the two conjugates are the object, at a remote distance from the lens, and the image in the posterior focal plane. The terms *aplanat* and *aplanatic* are now applied to systems free from both spherical aberration *and* coma, i.e., systems free from axial and extra-axial spherical aberration for a pair of conjugate surfaces. *Vide* also page 192 and footnote.

Fig. 78—Temporary mounting used by E. J. Cooper for Cauchoix's 14-inch objective

The telescope was erected at Markree Castle, Sligo, Ireland. In 1834 the objective was remounted on a large Fraunhofer-type equatorial made by Thomas Grubb.

(Royal Astronomical Society)

the Dorpat (now Tartu) Observatory, was entrusted to Wilhelm Struve who records that, when he first saw it, he was ' undetermined which to admire most, the beauty and elegance of the workmanship in its most minute parts, the propriety of its construction, the ingenious mechanism for moving it, or the incomparable optical power of the telescope, and the precision with which objects are defined.'[21] With this refractor, Struve undertook no less than a complete survey of the sky from the celestial pole to $-15°$ declination. At the remarkable rate of seven stars a minute, Struve examined about 120,000 stars. He took micrometrical measures of over 3000 double stars, of which number only 700 were known previously.[22] Many had a separation of only $1''\cdot 0$ or less and, as the theoretical resolving power of the object-glass was just half this value, it is clear that the excellence of the telescope and the skill of the observer must have been of a high order.

The focal length of the Dorpat objective was 14 feet and necessitated a tube composed of deal strips, joined and reinforced by iron diaphragms. This tube was covered by a mahogany veneer, highly polished so as to resemble burnished copper. To prevent flexure, two rods were arranged parallel to the length of the tube and terminated at the eyepiece end in two spherical metal counterweights.[23] The objective had a crown glass lens whose rear radius was shorter than the front radius of the succeeding flint. The two lenses were consequently separated by spacers at the edge which, in small achromats in simple cells, is rather inconvenient. Unless this is done, the main advantage of the Fraunhofer combination—aplanatism—is considerably impaired. Fraunhofer left no account of his methods of computing and making acromatic lenses; this information he probably regarded as a trade secret. For the Dorpat achromat, a glass fairly representative of his work, chromatic aberration was slightly over-corrected. The excess of blue in the image, although not serious, affected colour estimates of double stars and planetary markings. The combination was fairly free from coma and, while this correction was by no means so rigorous as in the modern aplanat, critical definition could be obtained over a field of from 2 to 3 degrees extent. This slight under-correction was not serious in visual work and, for field, definition, and colour correction, the Dorpat glass, in common with nearly all Fraunhofer's other objectives, surpassed the best English specimens and compares well with modern achromats.[24]

A striking feature of the Dorpat telescope was its equatorial mounting, another characteristic common to Fraunhofer's telescopes. While the name *equatorial* reminds us of Tycho's equatorial armillae, Scheiner's helioscope, Römer's equatorial instrument, and Short's large Gregorian equatorials, none of these instruments was an equatorial in the modern sense. Like Sisson's equatorial, they were all based on a polar axis, arranged parallel to the earth's axis. Tycho's armillae were merely graduated rings attached to a polar axis, Scheiner's helioscope, a telescope hinged to a polar axis, and Römer's equatorial, a predecessor to the small portable stands made by Short and Ramsden. Short's larger equatorials had a hollow conical polar axis, mounted in a heavy wooden trolley, with the telescope swung in a yoke or cradle at the top of the cone. The only feature which Fraunhofer borrowed from all these was the polar axis. This was of steel and was arranged with a declination axis like the lines of a capital T. The polar axis, the upright of the T, pointed to the celestial pole and was carried in bearings attached to the

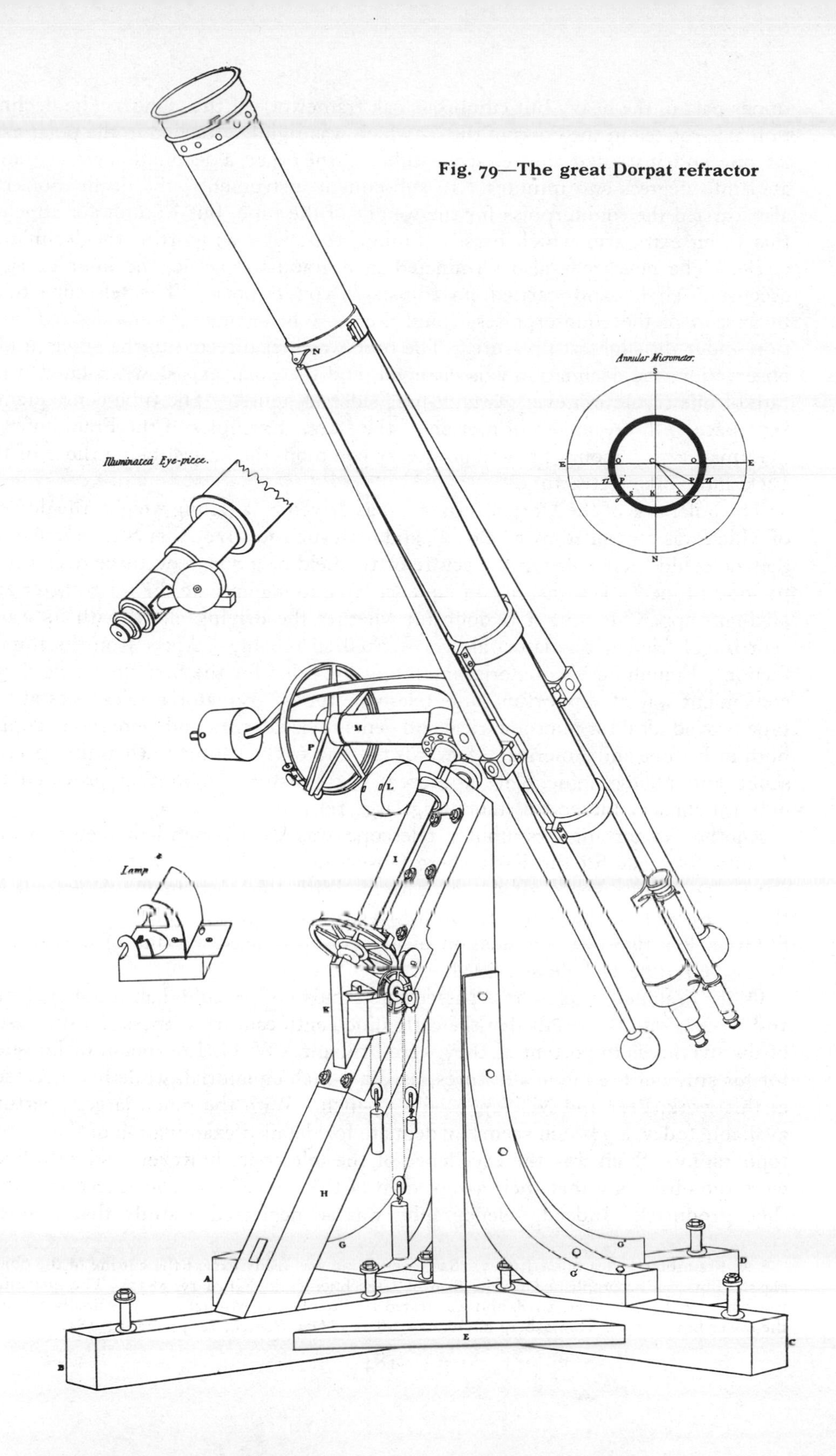

Fig. 79—The great Dorpat refractor

upper part of the heavy but cumbrous oak framework of the stand. The declination axis rotated in the cross of the T, which was rigidly attached to the polar axis. At one end it carried the telescope tube, at the other, a declination circle graduated into degrees and minutes. In subsequent instruments, the declination axis also carried the counterpoise for the weight of the tube, but Fraunhofer attached this to an extra arm which passed through the spokes supporting the declination circle. The polar axis also terminated in a graduated circle, the hour or right ascension circle, and carried an adjustable counterpoise. The telescope tube, by reason of the counterpoises, could therefore be swung into any desired position under the slightest pressure. The tube was first directed to the object under observation, the declination axis clamped, and the polar axis slowly rotated at the rate of one revolution every twenty-four sidereal hours. The tube consequently kept pace with the apparent motion of the stars. Examples of the Fraunhofer or ' German-type ' mounting are now very common, the largest being those of the Lick and Yerkes refractors.

The polar axis of the Dorpat refractor was driven by a falling weight, the descent of which was controlled by a ' clock ', and with such accuracy that Struve remarked that he could keep a star in the centre of the field of a 700 × eyepiece over a long period of time.[25] He was, in consequence, free to manipulate the micrometer and auxiliary apparatus, but it is doubtful whether the driving clock, with its crude centrifugal friction regulator, always worked so reliably. Apart from its imperfections, Fraunhofer's equatorial mounting provided for the first time a rigid and convenient way of supporting large telescope tubes. Equatorial telescopes of this type proved ideal for micrometrical and general visual work and were soon adopted both in Europe and America: Towards the end of the century, when the spectroscope and photographic plate appeared, the equatorial mounting provided the only satisfactory method of mounting large refractors.

Another important Fraunhofer telescope was the 6¼-inch heliometer[26] which Fraunhofer made for the Königsberg Observatory and which was not completed until 1829, three years after his death.* With this instrument, Friedrich W. Bessel measured the parallax of 61 Cygni, an angle of only 0″·3483 (1838). Struve's less rigorous examination of α Lyrae produced a value of 0″·2613, an excessive but nevertheless real value.[27]

Other Fraunhofer telescopes passed into hands no less able than those of Struve and Bessel, with the result that the early nineteenth century witnessed a succession of discoveries as important as they were brilliant. W. G. Lohrmann of Dresden, for his study of the moon's features, used a 5-inch equatorial, while his successors at this work, Beer and Mädler, used a 3¾-inch. With the much larger apertures available today, a 3¾-inch seems inadequate for detailed examination of the moon's topography. Such was the excellence of the telescope, however, and so diligent were the observers, that their 3-foot map of the moon[28] was considered the finest then produced. Indeed, selenography was so neglected a study that Mädler's

* At Fraunhofer's death, on June 7, 1826, everything was ready except the cutting of the object-glass. This was accomplished by the firm of Utzschneider in February, 1827. The instrument was completed by October, 1828, and was tested by Steinheil at Munich. It was finally set up in the tower built for it at Königsberg in October, 1829. (*Mem. R.A.S.*, **12**, pp. 52–53, 1842.)

exhaustive work tended to retard rather than encourage further studies. With an even smaller Fraunhofer telescope, Argelander made his photometric survey of all the stars in the northern hemisphere down to the ninth magnitude—the now classic *Bonner Durchmusterung*. He used a 3·4-inch, *f*/7 equatorially mounted Fraunhofer comet seeker, with a Kellner orthoscopic eyepiece* magnifying only 10 diameters.[29]

Two further inventions must be credited to Fraunhofer. The first is the suspended annular micrometer—a steel ring inserted in a disk of glass placed in the plane of the primary focus. The use of a glass plate in this position was, however, by no means new. According to Bion,[30] de la Hire in the seventeenth century suggested replacing the usual fibres employed for cross-wires by fine lines faintly ruled on a glass plate by means of a diamond. Fraunhofer always supplied a lightly ruled glass plate with his larger refractors. His other invention was a friction regulator for telescope drives. This consisted of a hollow brass cone, fixed over the balls of the governor. In diverging, the balls rubbed against the

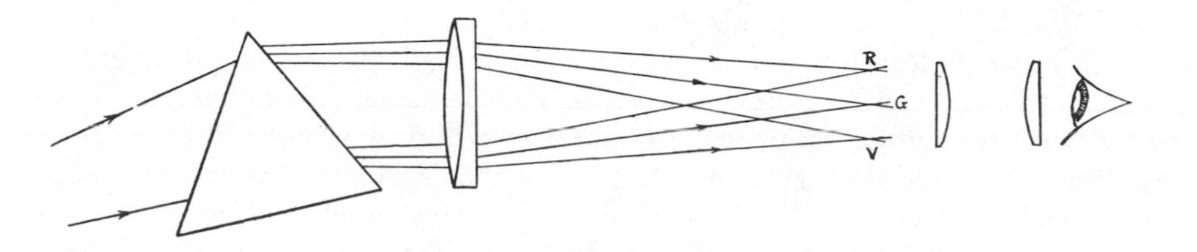

Fig. 80—Fraunhofer's arrangement for viewing the solar spectrum and its absorption lines

The sun's rays were almost parallel when they met the prism, since the slit in the window was some distance from the theodolite. This parallelism is now achieved by placing the slit at the focus of a convex lens—the collimator lens.

inner surface of the cone with increasing friction, the effect of which was to reduce their angular velocity. These two inventions were really two of many, for it can be said that all Fraunhofer's micrometers, eyepieces, telescope-drives, and mountings embodied some improvement upon older types.

Fraunhofer was the first to examine and map the absorption lines of the solar spectrum. He placed a flint-glass prism before the objective of a small theodolite and fed both with sunlight from a vertical slit in the window shutters of his room. He saw that the sun's spectrum was crossed by a series of vertical, dark lines of

*This compound eyepiece was introduced in 1849 by Carl Kellner, an optician of Wetzlar, who described it in a brochure entitled *Das orthoskopische Okular*. The eye-lens is an over-corrected pair with the flint component facing the eye. The field-lens, usually placed just within the anterior focal plane of the eye-lens, can be either plano-convex or biconvex. In low-power form, the system gives a usable field of about 50 degrees angular extent, and in this respect it is superior to the basic Ramsden eyepiece. The Kellner eyepiece is orthoscopic in the sense that it is fairly well corrected for the aberration known as distortion, that is, magnification is almost constant over the entire field. A similar eyepiece with both achromatic field- and eye-lenses was introduced by A. Steinheil of Munich. Later orthoscopic and monocentric forms by Steinheil and Zeiss, especially in high powers, are notably superior to the basic Kellner.

varying degrees of intensity—a series which, as W. H. Wollaston was the first to discover in 1802[31] (although he saw only a few lines), always kept the same relative order and position. These lines were, in consequence, excellent marks for finding the refractive index of glass for different colours. Fraunhofer selected eight lines, situated in the most important parts of the spectrum, and gave them the reference letters A, B, C, . . . H. Between the lines B and H he counted 574 lines. Of these, he measured the relative distances of the strongest and entered them in a figure (Fig. 81). The faintest lines were then inserted by eye estimation.[32]

Fraunhofer now constructed a 4½-inch refractor, which he fitted with an objective-prism of 37° 40′ apical angle.[33] He found the lines were present in every kind of sunlight, whether reflected from terrestrial objects or from the moon and planets. With a cylindrical lens to give breadth to their spectra, these lines appeared in stellar spectra, but in different arrangements.[34] From these and other observations, he concluded that the lines had their origin in the very nature of these bodies. Indeed, he came close to discovering their nature when he noticed that the two dark lines in the solar spectrum, which he had designated D, coincided with the two bright lines in the spectrum of his sodium lamp.[35]

With a view to learning more about the nature of light, Fraunhofer brought the powers of his 4½-inch telescope to bear on the phenomena of diffraction.[36] He first studied the diffraction pattern produced by a narrow slit placed at the focus of a lens and viewed through his telescope. He then thought he would accentuate the pattern by placing before the telescope objective a number of equally-spaced thin wires, all parallel to each other and to the collimator slit. Instead of seeing more pronounced maxima and minima, however, he obtained, when sunlight was passed through the collimator slit, a series of spectra, the bright central maximum being flanked on either side by ' first order ' and ' second order ' spectra. Similar appearances, but with greater dispersion, appeared when closely packed but regular diamond rulings in gold film on a glass plate replaced the wire grating. Upon interpreting these appearances in terms of the wave theory of light, Fraunhofer was able to determine the wavelength of light with considerable accuracy.* Prior to this, David Rittenhouse of Philadelphia (1785)[37] had commented on the dispersion produced by coarse gratings, whilst Thomas Young of London had investigated the colours of striated surfaces viewed by reflected light (1801).[38] It was left to Fraunhofer, however, to use the grating for wavelength determinations, that is, to discuss its effects quantitatively.†

In 1811, Fraunhofer had charge of forty-eight workpeople, and the manufacture of spectacle lenses about this time brought further additions to the staff. In February, 1814, Reichenbach left the Munich factory to establish his own workshop for the manufacture of precision instruments. Five years later, the Benediktbeuern optical shop was transferred to Munich and, in February, 1820, Utzschneider made Fraunhofer partner and director of the Optical Institute.[39]

For his discoveries in physical optics, Fraunhofer was made corresponding

* He established the relationship $\{(a + b) \sin \theta\}/k = $ a constant, where θ is the angle of diffraction for a particular line in a spectrum of order k and $a + b$ is the grating interval. This expression is true only when the light falling on the grating is incident normally. The constant is the wavelength of the light diffracted at the angle θ.

† *Vide* also pp. 283–284.

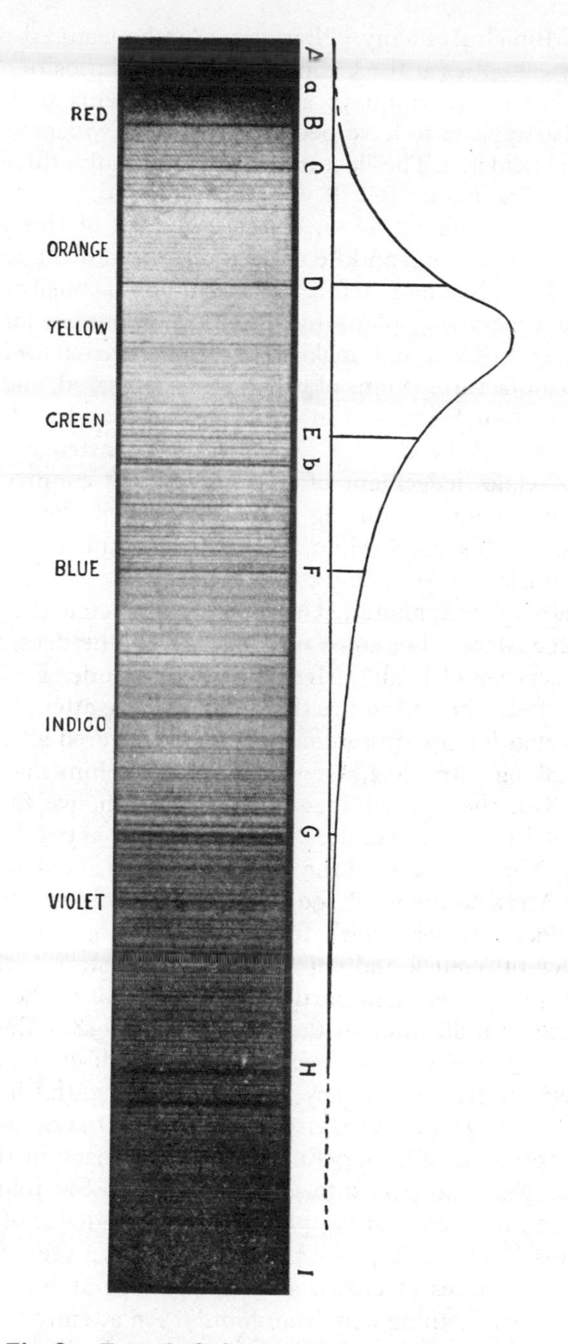

Fig. 81—Fraunhofer's map of the solar spectrum

A reduced copy showing the letters used to identify the more prominent absorption lines and (*right*) a curve to indicate the relative luminosity of different parts of the spectrum

member of the Munich Academy.[40] Early in 1825 he lectured publicly on mathematical and physical optics at the Cabinet of Natural Curiosities at Munich. This extra work, performed on Sundays and holidays, seems to have overtaxed his strength. He also appears to have been worried by a widening rift in his friendship with Utzschneider. The lectures lasted for only three months and, in September, 1825, Fraunhofer fell ill with a severe cold.

Fraunhofer died on June 7, 1826, at the early age of thirty-nine years. His constitution was never strong, and he seldom rested from the numerous activities that crowded his life. Not long before his death, he was busily engaged in making drawings and in supervising plans for a 12-inch refractor for Bogenhausen, in giving his Sunday lectures and making occasional visits to Benediktbeuern to attend to glass manufacture. Signs of tuberculosis appeared, and the sick man conducted his affairs from his bed. In this condition, and in his anxiety for the continuance of his work he chose F. A. Pauli to be instructed in his knowledge and methods.[41] Acknowledgement of his work by his country came just before his death. He was knighted during the exhibition of the Dorpat telescope in 1826 and relieved of his taxes while a citizen of Munich. He received a state funeral and was buried in Munich.

After the death of Fraunhofer, Utzschneider directed the Munich Institute assisted by George Merz. For some unknown reason he does not appear to have made use of the services of Pauli. Merz had worked under Fraunhofer since 1808 and, as a boy of fifteen, helped to fire the furnaces. He attended the school which Utzschneider started for his apprentices and there covered all the various aspects of instrument-making. In 1832, Utzschneider made him manager of the glassworks and, in 1839, the year of Utzschneider's death, we find Merz, with the monetary help of his friend Mahler, purchasing the entire business.[42] Mahler died in 1845, but Merz maintained the glassworks until about 1884. The business continued in the Merz family until 1903, when it was sold to Paul Zschohke. The glassworks at Benediktbeuern, with furnace building intact, were purchased by Prof. W. Küchler of Vienna and, in 1937, the monastery section was occupied partly as a theological college and partly by a contingent of the Labour Corps.[43]

When we realize that all through the period 1750 to 1825 the English opticians were faced with an ever increasing excise duty on both crown and flint glass, we can imagine their chagrin when they heard Struve's enthusiastic report on the Dorpat refractor. Sir David Brewster did not hesitate to criticize the Government in the strongest terms for allowing England's 'supremacy in the manufacture of achromatic telescopes'[44] to pass into German hands. Sir John Herschel visited Fraunhofer in 1824 but, save for supplying further evidence of the superiority of German glass, was unable to improve the position. As a great favour, Fraunhofer gave Herschel some pieces of crown and flint glass, but the Slough astronomer admitted that he learnt nothing new from him.[45] An attempt to come to an agreement with Guinand for starting a glass factory in England was frustrated by Guinand's death in 1824, and at the time his son was busy making alternative arrangements in France.

At last, but too late, the Government passed two resolutions. It decided, first, to finance research in glass manufacture and, second, to reduce the tax on flint glass by two-thirds the exorbitant 1825 rate of 98 shillings per cwt.[46] Research came first, and the Government undertook to bear all the expenses of furnaces, materials, and labour, and appointed the Council of the Royal Society to control the investigation. A committee was formed and Faraday was given the chemical part of the inquiry, Herschel the physical part, and George Dollond the task of figuring the lenses.[47] The inquiry began at an experimental glass-house and was continued in Faraday's laboratory at the Royal Institution, but the results were disappointing. From certain batches of glass, Dollond managed to make a telescope objective, but its performance was no better than that of others of standard type. Faraday used platinum moulding trays[48] and stirring rakes[49] in his researches, an expensive method compared with Guinand's fireclay rods. The electrical studies for which he later became so famous gradually monopolized more of his time and, although urged to continue, he finally left his optical work. Thus the foremost chemist of his day, financed by the Government and in control of a fine laboratory, failed to accomplish what a Swiss cabinet-maker had done with meagre resources. Faraday's feelings in the matter are best summed up in his remark to W. H. Smyth—' the best step to ensure improvement will be to take off the Excise duty.'[50]

After this further failure, other investigators turned to the only alternative left to them—the fluid lens. According to Smyth, Benjamin Martin constructed a telescope which, by the use of fluids, succeeded better than ' that of his friend Dollond.'[51] More positive success was achieved towards the end of the eighteenth century by Robert Blair, a Scottish naval surgeon and later professor of astronomy at Edinburgh University. Blair's experiments date back to 1787 when, with his limited instrumental resources at Merchiston, he examined one of Peter Dollond's standard object-glasses. He took a small, bright point object and observed the coloured fringes near the focus with high-power eyepieces. He saw, like Clairaut and others, that the ' achromatic ' objective was not free from colour and that the residue, or *secondary spectrum* as he called it, was quite prominent. After many years of patient experiment with fluid lenses, he abolished not only secondary spectrum but rendered the tertiary so insignificant as to place it beyond detection by the means at his disposal. Using formulae published by Huygens, he also tried to remove spherical aberration entirely and met with such success that he applied the term *aplanatic* to his lenses. He determined the partial dispersions of a great number of solutions; in some cases he enclosed them in separate cells, in others, where they would not decompose each other, he mixed them in one cell between two convex crown glasses. Solutions of metal salts proved in all cases more dispersive than crown glass. The most dispersive fluids were those in which hydrochloric acid and metals were combined.[52]

Blair's final fluid object-glasses consisted of a biconcave lens of hydrochloric acid combined with a metal and placed next to a plano-convex lens of crown glass. The liquid was introduced between the convex surface of the solid glass lens and a meniscus-plane or zero-power lens, the plane side of the crown lens facing the object. Blair's first successful object glass was of 3 inches aperture and 17 inches focus. According to the astronomer Thomas Dick, this glass was ' manifestly superior to one of Mr. Dollond's of 42 inches focal length.' Dr Robinson and John

Playfair tested the glass at night. 'They had most distinct vision of a star' Dick continues, 'when using an erecting eye-piece, which made this telescope magnify more than a 100 times; and they found the field of vision as uniformly distinct as with Dollond's 42-inch telescope magnifying 46 times; and were led to admire the nice figuring and centering of the very deep eye-glasses which were necessary for this amplification.'[53] In 1791, according to the Rev. T. W. Webb, himself a user of a fluid lens telescope, Blair constructed another 3-inch lens of the unusually short focal length of 9 inches.[54]

Opticians were sceptical regarding the permanency of the fluid lens. They argued that the corrosive action of the acids would have a deleterious effect on the lens surfaces, that the components were too easily deranged and that they could not be used for solar observations. Blair did not share their opinion and, in 1785 and 1791, he took out patents for his invention.[55] He even tried, with his son's help, to manufacture fluid lenses on a commercial basis, but the attitude of certain prominent opticians so discouraged him that the scheme was abandoned.

Peter Barlow, professor of mathematics in the Royal Military Academy, Woolwich, used an ordinary crown-glass objective which he corrected by a small fluid lens situated further down the tube.[56] He chose carbon bisulphide for the liquid, the transparency, remarkable dispersive properties (twice those of flint glass), and absence of colour of which were investigated by Brewster in 1813.[57]

On this principle, Barlow made a 3-inch telescope with which he resolved several close double stars. In 1827, he constructed a 6-inch and, two years later, one of 7·8 inches aperture. The latter was mounted in a tube 11 feet long and bore a magnification of 700 on some very close doubles, 'although the field' Barlow wrote,[58] 'is not then so bright as I could desire. . . . Venus is beautifully white and well-defined with a power of 120, but shows some colour with 300. Saturn, with 120 power, is a very brilliant object, the double rings and belts being well and satisfactorily defined, and with 360 power it is still very fine.'[59] He had no doubts of the permanency of his fluid lenses. In the case of his 3-inch telescope, made fifteen years before, 'no change whatever has taken place in its performance, nor the least perceptible alteration either in the quantity or quality of the fluid.'[60] The 7·8-inch he mounted on a heavy rotatable stand inside an observatory. He compared its performance with that of Herschel's 20-foot reflector and South's $11\frac{3}{4}$-inch Cauchoix object-glass. With a power of 700, the closest double stars in South and Herschel's 1824 catalogue appeared 'round and defined'. South's telescope defined Jupiter and Mars better than did Barlow's, 'but in this respect the superiority of the former instrument was by no means as great as he [Barlow] expected.'[61] Barlow was encouraged by the results of this comparison and wrote, in 1830, that he had strong hopes of constructing 'a telescope of two feet aperture and 24 feet in length.'[62]

Barlow attempted his largest and last lens in 1832, when the Royal Society stood the cost (£157) of an 8-inch fluid-type telescope. The optical work was executed by George Dollond and the completed instrument was sent to Herschel, Airy, and Smyth for examination. Their extensive reports[63] mention its good light-grasp but refer to offending secondary chromatic aberration with powers greater than 300. There was also marked spherical aberration and useful magnifications were limited to 150 and under. In brief, they considered the lens a failure. Barlow thought otherwise. 'The spherical aberration is perfectly corrected' he wrote in 1832,[64] 'the field is open and flat, with abundance of light,—all very desirable qualities in an astronomical telescope; with respect to colour, there is perhaps more outstanding on the violet side of the spectrum, than is generally found in the usual refractor, particularly towards the limits of the field.' Barlow appears, however, to have discarded the 8-inch. In fact, apart from designing a fluid achromatic negative lens,[65] he discontinued his work on fluid telescope objectives.

A more practical alternative to Barlow's arrangement was described in 1828 by Rogers in a paper to the Astronomical Society,[66] later the Royal Astronomical Society. He replaced the liquid concave lens by a compound crown-flint which, while it had no effect on the mean rays, shortened the focus for red and lengthened it for violet, so bringing all colours together. By this arrangement, a 3-inch flint disk sufficed to correct the colour of a 9-inch crown glass of 14 feet focus. Of course, an even smaller lens placed nearer the focus would have served the same purpose, but the steeper curves required to give it a greater power would have introduced too much spherical aberration to be even approximately corrected. Hence, when Wray and other opticians tried their hand at constructing this instrument, they chose a moderate ratio like 2: 1 or 3: 1 for the diameters of the objective and corrector. George Dollond first put the idea into practice with the 5-inch which he made for Barlow. Wilson of Glasgow apparently made a 4-inch before he was even aware of Rogers' plan.[67] Wray made another of 8¼-inch aperture and, although Herschel remarked that it was a 'very artificial and beautiful invention, highly deserving of further trial',[68] the *dialytic* telescope, as it was called, was generally neglected in England. The best examples were furnished by G. S. Plössl, an optician of Vienna, under Stampfer's direction.[69] He was introduced to the arrangement by Karl von Littrow, and many of his smaller instruments are reputed to have been of excellent workmanship. A 6½-inch went to Pulkowa to supplement the Dorpat refractor, while Dembowski at Naples used a 5½-inch on double stars. Schmidt at the Athens Observatory had an 8-inch, the performance of which does not seem to have been very good. Plössl's largest glass, an 11-inch of 11 feet 9 inches focus, went to a sultan and never had a chance of doing useful work.

While Barlow was mastering the difficulties of fluid lenses, the council of the newly founded Astronomical Society of London received a letter from Reynier of Neuchatel to the effect that Guinand was able to supply flint blanks up to 12 inches diameter.[70] The council thereupon invited Guinand to submit specimens of his glass, but the samples which arrived were disappointingly small. The largest blank, a 2-inch, was given to Charles Tulley, an optician of 4 Terrett's Court, Upper Street, Islington, with instructions to make with it a concave lens suitable for an achromatic object-glass. Tulley effected this and presented to the council an objective which a sub-committee consisting of Gilbert, J. Herschel, and Pearson considered 'trifling' in size but good in performance.[71]

Meanwhile, further flint blanks arrived from Guinand, one of which, a 7¼-inch disk, looked so promising as to warrant its working into a component for an object-glass. Tulley had by him a disk of French plate-glass which he worked into a convex lens, but the completed combination was unsuccessful. Tulley now,

' with a zeal and constancy for which he is entitled to much credit ',[72] obtained and worked a second convex component, this time of English crown glass. The resultant objective, of 6·8 inches clear aperture and 12 feet focus, was then mounted in a temporary wooden tube at Tulley's house, and G. Dollond, J. Herschel, and Pearson spent many hours testing it. The definition of Jupiter, Saturn, nebulae in Virgo and ' difficult ' doubles like Polaris, γ Leonis, ζ Cancri and ω Leonis with powers from 200 to 700 was considered extremely good. The observers attributed the good achromatism to the quality of the flint glass and referred the almost complete absence of spherical aberration to ' the goodness of its workmanship '. In concluding their report on this lens, the three astronomers paid tribute to ' the great pains bestowed on its workmanship by Mr. Tulley.'[73]

The council were now anxious to dispose of the object-glass but Tulley asked £200 for his work and there was, in addition, the cost of £20 16s. 6d. plus 700 francs due to Guinand for the flint blank. The council considered Tulley's charge excessive and did not hesitate to tell him so, adding that it was discouraging to further experiments.[74] Pearson, however, purchased the glass outright and so prevented what might have become a serious dispute. He got George Dollond to mount it on one of his new parallactic ladders, a simplified version of J. Sisson's equatorial, and never once had occasion to regret the expense.

It will be seen from this account that, by 1823, Charles Tulley was an optician of some standing. He had been a member of the Astronomical Society since its foundation in 1820 and, with George Dollond, one of the founders, was always consulted on optical matters. How Tulley came to the Islington business we do not know; presumably he served an apprenticeship under one of the London opticians—perhaps Peter Dollond—and then launched out on his own. It is probable that he set up in business between the years 1775 and 1785, using a room in his own house as a workshop and his garden as an observatory.

Tulley's optical work again came into prominence when he made, in 1822, a $3\frac{1}{4}$-inch, $f/14$ object-glass for Sir James South. This glass was based on the computation given by Sir John Herschel for an aplanatic doublet[75] and was, for the time, quite novel. Charged with a power of 300, the glass gave excellent results; ' what therefore, Mr. Herschel's theory told him would be good, Mr. Tulley's practice has declared so,' South wrote.[76] Herschel's paper was the first serious attempt to reduce outstanding aberrations by analytical methods. The result was a number of practical combinations of lenses for object-glasses in which spherical aberration was corrected, not only for very remote objects, but for those at a moderate distance. These objectives were of no particular importance in astronomy but found application in surveying instruments. In correcting for spherical aberration, Herschel so reduced coma that he could with justification christen his objectives *aplanatics.** A. E. Conrady has remarked[77] that Fraunhofer's coma-free object-glasses were designed along the lines of the Herschel condition and that it is practically certain that he anticipated Herschel, seeing that he used pages of

* It occurred to Herschel that an objective could be so designed that it could be tested on some object inside the workshop and yet used on distant objects. The resulting so-called *Herschel condition* requires that $n \sin(\theta/2) = kn' \sin(\theta'/2)$ which clearly does not permit the simultaneous correction for coma, i.e. that $n \sin \theta = kn' \sin \theta'$—the Abbe ' sine condition '. Whilst Herschel regarded his objective as ' aplanatic ' it was not, therefore, ' aplanatic ' in the Abbe sense.

Fig. 82—A Tulley telescope on its stand

(Science Museum, London. British Crown copyright)

print at convenient distances as test objects. For its success, the Herschel combination requires great accuracy in construction, for small errors in curvature produce comparatively large amounts of outstanding aberration.

Such was South's confidence in Tulley's ability that he entrusted him with the working of two large flint blanks which he received from Paris about this time. The first object-glass, a 5-inch of 7 feet focus, South considered to be the finest in existence. 'Proof of this' he wrote, 'will be found in the separation and measurement of the most minute double-stars, such as η and σ Coronae Borealis, in its sharp definition of the double ring of Saturn, and various others of the most delicate celestial objects.' 'Under favourable circumstances' he adds,[78] 'with a power of 600, the disks of the two stars of η Coronae ($0''\cdot9$) and of σ Coronae ($1''\cdot2$); of ξ Bootis ($1''\cdot2$) and ξ Orionis ($2''\cdot7$), are shown perfectly round, and as sharply defined as possible.' Before parting with this telescope, Tulley and his son compared it with a 6·8-inch Newtonian of 7 feet focus and observed the eclipses of Jupiter's satellites 'as well and as long as with the other Telescope.' 'I believe' Tulley wrote to W. Kitchener,[79] 'that the 7 feet Achromatic is the only Achromatic Telescope of that focus which has been made in this country of 5 inches aperture—the Magnifying powers were from 100 to about 500 times.'

The second glass of 5·9-inch aperture and $8\frac{1}{2}$ feet focus was mounted on a brass polar axis, provided with 3-foot brass declination and hour circles taken from Sisson's equatorial sector at Greenwich. South sold this telescope to W. H. Smyth for £220. South told Smyth that the objective was, in his estimation, Tulley's *chef-d'œuvre.* So large an aperture was almost unique in those days—'I deemed myself fortunate in securing the prize' Smyth records.[80] At his Bedford observatory, Smyth had the telescope tube remounted by George Dollond on a large polar axis, driven by a Sheepshanks clock-drive. 'On repeated trials' Smyth writes,[81] 'I find the instrument bears its highest magnifiers with remarkable distinctness, as is especially evinced by the roundness of small discs, the dark increase of vacancy between close double stars, and from particular portions of the moon when dichotomized; I have therefore reason to presume that the curves of the lenses are in exact chromatic and spherical aberration throughout.' Upon observations made with this instrument, Smyth founded his *A Cycle of Celestial Objects*, the first book to be written expressly for the amateur astronomer.

When he had completed his catalogue of nebulae and double stars, Smyth transferred his telescope to Dr Lee who had erected a private observatory at Hartwell House in Buckinghamshire. The two friends continued their work together, assisted by N. Pogson and J. Epps, the former becoming well known for his catalogue of variable stars. The telescope is now in the Science Museum, London.

According to Pearson,[82] Tulley determined the refractive indices of his glasses by a method first practised in England by L. C. Martin and well known to the Paris opticians. This was to grind and polish several glasses to the same radius of curvature and then to compare their respective focal lengths when held up to the sun. 'After having done this' Pearson writes, 'with a specimen of both crown and flint glass, this artist is able to compute his curves and, by his peculiar dexterity

Fig. 83—The Smyth equatorial

(By courtesy of the Director of the Science Museum, London)

of tact, to work the faces to such perfect figures, that art can accomplish with given specimens of glass.' We cannot but admire the patience of opticians a century ago, and yet deplore the fact that they were apparently content to proceed in this wasteful but time-honoured fashion. This procedure changed when Fraunhofer's new conception of refractive index and accurate determinations of partial dispersion became known in England. The continued scarcity of flint glass was felt so acutely that it even features in the obituary notice of Tulley which appeared in 1833. 'It is grievously to be lamented' the author writes,[83] 'that much of the labour of this invaluable artist was absolutely wasted, from the impossibility of procuring flint glass capable of forming an object-glass at all; lens after lens having been rejected by him before one could be found of requisite purity, even of moderate aperture.' We would add also that many crown glass components were discarded for similar reasons.

Charles Tulley was not a member of the Worshipful Company of Spectacle Makers and we are left to only one source of information as to his death—the *Memoirs of the Royal Astronomical Society.* He is sometimes referred to as the 'Senior Tulley', often as 'Mr. Tulley'. It is possible, therefore, to confuse his work with that of his two sons, Thomas and William, both of whom assisted their father. After the death of Thomas Tulley in 1846, we hear no more of the Islington business.*

Tulley's rival in object-glass manufacture was George Dollond, already referred to in connection with the mounting of Smyth's Bedford telescope. Dollond made two similar mountings, one for G. Bishop's observatory in Regent's Park (1836), and the other for Lord Wrottesley's observatory in Staffordshire (1843). Dollond ground and polished the object-glasses for these instruments from flint disks sent over by Guinand.[84] Bishop's telescope had an aperture of 7 inches and a focal length of 10¾ feet; powers ranged from 45 to 800. Lord Wrottesley's telescope of 7¾ inches aperture and of the same focal length as Bishop's was mounted on a mahogany polar axis, 14 feet long. The circles on both instruments were 3 feet in diameter. A similar but simpler stand was Dollond's parallactic ladder, the first example of which was made for Smyth.[85] Pearson had another at South Kilworth in 1827 for carrying a 12-foot telescope of 6·8-inch aperture—the Tulley telescope previously mentioned. In this design the polar axis took the form of a cradle within which slid a frame which carried the telescope. The parallel sides

* In 1951 the author visited Terrett's Court, now Terrett's Place, a group of houses wedged between 146 and 147 Upper Street, Islington. No. 4 no longer exists but reference is made to it in the Directories and Voters' Registers up to 1908.

A tombstone to William Tulley and his wife Sarah remains in the cemetery of the Parish and Borough Church of St Mary, Islington. From this and the burial register it is definite that William Tulley died on December 15, 1835, aged forty-six years. The burial register gives the date of the death of Charles Tulley as October 16, 1830—he was then sixty-nine years of age. William Tulley, therefore, was in business for only five years after his father's death. Thomas Tulley died in 1846 but the parish records suggest that the business closed down about 1843.

About 1824 the Tulleys made a series of achromatic microscope objectives at the suggestion of a Dr Goring. The first samples were achromatic cemented triplets, each a telescope objective in miniature. Later a doublet was placed in front of a triplet of somewhat shorter focus. This objective was considered an advance on the single doublet and triplet forms and superior to the best achromatics of Chevalier, a leading French optician. For further details *vide* Carpenter, W. B., 1901, *The Microscope,* 8th edition by Dallinger, i, pp. 354–355.

Fig. 84—George Dollond

(By courtesy of the Director of the Science Museum, London)

of the cradle formed, as it were, the ladder, the cross-pieces of the frame and the two ends, the rungs.[86]

George Dollond collaborated with Barlow in contributing two papers[87] to the Royal Society on what is now called the *Barlow lens*. In the early part of the eighteenth century, Wolfius proposed placing a concave lens between the objective and eyepiece of a telescope in order to give greater magnification of the image with slight increase in focal length. Barlow computed a concave achromatic lens which Dollond made and mounted in a telescope. Dawes and Smyth both comment favourably on the Barlow lens. It left the final image as much without colour when the lens was used as when not, and enabled the observer to alter the image size at pleasure. Another paper[88] communicated to the Royal Society by George Dollond concerns a spherical rock-crystal micrometer. This device was similar to that designed by A. M. de Rochon[89] whereby two images could be produced and made to coincide by moving an achromatized rock-crystal plate along the axis of a refracting telescope. In Dollond's micrometer the Iceland-spar crystal took a spherical form and took the place of the eye-lens of a Huygenian eyepiece. The angle through which the sphere had to be rotated to bring two separated stars into coincidence afforded a measure of their apparent angular separation.

After the death of Sir William Herschel in 1822, several London opticians followed his example of selling reflectors at reasonable prices. W. Harris of 50 High Holborn, and the brothers W. and S. Jones of 30 Lower Holborn, offered Gregorians and Newtonians up to 15 inches aperture. The tubes were mounted on 'improved' wood and brass stands, aperture ratios were larger than those of Herschel's mirrors, prices ranged from £100 to £1000.[90] Cornelius Varley of 51 Upper Thornhaugh Street, Tottenham Court Road, made useful modifications to existing stands, and J. Watson of 4 Saville Place, Lambeth, was noted for the excellence of his mirrors. Cuthbert, a microscope-maker, produced a range of small reflectors, and both S. Pierce and W. Cary, two of Ramsden's pupils, made large Newtonians and Gregorians. The Tulleys offered an ambitious range—Gregorians from 1 foot to 6 feet focus and 9 inches aperture, Newtonians from 7 feet to 10 feet focus and 12 inches aperture. Prices ranged from 6 guineas to £210 for the Gregorians and £105 to £315 for the Newtonians.[91]

Before visiting any one of these opticians, a prospective buyer of a reflector would probably read through William Kitchener's little book on telescopes.[92] Kitchener, to whom we are indebted for so much of our knowledge of Short, Ramsden, and the Dollonds, inherited some £70,000 which he spent on his three hobbies—optics, cookery, and music. During thirty years he collected over fifty telescopes and probably inspected many more, so that he was in a better position to judge the relative merits of different instruments than either the opticians who made them or the astronomers who used them. His book appeared in 1825 and covered, in consequence, the period when Peter Dollond was at the height of his career and when the opticians just alluded to were vying with one another to capture what remained of the telescope trade.

The following extracts are typical of Kitchener's many observations on reflectors.

> I have a Gregorian Telescope of 4 inches aperture, and 15 inches long, made by Mr. Watson, which shews the Belts of Jupiter and the Division in Saturn's Ring, as well as I have seen these objects in my 3½-foot Achromatic of 2¾-inch aperture. This Gregorian shews Double Stars with a sharply defined disk, like little planets. . . .
>
> I have a little Dumpy Gregorian of only Two inches aperture and barely Four inches focus, made by Mr. Cuthbert, which shews Jupiter and Saturn beautifully sharp, and defines some Double Stars in the neatest manner, with a power of 130; the separation of the Two Stars of Castor [4″·7, epoch 1825] is extremely distinct. . . .
>
> Mr. Tulley informed me, that with the Cassegrainian Reflector, which he finished in the year 1802, and which has an Aperture of 15 inches—that Planets were best seen with a power of 200, or 250 at the utmost; he shewed Saturn to me with those Powers, beyond which it was not so well defined. . . .
>
> In the course of the last 30 years, I have seen the following Cassegrain Telescopes—Short's Dumpy, 2 feet long, 6 inches diameter—Watson's Dumpy, 7½ inches long, 3 inches diameter—a 30-inch of 6 inches Aperture, made by Mr. Tulley, for Mr. Custano, which was afterwards in Mr. Wm. Walker's Observatory.

While the London opticians provided large telescopes for those who could afford them, amateurs were not wanting who could make equally effective instruments at a small fraction of the cost. Of these, James Veitch of Inchbonny, near

Jedburgh, should take first place, although his name and work are now almost forgotten. Veitch was a friend of Sir David Brewster and, we are told, 'the construction of telescopes was his most favourite occupation. The curves of his specula, and also of the lenses for achromatic object-glasses, he determined most carefully and laboriously. Much of his time was, however, wasted in mere mechanical work, in which he delighted, of tubes, stands, and other apparatus, which could have been better done by ordinary workmen.'[93] Veitch's optical work was known to a wide circle of scientific friends. Sir Walter Scott calls him 'one of the very best makers of telescopes, and all optical and philosophical instruments, now living', but says that Veitch 'prefers working at his own business as a ploughwright, excepting at vacant hours.'[94] Veitch's optical achievements were known to Sir William Herschel, who invited him to Slough, and, under Veitch's tutelage, Brewster as a young man made his first Newtonian telescope. Veitch was also an observer and the first in England to see the great comet of 1811.[95]

Another ambitious amateur of this period was John Ramage, a tradesman of Aberdeen, who spent over fifteen years experimenting with the casting and polishing of large specula. As early as 1806, Ramage made trials with the Gregorian form and, in 1810, made a 'front-view' Newtonian of 9 inches aperture. In 1817, he made a Newtonian of $13\frac{1}{2}$ inches aperture and 20 feet focus and followed this with three others, each of 25 feet focus with 15-inch mirrors. One of the latter he mounted in a manner similar to Herschel's large 20-foot telescope and offered to let the Council of the Astronomical Society test it at the Royal Observatory, Greenwich.[96] His offer was accepted and, in 1825, we find him in London supervising the erection and preliminary testing of his instrument. At first, the observers were impressed with the mirror's light-grasp, but closer examination revealed that the figure was far from perfect. A persistent zonal aberration limited its effective aperture to 11 inches and, with a cover over the offending zones, it still failed to give critical definition. Only the lower powers of the impressive range of eyepieces from 100 to 1500 could be used, and we are not surprised to find that the instrument was dismantled a few years later. Ramage claimed that the mounting was safe and convenient to use, but even this seems doubtful, for Sir David Brewster fell off the observing platform when it was at its highest elevation. Brewster, however, wrote that 'the apparatus for moving and directing the telescope is extremely simple and displays much ingenuity'[97]—Ramage was a fellow countryman, and this fact no doubt influenced his criticism.

Undeterred by the disappointing performance of his 25-foot telescope Ramage drew up plans for a telescope of 21 inches aperture and 54 feet focus. But he died suddenly in 1835 and the mirror, although made, was never mounted. Thomas Dick, who visited Ramage in 1833, says that he had another 25-foot telescope erected in an open space in Aberdeen. Cloudy weather prevented Dick from viewing stars through it, but he saw 'two or three large speculums, from 12 to 18 inches in diameter' which had a high polish. 'He told me, too' Dick writes,[98] 'that he had ground and polished them simply with his hand, without the aid of any machinery or mechanical power—a circumstance which, he said, astonished the opticians of London, when it was stated, and which they considered as almost incredible.' Of the great physical labour which Ramage must have expended over his mirrors a further impression may be gained from his own statement that it was not until after 'the experience obtained in casting and polishing upwards of a hundred 15-inch mirrors besides numerous smaller ones '[99] that he was able to improve standard processes. Ramage died in Aberdeen in December, 1835, aged fifty-two years.

A much less practical but nevertheless sincere experimenter was Charles, third Earl of Stanhope, who drew up plans for erecting telescopes of immense size and capable of magnifying many thousand diameters. At one time, the optician Varley revealed that he had cast mirrors for Stanhope far larger than any ever made previously and which he planned to mount in giant stands.

> His Lordship's vast design [Varley writes][100] was no less than the construction of a telescope of 384 feet in Length, with reflectors 6 feet in Diameter. . . . The observer may sit or stand in a warm room and, without ever changing his position, observe more than one half the horizon, the object appearing directly before him, however elevated it may be in the heavens; thus continuing in the easiest posture and without being exposed to the open air.

Needless to say, this fanciful instrument was never constructed and the realization of Stanhope's dreams went no further than the acquisition of a large quantity of mahogany for the stand. At a later date, Sir James South went through the papers which Stanhope bequeathed to the Royal Institution, but failed to find a single hint worth remembering.

The failure of Ramage's telescope and the inability of the London opticians to make reflectors greater than 12 inches aperture meant that, as late as 1840, Herschel's large 20-foot was still the finest and most useful instrument of its class. The ageing 40-foot remained the largest telescope, but such was its dilapidated state that John Herschel dismantled it in 1839. The large 20-foot was too valuable an instrument to lie idle and, after the elder Herschel's death, it was used by his son to review his father's nebulae. It was during this work that John Herschel decided to extend his survey to the southern hemisphere and, in 1833, he set sail with his family and instruments for the Cape of Good Hope.

John Herschel secured a house, named Feldhausen, at Claremont, on the eastern slope of Table Mountain, about six miles from Capetown and surrounded by an orchard near which he erected the 20-foot reflector. In the vicinity stood a little wooden observatory with a revolving roof for South's 5-inch Tulley equatorial (the telescope of which became the finder of the 30-inch Helwân, Egypt, reflector). With this instrument, he intended making micrometrical measures of the more interesting double stars. Herschel took three mirrors with him, one made by his father, another made under his father's supervision, and a third figured subsequently by himself. These he supported in the tube simply but effectively on several thicknesses of woollen cloth, the fibres of which acted as so many tiny springs which shared the weight evenly between them.[101] The same material supported the edge, for Herschel was opposed to the old method of pressing the mirror against a rim in front. The disadvantage of the new method lay in the very softness of the material, for as the inclination of the tube altered, so the mirror's surface became off-square with the tube. It was to check this displacement that Herschel introduced an interior collimating telescope into the tube.[102]

Fig. 85—Ramage's 25-foot reflector at Greenwich

(Royal Astronomical Society)

Fig. 86—The 20-foot reflector at Feldhausen, 1834

Generally speaking, the 20-foot telescope worked well. Compared with modern instruments of the same size, it was clumsy and, for one man unassisted, very tiring to use. This latter point should not be forgotten when we examine the thousands of observations that Herschel made with it. His fine monographs of the Orion nebula (M 42 and 43) and the irregular nebula round η Argus are, in themselves, eloquent testimony of his industry, skill, and patience. To represent these difficult objects, he would first form a framework by taking measures of the principal stars involved and then fill in details on subsequent nights. This task is difficult enough with a modern clock-driven equatorial, let alone with an instrument like the 20-foot pointed to a comparatively uncharted sky. Little wonder, then, that the delineation at the eyepiece of the ' overwhelming complexity ' of objects like the Magellanic Clouds proved too great a task for even Herschel, so that he drew them ' seated at a table in the open air, in the absence of the moon.'[103]

The Cape offered seeing conditions which were often superior to those commonly obtaining at Slough. ' The best nights ' Herschel wrote, ' occur after the heavy rains which fall at this season* have ceased for a day or two; and on these occasions, the tranquillity of the images, and sharpness of vision is such, that hardly any limit is set to magnifying power but what the aberrations of the specula necessitate.'[104] In the hot season, the sky was generally clear, but sometimes belts of

* The cooler months, May to October.

cloud would remain for several nights in succession or hot air currents make the stars ' tremble, swell and waver most formidably.'

Some idea of the light-grasp and resolving power of the 20-foot can be got from the monograph of the nebula surrounding η Argus, with its complicated convolutions and the multitude of stars scattered over it. The drawing of this object occupied several months and Herschel often despaired of ever being able to transfer its ' endless details ' to paper. Other objects were so remote and so faint that even the full 250 square inches of bright speculum metal only just revealed their existence. Certain nebulous masses, for instance, appeared ' only as a gentle radiance ' in the telescope's field. Upon this luminous background ' other objects of striking and indescribable form ' were scattered, each one to be measured, studied and recorded. As if this work was not enough, Herschel made naked-eye comparisons of the brightnesses of stars, observed Halley's comet at its 1835 apparition, and studied and described over 2000 double stars and nearly the same number of nebulae and clusters. With the 5-inch equatorial, he took micrometrical measures of another 417 doubles and gave much attention to the binaries λ Toucani, α Centauri, α Crucis, γ Virginis, and π Lupi. Other observations included a study of Saturn's satellites and, with the 5-inch, details of sunspots. Such was the foundation of modern sidereal astronomy in the southern hemisphere.

For solar observations, Herschel adopted his father's method of placing combinations of green and blue-green filters between his eye and the eyepiece. While these combinations often gave ' most agreeable ' views of the sun there remained the ever-present danger of the glasses cracking. John Herschel therefore suggested [105] the use of a glass primary speculum figured on its front face but unsilvered (silvering techniques in any case coming later with Steinheil and Foucault). Instead of the secondary flat of speculum metal he proposed using a surface of a glass prism of apical angle ' at least 28° 8′ '. The observer could then use a low-density green filter at the Newtonian focus since most of the light and heat passed *through* the prism. This was the basis of the so-called *Herschel solar eyepiece*, a popular item in the equipment of amateur solar observers and now used with achromatic objectives and coated primary mirrors.

Herschel's career as an observing astronomer closed when he left the Cape in 1838. He had not long been in Slough, before the 20-foot was almost forgotten and he was immersed in his early Cambridge investigations on the action of light on salts of silver. These researches, which date back to 1819 and which were independent of Fox Talbot's similar studies, consisted of ' fixing ' photographic images on sensitized paper by thiosulphate, thought by Herschel to be sodium hyposulphate. An interesting application of this process occurred in 1839, just before he decided to dismantle the then dangerous 40-foot telescope. Using one of his aplanatic lenses made by George Dollond, he obtained, after about two hours' exposure, a faint image which bore washing with ' hypo ' and which showed the main ladders and spars silhouetted against the sky. This photograph was taken on a sensitized glass plate, four years before J. N. Nièpce used glass plates covered by a superior albumen process. Fox Talbot's first photographs or, rather, ' photogenic drawings ', date back to 1834. The *Talbotype* or *Calotype* process was not introduced until 1840, one year after Herschel's photograph.

The career of the 40-foot closed as it had begun, with a concert in the tube. On New Year's Eve, 1839, the Herschel family entered the tube, then lowered on the ground, and sang a ballad written by Sir John.[106] The tube remained on the site for some years but eventually a tree was blown on it during a storm, destroying all but ten feet of the lower end. This meagre remainder rests in the garden at Observatory House (Fig. 60, page 131), while the speculum itself has hung for many years in the hall there.

REFERENCES

[1] *Biblio. Univ. des Sc.*, Feb.–March, 1824, 'Some Account of the late M. Guinand' by 'E.R.', p. 9.

[2] *Ibid.*, pp. 10–11.

[3] *Ibid.*, pp. 11–13.

[4] *Ibid.*, p. 17.

[5] Chance, W. H. S., *Proc. Phys. Soc.*, **49**, p. 433, 1937.

[6] *Ibid.*, p. 434. 'E.R.' however, says that Fraunhofer went to les Brenets and 'engaged M. Guinand to take a journey into Bavaria', p. 21. *Vide* also Rohr, M. von, *Trans. Opt. Soc.*, **27**, p. 283, 1926.

[7] *Vide* Ref. 1, pp. 13–16.

[8] Chance, *op. cit.*, p. 435.

[9] *Ibid.*, p. 436.

[10] *Ibid.*, p. 436.

[11] *Ibid.*, pp. 436–437.

[12] *Ibid.*, p. 437.

[13] *Vide* Ref. 1, p. 24.

[14] *Ibid.*, pp. 17–18, p. 15.

[15] Gautier, A., 1825, *Coup d'œil sur l'état actuel de l'astronomie pratique en France et en Angleterre*, pp. 255–259.

[16] Brewster, D., 1853, *A Treatise on Optics*, p. 502.

[17] Jeans, G., 1841, *Practical Astronomy*, p. 288.

[18] Cooper, E. J., 1851–56, *Catalogue of Stars near the Ecliptic, observed at Markree during the years 1848 to 1856.*

[19] Bauernfeind, C. Max von, 1887, *Gedächt. auf J. von Fraunhofer zur Feier seines Hundertsten Geburtstags*, p. 9. Rohr, M. von, *op. cit.*, p. 284.

[20] Fraunhofer, J., *Denkschriften de K. Acad. der Wissenschaften zu München*, **5**, pp. 193–226, 1815. Rohr, M. von, *Zeitschrift für Instrumentenkunde*, **46**, pp. 273–289, 1926.

[21] *Mem. R.A.S.*, **2**, pp. 93–94, 1826. Brewster, D., *Edinb. Jour. Sc.*, **5**, p. 105, 1826.

[22] Struve, F. G. W., 1837. *Mensurae Micrometricae.*

[23] Struve, F. G. W., *Mem. R.A.S.*, **2**, p. 95, 1826.

[24] Bell, L., 1922, *The Telescope*, pp. 82–83, 91–92.

[25] *Mem. R.A.S.*, **2**, p. 96, 1826.

[26] Bessel, F. W., *Astr. Nach.*, **8**, pp. 397–426, 1831. For technical details of this and subsequent heliometers *vide* Ambronn, L., 1899, *Handbuch der astronomischen Instrumentenkunde*, ii, pp. 552 620.

[27] Clerke, A., 1885, *History of Astronomy, during the Nineteenth Century*, pp. 46–47.

[28] Beer, W., and Mädler, J. H., 1837, *Der Mond....*

[29] Robbins, F., *J.B.A.A.*, **49**, pp. 210–216, 1939.

[30] Stone, E., Supplement to Bion's *Instruments de Mathématique*, 1758 (Stone's translation), cited by Martin, L. C., *Trans. Opt. Soc.*, **24**, p. 300, 1923.

[31] Wollaston, W. H., *Phil. Trans.*, **92**, i, p. 378, 1802.

[32] Fraunhofer, *op. cit.*, **5**, pp. 193–226, 1815. *Edinb. Phil. Jour.*, **9**, p. 296, 1823; **10**, p. 26, 1824, gives translation.

[33] *Ibid.*

[34] *Edinb. Phil. Jour.*, **10**, pp. 37–38, 1824.

[35] *Ibid.*, p. 39.

[36] Fraunhofer, J., *op. cit.*, **8**, pp. 28–34, 1821.

[37] Ford, E., 1946, *David Rittenhouse*, pp. 139–140.

[38] Young, T., *Phil. Trans.*, **92**, i, pp. 35 37, 1802.

[39] Chance, *op. cit.*, pp. 437, 441.

[40] *Ibid* pp. 437, 441.

[41] Chance, *op. cit.*, p. 442.

[42] *Ibid.*, p. 443.

[43] *Ibid.*, p. 443.

[44] Brewster, D., Ref. 21, p. 110.

[45] Herschel, J. F. W., Correspondence between Herschel, Fraunhofer, South and Tulley is preserved at Observatory House, Slough, as also are samples of Fraunhofer's glass. *Vide* also Rohr, M. von, *Trans. Opt. Soc.*, **27**, p. 290, 1926; **28**, p. 131, 1927.

[46] *M.N.R.A.S.*, **7**, p. 54, 1845; **9**, p. 147, 1849.

[47] *Phil. Trans.*, **120**, p. 3, 1830.

[48] *Ibid.*, p. 34.

[49] *Ibid.*, p. 17.

[50] Smyth, W. H., 1850, *A Cycle of Celestial Objects, including the Ædes Hartwellianae*, p. 159.
[51] *Ibid.*
[52] *Nicholson's Journal*, **1**, pp. 1–13, 1797.
[53] Dick, T. 1845, *The Practical Astronomer*, pp. 271–272.
[54] Webb, T. W., *Intellectual Observer*, **7**, p. 186, 1865.
[55] Tilloch, *Phil. Mag.*, **2**, p. 24, 1798–1799. Patent Nos. 1473, 1800.
[56] Barlow, P., *Phil. Trans.*, **118**, p. 105, 1828; **119**, pp. 33–46, 1829; **121**, pp. 9–15, 1831; **123**, pp. 1–13, 1833.
[57] Brewster, Ref. 16, p. 516.
[58] *Phil. Trans.*, **119**, p. 33, 1829.
[59] *Ibid.*, p. 40.
[60] *Ibid.*, p. 38.
[61] Barlow, P., *Proc. Roy. Soc.*, **3**, p. 14, 1830–1831.
[62] *Ibid.*, p. 15.
[63] *Mem. R.A.S.*, **2**, pp. 507–511, 1826.
[64] *Phil. Trans.*, **123**, p. 2, 1833.
[65] *Ibid.*, **124**, pp. 205–207, 1834.
[66] Rogers, A., *M.N.R.A.S.*, **1**, p. 71, 1828.
[67] Dick, *op. cit.*, p. 282.
[68] Webb, *op. cit.*, pp. 182–183.
[69] Rohr, M. von, *Trans. Opt. Soc.*, **26**, p. 291, 1926, points out that S. Stampfer and J. J. Prechtl tried to teach J. Voigtländer and G. S. Plössl the art of making good telescopes.
[70] *Mem. R.A.S..*, **2**, p. 507, 1826.
[71] *Ibid.*, p. 508.
[72] *Ibid.*, p. 509.
[73] *Ibid.*
[74] *History of the R.A.S.*, 1923, p. 16.
[75] Herschel, J. F. W., *Phil. Trans.*, **111**, pp. 111–167, 1821.
[76] *Journal of Science*, Royal Institution, **26**, p. 386, 1822.
[77] Conrady, A. E., 1929, *Applied Optics and Optical Design*, pp. 400–401.
[78] *Phil. Trans.*, **115**, p. 12, 1825.
[79] Kitchener, W., 1825, *The Economy of the Eyes*, ii, p. 296.
[80] Smyth, *op. cit.*, p. 153.
[81] Smyth, W. H., 1844, *A Cycle of Celestial Objects*, i, p. 339.
[82] Pearson, W., 1819, in Rees' *Cyclopaedia*, art. 'Telescope'
[83] *Mem. R.A.S.*, **5**, p. 386, 1833.
[84] Smyth, Ref. 50, pp. 154–155.
[85] Smyth, Ref. 81, p. 373.
[86] *Ibid.*, pp. 374–377.
[87] *Phil. Trans.*, **124**, pp. 199–207, 1834.
[88] *Ibid.*, **111**, pp. 101–103, 1821.
[89] Rochon, A. M. de, *Jour. de Phys.*, **53**, pp. 169–199, 1801. Maskelyne, N., *Phil. Trans.*, **67**, p. 799, 1777.
[90] Court, T. H., and Rohr, M. von, *Trans. Opt. Soc.*, **30**, p. 237, 1929, and Price Lists in the Court Collection, Science Museum, London.
[91] Kitchener, *op. cit.*, ii, p. 83, and Price Lists in the Court Collection.
[92] Kitchener, *op. cit.* Also, 1815 and 1818, *Practical Observations on Telescopes*; 1811, *A Companion to the Telescope.*
[93] Brewster, D., Ref. 16, p. 24.
[94] *Ibid.*, p. 25, cites letter from Sir Walter Scott dated 1818.
[95] *Ibid.*, p. 25.
[96] Ramage, J., *Mem. R.A.S.*, **2**, pp. 413–418, 1826. Brewster, Ref. 16, pp. 496–497. Pearson, W., 1829, *Introduction to Practical Astronomy*, ii, pp. 79–82.
[97] Brewster, D., 1831, *A Treatise on Optics*, p. 357.
[98] Dick, *op. cit.*, p. 310.
[99] *Mem. R.A.S.*, **2**, p. 413, 1826.
[100] *Journal Arts and Science*, **1**, p. 36, 1820.
[101] Herschel, J. F. W., 1847, *Cape Observations*, p. xiii.
[102] *Ibid.*, p. xiv; *Phil. Trans.*, **123**, p. 488, 1833.
[103] Herschel, Ref. 101, p. 144.
[104] *Ibid.*, p. xvi.
[105] *Ibid.*, pp. 436–437.
[106] Weld, C. R., 1848, *History of the Royal Society*, ii, p. 195.

CHAPTER X

> Many practical men whom I have spoken to think that after Fraunhofer's discoveries, the refractor has entirely superseded the reflector, and that all attempts to improve the latter instrument are useless.
>
> LORD ROSSE

SIR JOHN HERSCHEL'S *Cape Observations*, published in 1847, were followed by the results[1] of Struve's studies with the Dorpat refractor. While the latter confirmed the conclusions reached by the elder Herschel concerning stellar distribution and the general pattern of the system of stars, they also suggested that Fraunhofer's refractor was just as capable as a reflector of twice the aperture. In 1839, a formidable rival to the reflector appeared in the 15-inch equatorial for the Pulkowa Observatory. From America also came reports that Merz and Mahler were supplying a similar instrument for Harvard College Observatory. The successors of Fraunhofer seemed to have the manufacture of flint glass well in hand. Small wonder, therefore, that many English astronomers thought that the refractor was superseding the reflector and that attempts to improve the latter would be useless.

William Parsons, third Earl of Rosse, did not share this opinion; he saw no reason why a reflector far larger than Herschel's 48-inch could not be made. Certainly no one could have contemplated building the world's largest telescope, for such was his ultimate object, under more propitious circumstances.

Parsons was born at York on June 17, 1800, was educated at Dublin and Oxford and, while still an undergraduate, had represented King's County in Parliament.[2] Upon his father's death in 1841, he succeeded to the earldom, became sole owner of the ancestral estate of Birr Castle and, as it were, feudal lord of the village of Parsonstown. He possessed a considerable income, was master of his time, and had every facility for erecting a foundry and workshops. To these material blessings he added a practical knowledge of engineering and great interest in astronomy. The only drawback—the absence of skilled labour in Parsonstown—he remedied by training labourers on his estate. Under a capable smith named Coghlan, these men were eventually working lathes, building grinding and polishing machines, and casting metal.

Lord Rosse first experimented with compound specula, for he had little hope of casting large disks by ordinary foundry methods. The first of these, a 6-inch, consisted of two parts—a central adjustable disk mounted inside a fixed ring 1½ inches wide. Both plates were of speculum metal and were cemented to a flat brass base. They were then ground and polished together to a spherical figure, when the centre was moved back relative to the outer annulus by turning three screws let into the brass base. This caused the two images to coincide and reduced spherical aberration by about 50 per cent.[3] The next experiments were with solid specula, ribbed at the back as Herschel had once proposed,[4] and carefully annealed. The

Fig. 87—William Parsons, third Earl of Rosse

results were discouraging and Rosse had to admit that he could not cast a mirror of the modest size of 15 inches without adding more copper to the alloy.[5] The alternative was to cast specula in separate pieces and then to unite them by tinning the surfaces. This was found practicable and led to the following plan.[6] A deeply ribbed and comparatively light brass plate was first prepared by riveting and soldering together eight brass sectors. It was then turned smooth and flat on one side and this was then tinned. The plates of speculum metal were now arranged in their places on the brass base, and the composite disk was gradually heated in an oven. As soon as the tin fused, a resinous flux was applied and the sectors were moved a little until it was apparent that the tin had acted upon the speculum metal. The heat was then reduced over a period of about five days, when the speculum was ready to be ground.

The first mirror to be made in this way was a 15-inch of 12 feet focus. This was mounted roughly in a stand similar in design to Ramage's. With powers from 80 to 600, the new moon appeared 'very perfectly defined'. Polaris and ε Boötis (separations 18″·4 and 2″·6 respectively) 'and some other stars not requiring high powers' were well shown.[7] High powers were not used owing to difficulty in controlling the motion of the telescope. The 15-inch composite mirror was followed by a 24-inch and, finally in 1840, by one of 36 inches aperture. The latter was intended for a 26-foot tube, ready mounted on a large altazimuth stand. At the same time, Rosse managed to cast a 36-inch solid speculum. With the apparatus in working order, he was now able to make comparisons between the two mirrors. Dr T. R. Robinson summarizes the results as follows:[8]

Both specula, the divided and the solid, seem exactly parabolic, there being no sensible difference in the focal adjustment of the eyepiece with the whole aperture of 36 inches, or one of twelve; in the former case there is more flutter, but apparently

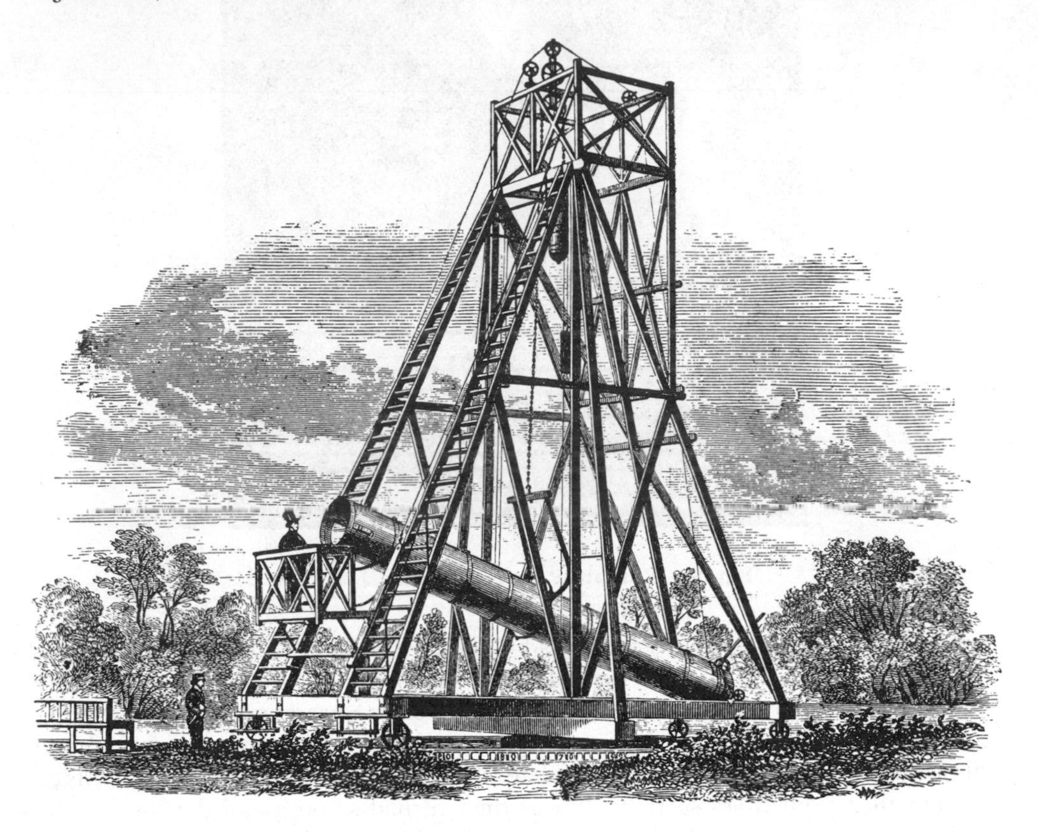

Fig. 88—The Earl of Rosse's 3-foot reflector

no difference in definition, and the eyepiece comes to its place of adjustment very sharply.

The solid speculum showed α Lyrae round and well-defined, with powers up to 1000 inclusive, and at moments even with 1600; but the air was not fit for so high a power on any telescope. Rigel, two hours from the meridian, with 600, was round, the field quite dark, the companion, separated by more than a diameter of the star from its light, and so brilliant that it would certainly be visible long before sunset.

γ Orionis, well defined, with all powers from 200 to 1000, with the latter a wide black separation between the stars; 32 Orionis and 31 Canis Minoris were also well separated. [Separations 2″·7, 1″·0, and 1″·5 respectively.]

These and other observations were made in bad weather with unsteady air almost every night. After midnight, when the mirrors seemed at their best, the sky became overcast—a foretaste of the general observing conditions at Parsonstown.

The difference between the two mirrors was most marked when bright stars were examined with high magnifications, the component plates of the compound speculum giving rise to a noticeable diffraction effect. This could be reduced by diminish ing the number and size of the joints, but there was an obvious limit to this. In any case, the solid speculum gave such excellent results that, when Rosse began to consider his next and largest instrument, he never gave composite mirrors another thought.

The mounting of the 3-foot reflector was similar to that of Ramage's 25-inch, except that the heavy movable parts were counterpoised and thus more easily manipulated. The telescope was used as a Newtonian, for Rosse found that an inclination of the axis of the mirror to the axis of the tube of only 11 minutes of arc introduced sufficient astigmatism to mar the definition. The correct inclination for the Nerschelian form in this case would have been over 3 degrees and, as Robinson pointed out, ' The observer's position at the mouth of the tube, must cause currents of heated air, which will materially interfere with sharpness of definition.'[9] When not in use, the mirrors were preserved from moisture and ' acid vapours ' by connecting their boxes to chambers containing quicklime, an arrangement devised by Robinson some years previously.

Robinson considered that Rosse had improved the brightness and sharpness of reflector images ' to an extent which even Herschel himself did not venture to contemplate'[10] .In his opinion, the 3-foot was ' the most powerful telescope that has ever been constructed '.[11] Yet he had to admit that so little had been published about the performance of the 48-inch, that no fair comparison could be made. Herschel seldom used a high power with his telescope and failed to mention the fifth and sixth stars in the trapezium of the Orion nebula, both readily visible in the 3-foot. One thing is obvious—that Rosse had already progressed further than Herschel in mastering the difficulties of casting, figuring, and mounting large mirrors.[12] He realized from the beginning that a speculum performed best at the temperature at which it was figured. To ensure this, he immersed the 36-inch disk in a cistern of water maintained at constant temperature. All working operations, even to final figuring, were performed by a machine driven by a small steam engine. The mirror, already mounted on a specially designed bed of weighted levers, rotated slowly in the cistern, with the water reaching almost to its surface and with the cast iron tool, counterpoised above, pressing lightly upon it. Two adjustable cranks imparted a slow side-to-side motion to the tool which also rotated under the imparted motion of the mirror and, with the aid of sand and emery, ground it to a spherical figure. The grinder was then covered with pitch and the speculum figured with rouge by increasing the length of stroke and rate of rotation. Here a new obstacle arose, for the pitch was deformed even under slight pressure and therefore polished the metal unequally. Rosse thereupon borrowed another idea of Herschel's—that of grooving the polisher into small squares. The intermediate grooves acted as channels for dispersing the rouge, accommodated any excess pitch and, at the same time, prevented the adhesion of the two surfaces. The softness of the pitch still prevented

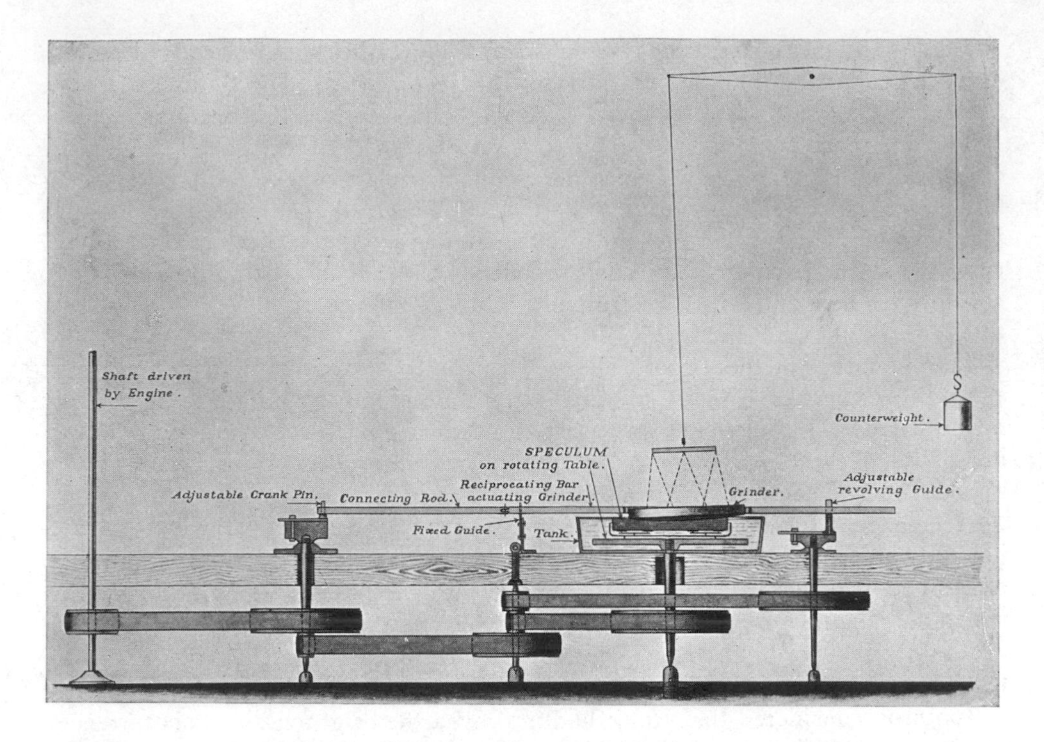

Fig. 89—Drawing of Rosse's machine for grinding and polishing 3-foot specula

(*Science Museum, London. British Crown copyright*)

the working of a good figure, which required that the medium holding the rouge particles should be as hard as possible. Rosse, therefore, applied the lap to the tool in two layers—a foundation of pitch and turpentine covered by a harder compound of flour and resin. Water was introduced to moisten the rouge and, as the final figure approached, the addition of a solution of soap and ammonia improved the surface brightness. During the last stages of the polish, the figure was tested by viewing a watch-dial placed at the centre of curvature, some 50 feet above the surface. 'The curvature' Rosse writes, 'I measure as Mudge did, by means of diaphragms; and when the surface is true, the separate portions of it, though the general figure may be indifferent, will define sharply, and their focus can be ascertained with great precision.'[13]

As soon as Rosse knew that the 3-foot reflector was a success, he drew up plans for an instrument of double the size. A foundry large enough to take three furnaces was erected, together with a mould, annealing oven, and crane and tackle for carrying crucibles from furnace to mould.[14] An alloy of 126·4 parts copper and 58·9 parts tin by weight was then formed into ingots and remelted at a low temperature so as to reduce porosity. To do this, the furnaces, together with the annealing oven, consumed over two thousand cubic feet of peat. The metal was contained in three cast iron crucibles, one for each furnace and each capable of holding $1\frac{1}{2}$ tons. As the metal melted, each crucible was lifted by the crane to three iron baskets placed at the side of the mould. Each basket turned on a pivot and was provided with a lever, the depression of which tipped the melt into the mould.

Rosse knew that the success of his 36-inch mirror casting had depended, to a large degree, on the construction of the mould. He had seen some small castings by Cuthbert, a London optician, and these had been cast in iron moulds and so chilled instantaneously.[15] These ' chilled ' castings were sounder and harder than sand-castings, although the technique adopted in making them was unsuitable for a 36-inch disk. Rosse eventually evolved a mould in which heat was abstracted rapidly and at equal rates from the lower surface of the metal. This consequently solidified in successive strata from the bottom upwards and a perfectly sound casting resulted.

The 72-inch disk was cooled in the same way. The bottom of the mould was formed by binding together layers of hoop-iron and turning the required convexity on them edgewise. It was thus porous to air although solid enough to prevent the metal from escaping. This laminated bottom was 6½ feet in diameter and 4 inches[16] thick and was set horizontal by means of spirit-levels. A wooden pattern, the exact size of the speculum, was then placed on the base; moist sand was rammed between this and the surrounding wooden frame, and the pattern was removed.[17] As the melt came into contact with the iron base plate it chilled into a hard, dense sheet; the upper surface, being in contact with air, naturally solidified last.

The casting took place on April 13, 1842, at 9 o'clock in the evening and in the open. The crucibles were two hours heating in the furnaces before the metal was introduced and this took ten hours to become sufficiently fluid for pouring. To Dr Robinson, the moment of casting, with its suspense and excitement, was like a scene on another planet.[18]

> Above, the sky, crowded with stars and illuminated by a most brilliant moon, seemed to look down auspiciously on their work. Below, the furnaces poured out huge columns of nearly monochromatic yellow flame, and the ignited crucibles during their passage through the air were fountains of red light, producing on the towers of the castle and the foliage of the trees, such accidents of colour and shade as might almost transport fancy to the planets of a contrasted double star.

While red hot and scarcely solid, the mirror was encircled by an iron ring and dragged by chains and windlass along a railroad to the centre of the annealing oven. Here it was left to cool for sixteen weeks, when it was removed to the grinding machine. Like the 36-inch, the mirror was placed directly on a bed of levers and the work of grinding and polishing proceeded as before. The definition of the watch-dial was again taken as a criterion of the figure's accuracy, while final tests were to be made with the mirror in its tube, directed to celestial test-objects. Unfortunately, before these were possible, the mirror was accidentally broken.[19] Another casting with a higher copper content had a surface covered with minute fissures.[20] Most of these were ground out over a period of nearly two months, the speculum was polished and passed the tests for definition. A third casting cracked in the annealing oven and a fourth was ' as porous as pumice-stone.'[21] The fifth and last, a perfect casting in every way, was figured and mounted within a month.[22]

Rosse planned to use the 6-foot as a Newtonian and he made several plane specula. These he ground and polished on the same large machine but with smaller polishers. He gives no details of his method and we can only assume that he found, contrary to general experience, that the preparation of an optical flat was a comparatively simple task. The test for flatness was not an accurate one. A distant object was viewed with a well-corrected achromatic telescope first direct and then by reflection in the mirror. Any distortion or falling off of definition was thus due to errors in the flatness of the mirror.[23]

Towards the end of 1842, Rosse began to erect the mounting,[24] the predominant features of which were two parallel piers of solid masonry, each 72 feet long and 56 feet high on the inner side. They were 24 feet apart and lay in the plane of the meridian. The west wall carried three observing galleries; the east wall, an iron semicircle 85 feet in diameter for guiding the tube in altitude. The tube was 56 feet long, 8 feet in diameter and made of deal planks hooped with iron rings and strengthened by diaphragms. The upper end carried the Newtonian flat and eyepiece mount, the lower terminated in a cubical wooden speculum-box, bolted to a massive cast iron equatorial mounting set in brickwork. The telescope was thus virtually a circum-meridian instrument with equatorial movements,[25] its range being from near horizontal through the zenith to the pole. The maximum lateral movement was limited to 15° or one hour of right ascension. Near the zenith, this range was extended a little, but still proved a serious handicap when it came to making regular observations of one particular object. To raise or lower the tube, the observer required the assistance of two workmen who turned a windlass on the ground; he then gave it the necessary slow motions by turning two small handwheels on the observing platform. Polar distances were indicated by an arm which was attached to the near horizontal declination axis of the mounting and which moved over a graduated arc of 6 feet radius.

The mounting of a mirror four tons in weight presented some difficulty. Rosse adopted the weighted lever and triangle system, which had answered well with the 3-foot, but with eighty-one bearing points instead of only twenty-seven.[26] These 'points' were actually ball-and-socket joints, fixed at the apices of twenty-seven cast iron triangles. The triangles were, in turn, carried by nine secondary and three primary triangles, the latter being mounted on the carrying frame or base. The eighty-one balls thus supported twenty-seven equal portions of speculum and, so long as this remained in a horizontal position, relieved it from all strain. But as soon as the tube was lowered, the weight of the supporting system alone—some 600 lb—on the primary triangle was so great that Rosse had to introduce a system of compensated levers. To prevent the mirror from moving laterally, it was also suspended in a ring which at the same time relieved some of the weight from the triangles. Lastly, the carrying frame was supported in the speculum box by three screws which also allowed the mirror to be set perpendicular to the tube. By placing the tube perpendicular and unscrewing these, four wheels under the frame touched a railroad which ran to the castle workshops.

Work on the mounting continued all through 1843 and well into 1844. By February, 1845, the first of the two mirrors (casting No. 2) had received a fair polish and was considered worth a trial in the tube. Lord Rosse, Dr Robinson, and Sir James South observed together, hoping to see the Orion nebula, but the sky clouded every time the nebula came within range. On February 13, however, and without an eyepiece, they saw the components of Castor ' far apart ' and the individual stars of M67.[27] Some days later, the figuring was carried a stage further and the mirror tried again. Several nebulae were observed and γ Leonis (2″·8) and ϵ Virginis (230″)* resolved with powers from 400 to 800. Regulus appeared ' neat and round, without appendages or flare.'[28] The brightness of the images surpassed expectation. When Jupiter entered the field, the effect was as if a coach lamp had been placed in the tube, and certain clusters appeared ' such as man before had never seen ' and which for ' magnificence baffled all description.'[29] With a power of 828, γ^2 Andromedae was seen as two separate stars, the separation being 0″·5 at that time. The moon's surface offered a profusion of minute craters, gullies, and streaks.

The mirror was now considered good enough for making regular observations of Herschel's nebulae. In this sphere, its light-grasp revealed that many objects listed as ' nebulous ', ' planetary ' or ' diffuse ', were masses of faint stars. In one nebula, M51 in Canes Venatici, curved filaments of nebulosity gave to the whole the appearance of a spiral.[30] Several planetary nebulae were also resolved into rings and spirals, ' indicating ' Robinson writes, ' modes of dynamic action never before contemplated in celestial mechanics.'[31]

Rosse supplemented the great telescope with many interesting pieces of apparatus. The tube had no finder, so he made a low-power eyepiece with a field of view of 31 minutes of arc and a field lens 6 inches in diameter.[32] Other eyepieces could be used in pairs, for the eyepiece mount had a sliding frame fitted with two adaptors; a low power could thus be replaced by a higher power merely by moving the slide. For drawing faint and extensive nebulosity, a micrometer was essential, but even with the mirror's great light-grasp, the fainter details of some nebulae would have been extinguished by even the faintest illumination of the field. When Rosse drew M51, he rubbed the micrometer threads with phosphorus.[33] An attempt to heat platinum wires electrically failed because the air also became heated and air currents spoiled the definition. After the introduction of the wet plate, Rosse photographed the moon, constructing a sliding clock-driven plate-holder which moved at the same rate as the moon's image. The resulting pictures were interesting, but too devoid of detail to be of practical value.[34] About 1869, a clock-drive was added[35] and proved indispensable when it came to drawing planetary markings.

The 6-foot telescope resolved so many nebulae that Rosse and Robinson came to the conclusion that ' no REAL nebula seemed to exist . . . all appeared to be clusters of stars '.[36] Bond's investigations in America with the 15-inch Harvard refractor, and his ' resolution ' of the Orion nebula, strengthened this belief and, among the ' resolved nebulae ', Rosse included both this object and the Andromeda nebula.† But, in 1864, Sir William Huggins turned a spectroscope to the Orion nebula and found, not an absorption spectrum, but the bright emission lines of glowing gas. After this discovery, Rosse designed a 70-lb spectroscope, a monster for those days, and saw for himself the bright lines of the Orion nebula.[37] He made a long study of this object and his final drawing, the result of many nights'

* A well-separated pair, but the fainter star is about magnitude 12.

† The Andromeda nebula was resolved photographically by Hubble in the 1920s with the 100-inch Mount Wilson reflector.

Fig. 90—The Earl of Rosse's 6-foot telescope

(By courtesy of the Director of the Science Museum, London)

observation, measured two feet square and was as faithful a representation as hand and eye could give.

The ' Leviathan of Parsonstown ', as our grandparents calledthe Rosse telescope, took three years to build and cost £12,000. Yet the only discovery of significance made with it was that of the spiral structure of certain nebulae. The causes of its meagre output are not difficult to find. From the first, the weather conspired against the instrument. Clouds and haze covered the sky for days on end and, when a break did occur, the observers were unable to take full advantage of it owing to the restricted lateral motion of the tube. They were, in fact, using a transit instrument and, once having seen an object, had to wait twenty-four hours before they could examine it again. The proximity of the Bog of Allen did not improve weather conditions,[38] and only on rare occasions could the full aperture be used to advantage. At other times, the 3-foot reflector bore equal magnifications, gave better performance and was much more manageable. ' When the air is unsteady ' Rosse writes, ' minute stars are no longer points, the diffused image is much fainter, and single stars, easily seen when the air is steady, are no longer visible.'[39] On looking through the reports of different astronomers on the performance of the 6-foot, one cannot but be struck by their repeated references to the bad state of the

Fig. 91—The second 3-foot Rosse telescope

Notice the counterpoised observing 'chair' giving access to the Newtonian focus.

(By courtesy of the Director of the Science Museum, London)

weather. When the mirror was ready for trial in the tube, for instance, the observers waited in vain to glimpse the Orion nebula. ' Unfortunately the whole month of February ' Dr Robinson writes, ' was of the worst astronomical character; and though the great speculum had only the imperfect polish already noted, it was kept in the tube as long as there were any hopes of seeing the great nebula in Orion. That, however, was always clouded while within its range.'[40]

In August, 1848, Sir George Airy visited Lord Rosse to witness some trials of the telescope and, on the 29th of that month, wrote to his wife:[41]

> The night was uncertain: sometimes entirely clouded, sometimes partially, but objects were pretty well seen when the sky was clear. . . . From the interruption by clouds, the slowness of finding with and managing a large instrument (especially as their finding apparatus is not perfectly arranged) and the desire of looking well at an object when he had got it, we did not look at many objects.

On the 31st he wrote to say that ' The weather here is still vexatious: but not absolutely repulsive ' and, in the same letter, ' The night promised exceedingly well:

but when we got actually to the telescope it began to cloud and at length became hopeless.'[42]

The definition of the two mirrors seems to have been fairly good, star disks with powers of 800 and 1000 appearing as points or very nearly so, and being free from rays and stray light. The second mirror (casting No. 5) was of $3\frac{1}{2}$ tons weight and ' considerably weaker ' than the first. Rosse writes that ' by carefully comparing the two specula at low altitudes, we have been made thoroughly sensible of the great importance of strength in preventing flexure. . . . I think there would still be sufficient gain to make it worth while to cast a third speculum considerably heavier than either of the others.'[43] Despite every precaution, it is doubtful whether the performance of both mirrors was as good at low as at high altitudes. As temperature dropped, so the mass of metal cooled, but at a slower rate, and, in cooling, changed its figure. This effect was probably masked or confused with faults due to the mirror supports.

South once stated that the mirrors *could* be uncovered and the tube trained on a given object in less than eight minutes.[44] This, to say the least, was an exaggeration, especially after Airy's statement. At least four workmen were required to assist the observer. One at the winch raised or lowered the tube, another gave it an eastward or westward motion, a third moved the gallery so as to keep the observer near the eyepiece, and the fourth looked after the lamps and attended to minor matters. Whilst the observer froze in the balcony 50 feet above ground, these individuals probably solaced themselves with thoughts far from astronomical.

The Rosse telescope would have been more effectively used had its maker been as interested in observing as he was in instrument-making. His special interest in the instrument seems to have ceased when the last nail was driven into it.[45] Then the troubled years during and following the great Irish potato famine, which began almost as soon as the telescope was completed, made other demands upon his time. For twenty-two years he took his place among the Irish peers in the House of Lords and, from 1848 to 1854, undertook the duties of president of the Royal Society. Yet the telescope never remained idle for long, being placed at the disposal of any astronomer whose work required the use of an exceptionally large aperture. Dr Robinson became associated with Lord Rosse before the 3-foot was constructed. For twenty years, Dr G. J. Stoney made regular observations of nebulae and clusters with the 6-foot in company with Lord Oxmantown, the fourth earl and Lord Rosse's eldest son.[46] Sir Robert Ball, author of popular books on astronomy, used it as a young man for the two years ending 1867. By then its novelty had, to some extent, died away, but Lord Oxmantown continued his observations until 1878. The observatory was dismantled in 1908, but not before several astronomers —R. Copeland, J. L. E. Dreyer, Otto Boeddicker, and others had put the Rosse telescopes to useful work.*

Between 1874 and 1876 the 3-foot was re-mounted on a fork-type equatorial (Fig. 91) and performed so favourably in comparison with its ageing companion that the latter fell into gradual disuse.[47] To supplement the 3-foot, Lord

* The 72-inch mirror was rescued from the neglected and somewhat derelict telescope in 1913 or 1914 and is now in the Science Museum, London. It is not known with certainty whether this is the mirror of casting No. 2 or of casting No. 5.

Oxmantown made an 18-inch reflector of 10 feet focus. This instrument, driven by a water-clock, was used primarily for micrometrical measures of star fields.[48]

In constructing the 6-foot reflector, Rosse demonstrated that the building of large telescopes could no longer be regarded as 'a perilous adventure, in which each individual must grope his way.'[49] His improved methods of casting specula, and his large grinding machines, solved many earlier difficulties and encouraged younger men to follow suit. Lassell, De la Rue, and Draper all caught his enthusiasm and, when they made their first specula, adopted many of his methods. Another valuable legacy was the warning that high powers and large apertures were wasted in a climate where clouds interrupted observations, and indifferent seeing conditions monopolized the greater part of the year. Unfortunately, the wisdom of this advice was ignored, and it was not until the end of the century that astronomers began to consider mountain-top observatories.

Before passing to the work of the next mid-nineteenth-century telescope-maker, William Lassell, we must mention James Nasmyth, inventor of the steam hammer. Nasmyth was a master engineer of Patricroft, near Manchester, and had the advantage over Rosse since he was an experienced founder and possessed, at his Bridgewater factory, all the appliances for casting specula. His first casting, an 8-inch of 32 parts copper, 15 tin and 1 arsenic, was cast in a sand mould and slowly cooled in an annealing oven. But, after many hours' grinding and polishing, the disk broke into fragments.[50] Nasmyth thereupon cast another of the same size, using the surplus metal in the crucible. This residue, when remelted and cast, seemed tougher and brighter than before. This time, he used a metal mould and obtained a perfect casting; when ground and polished he considered it worth mounting as a Newtonian. 'I was most amply rewarded' he writes, 'for all the anxious labour I had gone through in preparing it, by the glorious views it yielded me of the wonderful objects in the heavens at night. My enjoyment was in no small degree enhanced by the pleasure it gave to my father, and to many intimate friends.'[51] He tells, also, how in a fit of enthusiasm he used to observe from his small garden at the side of the Bridgewater Canal, clad only in his nightshirt and at all hours of the night. In this guise he was apparently seen one night by a boatman, who set the story going that a ghost was seen at Patricroft, moving among the trees with a coffin in its arms.[52]

Nasmyth's largest mirror was a 20-inch which he used as a Cassegrain, mounting it in a sheet-iron tube made to move on trunnions like a cannon. One trunnion was perforated and took the eyepiece, the rays from the small convex mirror being reflected in this direction by a third and plane mirror fixed at 45° to the axis of the tube. Tube and trunnions were mounted on a large turn-table and, by turning a small handwheel near his seat, Nasmyth could direct the telescope to any part of the sky without moving (Fig. 92).*

Upon his retirement to Penshurst, Kent, Nasmyth used this instrument for a detailed study of the sun and moon. He made many observations of sunspots, using high powers and choosing moments when the seeing was at its best. These led to the discovery of the so-called 'willow-leaves'.[53] The sun's surface appeared to be resolved into a compact pattern of luminous filaments which, however, later

* This telescope is now at the Science Museum, London.

observers have not confirmed. Nasmyth adopted the same methods when he observed the moon, hoping thereby to discover the cause of its craters, walled plains, and bright ray systems. The result was his volcanic theory—the suggestion that lunar craters were magnified examples, both in origin and appearance, of those on the earth.[54]

William Lassell was the first to apply Fraunhofer's equatorial mounting to large reflecting telescopes. His 9-inch Newtonian of aperture ratio *f*/12·5 was mounted in this way, and stood in a small observatory at Starfield, near Liverpool.[55] This instrument was not the only example of his handiwork, however, for from his schooldays until about 1844, he made several mirrors of about this size. In 1844, Lassell considered making a 24-inch reflector and, as a preliminary step, visited Parsonstown where he inspected Lord Rosse's workshops and watched part of the erection of the 6-foot reflector. Back in Liverpool, he made a grinding and polishing machine similar to Rosse's, but altered it a little later so that the tool imitated more closely his own particular hand movements.[56] Otherwise, the machine retained many of Rosse's original features. The mirror, for instance, rested on its bed of levers in a water cistern which rotated with it under the polisher. But, whereas in Rosse's machine the strokes were nearly straight, here they were circular. This motion was achieved through a gear train, so mounted that any point on the polisher described epicycloidal or hypocycloidal curves whose size could be varied at will (Fig. 93).

Lassell added a small quantity of arsenic to the composition of his specula, but their permanence and brilliancy were due more to the purity and correct relative proportion of tin and copper. Lassell must have realized this, for he omitted arsenic from the two 48-inch mirrors which he made later. He used cast iron moulds, the Rosse composite mould being too expensive and difficult to make. These he tilted and, after pouring in the molten metal at the lowest point, gradually restored to the horizontal so that the metal ran evenly to the other side. In the 24-inch, the weight of the mirror, 370 lb, was shared equally by eighteen brass disks. These were divided into nine pairs, each pair being connected by a bar, while the centres of the bars rested on the apices of a triangular plate. This plate was then supported at its centre by a large screw which passed into the end-plate of the tube.[57] Lassell found that this arrangement prevented sensible flexure of the mirror so long as temperature remained fairly constant and the tube was used near the vertical. Distortion which occurred in lower altitudes was overcome by casting a series of parallel bars or ridges on the back of the mirror, each rib acting as an area of support for the pads of a number of counterbalanced levers.[58] These exerted their maximum effort when the tube was horizontal and became less operative as it was elevated.

With the 9-inch, Lassell independently detected the sixth star of the trapezium of θ Orionis and, with a power of 450, satisfied both himself and the Rev. W. R. Dawes that Saturn's exterior ring consisted of two distinct annuli.[59] With the 24-inch, he discovered Triton, the satellite of Neptune. He first saw this minute body on October 10, 1846, but could not satisfactorily follow it until the next year, when further observation fully confirmed its existence.[60] In 1848, while making observations of Saturn, he saw a thirteenth-magnitude star near the plane of the rings.

Fig. 92—Nasmyth's 20-inch Cassegrain-Newtonian, *circa* 1845

Nasmyth is at the fixed eyepiece

This proved to be Hyperion, Saturn's eighth satellite in order of discovery, seen simultaneously by G. P. Bond* with the 15-inch Merz refractor at Harvard.[61] After a long and careful search, he confirmed the existence of Umbriel and Ariel,[62] two satellites of Uranus additional to the two discovered by the elder Herschel in 1787.[63]

In the autumn of 1852, Lassell took his 24-inch telescope to Valetta, Malta, where he hoped the better climate would enable him to extend his planetary studies to observations of nebulae. Here he continued his search for further satellites and

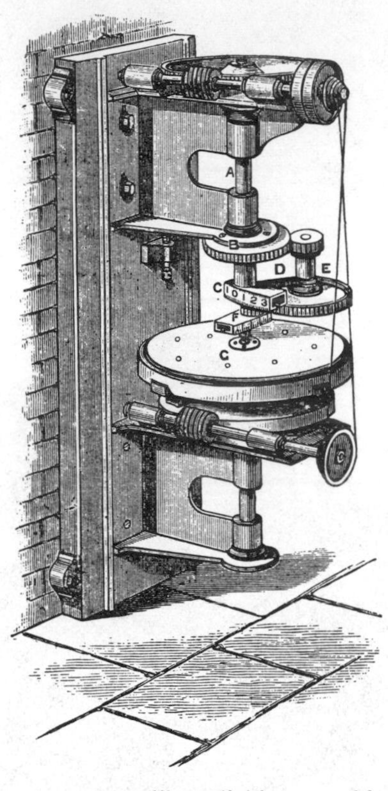

Fig. 93—Lassell's polishing machine

(From Lockyer's 'Stargazing', 1878)

made drawings of planetary and other nebulae. The search was, to use his own words, ' rather negative than otherwise ', but it showed that no further satellite of Uranus or Neptune could be detected without great increase in optical power. Writing of Uranus in January, 1853, he says ' I am fully persuaded that . . . he has no other satellites than these four.'[64] He met the need for a larger aperture in 1861, when he erected a 48-inch equatorial at Malta,† but the instrument failed to reveal further satellites. In 1949, however, Gerard P. Kuiper, second director of

* *Vide* p. 249.

† Constructed at Bradstones, near Liverpool, during 1859 and 1860.

Fig. 94—Lassell's 48-inch reflector at Malta

(*Royal Astronomical Society*)

the Yerkes and McDonald Observatories,* discovered a fifth satellite of Uranus† and a second satellite of Neptune, on photographs taken with the 82-inch reflector on Mount Locke, Texas.[65]

Lassell provided the 48-inch with two alternative mirrors, each about $4\frac{1}{2}$ inches thick and each weighing well over a ton. The tube, 37 feet long, was of original design, being made of a latticework of flat iron bars. ' The object of making the tube of this form ' Lassell writes, ' was to prevent the possibility of any currents of differently heated air in the tube; or of any inequality of the internal and external temperatures—it appeared to answer this end perfectly.'[66] The section near the mirror was also ' open ' but, a little higher up, a solid cradle enabled the entire tube to be rotated about the optical axis. The mount was of the fork variety and surmounted a heavy cone-shaped casting which formed the polar axis. Lassell thought that a driving clock of the size necessary to move so large an instrument would be inconvenient and unreliable. He therefore connected the polar axis through a gear system to a fly-wheel and winch handle. By turning the latter once every second, as indicated by a loud ticking clock, an assistant could keep an object in the field of view. The motion could be stopped, accelerated, and retarded at pleasure and, according to Lassell, could be ' continued for hours without being oppressive.'[67]

Progress in hour angle was read from a dial attached to the clock mechanism, whilst a second dial indicated decreasing eastern hour angle. On commencing a night's observation of, say, the Orion nebula, Lassell set the dials to the hour angle of the nebula and left the assistant to turn the handle. If, at any time, observations stopped owing to cloud, the assistant could rest until the sky cleared. He need then refer only to the dials and, by turning a quick-motion winch, bring up the telescope to correspond. In the meantime, Lassell had ascended the observing tower or ' colossal sentry-box ' by means of an internal staircase and had gone to the platform or ' storey ' nearest the eyepiece. As the telescope retreated from the tower, a second assistant moved the latter on its circular track. Lassell writes that the storey afforded ' abundant room for papers, micrometers, eyepieces, lamps, and any other small apparatus required; beside furnishing to the observer a most grateful shelter from the dew, and occasionally from inclement wind.'[68] Even so, there must have been times when Lassell found his position anything but comfortable. ' Almost all altitudes are equally convenient ' he writes, ' by the adaptation of the several attitudes of kneeling, sitting, or standing, none of them irksome when not continued too long.'[69]

Lassell's work at Malta left no doubt as to the value of large apertures for searching for new satellites. It showed, once again, that a large instrument performed better in an atmosphere whose ' quietude ' nearly equalled its transparency. But, if weather generally favoured his work, the winters were little better than those in England, and sometimes the telescope remained idle for weeks. On these occasions, a smaller instrument performed better and proved capable of following faint satellites once their presence was known. As Lassell once pointed out,[70] the brightness of a planet in a large telescope can reduce the sensitivity of the eye for

* Second director, that is, of the joint observatories.

† First photographed in February, 1948.

detecting faint satellites, especially those close to the primary. In addition, atmospheric disturbances were more pronounced in a large instrument. During bad weather, Lassell checked the motions of the telescope and polished and interchanged the mirrors, work which involved the complete re-adjustment of the optical components.

Two extracts from a letter from Lassell to W. H. Smyth[71] give some idea of the remarkable light-grasping powers of the 48-inch telescope.

> Sirius is an object I would challenge any one to see as I see it, for the first time, without an involuntary exclamation of wonder! With 260, it is an incandescent diamond, which I cannot describe. . . . The pencil of light transmitted through the eyepiece from this star, cast a brilliant spot on the wall of the observatory sufficient forcibly to attract my notice, without being at all prepared for it. . . .
>
> All the stars I see [in the Orion nebula] are individual, isolated, and rather unusually brilliant points, without apparently any connection with it. Examined under good circumstances, with a power of 1018, the brightest parts of the nebula look like masses of wool . . . one layer seemingly laying partly over another, so as to give the idea of great thickness or depth in the stratum.

Lassell stayed in Malta for three years and was assisted in the last two by A. Marth. Between them, they made observations of the planets and their satellites, drew up a catalogue of the places of 600 new nebulae and monographed many others. The following extracts from Lassell's *Miscellaneous Observations*[72] give some idea of the appearances of the Orion nebula, planets, and double stars.

> 1862, Jan 20. Observed the great Nebula of Orion, powers 231 and 285. I do not detect many new stars, but I gain a vivid impression of a scroll-like form in the brightest part of the Nebula, south preceding the Trapezium, as remarked by the late Professor G. Bond. I can, however, get no impression of resolvability, properly so called. The Nebula indeed rather retreats from the stars than concentrates around them.
>
> Jan. 22, 4h.30m. Observed Venus with 285. A fine sharp view, but the atmosphere quivers much; as indeed might be anticipated from the Sun shining all day on the telescope. The cusps are excessively sharp and fine. The surface of the planet is shady and mottled, without any decisive mark. The portions of the limbs near the cusps are sharper than any other parts of the circumference of the disk, and I am persuaded that the boundary of the unilluminated portion is wavy or serrated.
>
> Jan. 24. 7h.5m. Uranus. The four satellites well seen. The fainter ones, Ariel and Umbriel, nearly equal in brightness, but I can detect no others with various powers, even up to 1480. The fainter satellites, though incomparably less bright than Oberon and Titania, are yet conspicuous with powers from 500 to 1000.
>
> Jan. 25. 12h. 45m. Saturn appeared without his ring, excepting the portion crossing his disk, which appeared as a dusky line, very slightly but yet sensibly convex towards the south—doubtless the edge of the unenlightened side of the ring projected on the planet. All the eight satellites are visible, unless I am mistaken in what I take to be Hyperion.
>
> Sept. 27. The atmosphere being fine I turned the telescope on δ Cygni,* power 466. With full aperture and this power there are certainly a good many rays round the bright

* Dawes, in his 1867 Catalogue of Double Stars (*Mem. R.A.S.*, **35**, pp. 137–502, 1867) recorded the brighter star yellowish and its companion blue. Separation about 1″·6. This star is a binary with period 321 years.

star, but the small one is conspicuous among them all—constantly and steadily visible. The companion is much redder and duller in colour than its primary, and is separated from it a full diameter of the large star. With 760, I am forcibly reminded of the appearance of ε Boötis in the Two-foot Equatorial. . . . The hardness and roundness of the disks in this fine atmosphere are strikingly pleasing. I was never more struck with the conviction how necessary a pure tranquil sky is to the just performance of a very large telescope.

Upon his return to England, Lassell settled at Ray Lodge, Maidenhead, Berkshire. The 48-inch was never re-erected, but the 24-inch was used until failing eyesight and health forced him from his work. From 1870 until his death in 1880 he seldom used the telescope, and it was finally re-erected at Greenwich. His disappointment was keen when the committee appointed to consider the erection of a large telescope at Melbourne, N.S.W., decided to build a new instrument instead of accepting his offer of the 48-inch. The telescope was ultimately consigned to scrap-metal merchants. 'I may add' Lassell writes, 'that when witnessing the breaking up of the specula, I was not without a pang or two on hearing the heavy blows of sledge-hammers necessary to overcome the firmness of the alloy.'[73]

In 1840, a young man visited Nasmyth to consult him about appliances for making white lead. 'I was then busy with the casting of my 13-inch speculum' writes Nasmyth. 'He watched my proceedings with earnest interest and most careful attention. He told me many years after that it was the sight of my special process of casting a sound speculum that in a manner caused him to turn his thoughts to practical astronomy.'[74] The visitor was Warren De la Rue who, a few years later, astonished astronomers with his photographs of the moon, stars, and planets. With the joint assistance of Nasmyth and Lassell, De la Rue mastered the difficulties of speculum-making, even to the extent of figuring 13-inch disks. 'I usually succeed in producing thirteen-inch mirrors' he writes in 1852, 'which define the planets . . . in a manner rarely equalled, and never surpassed, by any of the refractors which I have yet had an opportunity of looking through. I am, however, free to confess that they [the refractors] defined a fixed star much more satisfactorily than my best mirrors.'[75]

De la Rue's largest reflector was a 13-inch equatorial of 10 feet focus, the work of his own hands except for the speculum casting, which Nasmyth made for him. This instrument he set up in his garden in Canonbury, London, and at first made a series of drawings of Jupiter, Saturn, and Mars.[76] In the autumn of 1852, he began to experiment with F. Scott Archer's newly invented collodion wet-plate process and, with exposures of only 10 to 30 seconds, obtained pictures of the moon large and clear enough to show the main surface features.* On moving to Cranford, Middlesex, five years later, he devoted his energies and a large part of his income to

* The first photographs of the moon were daguerreotypes taken in 1840 by J. W. Draper of New York (*vide* page 267). Using a heliostat and a flint-glass prism, Draper also took a daguerreotype of the solar spectrum (September, 1841).

The first daguerreotype of the sun was made by Fizeau and Foucault on April 2, 1845. An exposure of 1/60th of a second was used and two large sunspot groups were recorded.

The first daguerreotypes of a solar eclipse were attempted by Majocchi of Milan but he obtained impressions of only the partial phases. The first successful daguerreotype of a solar eclipse was taken by Berkowski with the 6¼-inch Königsberg heliometer on July 28, 1851.

Fig. 95—De la Rue's Kew photoheliograph

It is shown erected in a temporary observatory for photographing the 1874 transit of Venus

astronomical photography. The reflector was mounted in a covered observatory on a 15-foot pier, a clock-drive was added, and the floor of the observatory was high enough to allow him to use the space underneath as a darkroom.[77] By degrees, he so improved on the quality of his lunar photographs that they bore enlargement to several times their original size. The set of twelve which he published about 1865 was on 18-inch plates, and showed ' nearly every mountain or object of importance . . . [with] a minuteness of perfection that would scarcely have been deemed possible.'[78]

Between 1858 and 1866, De la Rue took a series of photographs of the crater Linné, to investigate Schmidt's allegations that it was undergoing changes.[79] He also produced stereoscopic pictures, taking two exposures of the moon after an interval of a few minutes and viewing the final prints through a stereoscope.[80] For this original work, he received the Gold Medal of the Royal Astronomical Society.

De la Rue also pioneered in solar photography, a subject which had become prominent following S. H. Schwabe's discovery of the periodicity of sunspots and Sabine's subsequent discovery of the identity of this period with that of magnetic disturbances. Sir John Herschel repeatedly advocated the construction of a photoheliograph for photographing the sun daily[81] and, at length, the Royal Society advanced the necessary funds. De la Rue designed the instrument, while the optician A. Ross was commissioned to make it under his supervision (Fig. 95). The most interesting feature was the objective, a 3½-inch achromat of 50 inches focus, corrected for violet photographic rays. Exposures of a fraction of a second were made possible by a spring-loaded roller-blind shutter which operated in the focal plane.[82]

The photoheliograph was installed under a dome at Kew Observatory, now used by the British Association for researches in terrestrial magnetism. Progress was at first slow; two years elapsed before most of the difficulties of this novel work were mastered, and the death of John Welsh, the superintendent, delayed research. In 1860, De la Rue took the instrument to Rivabellosa, Spain, to observe, with Secchi, the July total solar eclipse. This was the first time that collodion wet-plate photography had been used in eclipse work, and the results were encouraging. The prominences came out clearly, and the plates proved conclusively that these appendages had their origin in the sun.[83] One prominence, which De la Rue likened to a Turkish scimitar, reached a height of 70,000 miles. But the faint coronal light, easily seen by the naked eye, was conspicuously absent on all plates.

On his return to England, De la Rue erected the photoheliograph at Cranford. In 1861, it was again removed to Kew and placed in the hands of Balfour Stewart, the new superintendent. During the next ten years, De la Rue and Stewart took a series of plates which covered a complete solar cycle. In 1873, the instrument finally came to rest at Greenwich, where the daily records were continued, although with more up-to-date equipment.

In 1873, also, De la Rue's Cranford Observatory was dismantled and the 13-inch reflector passed to Oxford University Observatory. ' One use to which I should like to see my reflector applied ' De la Rue wrote to Charles Pritchard, ' is the determining whether or not the moon has a physical libration; for this purpose, photograms of this planet would have to be taken as often as practicable, and the original negatives measured by means of a properly constructed micrometer.'[84] This work, readily undertaken by Pritchard, made clear that photographs were susceptible of very accurate measurement. Pritchard thereupon tested the photographic method on stellar parallax determinations, taking 61 Cygni as a test case. The parallax of this star* had, by 1887, been determined many times. From the reduction of no fewer than 30,000 bisections of star images on 330 photographic plates, Pritchard obtained eight independent determinations of the parallax of this double star.[85] These determinations, in good agreement among themselves and with earlier results, encouraged Pritchard to proceed further. In 1889 and 1892 he was able to publish the parallaxes of twenty-nine stars, some of the second magnitude. We shall see in Chapters XV and XVII how F. Schlesinger, using large American refractors, greatly improved this important photographic method.

* A triple system, with two components of magnitude 5·6 and 6·3 and the third unseen but detected through the perturbations in the motions of the two bright stars.

REFERENCES

1 Struve, F. G. W., 1847, *Études d'Astronomie Stellaire.*

2 Ball, R. S., 1907, *Great Astronomers*, p. 272.

3 *Edinb. Jour. Sc.*, **7**, p. 25, 1828. *Scientific Papers of Wm. Parsons, Third Earl of Rosse* (*S.P.P.*), 1926, pp. 1–4, p. 99.

4 Dreyer, J. L. E., *Scientific Papers Sir W. Herschel*, Introduction p. lvi.

5 *S.P.P.*, p. 10.

6 *Ibid.*, p. 11; *Edinb. Jour. Sc.*, **9**, p. 136, 1830.

7 *S.P.P.*, p. 11.

8 *Ibid.*, p. 17.

9 *Ibid.*

10 *Ibid.*, p. 15.

11 *Ibid.*, p. 19.

12 For full account of casting, polishing and testing 36-inch specula *vide Phil. Trans.*, **130**, pp. 503 ff., 1840. *S.P.P.*, pp. 80–104.

13 *S.P.P.*, p. 100.

14 For full account of moulds and of the casting and annealing of 72-inch specula *vide Phil. Trans.*, **151**, iii. pp. 681–745, 1861; *S.P.P.*, pp. 125–133.

15 *S.P.P.*, p. 130.

16 *Ibid.*, p. 127

17 *Ibid.*, p. 21.

18 *Ibid.*, p. 21.

19 *Ibid.*, p. 128

20 *Ibid.*

21 *Ibid.*, p. 129.

22 *Ibid.*

23 *Ibid.*, p. 101.

24 For description of mechanical parts *vide S.P.P.*, pp. 143–145, p. 150 and Plate 24.

25 Or nearly so; the mounting was a universal joint with its primary axis directed east to west. *Vide S.P.P.*, p. 144, for alterations introduced by the changing position of the suspension chain.

26 *S.P.P.*, pp. 133–135.

27 *Ibid.*, p. 25.

28 *Ibid.*, p. 26.

29 *M.N.R.A.S.*, **9**, p. 120, 1849.

30 *S.P.P.*, p. 36, pp. 114–115.

31 *Ibid.*, p. 56.

32 *Ibid.*, p. 34.

33 *Ibid.*

34 *M.N.R.A.S.*, **40**, p. 231, 1880.

35 Ellison, M.A., *J.B.A.A.*, **52**, p. 270, 1941–42. Lord Rosse died in 1867 and the driving clock was added by the fourth earl.

36 *S.P.P.*, p. 29.

37 *Ibid.*, pp. 203–206. Ball, W. V., 1915, *Reminiscences and Letters of Sir R. Ball*, pp. 68–69.

38 Ball, Ref. 37, p. 66.

39 *S.P.P.*, p. 113.

40 *Ibid.*, p. 25.

41 Airy, G. B., 1896, *Autobiography*, p. 198.

42 *Ibid.*, p. 199.

43 *S.P.P.*, p. 130.

44 *Ibid.*, p. 25.

45 Ball, Ref. 2, p. 287.

46 Ball, Ref. 37, p. 65.

47 *M.N.R.A.S.*, **69**, p. 251, 1909.

48 Ellison, *op. cit.*, p. 270.

49 *S.P.P.*, p. 14.

50 Smiles, S., 1883, *James Nasmyth: an autobiography*, p. 406.

51 *Ibid.*, p. 408.

52 *Ibid.*, p. 329.

53 Nasmyth, J., *Mem. Lit. Phil. Soc. Manchester*, ser. 3, i, p. 407, 1862.

[54] Nasmyth, J., and Carpenter, J., 1874, *The Moon*, 2nd edit.
[55] *Mem. R.A.S.*, **12**, pp. 265–272, 1842.
[56] *Ibid.*, **18**, pp. 1–15, 19–20, 1850. *M.N.R.A.S.*, **8**, p. 197, 1848; **9**, p. 29, 1849; **13**, p. 43, 1853.
[57] *B.A. Report*, pp. 180–181, 1850; *Mem. R.A.S.*, **18**, p. 17, 1850.
[58] *B.A. Report*, p. 182, 1850.
[59] *M.N.R.A.S.*, **6**, p. 11, 1843; *Mem. R.A.S.*, **18**, p. 196, 1850.
[60] *M.N.R.A.S.*, **9**, p. 87, 1849. Denning, W. F., 1891, *Telescopic Work for Starlight Evenings*, pp. 223–224.
[61] *Mem. R.A.S.*, **18**, pp. 21–25, 1850.
[62] *M.N.R.A.S.*, **11**, p. 248, 1851.
[63] *Phil. Trans.*, **77**, pp. 125–129, 1787.
[64] Smyth, W. H., 1860, *A Cycle of Celestial Objects, including the Ædes Hartwellianae*, p. 85.
[65] *J.B.A.A.*, **60**, p. 40, 1950.
[66] *Mem. R.A.S.*, **36**, p. 1, 1866.
[67] *Ibid.*, p. 3.
[68] *Ibid.*, p. 2.
[69] *Ibid.*, p. 3.
[70] *M.N.R.A.S.*, **11**, p. 26, 1851.
[71] Smyth, *op. cit.*, p. 165.
[72] *Mem. R.A.S.*, **36**, pp. 33–35, 1866.
[73] *Observatory*, **1**, p. 179, 1877.
[74] Smiles, *op. cit.*, pp. 327–328.
[75] *M.N.R.A.S.*, **13**, p. 44, 1853.
[76] *History of the R.A.S.*, 1923, p. 155.
[77] *Ibid.*
[78] *Intellectual Observer*, **3**, p. 228, 1863.
[79] Clerke, A., 1885, *History of Astronomy during the Nineteenth-Century*, p. 312–314.
[80] *M.N.R.A.S.*, **19**, p. 40, 1859.
[81] Ref. 76, p. 157.
[82] *M.N.R.A.S.*, **19**, p. 357, 1859.
[83] Ref. 76, p. 156; also Clerke, *op. cit.*, p. 214.
[84] Pritchard, A., and Turner, H. H., 1897, *Charles Pritchard, Memoirs of his Life and of his Astronomical Works*, p. 267.
[85] *Ibid.*, p. 302.

CHAPTER XI

The circles that have lately been constructed by instrument makers of our own times, may be said to be micrometrical all round their circumference, and to leave nothing more to be hoped for on the score of accuracy both of construction and division.

WILLIAM PEARSON

THE death of Langlois about 1740 was followed by a gradual decline in French instrument-making. His successor, Charité, did his best to maintain its prestige by constructing instruments modelled on Ramsden's, but he had neither the money nor equipment to develop them further. J. D. Cassini's* predicament at the Paris Observatory can well be imagined. Along with minor instruments, he possessed only a few Langlois quadrants and two or three mediocre achromatic telescopes. He tried in vain to get Canivet's successor, Lennel, appointed to the observatory, but had to rely instead on Doyen, a less experienced workman. Lennel, however, offered to repair several of the older instruments, and Cassini speaks highly of his ability. The prospect for Cassini brightened somewhat when, about 1775, in response to his repeated appeals, the French Academy voted 240,000 francs for reconditioning the observatory. One stipulation must have caused him some uneasiness—it required the instruments to be made by French artists who, as he well knew, lacked the experience and skill of their English counterparts. 'Ramsden and the Dollonds are geometers and scientists' he once wrote; 'our best artists are only workmen.'[1] The Academy's idea was, of course, to encourage French artists, whose interests no one could have had more at heart than Cassini. He therefore ordered a large mural quadrant from Charité, a 3-foot meridian circle from Lenoir, 'the most skilful artist known for the construction of these instruments', and a Sisson-type equatorial refractor from Megnie, who had just won the Academy's prize for astronomical instruments.[2]

Cassini's misgivings as to the reliability of the Paris instrument-makers were soon realized. After some months, when the work should have been well in hand, Megnie had not started the equatorial and Lenoir was asking for 1500 francs in advance and all the necessary metal. The latter also overlooked the fact that he did not belong to the all-powerful Corporation of Master Founders whose trustees, just after the work had started, descended on his workshop, seized his tools and castings, and charged him 36 francs for their trouble.[3] Only the prompt intervention of the police stopped what was tantamount to robbery, and Cassini, to free instrument-makers from further tyranny, used his influence with de Breteuil to get them State recognition. He was successful and, in 1787, Lenoir, Carroché, Charité, Fortin, and a few others became the first members of the new Corporation of Mathematical and Optical Instrument-Makers.

* J. D. Cassini, often called Cassini IV, great-grandson of J. D. Cassini (Cassini I), the first director.

By 1790, Lenoir was able to show Cassini a circle ready to receive the graduations and final adjustments. Charité was still struggling with the large quadrant, and it was becoming apparent that both he and Lenoir had too much smaller work on their hands, work on which they depended for a living. ' Our artists are poor ' Cassini complained to de Breteuil,[4] ' none of them is able to make the necessary outlay for machinery and for procuring proper means for making accurate instruments with promptitude and certainty; this need the Government should supply. Establish at an Observatory a large workshop where artists can find the best machinery and where they can train others.' Such an establishment, he declared, could not fail to produce all the necessary instruments with minimum delay.

De Breteuil again came to Cassini's assistance and arranged for Charité and his family to reside permanently at the observatory. Cassini, however, was not amenable to this and, rather than consider any alternative, Charité refused the offer. Lalande then recommended Megnie for the post, and Cassini forthwith had him installed in the observatory ' to prepare all the necessary tools and equipment.' All went well until October, 1786, when Megnie suddenly vanished, taking with him nearly 20,000 francs and leaving Cassini to face the creditors.[5] Cassini never completely recovered from this last blow to his cherished plans and, as far as we know, took little further interest in French instrument-making. Even his efforts to establish a suitable glassworks in Paris failed, the downfall of de Breteuil at the time taking away the necessary financial support. The observatory workshops were dismantled, the foundry and tools were used for making guns. Two years later, Paris was locked in the bitter struggle of the Revolution.

In 1788, Cassini went to London, for George III was anxious to include France in Roy's triangulation. Accompanied by Méchain, Legendre, and Carroché, he visited Ramsden's workshops where he saw instruments that surpassed all those of his countrymen:

> The fertility of this artist's genius [he wrote to de Breteuil],[6] the perfection of his achievements and his consumate experience in his art, compel me to acknowledge that for now and for a long time it will be difficult to attain or imitate his work. In all honesty I must warn you that our artists, to make much inferior instruments, will cause you to spend twice the cost of Ramsden's masterpieces and, in approaching this artist for our instruments, we shall serve the double purpose of saving the King's money and acquiring models for our artists to copy.

De Breteuil acted on this advice and, a little later, Ramsden received orders for a transit instrument and an 8-foot quadrant. He was also asked if he would accommodate two French instrument-makers in his workshops to study his methods. With characteristic adroitness, Ramsden replied that, while he himself would welcome their company, he was afraid their presence in the workroom would make his assistants jealous and therefore hinder the progress of the work.

We are not surprised to find that the French transit instrument, ordered from Ramsden in February, 1788, and promised for August of that year, failed to arrive. M. Restif, secretary to the French ambassador in London, made repeated visits to Ramsden's workshops, but to no avail. In February, 1790, Count Brühl, ambassador of Saxony in London, wrote that all the larger parts were complete and merely

Fig. 96—Edward Troughton

Cooke, Troughton, and Simms Ltd)

required assembling. He consoled Cassini with the observation that the Duke of Saxe-Gotha had paid Ramsden for a meridian telescope which had lain unfinished in his workshop for five years.[7] In January, 1791, A. M. de Rochon wrote that he had not ' ceased to torment Ramsden but had received in return only promise after promise.'[8] In 1793, when Cassini left the Paris Observatory, the transit was still unfinished although half paid for. Seven years later, Ramsden died, leaving Berge to complete the instrument, so that it did not arrive in Paris until September, 1803.

Ramsden's demise had comparatively little effect on the high standard of British instrument-making. To some extent, this was due to the efforts of Berge, but the main reason was that another firm had arisen to take his place. ' The brothers Troughton ' Cassini wrote, ' are the best artists after Ramsden.'[9]

Edward Troughton, the better known of the two brothers, was born at Corney, Cumberland, in 1753. He spent his boyhood on his father's farm, but in 1770 joined his eldest brother, John, in London. John Troughton was then in Surrey Street, Strand, managing a small instrument-making business left him by his uncle, Edward Troughton.[10] In 1782, the two brothers established themselves at 136 Fleet Street, under the sign of ' The Orrery ', as successors to Benjamin Cole. ' The Orrery ' was even then an old and well-known business. Benjamin Cole's father (of the same name) took over the business from Thomas Wright in 1748, and there is evidence to suggest that Wright succeeded John Rowley.[11] If this connection is true, then ' The Orrery ' establishment was of over eighty years' standing, for Rowley probably set up in Fleet Street before 1700.[12] That the business was of good repute will be judged from the fact that Rowley was ' Master of Mechanicks ' to George I. Rowley made, besides surveying instruments and globes, some of the first orreries, while Wright (who for a time lived with Graham) was ' Mathematical Instrument Maker in Ordinary ' to George II.[13]

Edward Troughton worked with his brother at 136 Fleet Street, first as an apprentice and then as a partner. Both brothers suffered from the family trait of colour-blindness and were unable to specialize in lens-work.[14] This they usually delegated to George Dollond or the Tulleys, concentrating instead on the improvement of graduated instruments. At first, they manufactured simple surveying and drawing instruments but, in 1778, John Troughton made himself a dividing engine[15] on the plan of Ramsden's second machine, and began to graduate sextants and protractors. A little later, his brother devised a new way of dividing and testing arcs, ' but as my brother ' he writes,[16] ' could not readily be persuaded to relinquish me a branch of business in which he himself excelled, it was not until 1785 that I procured my first specimen by dividing an astronomical quadrant of two feet radius.'

Among Edward Troughton's first instruments was a portable quadrant for Bilboa, Spain.[17] He made it at a time when quadrants were giving place to circles and it was thus not only the best but the last of its kind. There were two 3-foot graduated arcs, united by tubular cross-pieces, and two telescopes, one for each arc. Also of this early period was an improved but smaller version of Ramsden's Shuckburgh equatorial.[18] Compared with modern standards, the instrument was flimsy and unstable, with many component parts to work loose. At the time, it was generally considered the best example of its type in finish and workmanship, and superior to Ramsden's stands.

For stability combined with accuracy, Troughton's altazimuth circles were unsurpassed. The circle made for Count Brühl in 1792 ' probably exceeded in accuracy every instrument that preceded it.'[19] Another circle, Troughton's largest, was intended for the Royal Academy at St Petersburg, but the order was countermanded when the instrument was almost complete for fear that it might undergo the fate of Cary's Moscow transit, which was taken by the French at Moscow in 1812. The ever-liberal Pearson then purchased it for his private observatory at South Kilworth, Leicestershire, and asked Tulley to supply the $3\frac{1}{4}$-inch objective.[20] A third circle, known as the ' Lee Circle ' because it eventually came into the hands of Dr Lee of the Hartwell Observatory, was considered the ' most perfect specimen of simplicity in design.'[21] Troughton made it in 1793 for his friend Lowe for £120, the bare cost of materials and labour. The circles were 24 inches in diameter and the telescope 30 inches long. By far the most successful altazimuth, however, was the 30-inch circle[22] which John Pond set up at his house at Westbury, Wiltshire, for which reason the instrument became known as the ' Westbury Circle '.

Not only was the Westbury circle particularly accurate, but Pond was an observer of energy and skill. By measuring the declination of certain fixed stars, he concluded that Bird's quadrant, still in use at Greenwich, could not longer be trusted.[23] Maskelyne and Pond thereupon consulted Troughton, who designed what he considered to be a far more accurate and permanent instrument. This, unfortunately, was a 6-foot mural circle of design so unconventional that the Council of the Royal Society had difficulty in coming to a decision.[24] Troughton turned a deaf ear to all objections until his plan was adopted and, six years later, the great circle took the place of Bird's quadrant. Pond had, by then, succeeded Maskelyne at Greenwich. A born observer, keen and thorough to a degree, Pond found in Troughton both friend and invaluable collaborator. When Pond drew up plans for refurnishing the observatory, Troughton came forward with new and ingenious ideas which Pond was generally willing to consider. More important was the advice and technical ability which Troughton placed at his disposal when he began to investigate, with a view to eliminating, all possible sources of instrumental error.

The new mural circle took the form of an immense brass ' cart-wheel ', fixed at right angles to a 4-foot-long conical axis of the same material. This axis rotated in bearings, fixed at each end of a large semicircular hole cut in the thick supporting wall, so that the circle was close to it and consequently in the plane of the meridian. The limb consisted of two rings, carried by solid brass radial spokes.[25] The second and outer ring had a narrow platinum band which took the divisions. Six micrometer-microscopes read to $0''\cdot1$ and, with all six in use, Pond found the errors of division to be almost insensible. But taking six readings at every observation caused unnecessary delay, and Pond usually used only the two horizontal microscopes.[26] The telescope, of 4 inches aperture and 74 inches focus, turned on its axis, but it could be clamped to any desired position on the circle and so made to rotate with it.[27]

The mural circle was designed to measure the polar distances of stars instead of their zenith distances as with a quadrant. By taking the celestial pole as zero point, measured polar distances became independent of plumb-line and spirit-level settings, also of the latitude of the observatory. To compare measured polar distances with zenith distances, Troughton made a reflecting zenith telescope,[28] but Pond seldom used the instrument.

A ready check on the accuracy of Troughton's circle was provided when a similar circle by Thomas Jones was sent to Greenwich for trial before being shipped to the Cape. Pond found the Jones circle so useful that he obtained permission to retain it at Greenwich.[29] ' The two circles ' we are told, ' are placed in the Observatory, with their axes in a right line, where, with their faces opposed to each other, at about seven feet distance, they seem to regard each other as antagonists; yet is there the same cordiality between them, as there has subsisted between their respective makers for many years.'[30] In 1851, Jones' circle was transferred to the observatory of Queen's College, Belfast.

Troughton's circle was mounted and ready for use in June, 1812; a year later, Pond presented to the Royal Society a catalogue[31] of the polar distances of forty-three stars determined with it and which Bessel pronounced ' the *ne plus ultra* of modern astronomy '

Pond next required a transit instrument from Troughton so that he could determine the right ascensions of stars with the same degree of accuracy as the mural circle measured polar distances. Troughton utilized Peter Dollond's 5-inch achromatic object-glass, lying idle at the time,[32] and the transit thus became larger than was intended. For the rapid location of stars with this instrument, Troughton fixed two graduated polar-distance semicircles at the eye-end of the tube. Each semicircle carried a movable index arm with a level at one end and a vernier at the other, the arm, by reason of the weight of the spirit-level, acting as a plumb-line.[33] Once the tube was depressed below the horizon, however, the semicircles became redundant and had to be replaced by 2-foot circles. This depression became necessary when Pond introduced the artificial horizon, an accessory supposed to have originated with the elder G. Adams about 1750.[34] The artificial horizon was previously used only as ancillary to the sextant; the idea was to view a star direct and then by reflection in the mercury, half the difference in readings being the star's apparent altitude. Pond saw that this method provided an easy means of ascertaining the vertical and checking axis flexure of both meridian circle and transit instrument.

When Troughton designed the Greenwich mural circle, he was fully aware of the advantages of the transit circle, which allowed observations in right ascension and declination at one and the same time. In England, the transit circle was a neglected instrument, the only recent example being Cary's Wollaston circle, the axis of which was too slender and the circle of which too small to render it as effective as it might have been.[35] In 1806, Troughton made a reversible transit circle for Stephen Groombridge, the first accurate instrument of its type in the country. The telescope, of 3½-inch aperture and 5 feet focus, was fixed in the centre of a 3-foot axis, supported at both ends by a stone pier and fitted with two 4-foot circles.[36] For no apparent reason, Troughton disliked this type of instrument and broke up a second after he had already spent £150 on it. 'I was afraid' he said, 'I might grow covetous as I grew old and so be tempted to finish it, and I don't think it is a good kind of instrument.'[37]

Groombridge found his transit circle both easy to use and remarkably reliable. His observatory at Blackheath adjoined his dining-room, and it is said that he often left his dinner, made an observation, and then returned to his meal.[38] With this circle, he fixed the positions of over 4000 circumpolar stars, among their number one with a large proper motion of 7″ a year, since known as the 'Runaway Star' or Groombridge 1830.[39] Some years later, the instrument was purchased by South for his Kensington Observatory, where it remained until 1870.

A description of the many minor improvements which Troughton incorporated in his larger instruments would be tedious and outside the scope of this work. From the brief description of the Greenwich instruments, some idea will have been gained of the devices which Troughton employed. These instruments were but few of the hundreds which left his hands and which found their way into observatories the world over. His mural circles were established at the observatories of Paris, the Cape, St Helena, Madras, Cracow, Edinburgh, Brussels, Cadiz, Armagh, and Cambridge.[40] He made, besides, a large number of small instruments for navigators and surveyors, among them sextants and reflecting circles,[41] divided for him by Fayrer of Pentonville. Many items, including some of his larger instruments, eventually became lumber in the observatories to which they were sent. Once again we have the frequent picture of only a few instruments out of many hundreds contributing anything of permanent value to astronomy.

Troughton died on June 12, 1835, at his premises in Fleet Street which, for sixty

Fig. 97—Troughton's 4-foot transit circle made for Groombridge in 1806

(*Science Museum, London. British Crown copyright*)

years, served as workshop, home, and observatory. Towards the end of his life, increasing pressure of work and his absorption in it made him a recluse, and he left his workshop only to visit friends or to attend scientific meetings. 'A man of simple and frugal habits,' we are told, 'he never married and, it is said, seldom left his back parlour in Fleet Street, where he sat with ear trumpet in hand, wearing clothes stained with snuff and a soiled wig.'[42] To outsiders, this isolation, with his growing deafness and terse, outspoken manner, must have often been taken for unsociability but, by his friends, he was welcomed as much for his company as for his knowledge. In 1826, he took William Simms into partnership.[43] At no time were his relations with the Simms family and the members of his profession other than cordial, and he could well afford to be liberal in his communications both to them and to the learned societies. By the time his rivals—and he considered Ramsden the only serious one—had adopted his suggestions, he had either made others or was improving something else. It was impossible to keep pace with his prolific output. He was elected Fellow of the Royal Society in 1810, and received the award of the Copley Medal for his improvements to instruments and his paper, *A Method of Dividing Astronomical and other Instruments by Ocular Inspection*, communicated to the Royal Society[44] by Maskelyne. In 1779, he was enrolled Freeman of the town of Lancaster and, in 1822, Fellow of the Royal Society of Edinburgh. About 1830, after a visit to Paris, his work was recognized by the King of Denmark in the form of a gold medal.[45] He regularly attended the meetings of the Astronomical Society—of which society he was an original Fellow—and his name appears on the Council at its first meeting on March 10, 1820.[46]

So far as we know, only one incident interrupted the quietude of his later years (about 1829), and that through no fault of his own. From time to time, Troughton and Simms made instruments and carried out repairs for South, whose collection of astronomical instruments was little short of princely. When South purchased Cauchoix's 11¾-inch object-glass, he gave Troughton and Simms the task of mounting it. The mount took the form of an English equatorial which, much to South's annoyance, was from the first slightly unsteady. Troughton offered to correct this and, with the assistance of the Rev. R. Sheepshanks, did so by alterations to the bearing of the lower pivot. South, however, put every obstacle in their way and grew more and more hostile towards Troughton and Simms. He refused to pay the bill and drove Troughton to take proceedings. The verdict came out in the instrument-maker's favour, whereupon South, in a moment of rage, smashed the instrument to pieces.* Fortunately, he spared the objective, which he afterwards presented to Dublin University for the Dunsink Observatory.[47]

When Troughton took William Simms into partnership, the latter was thirty-three years of age, the former, seventy-two and only too willing to share his heavy responsibilities with a younger man. The arrangement lasted five years and proved highly successful, after which period Troughton retired from active work but continued to live with the Simms family.

* South paid about £800 for the mounting. He purchased the Groombridge circle and obtained the Westbury circle for only £50. He got Simms to renovate the latter instrument at a cost of £140. Trouble arose over the work and in March, 1833, Simms accused South of selling the instrument to J. Scott, a London physician, for the apparently exorbitant sum of £400. In 1861, according to Charles Babbage, the instrument was sold at a public auction for £17. 10*s*. 0*d*.

Fig. 98—William Simms

(Cooke, Troughton and Simms Ltd)

There were then two Simms businesses in London. The father of the above-mentioned Simms, William Simms senior, moved from Birmingham to London about the year 1793 and set up as mathematical instrument-maker at 1 Bowman's Buildings, Aldersgate Street. Upon his retirement in 1820, the business passed to two sons, James and George, who carried on at Broadway, Blackfriars, for a further fourteen years.[48] The eldest son, William, was apprenticed to Bennett,[49] one of Ramsden's workmen, and later opened on his own. He was a clever engineer and lived through many of the sweeping changes in his profession due to the Industrial Revolution. He was quick to apply the higher standards of precision required in engineering and assisted the commission set up to investigate the restoration of the Imperial Standards of Length, destroyed by fire in 1835. His self-acting dividing machine,[50] now in general use, was but one of his many inventions. It reduced the work of weeks to the work of hours and was used to graduate many important large circles. Simms was an Associate of the Institute of Civil Engineers, Fellow of the Royal Society, and an active member of the council of the Astronomical Society.[51]

Before he joined Troughton, Simms made a large theodolite for Colonel Colby, Superintendent of the Ordnance Survey, and assisted in the construction of the Cambridge mural circle and reflex zenith tube. Under the joint name 'Troughton and Simms', he made the 8-inch Liverpool and Madras equatorials (object-glasses by Merz), the 36-inch theodolite and zenith sectors for Sir George Everest's great Indian survey, and the 36-inch transit circle of 8 inches aperture for Lord Lindsay's observatory at Dunecht, Scotland. Of these, the Indian survey theodolite excited the greatest admiration and, at the time, was considered the most complete and perfect theodolite ever constructed.[52] Simms made it to the design of Col. A. Strange, a geodesist who introduced many improvements into the design of surveying instruments. With its 36-inch telescope of 3 inches aperture, it enabled angle measures to be taken to an accuracy of 0″·1. Simms also made instruments for other surveys, and telescopes and meridian circles for observatories at home and abroad. He undertook, besides, the construction of special pieces of apparatus for private and professional astronomers, graduated circles for the Northumberland and Greenwich equatorials, made driving clocks, spectroscopes, sextants, eyepieces, micrometers and, when he could get the glass, object-glasses.

William Simms died on June 21, 1860, at Carshalton. In 1836, he had been joined by his nephew, William Simms (son of James Simms) who, in 1860, entered into partnership with a cousin of the same name. The Fleet Street business had, by 1860, moved to larger premises at No. 138, and later the factory part was transferred to Charlton, Kent. At this point, however, we must for the present leave the Simms family or, more particularly, the firm of Troughton and Simms.

Pond died in 1836, leaving his post at the Royal Observatory, Greenwich, to George Biddell Airy, a man of rare and outstanding genius who possessed a wide knowledge of optical instruments. As a student at Cambridge, Airy distinguished himself by his ability at mathematics and optics. As Plumian professor of astronomy, he designed and supervised the erection of the Northumberland equatorial,*[53] the largest and best-mounted instrument in the country. At Greenwich, he at once set about remodelling the observatory on the same lines that had proved so successful at Cambridge. He undertook no less a task than the regular observation of the places of the moon, planets, and asteroids, and the reduction of all the Greenwich observations made since Bradley's time. So intense a programme necessitated first, the gradual replacement of Pond's instruments by larger ones, and second, the recruitment of a much larger staff.[54]

To obtain more frequent observations of the moon's place, particularly in those parts of its orbit where it could never be observed with a meridian instrument, Airy designed an altazimuth in the form of a large theodolite. He introduced three important principles of construction. Firstly, to produce as many parts as possible in a single casting, secondly, to avoid using screws for joining the parts, and thirdly, to reduce the necessity for adjustment to a minimum. The 3-foot horizontal azimuth circle, for example, with its spokes, hub and fixing brackets, was cast in one piece. The 3-foot vertical circle was made in the same way and rotated with the $3\frac{3}{4}$-inch telescope between two solid, vertical supports. Bearings, circles and general framework were all of gunmetal, a copper–tin alloy akin to speculum metal, harder than brass and with fine casting properties. Both circles were silver inlaid and carried major divisions down to 5 minutes of arc. The instrument was supported at the base by a conical gunmetal bearing, fixed to a stone pillar; the upper pivot was held by a triangular framework of metal bars.[55] Considering the many improvements embodied in it, the instrument was not particularly successful—it

* Object-glass by Cauchoix (*vide* p. 180). This object-glass was refigured by F. J. Hargreaves in 1937 (*J.B.A.A.*, **48**, pp. 54–55, 1938).

Fig. 99—Airy's altitude and azimuth instrument, 1847

(Science Museum, London. British Crown copyright)

lacked the rigidity of a meridian mounting. At first, Airy thought it hardly, if at all, inferior to Troughton's mural circle and thought it 'the most important addition to the system of the observatory that has been made for many years.'[56] As with other instruments of this type, the altazimuth enabled Airy to observe the moon's place when it was but a few degrees from the sun. During the course of a year, he found that the instrument enabled him more than to double the usual number of meridian observations.[57]

In 1848, Airy designed a transit circle and so dispensed with both Troughton's transit instrument and the mural circle. The former had become slightly unstable through continual use, and could no longer be relied upon to give consistent readings. The mural circle could no longer be trusted—observations revealed an error in the graduations to the extent of 6 seconds of arc. The telescope of the new instrument was of 8 inches aperture so as to bring within range asteroids and faint stars. Simms made the object-glass for £300 and graduated the circles; Ransomes and May made the rest.[58]

Messrs Ransomes were agricultural engineers, but Charles May, F.R.S., head of the engineering department, had a great interest in astronomy. He built himself an observatory, did most of the mechanical work for Airy's altazimuth, and assisted Airy in designing the transit circle.[59] He did much to develop the Ipswich firm into a large general engineering concern.

Airy's transit circle has now been in use for a century. In conjunction with a chronograph it formed, until 1922, the instrumental basis of time-keeping at Greenwich. The circle is graduated on a silver strip and can be read by six equidistant micrometer microscopes; using all six and taking their mean, the observer can *read* to $0''.01$, but, in practice, instrumental and personal errors considerably raise this limit. Instrumental errors can now be assessed with a high order of accuracy, but personal errors—'personal equation'—vary with the physical condition of the observer and the brightness of the star observed. Personal equation can, however, be almost eliminated by using a moving-wire micrometer, a device first introduced by Professor A. S. Herschel.[60] Instead of observing the passage of a star across a number of fixed wires, the star is bisected by a single wire, the motion of which is so adjusted that the star remains bisected throughout its travel across the field of view. At certain points, electric contacts are automatically closed and records are made on the chronograph. Small 'impersonal-micrometer equations' still persist, but these remain nearly constant for any one observer and can, therefore, be allowed for.

While Troughton enjoyed almost the monopoly of the instrument trade in Great Britain, a new school of instrument-making was being developed in Germany, with origins in the Munich workshops of Fraunhofer and Reichenbach. When Reichenbach left the Munich Institute, he concentrated on the design and manufacture of divided optical instruments. As an engineer, he had always been interested in mechanical problems and foundry methods. Cassini IV, like Airy, had hit upon the idea of dispensing with the many small pieces which made up large instruments by casting them all in one piece.[61] His first casting, a small 22-inch quadrant, was successful, as was that of a complete 5-foot circle made later. At the time of Megnie's disappearance, Cassini had prepared moulds for casting a $7\frac{1}{2}$-foot quadrant,

Fig. 100—The Airy transit circle, 1851

his plan being to make an instrument like Bird's at the Paris Military School. Reichenbach pursued the same idea, but quite independently. As he had been an artillery officer, he was familiar with the manufacture of large castings. This knowledge he turned to good account, constructing theodolites and transit circles unexcelled in rigidity and accuracy. He made, for example, a portable coudé transit[62] in which rays from the object-glass were reflected along the horizontal axis to the eyepiece. The observer was consequently confronted all the time by the graduated altitude circle and could keep his head still for all elevations of the tube. Certain modern theodolites are made on this principle.

Reichenbach's methods were adopted by his successor, G. Ertel, constructor of the Pulkowa vertical circle, the 9-inch meridian telescope for the U.S. Naval Observatory, Washington, and transit circles for observatories at Christiania, Glasgow, Markree, and Warsaw.

Another important engineering firm of this period was that of the Repsolds of Hanover, specialists in transit circles and large telescope mountings. The founder was Johann Georg Repsold, friend of Gauss, Bessel and Schumacher, who was accidentally killed during a fire in Hamburg in 1830.[63] His two sons, Georg (1804–1885) and Adolph (1793–1867), provided mountings for many of Merz's larger objectives. The most outstanding example of their work was the 7½-inch heliometer which Adolph Repsold went over in person to install at the Oxford Observatory in 1848. If not the finest precision instrument of its time, it was the only large heliometer in England. Merz figured and bisected the lenses, the Repsolds designed and made all the mechanical parts. The mounting was the same as that of a similar instrument sent to Christiania, Norway. It was of the German form, but differed from Fraunhofer's in having the hour circle at the upper end of the polar axis instead of at the lower. This admitted an hour circle of 2½ feet diameter; the declination circle was of the same size. The hollow steel polar and declination axes were about 4 feet long and rotated in gunmetal collars. The bright, lacquered brass tube was over 10 feet long, the declination circle read to 1 second of arc by micrometer microscopes. The entire instrument was driven by a clock governed by centrifugal balls, a device similar to that seen on old-fashioned stationary steam engines. There were, besides, two important innovations. First, the semi-lenses were mounted in cylindrical slides so that separation of the components did not affect the definition. This was one of the disadvantages of Bessel's instrument, in which the lenses were mounted in parallel slides.* Second, the scales at the objective end were illuminated by passing electricity from a battery through thin platinum wires and were read by long low-power micrometer microscopes at the eyepiece end.[64]

Lindsay's 4-inch Repsold heliometer achieved remarkable results in the hands of

* 'I have found that the indistinctness of the images' Bessel writes, 'arising from the oblique passage of the pencil, is not observable even for the extremity of the separation; this is probably a consequence of the construction of the object-glass, which, as Fraunhofer told me, is so made, that the aberration for excentrical pencils is compensated for. Another means for producing the same result, which I owe to Fraunhofer, would be to make the separation of the object-halves, not in a plane, but in a cylindrical surface, whose axis passes through the focus of the telescope; but he preferred the plane, since the construction of the mechanism of the separation would be simpler.' Quoted by Main, R. *Mem. R.A.S.*, **12**, pp. 53–54, 1842.

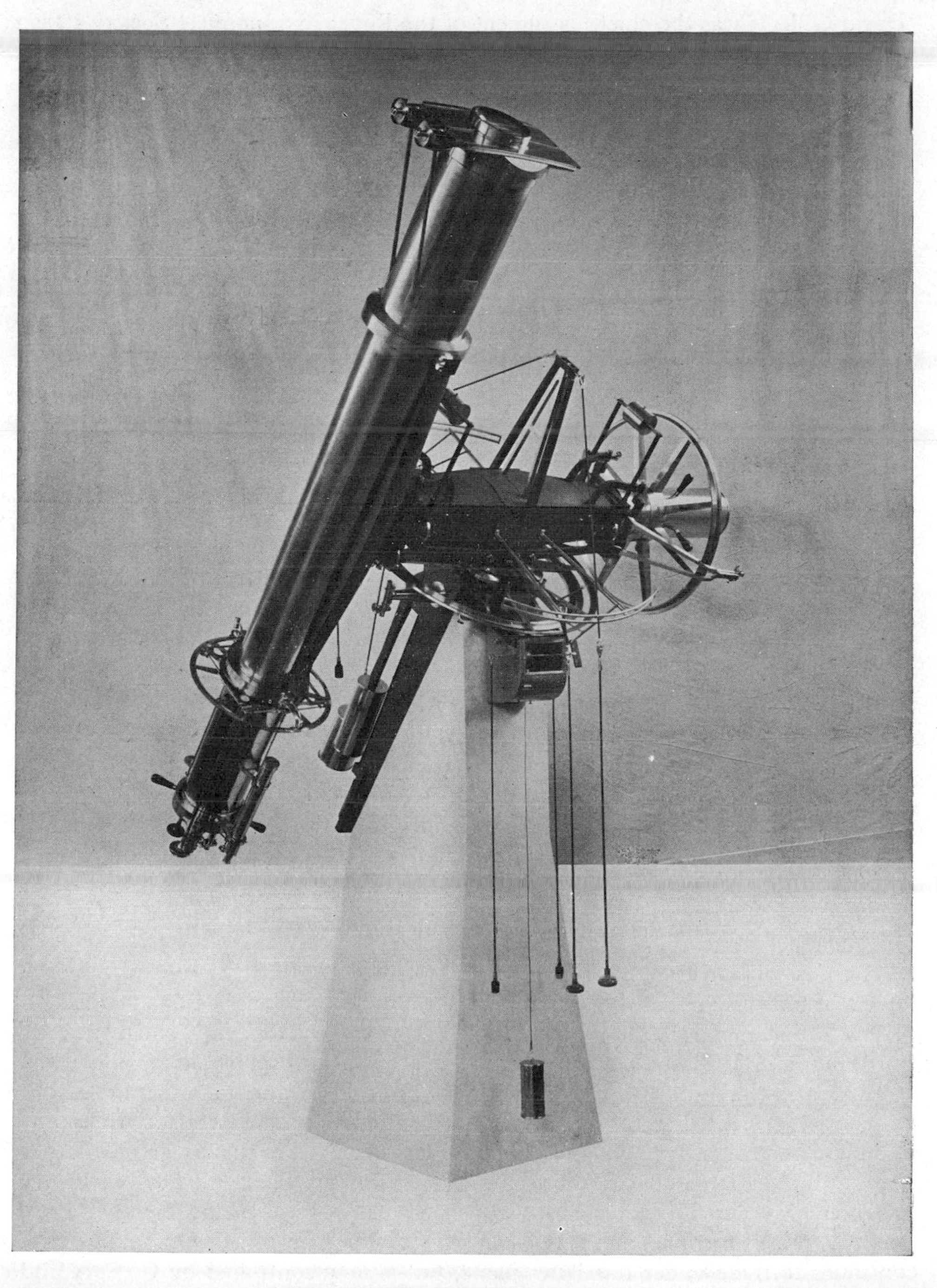

Fig. 101—The Oxford heliometer

(*Science Museum, London. By permission of the Trustees of the Radcliffe Observatory, Oxford*)

David Gill. It was the chief instrument of the Royal Astronomical Society's 1877 expedition to Ascension Island where Gill measured the parallax of Mars and hence the solar parallax. Years later, when Gill was H.M. Astronomer at the Cape, it performed useful work among the stars of the southern sky until replaced by a similar instrument of almost double the size. This new 7-inch heliometer was also the work of the Repsolds. It was completed in 1887 and was considered by Gill to be 'the most powerful and convenient instrument for refined micrometric research in existence.'[65] Under Struve's supervision, the Repsolds also made a prime vertical and meridian circle for the Pulkowa Observatory. They were, without question, responsible for the majority of heliometers in Europe and America during the nineteenth century. Heliometer observations were, however, slow and tedious and, towards the end of the nineteenth century, largely gave way to photographic methods. The Repsold business ended in 1919, after 120 years' valuable work, with the death of Johann A. Repsold.*

The Repsolds by no means monopolized astronomical instrument-making in Germany. The firms of C. Bamberg, J. Wanschaff, and Pistor and Martins, all of Berlin, were noted for the quality and variety of their theodolites and meridian instruments. The Steinheils, Fauth (later G. N. Saegmüller), G. Heyde, K. Fritsch, Reinfelder, and Hertel, all manufactured German-type equatorial mountings. Bamberg introduced equatorials which, by the use of special counterpoises, combined both equatorial and altazimuth characteristics. Examples of their work are described and illustrated in Ambronn, L., 1899, *Handbuch der astronomischen Instrumentenkunde.*

In France, the work of Reichenbach and Repsold was imitated by Gambey who, early in the nineteenth century, furnished the Paris Observatory with new instruments, including a mural circle and transit instrument. His contemporary, Marc Secrétan, founder of the present Paris firm of this name, outlived him by twenty years. Secrétan became associated in business in 1845 with N. M. Lerebours, son of J. N. Lerebours and, after his colleague's death in 1855, carried on alone.[66] Among other refractors, Secrétan made a 6-inch equatorial for Madras Observatory and a 12½-inch for Paris. He was followed in business by August and Georges Secrétan.

The flow of transit circles of French manufacture was further sustained by the efforts of F. W. Eichens, who set up in business in 1867 and made transit circles for several French observatories, among them Paris (1878), Abbadia, Hendaye (1879), Bordeaux (1881), Marseilles, and Lyons. He was also responsible for equatorial and siderostat mountings for Foucault's silver-on-glass mirrors (Chapter XIII). P. Gautier, his collaborator and successor, continued to make transit circles and also plain and coudé-type equatorial mountings (the latter to the design of Maurice Loewy) for the large achromatic lenses of the brothers Paul and Prosper Henry (Chapters XV and XIX). Gautier was succeeded in business by Georges Prin. In 1934, the Prin connection was taken over by the firm of Secrétan. The firm continues to trade under this latter name, but is now controlled by C. Épry and Jacquelin.

* J. A. Repsold was the author of a finely-illustrated book entitled *Zür Geschichte der astronomischen Messwerkzeuge von Purbach bis Reichenbach,* 1908.

REFERENCES

[1] Wolf, C., 1902, *Histoire de l'Observatoire de Paris*, p. 292.
[2] *Ibid.*, p. 275.
[3] *Ibid.*, p. 276.
[4] *Ibid.*, p. 277.
[5] *Ibid.*, p. 282.
[6] *Ibid.*, p. 288.
[7] *Ibid.*, p. 296.
[8] *Ibid.*, p. 296.
[9] *Ibid.*, p. 292.
[10] *Mem. R.A.S.*, **9**, p. 283, 1836; *Dict. Nat. Biog.*, 1893, art. ' Troughton '.
[11] Taylor, E. Wilfred, and Wilson, J. Simms, 1949, *At the Sign of the Orrery—The Origins of the firm of Cooke, Troughton and Simms Ltd*, pp. 15–17. T. Wright of Fleet Street must not be confused with T. Wright, the Durham astronomer.
[12] *Ibid.*, pp. 7–11.
[13] *Ibid.*, p. 13.
[14] *Mem. R.A.S.*, **9**, p. 288, 1836.
[15] J. Troughton's dividing engine, completed in 1778, is now in the Science Museum, London. *Vide* also Ref. 11, p. 25.
[16] Brewster, D., *Edinburgh Encyclopaedia*, **10**, 1830, art. ' graduation '.
[17] Pearson, W., 1829, *Introduction to Practical Astronomy*, ii, pp. 554–558.
[18] *Ibid.*, pp. 519–524.
[19] *Ibid.*, p. 429.
[20] *Ibid.*, p. 434.
[21] Smyth, W. H., 1844, *A Cycle of Celestial Objects*, i, p. 333.
[22] Pearson, *op. cit.*, pp. 429–433.
[23] *Phil. Trans.*, **96**, ii, pp. 420–454, 1806.
[24] *Mem. R.A.S.*, **9**, p. 287, 1836.
[25] Pearson, *op. cit.*, p. 476.
[26] *Ibid.*, p. 478.
[27] *Ibid.*, p. 477.
[28] *Ibid.*, pp. 479–480.
[29] *Mem. R.A.S.*, **10**, p. 359, 1838.
[30] Pearson, *op. cit.*, p. 475.
[31] Pond, J., *Phil. Trans.*, **103**, pp. 75–76, 1813.
[32] Pearson, *op. cit.*, p. 366.
[33] *Ibid.*, p. 369.
[34] Martin, L. C., *Trans. Opt. Soc.*, **24**, p. 298, 1923.
[35] Pearson, *op. cit.*, p. 402.
[36] *Ibid.*, pp. 402–405.
[37] *Mem. R.A.S.*, **9**, p. 287, 1836.
[38] Brown, B. J. W., *J.B.A.A.*, **42**, p. 213, 1932.
[39] Airy, G. B., 1810, *Catalogue of Circumpolar Stars reduced from Observations of S. Groombridge.*
[40] *Mem. R.A.S.*, **9**, p. 287, 1836.
[41] *Ibid.*, pp. 284–285.
[42] *Dict. Nat. Biog.*, 1893, art. ' Troughton '.
[43] *M.N.R.A.S.*, **21**, p. 105, 1861.
[44] *Phil. Trans.*, **99**, p. 105, 1809. Troughton's original dividing engine, improved by a self-acting mechanism added by Simms, is in the Science Museum, London. It was used continuously from 1793–1920.
[45] *Vide* Ref. 42.
[46] *History of the R.A.S.*, 1923, p. 8.
[47] *Ibid.*, pp. 52–55.
[48] From information received from Mr W. Taylor of Messrs Cooke, Troughton and Simms Ltd in 1946.
[49] *M.N.R.A.S.*, **21**, p. 105, 1861.
[50] *M.N.R.A.S.*, **5**, p. 291, 1843; *Mem. R.A.S.*, **15**, pp. 83–90, 1846.
[51] *M.N.R.A.S.*, **21**, p. 106, 1861; *Dict. Nat. Biog.*, 1893, art. ' Simms '.
[52] Stanley, W. F., 1914, *Surveying and Levelling Instruments*, pp. 298–306.
[53] Airy, G. B., 1844, *Account of the Northumberland Equatorial and Dome, attached to the Cambridge Observatory.*
[54] *Vide* Ref. 46, pp. 71–72.
[55] Airy, G. B., 1896, *Autobiography*, p. 236.
[56] *Greenwich, Royal Observatory Appendices*, 1867, i.
[57] Maunder, E. W., 1900, *The Royal Observatory, Greenwich*, pp. 207–211.
[58] *Greenwich, Royal Observatory Appendices*, 1867, ii.
[59] *M.N.R.A.S.*, **21**, pp. 101–102, 1861.
[60] *M.N.R.A.S.*, **31**, p. 239, 1871; **32**, p. 18, 1872. From an idea apparently due to Wheatstone, *M.N.R.A.S.*, **24**, p. 159, 1864.
[61] Wolf, *op. cit.*, pp. 281–282.
[62] Repsold, J. A., 1908, *Zür Geschichte der astronomischen Messwerkzeuge von Purbach bis Reichenbach*, p. 101, fig. 142.
[63] Repsold, J. A., 1896, *Vermehrte Nachrichten über die Familie Repsold.*
[64] Gunther, R. T., 1923, *Early Science at Oxford*, ii, pp. 330–331.
[65] Gill, D., 1913, *A History and Description of the Royal Observatory, Cape of Good Hope*, p. cxlviii.
[66] From information provided by the firm in 1947.

CHAPTER XII

> Refractors have always been found better suited than reflectors to the ordinary work of observatories. They are, so to speak, of a more robust, as well as of a more plastic nature. They suffer less from vicissitudes of temperature and climate. They retain their efficiency with fewer precautions and under more trying circumstances. Above all, they co-operate more readily with mechanical appliances, and lend themselves with far greater facility to purposes of exact measurement.
>
> AGNES CLERKE

TOWARDS the middle of the nineteenth century, a wave of observatory building spread through the United States of America which led to the installation of many interesting astronomical instruments. In 1825, President John Quincy Adams told Congress that, as an American, he felt no pride in the fact that, while Europe could boast of upward of one hundred and thirty ' lighthouses of the skies ' there was not one ' throughout the whole American hemisphere.'[1] Seven years later, Airy remarked[2] that he did not know of a single public observatory in the United States. Yet in 1840, as W. I. Milham has shown,[3] there were eleven equipped observatories in America, eight of which were erected by educational institutions and must, therefore, be classed as public observatories.

In 1830, the largest refractor in the States was the 5-inch Dollond achromatic which Elias Loomis and Denison Olmsted used in one of the towers of a college at Yale University.[4]* Six years later, a 6-inch Lerebours achromatic was installed at Wesleyan University, Middletown, Connecticut.[5] In 1844, O. M. Mitchel purchased an 11-inch glass for the Cincinnati telescope[6] and, a few years later, Harvard College Observatory possessed a 15-inch refractor. In 1862, an American-made object-glass of 18½ inches diameter, then the largest in the world, was mounted at the old University of Chicago.[7] America made a late but extremely vigorous entry into observational astronomy.

American instrument-making did not start in earnest until about 1850, before which time astronomers had to rely on European instruments. David Rittenhouse appears to have been the only American astronomer who was capable of or interested in making his own instruments. In 1749, when only seventeen years of age, Rittenhouse left working on his father's farm and took up clockmaking.[8] As he became experienced, he constructed many fine clocks, two large orreries, and various surveying and astronomical instruments. He achieved fame as a practical astronomer by his participation in the 1769 transit of Venus observing programme.

* American-made reflectors about this time ranged in size up to 10-inch aperture. They were the work of Amasa Holcomb, a surveyor of Southwick, Massachusetts who, in 1835, received a medal from the Franklin Institute for a serviceable telescope-mounting. In 1838, the largest telescope in the United States was a 12-inch reflector built by E. P. Mason of Yale and H. L. Smith of Hobart (*Sky and Telescope*, 1, p. 21, March, 1942).

For some months before the transit, Rittenhouse worked at Norriton, near Philadelphia, on the erection of a temporary log observatory. He made for it a transit telescope, an equal-altitude instrument, and an 8-day clock which 'does not stop when wound up, beats dead seconds, and is kept in motion by a weight of five pounds.'[9] To this equipment he was able to add a 2½-foot Sisson quadrant, his own 36-foot refractor, a 42-foot refractor and a 2-foot Nairne Gregorian reflector fitted with a Dollond micrometer. The last two instruments were destined for Harvard College and arrived from England only just in time for the transit. The Norriton observatory, although a temporary structure, was unique inasmuch as it contained American-made instruments. It was dismantled soon after 1770, when Rittenhouse moved to Philadelphia.[10]

When C. Mason and J. Dixon began their famous boundary surveys in 1764, they brought their own instruments with them from England. The work of the United States Coast and Geodetic Survey was considerably delayed until 1811, when F. R. Hassler was sent to Europe to purchase suitable instruments.[11] Lieutenant C. Wilkes' observatory on Capitol Hill, Washington, was a small wooden building and at one time contained a 4-inch Troughton transit instrument, a 4½-inch Dollond achromatic and several smaller British instruments.[12] The scientific Bishop James Madison observed from the College of William and Mary with a small fixed transit and a Short reflector.[13] Nearly every instrument in Dana House, the first observatory of Harvard College (1839), was of London manufacture.

About 1815, William C. Bond was in London inquiring about the cost of a telescope for Harvard, but the price asked was so high that the project was abandoned.[14] In 1824, Joseph Caldwell, president of the University of North Carolina, took $6000 to Europe to purchase books and scientific apparatus.[15] The $1200 which Sheldon Clark donated to Yale University went to Dollond for the 5-inch refractor[16] previously mentioned. In 1835, Willbur Fisk, first president of the Wesleyan University, Connecticut, was in Europe buying a Lerebours refractor and a 15-inch altazimuth telescope from Troughton and Simms.[17] Thus the equipping of new observatories was an expensive adventure and necessitated a visit to the opticians of either Paris or London. Except for a few clockmakers and a certain R. Patten of New York (who seems to have constructed mainly navigation instruments), Rittenhouse was almost alone in astronomical instrument construction.

The question: Which was the first American observatory? is effectively answered in two little books (accompanied by an extensive bibliography) by Professor W. I. Milham.[18] The question depends on the meaning attached to the term *observatory*. Milham shows that the oldest extant astronomical building in America is the Hopkins Observatory of Williams College, Williamstown, Massachusetts. This solid, stone-built observatory was founded by Professor A. Hopkins. He worked harder than any labourer in its erection, visited England to procure the necessary apparatus, and formally opened the building in 1838.[19] The equipment consisted of a small portable telescope, a 10-foot Herschel reflector, a Troughton and Simms transit instrument and a sidereal clock by Molyneux.[20] If by the term *observatory* we are to infer a permanent building, fitted with fixed apparatus regularly used for instruction and research, then the Hopkins Observatory is the oldest observatory in America.[21]

In 1842, the young Professor Ormsby MacKnight Mitchel was obliged to visit Europe for a telescope worthy of the newly founded Cincinnati Astronomical Society. England had nothing large enough to offer, Cauchoix gave a delivery date four to five years ahead. Only at the Fraunhofer Institute in Munich did Mitchel find an object-glass ready for mounting. ' This was the glass in search of which I had traversed the ocean and the land ' said Mitchel. ' Its magnitude (12 inches full diameter) and powers were beyond anything I had dared to hope or anticipate.'[22] The lenses were the work of Merz and Mahler who, after Fraunhofer's death, completed not only the Königsberg heliometer, and the 9·6-inch Berlin and 11·2-inch Munich equatorials, but turned to fresh projects. The Cincinnati object-glass cost $9000, a sum well in excess of that collected by public subscription, but the prize was too valuable to lose. Mitchel's enthusiasm gradually reloosened public purse-strings and the deal eventually went through.[23]

In 1843, at the instigation of Mitchel, ex-President Adams travelled over a thousand miles in the most inclement weather to lay the cornerstone of the observatory.[24] The ceremony took place in heavy rain before an umbrella-concealed audience: it was a wet, muddy, yet exciting affair. Several years elapsed before the building was completed and the telescope installed. Mitchel himself worked with shovel and trowel so as to reduce the expense of labour. Cincinnati had its telescope but Mitchel's financial troubles increased until death by disease in the Civil War closed a life at once tragic and triumphant.[25]

Thirty years after Adams' dedication, the Cincinnati refractor was moved to a more favourable site on Mount Lookout. Here the main instrument is a 16-inch Clark telescope, but the 11-inch, despite its hundred years, is still in use, although mainly for instructional purposes of the University of Cincinnati.

In 1839, Merz and Mahler made a 15-inch equatorial refractor for the Imperial Russian Observatory at Pulkowa.[26] In 1839 also, the Harvard College Observatory was founded—or rather, W. C. Bond was installed in Dana House, one of the dwellings owned by the College authorities.[27] The instrumental equipment was scanty. Most of it belonged to Bond himself and the building seems to have been ill-adapted for astronomical observations. When the great comet of 1843 appeared, the citizens of Cambridge clamoured for information which Harvard, owing to the absence of a good telescope, was quite unable to give. Townsfolk and officials thereupon pooled their resources, bought a building site and decided to erect a telescope as large as, if not larger than, that at Pulkowa.[28] The result of their negotiations with Merz and Mahler was another 15-inch equatorial. This instrument, mounted in 1847, formed the central feature of the observatory, and is to-day the largest telescope at the observatory's Cambridge headquarters.

Both the Pulkowa and Harvard telescopes had steel polar and declination axes and carried the usual divided circles which, on account of their height above the observatory floor, were never easily read. By reason of their size and weight, both instruments rested on massive stone piers, deeply set in the ground. Both had wooden tubes strengthened by iron diaphragms and veneered with polished mahogany.[29]

The two telescopes cost several thousand pounds apiece and the general public, quite naturally, looked forward to large returns for their money. The Pulkowa

Fig. 102—The 15-inch Harvard equatorial

telescope was entrusted to Otto Struve, son of Wilhelm Struve, who straightway discovered 500 new double stars, many with separations of less than 1″.0.[30] The atmosphere was particularly clear and Struve frequently observed with powers of 1000.[31] The Pulkowa Observatory, which stands on a hill 160 feet above the surrounding plain, was destroyed by the Germans in 1941. It has been rebuilt since the war, and the new observatory was dedicated in 1954.

The Harvard telescope was entrusted to W. C. Bond, first director of the new observatory, and to his son, G. P. Bond, the second director. Between them, the Bonds kept the instrument busy on observations of Saturn, Mars, and Donati's and other comets. It had been in use for only a year when the younger Bond discovered Hyperion, Saturn's eighth satellite,[32] while in 1850, simultaneously with the Rev. W. R. Dawes in England, he discovered the inner or 'crepe' ring of that planet.[33]

Soon after the introduction of the daguerreotype, the Bonds followed Draper's example and, with exposures of about 20 minutes, took a series of lunar photographs.[34] In 1850 also, J. A. Whipple, a professional photographer, assisted the elder Bond in taking the first photographs of stars—daguerreotypes of α Lyrae and α Geminorum.[35] The latter star gave an image distinctly elongated, proof of its duplicity. Seven years later, with a better clock-drive and with faster collodion plates, Whipple and Black obtained images of Mizar and Alcor, second- and fourth-magnitude stars respectively, with only 80 seconds' exposure.[36] G. P. Bond found that his measurements of the separations and angles of position from a number of plates of Mizar as a double star were in close accord with the micrometrical observations of Struve. Furthermore, the photographs provided a means of comparing the apparent brightnesses of stars. With the same exposure time, a bright star gave a larger photographic image than a faint star. Alternatively, a bright star produced a visible trace before a faint star.[37] Bond fully appreciated the photometric potentialities of the new method, investigated the relationship between growth of photographic image and exposure time, and concluded that his 'proposed system of magnitude determination would have an unquestionable advantage over that in common usage, . . . provided that the chemical action of starlight should be found . . . energetic enough to furnish accurate determinations of its amount.'[38] Bond did not, indeed he could not, foresee the peculiarities and disadvantages of the photographic method for brightness determinations. Difficulties multiplied rapidly enough when dry-plate astronomical photography was exploited, to be reduced only by the concerted efforts of many astronomers over many years.

Until 1869, when Thomas Cooke built the Newall telescope, the Pulkowa instrument was the largest of its type in Europe. In America, the Harvard refractor held its own until 1862, when Alvan Clark completed an 18½-inch equatorial for the old University of Chicago. For fifty years, the 'Great Refractor'—as the Harvard telescope was affectionately called—remained the principal instrument of that observatory. Interest in both telescopes waned considerably when astronomical photography became established; the objectives were corrected for visual rays only and photographs taken with them were never critically sharp. The mountings, although of solid construction, were not steady enough to permit long exposure times.

In 1846, Airy at Greenwich, religiously 'grinding' the Greenwich meridian with Simms' transit circle, began to feel that the existing 6·7-inch Sheepshanks equatorial was no match for the giants leaving Munich. Almost apologetically he designed and installed an English yoke-type equatorial made by Ransomes and May.[39] Merz made the 12¾-inch object-glass[40] of 18 feet focus and supplied all the small accessories. Airy stipulated that, if the new telescope was to be used for the observation of faint nebulae and close double stars[41], it should in no way interfere with the routine work of the observatory.

Mahler died in 1845 but Merz and his sons continued to run the Munich establishment which now enjoyed world-wide reputation for high-quality optical work. C. May and W. R. Dawes in England purchased Merz object-glasses, 7-inch glasses went to the Cape of Good Hope and Sydney (N.S.W.) observatories, and the City of Liverpool purchased an 8¾-inch glass. Dunér at Lund, Schiaparelli at Milan, Secchi at Rome, and Newcomb at Washington, all worked with 8- to 9½-inch Merz object-glasses. H. L. d'Arrest at the observatory of Copenhagen University used an 11-inch Merz object-glass mounted in 1858. The firm came to an end in 1903 with the death of Jacob Merz.[42] By this date, further Merz objectives up to and exceeding 19 inches aperture had found a ready home in

Fig. 103—Thomas Cooke

(Cooke, Troughton and Simms Ltd)

European observatories. The business was then sold to Paul Zschohke,[43] as mentioned previously in our account of Fraunhofer.

Returning to England, we find a formidable rival to the Continental artists in the person of Thomas Cooke. Cooke was born in 1807 at Allerthorpe in the East Riding of Yorkshire. His parents were so poor that, after two years' education at an elementary school, he had to help his father at the shoemaker's bench. With the intention of going to sea, young Cooke studied navigation but, in respect for his parents' entreaties, he reconsidered his future. At the age of twenty-two he moved to York where, for the next seven years, he worked as assistant schoolmaster and also gave private lessons. He studied mathematics and optics and, after many preliminary attempts, made a small achromatic telescope. The performance of this instrument encouraged him to make others and at length, chiefly at the instigation of his friends, he opened a small optical business at 50 Stonegate, York.[44]

After a few years in Stonegate, Cooke moved to larger premises in Coney Street. Here he concentrated on the construction of equatorially mounted refractors and, as the years passed by, ever larger telescopes left his workshop. Among his first orders was one for a 4½-inch equatorial which he executed so perfectly that his sponsors had no further doubts of his success as a manufacturing optician. In 1851 he made an equatorial of the then considerable aperture of 7 inches which Piazzi-Smyth used at Teneriffe.[45] This was followed by a 9½-inch for J. F. Miller of Whitehaven, one of many similar orders that almost inundated his small workshop. To cope with the demand, Cooke began in 1855 to erect the first telescope factory in England—the Buckingham Works at Bishopshill, Yorkshire. He started with a few workmen and one apprentice and was later joined by his two sons, Frederick and Thomas. At this period Cooke's interests covered the building of steam engines and the construction of telescopes and surveying instruments. In 1860 he constructed an elaborate equatorially mounted 5¼-inch achromatic for H.R.H. the Prince Consort, and the instrument was duly erected at Osborne House, Isle of Wight.[46] The early years at Bishopshill saw the construction of nine equatorials of from 8 to 10 inches aperture and some twelve others of from 5 to 8 inches aperture.

Cooke's rapid progress was due in good measure to his being able to obtain large disks of optical glass. These disks were provided by the newly founded firm of Chance Brothers, Birmingham. About 1828, H. Guinand, son of P. Guinand, founded a glassworks at Choisy-le-Roi which, after 1885, came under the supervision of Charles Feil, Mantois, and Parra.[47] During the revolution of 1848, however, George Bontemps, a collaborator with H. Guinand, left France and imparted his technical knowledge to the brothers Chance. He also brought to England a number of expert French and Belgian sheet-glassmakers so that, within a short space of time, English optical-glass production equalled and then surpassed that established in Germany and France.[48]

Cooke's fame centres around the careful design and execution of his medium-sized mountings and the high optical quality (by modern standards) of his object-glasses. His telescopes were eagerly sought by the wealthier English amateurs—J. Fletcher possessed a 6-inch Cooke refractor and mounted a 9·4-inch,[49] while J. N. Lockyer had a 6½-inch in his garden in south London.[50]

Cooke's largest telescope was the 25 inch Newall refractor, for some years the largest refractor in the world but now surpassed in England by the 26- and 28-inch telescopes of the Royal Greenwich Observatory. The Newall refractor took seven years to build and undoubtedly hastened the death of its maker for, worn out by the anxiety of so immense an undertaking, Cooke died in 1868, a year before the instrument was finished.

The project began when, at the Exhibition of 1862, R. S. Newall, a wealthy amateur astronomer, saw two large disks of optical glass on the stand of Messrs. Chance Bros.[51] These Cooke offered to work into an objective and to mount on a large equatorial stand of his own design. Together, the two lenses weighed 146 lb and, to avoid their flexure during polishing, Cooke floated them on mercury.[52] In the 32-foot tube they rested on three equidistant supports, while three counter-poised levers acted through the cell direct on the glass. The focal length of the lens was 29 feet and could be increased by a Barlow lens arranged to slide on a brass framework inside the tube. There were two finders of 4 inches aperture and a third of 6½ inches. A comparatively small pendulum clock, housed in the upper part of the 19-foot-high cast iron pillar, regulated the 7-foot driving sector of the polar axis. Right ascensions were indicated on a 26-inch hour circle, read from the floor by means of a small diagonal telescope attached to the pillar—an innovation due to Cooke and still used on large instruments.[53]

Newall mounted the telescope in his private grounds at Gateshead in the hope that its great optical power would lead to further discoveries in astronomy. A more

Fig. 104—The Newall refractor

(*From Lockyer's 'Stargazing'*, 1878)

unfavourable site for an instrument of this size would be difficult to imagine. During fifteen years, Newall had only one night in which he could use the full aperture with advantage. He was often absent on business for long periods, which not only prevented regular observation but caused him to miss most of the few breaks of fine weather. 'Atmosphere has an immense deal to do with definition' Newall wrote to Denning in 1885. 'I have had only one fine night since 1870! I then saw what I have never seen since.'[54] After several years' sporadic use, Newall offered to loan the instrument to the Cape Observatory, where the largest telescope was a 7-inch Merz and where D. Gill was anxious to determine the parallax of α Centauri. But the cost of dismounting and transporting the entire apparatus was considered prohibitive, and the advisory committee set up to consider the project thought it better to spend the money on a new instrument.[55] In 1890, the Newall telescope went to what was then the Solar Physics Observatory, Cambridge, where Newall's son, H. F. Newall, was appointed Newall Observer. The instrument is now little used.

Cooke was the first English optician to employ factory methods. He was a first-rate engineer and a skilled practical optician. Besides casting all his own brass, he made his own machine tools. He used the most up-to-date machinery, made a 3-foot automatic dividing engine and, just before his death, started a 4-foot. He was an expert horologist and made improved clocks for churches and public institutions. After his death, this work was continued by Thomas and Frederick, who specialized also in surveying instruments and general observatory apparatus. They were the first to suggest the use of papier-mâché coverings for observatory domes, instead of the customary heavy copper sheeting.[56]

Two further large English refractors of this period should be mentioned. The 1862 Exhibition contained, besides the Newall disks, a 21-inch aperture equatorial of 28½ feet focus. This was the work of J. Buckingham, who soon after installed it in a large observatory on Walworth Common, near London.[57] Here again, cloudy weather prevented regular observations, and Buckingham was frequently absent from home. Apart from sporadic planetary observations,[58] the instrument contributed little to astronomy. It was later erected at the Edinburgh City Observatory on Calton Hill, where it stood until well into the present century.* The second telescope, the 24-inch Craig instrument, was a complete failure. It was constructed for the Rev. Mr Craig, vicar of Leamington, by Messrs Rennie under the supervision of a Mr Gravatt, F.R.S., and stood on a piece of land just south of Wandsworth prison. From a contemporary volume,[59] we learn that 'the object-glass and all the optical work were executed by Mr Thomas Slater.' According to Sir David Brewster[60] and contemporary prints, this impressive glass was mounted at one end of a 75-foot cigar-shaped metal tube, increased to 85 feet by dewcap and eyepiece, and 13 feet wide at a point 24 feet from the objective. This tube was slung in chains from the side of a solid brick tower, the eyepiece resting upon a light wooden framework which ran on a circular railway. The object-glass was of poor

* Most of the work was probably done under Buckingham's supervision. N. G. Matthew, director at Calton Hill, suggests that W. Wray made the object-glass. The telescope was dismantled in 1926 and only the object-glass remains; its full powers could seldom be employed owing to unsuitable observing conditions.

Fig. 105—The Craig telescope, 1852

quality. Spherical aberration was so pronounced that the central part of the objective had to be stopped out. The crude structure was dismantled after a few years' use, but not before it had formed a strange landmark for the residents of Wandsworth and passengers travelling on the railway nearby.

The fact that their country had to depend on Europe for large telescopes like the Cincinnati and Harvard refractors must have been a sore point to many practically minded Americans. The mechanical work looked fairly straightforward, but a glass of large size seemed impossible to manufacture unless one knew the secrets of the Munich workshops. Alvan Clark probably felt in this mood when he heard of the Harvard refractor's remarkable performance. He was a portrait painter by profession, and a few attempts at mirror and lens grinding convinced him that the figuring of a large glass would be a long and difficult task. Such was his interest in the new telescope, however, that he asked Bond if he might be allowed to look through it. His request was granted, and the experience, although of only a few minutes' duration, marked the turning point of his career.

> I was far enough advanced in knowledge of the matter [he wrote later][61] to perceive and locate the errors of figure in their 15-inch glass at first sight. Yet these errors were very small, just enough to leave me in full possession of all the hope and courage needed to give me a start, especially when informed that this object-glass alone cost $12,000.

Clark's next step was to close his studios and to master the art of figuring by working on old lenses. This done, he ground and polished glass blanks, eventually completing a $5\frac{1}{4}$-inch achromat and then one of 8 inches. By the tests to which he subjected these two objectives, he knew they were as good as, if not better than, those of European manufacture. But, although he tried to sell others like them,

Fig. 106—Alvan Clark

(*Yerkes Observatory*)

the new venture made little financial headway. In 1851, after seven years' comparative obscurity, Clark wrote to W. R. Dawes at Haddenham, Kent, telling him about the close double stars he had seen with his telescope.[62] The following year, he wrote to say that he had discovered two doubles, one of which was 8 Sextantis. That same year he finished a new telescope of 7½-inch aperture and informed Dawes that it had resolved 95 Ceti (0″·7 separation), a star which he (Clark) had just discovered to be double. Dawes, by this time, could scarcely restrain his enthusiasm. He sent Clark a list of some of Struve's double stars, to see if the telescope could resolve them,[63] and added that he would like to make a purchase. Clark examined the stars, sent back a report, and the delighted Dawes bought the glass.* He afterwards bought four others from Clark,[64] one of which, an 8-inch ' which has afforded me some of the finest views of Saturn I have ever enjoyed ', passed into the hands of William Huggins, later Sir William Huggins. It was with this lens, mounted by Cooke, that Huggins did most of his pioneer work in astronomical spectroscopy.

* Dawes purchased the 7½-inch in 1854. At Haddenham, he observed with an 8¼-inch refractor which he described in *M.N.R.A.S.*, **20**, pp. 60–62, 1859–60. This instrument is now at the Temple Observatory, Rugby School. In 1865, Dawes added an 8-inch Cooke refractor to his equipment. Clark charged £200 for an 8-inch objective.

In the summer of 1859, Dawes invited Clark to London and introduced him to the leading English astronomers, among them Lord Rosse and Sir John Herschel.[65] This visit did much to give Clark the publicity he needed. The news of his telescopes, supported by Dawes' frequent reports of discoveries made with them, soon spread from England to the Continent and thence to America. When he arrived home, Clark was besieged with orders.

The old workshop was neither large enough nor suitably equipped and, in 1860, the little firm of Alvan Clark and Son began to look for larger premises. The same year, Dr F. A. P. Barnard, head of the University of Mississippi, ordered an 18½-inch refractor from the Clarks[66]—a glass 3½ inches larger than that of the Harvard refractor. Alvan Clark accepted the immense task. He sold his house and, with the proceeds and money from his earlier sales, bought a piece of land in Cambridge, Massachusetts.[67] On this he built the first American telescope factory. In this venture, as in his subsequent undertakings, he was accompanied by his two sons, George Bassett Clark and Alvan George Clark,[68] who between them designed the grinding and polishing machines. A fireproof safe was constructed, so that the lens could be wheeled into safety during the night, and an alarm connected with Alvan Clark's bedroom. Underneath the factory was a tunnel, 230 feet long, for testing lenses with artificial stars.[69] In this tunnel, vibration, dust, and temperature changes were reduced to a minimum. At his own residence nearby, Clark had a private observatory furnished with one of his large refractors and surmounted by a rotatable dome.

The two 18½-inch disks arrived from Messrs Chance Bros in 1861 and, on January 31 of the following year, Alvan G. Clark began to test the finished lens on certain bright stars. On turning the rough, experimental tube towards Sirius he saw not one star, but two; Sirius was accompanied by a relatively faint companion. The news was sent to Bond who, a few days later, picked up the companion with the 15-inch.[70] Thus a Cambridge telescope-maker confirmed Bessel's suspicion that the observed irregularities in the proper motion of Sirius were due to the presence of a 'dark' companion. For this important discovery, the younger Clark received the Lalande prize of the French Academy.

Unfortunately for Mississippi University, the new lens never reached the South owing to the outbreak of the Civil War. The telescope passed instead to the newly founded Chicago Astronomical Society. It was installed in the Dearborn Observatory, then located at the old University of Chicago, to be used by Professor Hough for planetary observations. S. W. Burnham, the famous American double-star observer, also worked with this telescope. In 1889, the instrument was transferred to the Dearborn Observatory of Northwestern University, Evanston, Illinois, to which it was deeded in 1929. Clark had meanwhile changed the mounting; the original is preserved in the Adler Planetarium, Chicago.

In 1870, the federal government asked Simon Newcomb to negotiate with the Clarks for a telescope for the U.S. Naval Observatory. The cost was to be limited to $50,000 and the instrument was to be the largest and best that this sum could provide. Newcomb says[71] that, although Clark had by this time risen to some prominence as a telescope-maker, 'his genius in this direction had not been recognized outside of a limited scientific circle. The civil war had commenced just as he had completed the largest refracting telescope ever made, and the excitement of the contest . . . did not leave our public men much time to think about the making of telescopes.' If there had been any doubt at Washington as to Clark's ability, this was dispelled when he refigured the lenses of the other observatory instruments. ' The result of this work ' Newcomb concludes, ' was so striking to the observers using the instruments before and after his work on them, that no doubt of his ability could be felt.'[72] Clark was promised the contract and, with Newcomb and Asaph Hall, made a second journey to England to examine and report on the Newall telescope, on which the Washington instrument was to be modelled.

After a number of unsuccessful trials, Messrs Chance Bros provided the 26-inch disks for the Washington telescope. These arrived at Cambridge in 1871 and took a year and a half to grind and polish. The front crown took the form of an equi-biconvex lens, nearly two inches thick; the flint was separated from it at the centre by only 0·03 inch and had a central thickness of just under an inch. Together, the two components weighed nearly 110 lb, gave a focal length of over 32 feet, and cost $6000.[73]

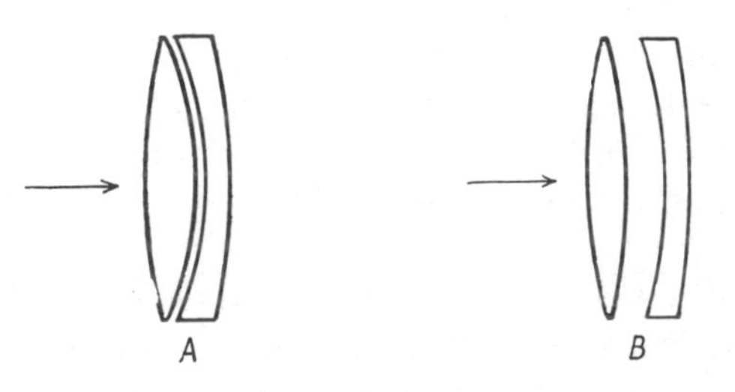

Fig. 107—*A* **Fraunhofer object-glass**
B **Clark object-glass**

The curves in both cases are exaggerated

According to Dawes, a lens of 26 inches aperture should have a resolving power of 0″·16, a limit which Newcomb and Hall almost reached in their measurements of certain very close double stars. The remarkable light-grasp of the Clark lens was demonstrated in 1877, when Hall discovered Deimos and Phobos, the two small satellites of Mars.[74] The tube was remounted in 1893 by Warner and Swasey, but the old Clark mount is still in use, having been adapted to carry several photographic telescopes. The U.S. Naval Observatory has always been devoted to the astronomy of position—it is the only American observatory in which the transit circle still reigns supreme. The 26-inch telescope, until recently used mainly for the minor role of satellite observations, is now being employed for photoelectric studies and for micrometrical work on double stars.

In spite of the outstanding capabilities of the Washington refractor, and of the 26-inch which Clark made ten years later for Leander J. McCormick, several astronomers were inclined to doubt the utility of glasses of this size. The Washington telescope had been in use only six years when the rumour spread that it had deteriorated.[75] Other observers had seen objects which were quite invisible in the 26-inch and, after they knew the Martian satellites existed, detected them with quite moderate apertures. The outer satellite, for example, was seen with the Paris 10-inch and Greenwich $12\frac{3}{4}$-inch refractors and even with W. Erck's $7\frac{1}{2}$-inch Cooke. With an $8\frac{1}{2}$-inch Merz refractor, Schiaparelli discovered *canali* on Mars and saw more details on that planet than had Hall. The 26-inch showed very little either of the *canali* or their alleged duplication, and it was feared that the glass had bent slightly under its own weight. Nothing of the kind had occurred, however, and we now know that the duplication, at least, is more illusion than fact. At a later date, Burnham often failed to see the companion of Sirius in the 40-inch Yerkes telescope when the air was disturbed, although it was visible in a 6-inch refractor under good conditions. Despite Denning's satirical remark that 'if all the data derived with such means [small telescopes] are to be absolutely accepted, then large telescopes . . . may as well be packed away in the lumber-rooms of our observatories',[76] trained and sensitive eyes have certainly seen details with small instruments which have been only just visible in larger telescopes. Denning himself was one of these keen-eyed astronomers, the 'eagle-eyed' Dawes another, also Webb and Ward. Considerable allowance must always be made when the objects sought are known to exist. Thus, Ward appears to have glimpsed the two outer satellites of Uranus with a 4·3-inch Wray telescope, an observation which, at the time (1876) led to much criticism. Webb states that these faint objects were seen by Huggins with an 8-inch refractor and by Sadler with a $6\frac{1}{2}$-inch reflector, but that he himself could see only one satellite in his $9\frac{1}{3}$-inch reflector.[77]

REFERENCES

1 Marshall, R. K., quotes in *Pop. Astr.*, **51**, p. 71, 1943.
2 Newcomb, S., 1906, *Sidelights on Astronomy*, p. 282.
3 Milham, W. I., 1938, *Early American Observatories*, p. 52.
4 *Ibid.*, p. 30.
5 *Ibid.*, p. 38.
6 Marshall, R. K., *op. cit.*, p. 73.
7 Clark, A., *Sidereal Messenger*, **8**, pp. 114–115, 1889.
8 Milham, *op. cit.*, p. 14.
9 *Ibid.*, p. 20. *Vide* pp. 16–21 for copy of the account of the 1769 transit of Venus observed from the Norriton Observatory, *Trans. Am. Phil. Soc.*, Vol. 7, 1810.
10 *Ibid.*, p. 22.
11 *Ibid.*, p. 9.
12 *Ibid.*, p. 12.
13 *Ibid.*, pp. 28–29.
14 *Ibid.*, pp. 26–27.
15 *Ibid.*, p. 32.
16 *Ibid.*, p. 30.
17 *Ibid.*, p. 38.
18 Milham, *vide* Ref. 3 and Ref. 19.
19 Milham, W. I., 1937, *The History of Astronomy in Williams College*, p. 9.
20 *Ibid.*, p. 12.
21 *Ibid.*, p. 25.
22 Black, R. L., quotes in *Pop. Astr.*, **52**, p. 73, 1944.
23 *Ibid.*
24 *Ibid.*, pp. 71–74.
25 *Vide* Ref. 1, p. 74.
26 Struve, F. G. W., 1845, *Description de l'Observatoire de Poulkova*.
27 Bailey, S. I., 1931, *History and Work of the Harvard Observatory*, pp. 17–18.
28 *Ibid.*, pp. 23–24.
29 *Ibid.*, pp. 39–40; *Vide* also Bond, W. C., *Mem. Am. Acad.*, new ser., **4**, pp. 177–188, 1849, for description of the observatory.
30 Dyson, F., *Trans. Opt. Soc.*, **24**, p. 61, 1923.
31 *Vide* observations in the *Mensurae Micrometricae*, 1837. The maximum power provided was 2000.
32 *Mem. R.A.S.*, **18**, pp. 24–25, 1850.
33 *Am. Jour. Sc.*, **12**, ser. 2, pp. 133–134, 1851. If these discoveries are to be assigned to a single person then they must be credited to G. P. Bond, but both observers worked in close co-operation and the honour belongs to W. C. Bond also. *Vide* Ref. 34, pp. 251–254.
34 Holden, E. S., 1897, *Memorials of W. C. Bond and G. P. Bond*, p. 262.

[35] Bailey, *op. cit.*, p. 116; Holden, *op. cit.*, pp. 262–263.

[36] Bailey, *op. cit.*, p. 117; *Astr. Nach.*, **47**, p. 1, 1857.

[37] *Astr. Nach.*, **48**, p. 1, 1858.

[38] *Ibid.*, **49**, p. 81, 1859.

[39] Airy, G. B., 1896, *Autobiography*, p. 182, p. 224.

[40] *Ibid.*, p. 231, p. 243.

[41] *Ibid.*, p. 182.

[42] Chance, W. H. S., *Proc. Phys. Soc.*, **49**, p. 443, 1937.

[43] *Ibid.*

[44] *M.N.R.A.S.*, **29**, p. 130, 1869, Obituary Notice. Taylor, E. Wilfred, and Wilson, J. Simms, *At the Sign of the Orrery—The Origins of the firm of Cooke, Troughton and Simms, Ltd*, p. 43.

[45] *M.N.R.A.S.*, **29**, p. 132, 1869.

[46] Taylor, E. Wilfred, and Wilson, J. Simms, *Vide* Ref. 44, p. 45.

[47] Rohr, M. von, *Trans. Opt. Soc.*, **28**, p. 140, 1927.

[48] *Ibid.*, p. 140. *Vide* Holden, *op. cit.* (Ref. 34), pp. 124–127 for reference to large disks seen at Messrs Chance Bros, Birmingham, by G. P. Bond in 1863.

[49] *M.N.R.A.S.*, **25**, pp. 241–243, 1865; **40**, p. 194, 1880.

[50] Lockyer, T. M., and W. L., 1928, *Life and Work of Sir N. Lockyer*, p. 10. Lockyer began with a 3¾-inch Cooke refractor in 1861, but soon changed it for the 6-inch. For Lockyer's relations with Cooke, *vide op. cit.*, pp. 8–10.

[51] *M.N.R.A.S.*, **50**, p. 166, 1890.

[52] *Ibid.*, **29**, p. 132, 1869.

[53] *Ibid.*, **30**, p. 113, 1870.

[54] Denning, W. F., 1891, *Telescopic Work for Starlight Evenings*, p. 25.

[55] *M.N.R.A.S.*, **40**, p. 236, 1880.

[56] *Vide* early catalogues of Messrs Cooke and Sons.

[57] *M.N.R.A.S.*, **30**, p. 114, 1869.

[58] *Ibid.*, **33**, p. 227, 1873.

[59] *The Family Tutor—Vide J.B.A.A.*, **47**, p. 90, 1936.

[60] Brewster, D., 1853, *A Treatise on Optics*, p. 506.

[61] Clark, A., *Sidereal Messenger*, **8**, p. 113, 1889.

[62] *Ibid.*, p. 113.

[63] *Ibid.*, p. 114.

[64] *Ibid.*, p. 114.

[65] *Ibid.*

[66] *Ibid.*, pp. 114–115.

[67] Bates, R. S., *The Telescope*, **7**, p. 78, 1940.

[68] Clark, *op. cit.*, p. 115.

[69] Bates, *op. cit.*, p. 78.

[70] Clark, *op. cit.*, p. 115.

[71] Newcomb, S., 1903, *Reminiscences of an Astronomer*, p. 129.

[72] *Ibid.*, pp. 129–130.

[73] Washington, United States Naval Observatory, *Astronomical and Meteorological Observations*, Appendix 1, pp. 7–9, 1877 (E. S. Holden).

[74] *M.N.R.A.S.*, **37**, p. 443, 1877; **38**, p. 205, 1878; **39**, p. 306, 1879.

[75] Newcomb, *op. cit.*, pp. 143–144.

[76] Denning, W. F., *Nature*, **52**, p. 232, 1895.

[77] Webb, T. W., 1896, *Celestial Objects for Common Telescopes*, p. 204.

CHAPTER XIII

> My experience in the matter, strengthened by the recent successful attempt of M. Foucault to figure such a surface more than 30 inches in diameter, assures me that not only can the four and six feet (speculum metal) telescopes of those astronomers be equalled, but even excelled. It is merely an affair of expense and patience.
>
> HENRY DRAPER

BY the middle of the nineteenth century the refractor was more than ever before the basic instrument in both private and national observatories. A census of observatory instruments at this time shows that, out of forty-eight British observatories, thirty-two possessed an equatorial refractor, eight had altazimuth refractors, while only seven possessed a reflector. Out of a total of forty-seven telescopes, fourteen were 'driven' by clock mechanism.[1] Amateurs like Rosse, Nasmyth, Lassell, and De la Rue were more exceptions than the rule. They had the time, the money and the practical skill to build large reflectors, the powers of which were seldom fully utilized in the skies of the British Isles. The age-long drawbacks of speculum metal were little nearer solution than when Edwards experimented with different alloys. Alternative reflecting surfaces had been tried from time to time, but without success.

In the eighteenth century, Caleb Smith proposed a reflecting telescope with a meniscus lens silvered on its convex surface for the main mirror.[2] The dispersion produced by the double passage of light through the glass was corrected by giving the reflecting prism concave surfaces, but optical difficulties prevented its manufacture. Sir William Herschel once made a mirror of 'Tussi's compound', a white porcelain, but its reflectivity was low.[3] Platinum and silver are expensive metals, although Carroché made a few platinum mirrors of up to 8 inches diameter. Silver tarnishes but can be given a high initial reflectivity; Jamin found that a freshly prepared silver surface reflects 50 per cent more light than does one of speculum metal.[4] G. B. Amici of Modena, maker of several 10- to 12-inch speculum metal mirrors, also worked with some silver ones. Rosse tried to polish optical flats of silver and made unsuccessful attempts to deposit silver electrochemically on the surface of a metal mirror.[5] He also coated a glass plate with silver precipitated from a solution of saccharate of silver, preserving the film with a thin shellac varnish.[6] While still a student at Cambridge, Airy sketched a plan for constructing a silvered-glass telescope[7] but was unable to procure good reflecting films. The Tulleys also tried their hand with mercury amalgam films, but with little success.[8]

At the 1851 Great Exhibition held in London, globes and vases were displayed silvered inside according to a process patented by Messrs Varnish and Mellish.[9] The vessels had been filled with a solution of silver nitrate and grape sugar so that silver films were chemically deposited on the interior surfaces. This process was revived a few years later by the German chemist, Justus von Liebig.[10] In its essentials, Liebig's method is the same as that used today. First, the surface to be treated must be chemically clean; it must be swabbed with nitric acid or a solution of caustic potash,* well rinsed in water and then, so that no part will become dry, completely immersed in distilled water. Liebig prepared two basic solutions. The first, a silvering solution, contained nitrate of silver to which he added caustic potash, which caused the liquid to turn brown with precipitated silver oxide. He added ammonia to this, drop by drop, so that each time the liquid was stirred it remained brown. Still stirring, he added more drops until all the precipitate had just entered the solution. The latter now contained its original silver nitrate in a form which was ready to give up its silver under the influence of a suitable reducing agent. For the latter, Liebig chose an ordinary sugar solution. This he added to the first and then poured the whole over the mirror. Metallic silver was thus evenly and slowly deposited on the glass surface—a film formed which was feebly transparent yet firm enough to take a good polish. After washing and drying the mirror, the surface was burnished by light circular strokes with soft chamois skin. This compacted the silver grains and made the surface less liable to tarnish.

The silvered-glass mirror brought both advantages and disadvantages. A glass disk can be ground and figured more easily than hard speculum metal. It is much lighter and therefore requires a less massive support system. When freshly formed, a silver surface reflects better throughout the visible spectrum than does speculum metal, but its reflectivity drops sharply in the ultra-violet. On the other hand, a silver film tarnishes quickly in a damp atmosphere, and is easily damaged. Compared with all the trouble of repolishing metal mirrors, however, its replacement is a comparatively simple and inexpensive operation.

Liebig did not apply his process to astronomical mirrors. This was first done in 1856 by Carl August von Steinheil of Munich and Léon Foucault, a French physicist. Steinheil made a 4-inch telescope of 8 feet focus which was stated to have taken a power of 100, that is, 25 to one inch of aperture.[11] He also advertised reflectors up to $9\frac{1}{2}$ feet focus and $12\frac{3}{4}$ inches aperture.[12] In the following year, and quite independently, Foucault silvered a 33-cm and smaller mirrors which Eichens mounted as Newtonians on ' fork-type ' equatorial stands[13] (Fig. 108). Foucault's largest mirror, of 80 cm aperture and 4·5 m focus, was made from a glass disk supplied by the St Gobain glassworks.[14] Mounted equatorially by Eichens, this mirror was used on planetary studies by J. Chacornac at the Paris Observatory and, later, at Ville-Urbanne near Lyons.[15] In 1864, the telescope was installed at the Marseilles Observatory.

Foucault achieved notable success with silver-on-glass optical flats, which he intended for use with fixed telescopes. He designed the first large siderostat, constructed by Eichens and Gautier in 1868, the year of Foucault's death.[16] The performance of this instrument led Lord Lindsay to order another from Eichens for his observatory at Dun Echt.[17] But Foucault's greatest contribution to the reflecting telescope was his invention of a simple yet accurate testing technique.[18] Possessed of this ace card, he was able to produce mirrors equal in definition to the finest objectives. He first ground them to a good spherical figure and gave them a

* Liebig used a solution of cyanide of potassium.

Fig. 108—Foucault's 13-inch reflector

Mounted by Eichens

fair polish. Parabolization was effected by abrading the centre portion more than the outer with polishers of increasing size and using straight strokes across every diameter. For polishers he used plano-convex lenses of curvature slightly less than that of the mirror and covered with paper instead of pitch.[19] The spherical surface was then carried through a series of ellipsoidal surfaces to a paraboloidal one and, which was so important, was tested at each stage of the work.

To understand Foucault's method of testing, consider a spherical mirror, checked to a gauge for curvature and brought to a fair polish. An artificial star in the form of an illuminated pin-hole is placed at the centre of curvature and the mirror is arranged so that the reflected image focuses alongside the pin-hole. A vertical straight edge is now introduced into the image and the eye behind it sees the shadow of the edge lying across the mirror. If the mirror be spherical and the straight edge exactly at the centre of curvature, the surface darkens over uniformly and gives the impression of being perfectly plane. But if there is any error in the curvature, with some zones having a greater and some a lesser radius, some of the reflected rays will miss the conjugate focus and cross either ahead of, or behind it. The edge will, therefore, catch them first or last as the case may be and the offending zones will be first darkened or remain bright after the light elsewhere is extinguished. The figure of the mirror is thus made visible to the observer who, from his position behind the straight edge, can explore the surface bit by bit and detect the slightest variation from sphericity. So delicate is the test that the expansion due to heat communicated to the glass by the touch of a finger can be clearly seen, also the motions of air currents between pin-hole and observer.

Foucault became familiar with the various shadowgraphs or focographs of a mirror as it went through its different polishing stages. Amateurs now test their small mirrors with pin-hole and knife-edge always at or near the centre of curvature. Foucault first placed them both in this position and then carried the mirror through a series of ellipsoidal forms, observing the image at each stage by moving the pin-hole towards the mirror and retreating with the knife-edge. The limited length of the workshop soon put a stop to this procedure and he performed the last stages by an empirical process, eventually resorting to trials in the telescope tube.[20] The same method is applicable to object-glasses and enables the operator to detect the smallest imperfections of polish. Both for mirrors and object-glasses it is essentially a laboratory test and highly accurate. As practised by Foucault, the process left a slightly under-corrected mirror, but under-correction is desirable to compensate for the usual over-correction which appears when a large mirror is cooling to night air temperature.

Silver-on-glass mirrors soon became popular in Europe and, with one or two exceptions, entirely replaced metal specula. One exception, and a conspicuous one, was the 4-foot Melbourne telescope—the last large reflector to be provided with metal mirrors. This instrument was built in 1862 in response to a call from the authorities of Victoria, Australia, for a powerful telescope for studying southern nebulae. The plan met with the full support of the Royal Society, who appointed a committee to consider what form the instrument should take. Thomas Cooke, the only English manufacturer likely to undertake the construction of a large refractor, had yet to prove his ability by making the Newall telescope. Merz asked

Fig. 109—The 48-inch Grubb reflector erected at Melbourne, Australia, in 1867

The drawing does not show the sliding-roof observatory which eventually housed the instrument

£9000 for a 30-inch objective and £20,000 for a complete equatorial, a sum far beyond that which the committee thought proper to give.[21] Lassell's offer of his 48-inch was turned down, as a more manageable instrument was preferred. The committee also rejected the idea of silver-on-glass mirrors owing to the uncertainty of silvering large disks. 'The silver film' Robinson wrote, 'would tarnish faster than speculum metal; and though on a small scale it is easily renewed, the manipulation of a [glass] speculum of such size, and probably of 5 cwt., would present considerable difficulty. Nor is it impossible that the film might break up under considerable changes of temperature, such as occur at Melbourne, or be spotted by rain or dew.'[22] However justified these remarks were at the time, we know now that the rejection of the silver-on-glass principle led to the lamentable failure of the instrument. The committee finally decided on a 4-foot Cassegrain, to be made by Thomas Grubb, F.R.S., owner of the Charlemont Bridge Works, Dublin.

Grubb was engineer to the Bank of Ireland, but he already had a reputation for his skill in making telescopes. In 1835 he had constructed a 15-inch clock-controlled Newtonian-Cassegrain for the Armagh Observatory in which, for the first time, a compound triangular system of balanced levers supported the main mirror.[23] Rosse used similar supports for the primary mirrors of his large telescopes and stated that he had availed himself of the suggestion of 'a clever Dublin artist, Mr Grubb.'[24] Robinson spoke well of the Armagh telescope and found that it resolved several difficult double stars.[25] Beside his own private observatory Grubb had built a factory for the manufacture of optical instruments and machine tools—a factory large enough to accommodate the plant for a telescope of the size contemplated.

Fig. 110—The mounting of the Melbourne telescope

Apart from Grubb's work with the Cassegrain, however, the plan was new. In the eyes of the committee this optical system had none of the disadvantages ascribed to it by Newton. Robinson wrote that, besides other advantages, 'the tube is shorter, therefore lighter, and less acted on by the wind; second, the magnitude of the second image gives facility for micrometer measures; it is also flatter than in any other telescope; third, the errors of the small speculum tend to correct those of the large. . . . Fourth, the greatest of all is the facility which it offers to the observer. The eyepiece is near the ground, and travels in a spherical surface of some 7 feet radius while the telescope sweeps the whole sky; the observer has to move but little, and the observing chair is light and easily managed.'[26]

Tube and speculum together weighed three tons and were carried by the polar axis, supported between two masonry piers embedded in concrete. Grubb's plan was to allow the axis to rest on its bearings with just sufficient weight to ensure contact. The remainder of the load was supported by carefully arranged counterpoises, so that a force of 12½ lb at a leverage of 20 feet sufficed to turn the polar axis.[27] Yet with all the care Grubb lavished on the design and construction of the various parts, the great instrument lacked rigidity and the adjustments seldom remained permanent. The greatest disadvantage was the metal mirror. For some months the specula retained their lustre but, in 1877, R. L. J. Ellery, director of the observatory, stated that they would soon require repolishing. This could be done only by sending both mirrors to Ireland. With commendable energy, Ellery taught himself the work and actually refigured the surface himself, but he lacked means for carrying out accurate tests.[28] The mirrors never performed well again.

> I consider the failure of the Melbourne reflector [Ritchey writes][29] to have been one of the greatest calamities in the history of instrumental astronomy; for by destroying confidence in the usefulness of great reflecting telescopes, it has hindered the development of this type of instrument, so wonderfully efficient in photographic and spectroscopic work, for nearly a third of a century.

The mounting and tube are now (1953) suitably housed at the Commonwealth Observatory, Mount Stromlo, Canberra. Considerably modified and modernized, they await Schmidt-type optics with a 50-inch glass spherical mirror.

Henry Draper, professor of natural science in the New York University, introduced Americans to the silver-on-glass reflector. He was the first to exploit its photographic possibilities, for he shared his father's interest in photography. Dr J. W. Draper had succeeded where Daguerre had failed—in the photography of the moon.

> There is no difficulty in procuring impressions of the moon by the daguerreotype [he wrote in 1840].[30] By the aid of a lens and a heliostat, I caused the moonbeams to converge on a plate, the lens being three inches in diameter. In half an hour a very strong impression was obtained. With another arrangement of lenses I obtained a stain nearly an inch in diameter, and of the general figure of the moon, in which the places of the dark spots might be indistinctly traced.

Henry Draper decided to make himself a reflecting telescope after seeing Rosse's reflector at Parsonstown. His private observatory was established at Hastings-on-Hudson, some twenty miles north of New York, where he began operations by casting a 15½-inch metal speculum. This he ground and polished on a machine almost identical with Rosse's, and in between times mastered photographic processes. But although he experimented with this machine for nearly a year, it persisted in grinding the surface in zones of different curvature. These he smoothed out by corroding away the defective parts with hydrochloric acid and by electrolysis. In the latter case, he isolated the offending zones, introduced acidulated water between grinder and speculum and passed an electrical current through the combination. As a result, tin and copper were removed from the uncovered zones. On another occasion, when he almost despaired of turning his disk into a good mirror, he thought of electroplating a brass mirror with speculum metal. The fracture of the original disk, however, put a stop to further experiments with metal.[31]

About this time, Sir John Herschel drew Draper's attention to the new silver-on-glass reflector. Draper accordingly improved his polishing machine and, while awaiting further details of the process, experimented for himself with different solutions. During the next three years and in hours snatched from sleep and professional duties, he polished more than one hundred mirrors of from $\frac{1}{4}$ inch to 19 inches diameter before he could produce large surfaces with speed and certainty.[32] The best of these, two 15½-inch disks, were completed during the winter of 1862 and, between 1858 and 1866, he used them and others to take hundreds of photographs of the sun and moon.[33] Some of the lunar images were good enough to bear enlargement to 36 inches and, in one instance, to 50 inches diameter.[34]

Draper experimented with no fewer than seven polishing machines made on principles recommended by Rosse and Lassell. The source of power was a treadwheel which he operated himself and on which he reckoned he sometimes walked over ten miles in five hours.[35] He complained that his mirrors had turned-down edges, for which he blamed his machine and, to avoid edge trouble, adopted Foucault's small polisher method of parabolization, only to find his surfaces 'diversified with undulations like a ruffle.'[36]

Draper's best 15½-inch mirrors bore powers of 1200, divided the difficult double star γ^2 Andromedae and showed the companion of Sirius and the sixth component of θ Orionis.

> As an example of light collecting power [he writes],[37] Debillisima between ϵ and 5 Lyrae is found to be quintuple, as first noticed by Mr Lassell. In the 18⅔-inch specula of Herschel, it was recorded as only double, and, according to Admiral Smyth, Lord Rosse did not notice the fourth and fifth components. Jupiter's moons show with beautiful disks, and their difference in diameter is very marked. As for the body of that planet, it is literally covered with belts up to the poles. The bright and dark spots on Venus, and the fading illumination of her inner edge, and its irregularities are perceived even when the air is far from tranquil. Stars are often seen as disks, and without any wings or tails.

Finding Foucault's silvering process uncertain, Draper adopted one with tartrate of potash and Rochelle salt, introduced by Cimeg in 1861 for backing ordinary looking-glasses.[38] He modified it so that the silver could be polished on its free surface and, in this way, never failed to secure 'bright, hard and in every respect perfect films.' For his diagonals he used silvered plate-glass,[39] but later found that optically worked surfaces gave better results. A 'directional' weakness in the plate-glass of this period—a defect absent in modern glass of this type—caused him to reject several otherwise fine mirrors. He found later that there was one diameter on either side of which a large mirror might stand without harm. He turned one of his 15½-inch mirrors through 90° from this position and could hardly believe that the surface was the same until further rotation by the same amount restored the original definition.[40] This defect was quite independent of any irregularity on the edge of the disk or in the mode of support.

Draper also studied the effect of temperature on a mirror's figure. He realized that the temporary distortion produced by the warmth of the hand would be made permanent if the surface was repolished in this condition. In fact, surfaces were so susceptible to temperature changes that even the smallest increase or decrease proved injurious during the delicate process of final figuring. 'A current of cold or warm air, a gleam of sunlight, the close approach of some person, an unguarded touch, the application of cold water injudiciously will ruin the labour of days.'[41]

Draper's telescope was mounted as an altazimuth in its own wooden observatory. The tube, 12 feet long and 18 inches diameter, consisted of walnut staves hooped together with brass bands. It was supported in an elaborate wooden frame and in such a way that the eyepiece remained stationary for all tube positions. The mirror support was similar to one introduced by Foucault—the disk rested on an india-rubber air cushion which pressed it against a semicircular tin-plate rim. A rubber tube ran from the air cushion to the observer who, by pressing the bulb, adjusted the inflation to that necessary just to keep the mirror against its retaining rim.[42] To prevent air currents within the tube, Draper opened up the lower part by drawing aside black velvet curtains that normally shut in the mirror.[43] The currents persisted, however, and at one time he covered the aperture with a sheet of thin plate-glass. On another occasion, when the currents seemed to be caused by heat from his body, he enclosed the opening in the dome by a conical muslin bag which fitted over the telescope's mouth—but without much improvement.[44]

The choice of an altazimuth mount meant that Draper had to provide the eyepiece of his telescope with a sliding plate-holder. The plate was then given a motion in the apparent direction of the moon's path by means of a sand-clock, the actuating weight of which was supported by a column of sand which ran out through a variable orifice below.[45] Later, he found a water clepsydra gave better results, even over four minutes—several times longer than the exposures usually required. For solar photography, Draper used an angular spring shutter, attached to the eyepiece and moved past the opening by means of a stout rubber band.[46]

At the close of 1867, Draper began the construction of a 28-inch Cassegrain reflector. Up to April 1, 1869, he ground and polished the primary mirror forty-one times and it was not until June, after a spell of $17\frac{1}{4}$ hours' polishing, that he considered the figure satisfactory.[47] With mirrors silvered and accurately aligned, the new telescope gave less critical images than those seen with the $15\frac{1}{2}$-inch, whereupon the 28-inch mirror again found itself on the polisher. It was not until June, 1872, that the large telescope was regarded as almost perfect.[48] Larger photographs of the moon (5 inches diameter) were now possible and, during August, 1873, this object was photographed on every clear night.[49]

Draper's contributions to spectroscopy are discussed in the next chapter. We can mention here the addition, in 1875, of a 12-inch Clark refractor to the axis of the 28-inch reflector and Draper's use of the dry plate on September 30, 1880, to secure the first photograph of the Orion nebula.[50] His first successful photographs of the nebula were taken in March, 1881, with exposures of the order of 104 minutes, after devoting much time, thought, and labour to the action of the driving clock and photographic manipulation of a new 11-inch achromatic telescope.[51] This fourth instrument replaced, in the spring of 1880, the 12-inch Clark. It was made by the Clarks for the Lisbon Observatory and was valuable owing to the fact that it was provided with a photographic corrector. On the finest plate of the Orion nebula taken with the 11-inch, stars of about the fourteenth magnitude were recorded.[52] We see presently how closely A. A. Common was following a similar and no less successful line of inquiry in England.

Draper's publication of his method of mirror-making and technique in astronomical photography was an important landmark in the history of American telescope-making. For the first time, amateur astronomers were invited to make their own instruments and to see celestial objects with definition and magnification equal to those of the larger refractors of the professionals.

Draper's work influenced John A. Brashear, a millwright of Pittsburgh. Brashear had been interested in astronomy from his boyhood, but not until he was thirty-two could he afford to buy books and materials for making telescopes. His first workshop was a wooden coal-house, his equipment a small, second-hand lathe driven by a steam engine. He decided to make a 5-inch achromatic objective, but for some time experienced nothing but disappointment. He was ignorant of the various processes used in lens-making and knew little about refractive indices and dispersive powers. He was able to compute the curves roughly, however, also to make the grinding and polishing tools. With the assistance of his wife, he eventually finished the lens. The task took three years and then only because he had worked until past midnight on some occasions.[53]

Brashear's next step was to show the lens to Professor Langley of the old Allegheny Observatory. Langley was impressed by its size and performance, allowed Brashear to view Saturn through the 13-inch Clark equatorial,* and gave him many words of encouragement.[54] Brashear left the observatory carrying Draper's paper under his arm and, shortly afterwards, began operations on his first mirror. Two 12-inch disks were cut from square plates, emery was washed for the different grades of abrasive, and grinding and polishing tools were prepared. This work, as before, had to be done after a strenuous day at the mill. Draper's paper proved invaluable. The expense of Cimeg's silvering process was avoided by using a process just published in *The English Mechanic*.[55] This gave a good deposit but the addition of the necessary heat caused the mirror to crack from edge to centre. Undaunted, Brashear made another mirror and experimented for himself with silvering solutions. He found one—a modification of Burton's method in *The Scientific American*—which he subsequently described in *The English Mechanic*.[56] This is the now popular *Brashear process*, since used by amateurs the world over for its cheapness, simplicity, and reliability.

In the Brashear process, part of the necessary ammonia is added before potassium hydroxide, which is then followed by the remainder. This avoids, to an appreciable extent, an excess of ammonia. In Liebig's process, silver oxide precipitated by potassium collects in compact grains which are only slowly dissolved in ammonia. The cleared solution, with its excess of ammonia, slows down the reduction of silver by sugar. Brashear's reducing solution was a mixture of rock-candy, nitric acid, alcohol, and distilled water. Best results were obtained when the solutions were used at the same temperature as that of the mirror. In hot weather, Brashear used ice to keep the temperature below 18°C. If deposited at a higher temperature,

* The original Fitz object-glass had been stolen and abused. Upon its recovery it was refigured by Clark.

Fig. 111—George H. With

(Hereford City Library, Museum, Art Gallery, and Old House)

the coat was apt to be soft and there was danger of the formation of explosive silver fulminate.

In England, the silver-on-glass mirror owed its adoption and rapid development to two enterprising telescope-makers, G. H. With (Fig. 111) and G. Calver, and to amateur astronomers like Bird and H. Cooper Key.

George H. With was born in 1827. For twenty-five years he was Master of the Blue Coat School, Hereford, and for pastime, made mirrors in a small workshop at his residence in Bath Street.[57] He was, unfortunately, unaware of Foucault's test, but nevertheless contrived to manufacture hundreds of mirrors of all sizes up to 18 inches diameter. He presumably tested their figures in the later stages in the telescope on bright stars, being guided by the appearance of extra-focal images under high magnification. Some of his mirrors were of high quality and are still doing useful work, although nearly a hundred years old. Several have been re-touched by others and all have been resilvered frequently, so that it is now difficult to assess their performance when they left their maker's hands. With himself admitted that some were over-corrected and that they were not all of the same high standard. On the other hand, he would often label a mirror ' absolute perfection ' or ' wonderful perfection ', and mention one or two stars which it could separate.[58] One such mirror, recently come to light—a 6½-inch of 67 inches focus—has a label to the effect that it ' easily divides γ^2 Andromedae and μ^2 Boötis ' and ' carries a power of 1000 easily on stars '.[59] But, apart from With's own remarks, the work of outstanding amateurs like T. W. Webb, W. F. Denning, and T. E. Espin is testimony enough to their fine performance. With's largest reflector, an 18-inch bequeathed to the British Astronomical Association by N. E. Green, an assiduous planetary observer, for many years did valuable work at the observatory of the late Rev. T. E. R. Phillips at Headley, Surrey; it is now housed in the private observatory of D. W. Millar at Keston, Kent.

Of With's first mirrors we know little, but the following account from the *Monthly Notices of the Royal Astronomical Society* for 1863 refers to early 5½-inch and 6¼-inch mirrors.[60]

> Mr. With, of Hereford, in his small intervals of leisure during the last three months, has completed three or four glass specula of the highest class, 6¼-inches aperture and 6 feet focus, and 5½-inches aperture and 4½ feet focus. With the larger telescope he is now able to elongate the disk of the small companion of γ Andromedae readily. Its definition of a large fixed star is remarkably fine, giving a perfectly round clean disk, surrounded at a little distance by one faint concentric ring, and without any other appendage or false light whatever. This was his first attempt. One of his 5½-inches specula is, however, slightly superior to this, and is presumed to be perfect: not the least trace of error of figure can be detected over the entire surface after the strictest scrutiny.

Comparisons between With's mirrors and the object-glasses of the time gave rise to a brisk interchange of letters among observers, some of which were published in the *Astronomical Register*.[61] Test-objects were double stars, the interest in which waxed very great halfway through the nineteenth century, following the publication of catalogues of these objects by Sir John Herschel, Darby, Webb, Gledhill, Dawes, and others.

During the course of his double-star work, Dawes made a careful study of resolving power and, in 1867, with the publication of his last catalogue of double stars,[62] gave a table in which he showed what aperture was necessary to resolve two stars of given apparent separation. This table was based on the fact that the diameters of star disks varied inversely as the aperture of the telescope and that a 1-inch objective ' divided ' stars of the sixth magnitude $4''\cdot56$ apart. At the same time, Dawes pointed out that ' tests of separation of double stars are not tests of excellence of figure ' as was the general belief.

In connection with these important observations we must mention that Sir William Herschel and others* had been aware of the fact that a good telescope images a star as a disk of light (the spurious disk). In 1813, Thomas Young commented on ' some lines of light and shade, or of colours ' which appeared round small objects viewed through a microscope. Young interpreted these colours in terms of his wave theory of light and concluded that ' their existence and their dimensions depend on the aperture of the microscope, and not on the magnitude

* *Vide* pp. 37, 58, 138.

of the particles in its focus.'[63] A few years later, both Fraunhofer and John Herschel[64] commented on the spurious disk and mentioned a surrounding system of faint rings. Both disk and ring system increased in size as the aperture of the telescope was reduced by smaller and smaller circular diaphragms. Both observers, in common with Young, recognized that the appearances seen were diffraction effects.

In 1831, G. B. Airy published the results of his mathematical investigation of the diffraction effects produced by circular and rectangular object-glasses.[65] He concluded that the image of a star formed by a circular objective takes the form of a strong central condensation of light surrounded by a series of rings of increasing diameter, but of low and rapidly diminishing brightness. The diameter of the spurious disk depends only on the wavelength of the light employed and the sine of the convergence angle under which the extreme marginal ray meets the central ray at the focal point. If the centre of the spurious disk of one star falls on the first dark ring of another star (of similar brightness), an observer with normal vision sees the stars almost separated. Simplified theory gives the least angular separation resolvable as 4″·5 divided by the aperture in inches, a relationship very close to that suggested by Dawes.

The optician Dallmeyer found from his measurements that 4″·33 divided by the aperture in inches gave resolving power, but that this depended upon the character of the seeing.[66] Dawes considered γ^2 Andromedae could not be separated in a refractor with an aperture less than 7¾ inches. Browning replied[67] that he had a 6½-inch mirror by With which divided this star 'as does also another of 6¼-inch in the possession of Mr Slack.' He also referred to a 6-inch metal speculum in America which divides γ^2 Andromedae and concluded, 'It would appear, then, that the superior dividing power of reflectors over refractors must beadmitted.'

With collaborated with John Browning who[68], early in the seventies of the last century, owned a flourishing scientific instrument business at 63 Strand, London. Browning was at that time making spectroscopes; assisted by With he also did much pioneer work in popularizing the reflector.[69] He mounted With's mirrors on cheap but substantial altazimuth and equatorial stands, thereby bringing the instruments within reach of the pocket of the ordinary amateur.

A With–Browning equatorial was used by Major J. F. Tennant at the Indian eclipse in August, 1868. The mirror was of 9 inches aperture, and the mount was driven by a clock similar to that employed on De la Rue's 13-inch equatorial. 'It is only fair to say that Mr Browning had never made such a clock before this' Tennant writes.[70] The account of Tennant's preparations before the eventful day, and of his experience during totality, is of interest as revealing the great need for improvement in the plates and clock-drives of the period.

> The nitrate of silver had been fused by myself in Mr. De la Rue's laboratory [writes Tennant][71] but when all other difficulties had been overcome, it seemed as though a bath made after his formula would not give clean pictures without a very minute portion of nitric acid. Several baths were made with different batches of distilled water, and with special precautions, still with the same result. . . . Still the pictures lacked definition, and though I knew the clock was not going well, I was satisfied that the focussing was wrong. The ground glass was useless for this purpose, except as a

rough guide, and I found the only way to focus on the collodion surface was by noting small specks and dust in it. . . . The clock all this time had been giving trouble. I tried everything short of taking it down and remounting it entirely, but without effect. . . . [During totality the clock stopped] though no great evil resulted, as Sergeant Phillips had, with great forethought, placed a man to start it again if it played tricks.

With's advice, always freely given, proved invaluable to others, and his example encouraged the Rev. H. Cooper Key to try his hand at figuring an 18-inch disk. Key succeeded, not without With's help and, after describing his method to the Royal Astronomical Society,[72] settled down to observing the moon and nebulae. One of Key's mirrors, a 12-inch of 10 feet focus, was purchased in 1866 by David Gill for his observatory in Skene Terrace, Aberdeen.[73] Bird also tried to emulate the example set by With and Key; after being introduced to mirror-making by Webb, he made himself 9-inch and 12-inch mirrors.[74]

The first large silver-on-glass reflector of this period was the comparatively useless 47-inch Newtonian which A. Martin,* Eichens, and Gautier completed in 1877. This instrument stood for many years under a movable housing in the grounds of the Paris Observatory. A spiral staircase which ran on a circular track enabled the observer to reach the eyepiece. Some years ago the mirror was refigured by A. Couder of the Paris Observatory and it now rests in a new mounting at Saint Michel, Haute-Provence.[75] This reflector is the largest telescope in France. The mounting was made by Secrétan (now Secrétan, Ch. Épry and Jacquelin).

George Calver of Widford, Chelmsford, made a large number of silver-on-glass mirrors, many of which are still in regular use. He made numerous 12-inch and smaller equatorial reflectors, an 18-inch for Klein at Stanmore, North London, and a 24-inch for the Rev. T. E. Espin at Tow Law, Durham. His trade-plate on a telescope, however, does not necessarily imply that it was made entirely in his workshop; many mounts and accessories were made for him by Messrs Ottway of Ealing. Calver aspired to quite large mirrors. ' Mr. Calver has recently figured a 50-inch mirror for Sir H. Bessemer ' Denning wrote in 1891, ' but the mounting is not completed; and he is expecting to make other large reflectors, viz. one over 5 feet diameter and another over 3 feet.'[76]

Calver is best known for the 36-inch silver-on-glass mirror which he made for Dr A. A. Common of Ealing, London. For some years Common had been engaged in astronomical photography with a 5½-inch refractor. With the intention of making a larger instrument, he purchased two 17-inch blanks for a refractor, but then changed his mind and ordered an 18-inch mirror from Calver.[77] He designed and constructed the mounting himself and, when the instrument was completed, made observations of Mars and Saturn.[78] But 18 inches was still too small an aperture for his requirements and, a little later, Calver received an order for a mirror of twice the size. At first he had doubts as to whether so large a disk could be properly annealed. The first broke into ' hundreds of fragments ' under the tool but a second disk, 4½ inches thick and weighing nearly 3 cwt, was successfully ground by machinery.[79]

* Inventor of a silvering technique somewhat similar to Brashear's. It involves the use of a predetermined weight of nitrate of ammonia.

Fig. 112—Tennant's 9-inch photographic reflector

Made by With-Browning, 1868

(*Royal Astronomical Society*)

Calver made five different machines, one of which was based on the same principle as Lassell's—' an excellent principle ' he wrote, ' but I have long since come to the conclusion that no machine can do the final work like the trained hand, and I was gratified to learn, when in conversation with Professor Draper, that his experience agreed precisely with my own on this point.'[80] Like his predecessors, Calver stressed the importance of making the polisher to the correct figure, for from this the glass received *its* figure. In polishing the 36-inch he made everything subservient to the form of the polisher—its size and number of strokes, the very machine itself.

> During the polishing and figuring [Calver wrote][81] I carefully studied the behaviour of the disc—for flexure—for distortion of figure by contraction and expansion during changes of temperature—and I found the disc as perfect as a 6-inch one. . . . The work of correcting was tedious and trying, especially in the later stages, when for every few minutes' polishing the whole preparations for testing had to be repeated, and the settling of the mass into its normal state had to be patiently waited for, and often days passed before further advance could be made.

Calver tested the figure at the centre of curvature by Foucault's method, and sometimes on small letters in daylight. Every care was taken to exclude draughts from the testing corridor and to keep the temperature constant. Common resilvered the mirror himself, suspending it face downwards in the silvering bath with the aid of a large suction disk.[82] The progress of deposition was more difficult to observe than with the mirror face up, but the resultant film was free from muddy precipitate.

Common mounted his mirror as a Newtonian in July, 1879. The mounting took the form of a fork-type equatorial with a latticework tube and the heavy polar axis supported by a large drum floating in a reservoir of mercury. When not in use, the instrument could be covered by a movable wooden housing, otherwise it was entirely in the open.[83] It was used chiefly as a photographic telescope and, as such, gave scope for much experimental work with different exposures on the dry gelatine silver-bromide plates introduced in 1871. The first attempt to photograph the Orion nebula in January, 1880, was a failure. Stars were in focus, but irregularities in the clock-drive made them appear as short lines, while a faint stain represented the nebula.[84] In June of the following year, with an improved drive and more sensitive plates, Common took the first successful photographs of Comet 1881b.[85] On March 17, 1882, he photographed the faint filaments of the Orion nebula—a photograph which caused great excitement at the May meeting of the Royal Astronomical Society.[86] A year later, Common had so improved the driving mechanism that he was able to give exposures of the unprecedented length of 1 hour 30 minutes. They were made possible by placing the slides in a movable frame attached to the telescope. An eyepiece was attached to the carrier slide of the photographic plate and, by constantly watching the star image when placed on the eyepiece cross-wires and moving the screws to maintain it there, almost circular star images were obtained.[87] With this device, Common obtained a splendid photograph (for those days) of the Orion nebula which earned him the Gold Medal of the Royal Astronomical Society. ' The success of these long exposures with this

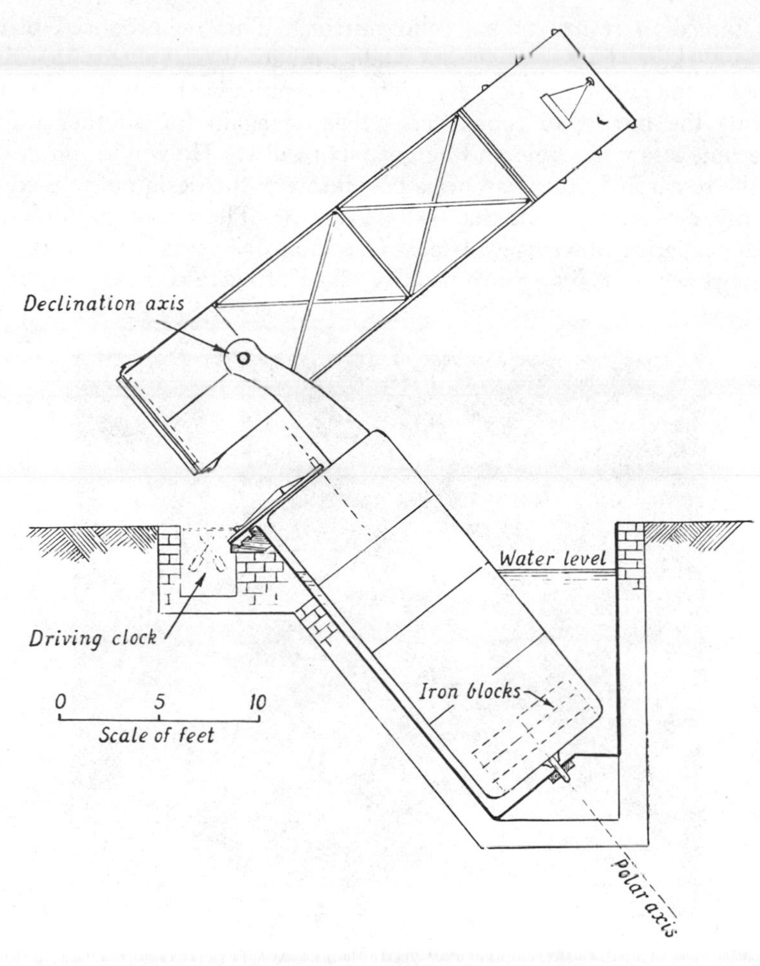

Fig. 113—Common's 5-foot reflector, 1891

powerful instrument' Edwin Dunkin remarked as he made the award, 'has opened out a new field of research by which the accumulating effect of the light of faint stars, too faint even for observation to the eye, has been registered on the photographic plate.'[88]

Common followed this achievement by taking photographs of Jupiter and Saturn, and of other nebulae besides that in Orion. At the same time, he drew up plans for a reflector of 5 feet aperture.[89] Nearly every part of this impressive instrument was made in a workshop adjoining his house. He used the instrument first as a Newtonian, but one night he nearly had a serious fall from the observing stage and there and then decided never to mount it again.[90] Calver's previous experience, when drilling a hole through the 3-foot mirror, turned Common against using the 5-foot as a Cassegrain, but this form offered so many advantages that he had a mirror cast with a hole already in it. This disk turned out to be so full of streaks that it was almost impossible to figure the surface with any degree of accuracy, and Common was forced to return to his solid mirror. This he proposed using as an ' oblique ' or ' skew ' Cassegrain,[91] but when a practical trial of this idea also failed, he returned to the faulty casting. By a laborious process of polishing for a minute, then leaving the mirror to cool, then polishing again for another minute, the streaks became less noticeable and he got fair results. He would, no doubt, have followed the research further had not a new interest, the design of telescopic sights for the Army, completely engrossed his last years.[92] The 5-foot, therefore, scarcely ever left the experimental stage. It cost Common five years' hard work and many anxious hours without giving any positive signs of success. He was an engineer

Fig. 114.—Common's workshop

His polishing 'machine' is a conspicuous feature

(*Royal Astronomical Society*)

by profession, but even his skill and experience could not make amends for the inadequate facilities of his workshop (Fig. 114).

Besides making telescope mirrors of various sizes,* Common did much pioneer work with the coelostat. This device was an improved form of siderostat which reduced to apparent rest, not only one star, but the whole sky. It was first conceived by the Frenchman, K. August, in 1839, but the name *coelostat* was not applied to it until 1895.[93] In 1895, G. Lippmann[94] drew attention to the fact that a plane mirror rotating in forty-eight hours about an axis in its plane, placed parallel

* A 36-inch Common mirror is at present in store at the Observatories, Cambridge, England. About 1889, Common made a mounting and figured a 30-inch mirror for Lockyer's private observatory at Westgate-on-Sea. This instrument was later moved to South Kensington and is now stored at the Norman Lockyer Observatory, Sidmouth. Common also made a 36-inch reflector for Edward Crossley's private observatory at Bermerside, Halifax. Crossley later presented this telescope to the Lick Observatory (*vide* p. 314). The 60-inch mirror is still in use (*vide* p. 395).

to the earth's axis, possesses properties similar to those of an equatorial. The direction of the horizontal telescope is dictated by the declination of the stars under observation, while the initial position of the mirror depends on the hour angle. The telescope is, in this way, restricted to one zone of stars and has to be moved bodily over considerable arcs to reach other declinations. The disposition of the mirror also means that light from circumpolar stars meets the surface almost at grazing incidence and makes their observation difficult; in other positions the diameter of the reflected beam varies greatly.

S'Gravesande's *heliostate* was developed by Gambey (1831), Silbermann (1844), and Foucault (1869).[95] It is around this last date that the name *siderostat* first appears. In Foucault's form, the mirror is carried in a fork movable about a vertical axis and actuated through connecting links by a polar drive. Like the coelostat, which followed some years later, it can pass sunlight or starlight into a horizontal

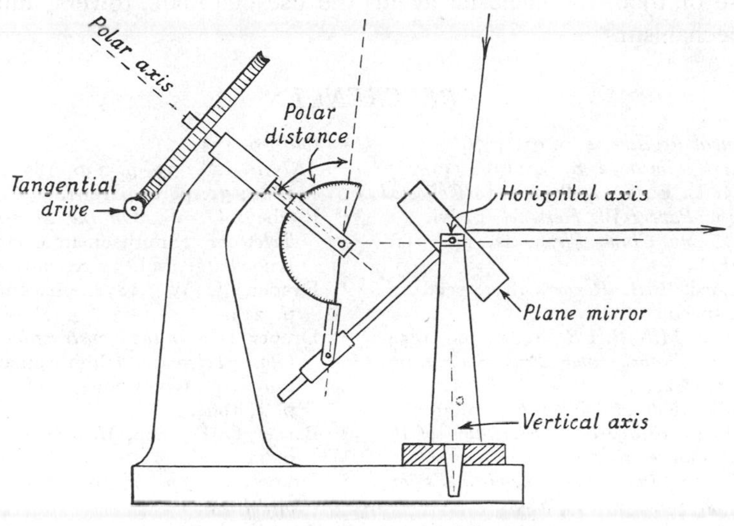

Fig. 115—The Foucault siderostat

telescope but at the expense of causing the field of view to rotate slowly. For the comparatively short exposures used in photographing the sun, this field rotation is inconsequential. A Foucault-type siderostat by Eichens was used by Gill at Mauritius in conjunction with a 40-foot Dallmeyer horizontal telescope for observing the 1874 transit of Venus.[96] On the same occasion, and at the quite independent suggestion of Joseph Winlock, third director of the Harvard Observatory, a similar arrangement was used by astronomers of the U.S. Naval Observatory, Washington.[97]

Following Lippmann's description of the coelostat, Common designed and made one in time for H. H. Turner to take it, after preliminary trials, to the Japanese eclipse of August, 1896.[98] The same coelostat was used by Turner for the Indian eclipse early in 1898. The 16-inch mirror fed two cameras placed side by side, each provided with its own lens and photographic plate.

The pioneer work of Common, Turner, and J. Stoney showed that the horizontal photoheliograph was superior to large mounted instruments. The rotation of the field characteristic of the siderostat decided Schaeberle, during the 1893 total solar eclipse, to fix the tube, not horizontal, but pointed directly at the sun during totality.[99] The Clark object-glass, of 5 inches aperture and 40 feet focus, was mounted on one pier, a moving photographic plate on another. Clock mechanism and inclined guides so regulated the motion of the plate that it kept pace with the sun's image. The long focal length rendered optical aberrations inappreciable and the resulting photographs were so excellent that the same equipment was used for the 1896 Japanese and later expeditions from the Lick Observatory.

When long tubes are used it is better to use a coelostat and to keep the telescope horizontal. Unless a second mirror is employed a previous calculation of the exact position of the tube in azimuth at the moment of totality is necessary. This presents no difficulty and, with horizontal cameras of 100 feet focal length, as in the American eclipse of 1900, the coelostat avoids the use of pillars, towers, and elaborate driving mechanism.

REFERENCES

1 *Astronomical Register*, **4**, p. 91, 1866.
2 *Phil. Trans., Abridged*, **8**, p. 113, 1740.
3 Dreyer, J. L. E., 1912, Preface to *Collected Scientific Papers W. Herschel*, p. lvi.
4 Jamin, J., *Ann. Chim. Phys.*, **22**, pp. 311–327, 1848.
5 Rosse, Lord, *B.A. Report*, 1851, sect. 12, 'Plain specula of silver'.
6 Rosse, Lord, *M.N.R.A.S.*, **14**, p. 199, 1854.
7 Airy, G. B., *Trans. Camb. Phil. Soc.*, **2**, pp. 105–118, 1822.
8 Airy, G. B., 1896, *Autobiography*, p. 38.
9 Tallis, 1852, *History and Description of the Crystal Palace*, p. 82.
10 Liebig, J. von, *Annalen der Chemie und Pharmacie*, **98**, pp. 132–139, 1856.
11 *Augsburger Allgemeine Zeitung*, March 24, 1856, quoted by Steinheil, *M.N.R.A.S.*, **19**, pp. 56–60, 1859.
12 Webb, T. W., *Intellectual Observer*, **3**, p. 129, 1863.
13 Foucault, L., *Comp. Rend.*, **44**. p. 339, 1857; Ref. 18, p. 221. Webb, *op. cit.*, p. 129.
14 Webb, *op. cit.*, **1**, p. 380, 1862.
15 *Ibid.*, **8**, p. 370, 1866.
16 Lockyer, J. N., 1878, *Stargazing*, p. 343.
17 *Ibid.*, pp. 344–348.
18 Foucault, L., *Annales de l'Observatoire de Paris*, **5**, pp. 197–237, 1859.
19 *Ibid.*, p. 212 ff.
20 *Ibid.*, p. 221.
21 *Phil. Trans.*, **159**, p. 128, 1869.
22 *Ibid.*, p. 132.
23 *Proc. Roy. Soc.*, **135**, Obituary Notice, p. iv, 1932.
24 Rosse, Lord, 1926, *Scientific Papers*, p. 101.
25 *Phil. Trans.*, **159**, p. 132, 1869.
26 *Ibid.*, p. 133.
27 *Ibid.*, p. 151.
28 *M.N.R.A.S.*, **33**, p. 230, 1873; **49**, p. 341, 1889; **51**, p. 231, 1891.
29 Ritchey, G. W., *On the Modern Reflecting Telescope* (Smithsonian Contributions to Knowledge, vol. 34 reprint), p. 47, 1904.
30 Draper, J. W., 1878, *Scientific Memoirs*, p. 213.
31 Draper, H., *On the Construction of a Silvered Glass Telescope* (Smithsonian Contributions to Knowledge, vol. 14 reprint), p. 2, 1864.
32 Barker, G. F., 1888, *Memoir of Henry Draper*, p. 12.
33 Draper, H., *op. cit.*, p. 47; Barker, G. F., *op. cit.*, p. 12.
34 Draper, H., *op. cit.*, pp. 1, 27, 52, 55.
35 *Ibid.*, p. 21.
36 *Ibid.*, pp. 23, 24.
37 *Ibid.*, p. 26.
38 *Ibid.*, p. 3.
39 *Ibid.*, p. 27.
40 *Ibid.*, p. 7.
41 *Ibid.*, p. 9.
42 *Ibid.*, p. 30.
43 *Ibid.*, p. 31.
44 *Ibid.*
45 *Ibid.*, pp. 33–40.
46 *Ibid.*, pp. 40–41.
47 Barker, G. F., *op. cit.*, p. 13.
48 *Ibid.*
49 *Ibid.*, pp. 13–14.
50 *Ibid.*, pp. 15–16.
51 *Ibid.*, p. 30.
52 *Ibid.*, p. 30.
53 Brashear, J., 1925, *Autobiography*, pp. 32–34.
54 *Ibid.*, pp. 35–37.

[55] Possibly *English Mechanic*, **29**, p. 398, 1879 (Blacklock) or **30**, pp. 415–416, 1880 (Various authors).

[56] Brashear, J., *English Mechanic*, **31**, p. 327, 1880.

[57] *Observatory*, **27**, p. 416, 1904.

[58] *J.B.A.A.*, **51**, pp. 206, 332, 1941.

[59] *J.B.A.A.*, **52**, p. 71, 1942; **51**, p. 332, 1941.

[60] Key, H. C., *M.N.R.A.S.*, **23**, p. 202, 1863.

[61] *Astronomical Register*, **3–7**, 1865–1869.

[62] *Mem. R.A.S.*, **35**, p. 159, 1867.

[63] Young, T., 1813, *Introduction to Medical Literature*. Peacock, W., 1855, *Miscellaneous Works of Thomas Young*, Vol. 1, p. 351.

[64] Herschel, J. F. W., 1827, *Light*, Art. 766 ff.

[65] Airy, G. B., *Trans. Camb. Phil. Soc.*, **5**, pp. 283–291, 1834. *Mathematical Tracts*, 1831, 2nd edition, pp. 321 ff.

[66] *Astronomical Register*, **5**, p. 129, 1867 (Account of R.A.S. meeting of May 10, 1867).

[67] *Ibid.*, p. 111.

[68] *M.N.R.A.S.*, **90**, pp. 359–362, 1930.

[69] Browning, J., 1876, *A Plea for Reflectors*.

[70] *Mem. R.A.S.*, **37**, p. 29, 1868.

[71] *Ibid.*, pp. 30–31.

[72] Key, *op. cit.*, pp. 199–202.

[73] Forbes, G., 1916, *David Gill*, pp. 37–38.

[74] Webb, T. W., *Intellectual Observer*, **3**, p. 217, 1863.

[75] Lockyer, J. N., *op. cit.*, pp. 314–315. Danjon, A., and Couder, A., 1935, *Lunettes et Télescopes*, p. 704.

[76] Denning, W. F., 1891, *Telescopic Work for Starlight Evenings*, p. 16.

[77] *M.N.R.A.S.*, **64**, p. 274, 1904.

[78] *Ibid.*

[79] Calver, G., *M.N.R.A.S.*, **40**, pp. 17–20, 1880.

[80] *Ibid.*, p. 20.

[81] *Ibid.*, p. 19.

[82] *M.N.R.A.S.*, **42**, pp. 79–82, 1882.

[83] Common, A. A., *Mem. R.A.S.*, **46**, pp. 173–182, 1881.

[84] *M.N.R.A.S.*, **44**, pp. 221–222, 1884.

[85] *Ibid.*, p. 222.

[86] *Ibid.*

[87] *Ibid.*, **49**, p. 297, 1889.

[88] *M.N.R.A.S.*, **44**, p. 222, 1884.

[89] *Mem. R.A.S.*, **50**, pp. 113–204, 1892.

[90] *M.N.R.A.S.*, **64**, p. 277, 1904.

[91] *Ibid.*, **45**, pp. 24–25, 1885.

[92] *Ibid.*, **64**, p. 277, 1904.

[93] Plummer, H. C., *M.N.R.A.S.*, **101**, p. 168, 1941.

[94] Lippmann, M. G., *Compt. Rend.*, **120**, p. 1015, 1895. *Observatory*, **18**, pp. 301–303, 1895.

[95] *Vide* Ref. 93.

[96] Forbes, *op. cit.*, pp. 62–65.

[97] Newcomb, S., 1903, *Reminiscences of an Astronomer*, pp. 163–166.

[98] *M.N.R.A.S.*, **64**, pp. 277–278, 1904.

[99] Dyson, F., and Woolley, R., 1937, *Eclipses of the Sun and Moon*, p. 60.

CHAPTER XIV

> The science of the *motions* of the stars is only a part of modern sidereal astronomy. Within the last quarter of a century, a science of their *nature* has sprung up and assumed surprising proportions; a science the reality of which confounds forecast, yet compels belief. Sidereal physics has a great future in store for it. Its expansiveness in all directions is positively bewildering.
>
> AGNES CLERKE

OVER half a century elapsed before a full explanation of the dark lines observed in the solar spectrum by Wollaston was forthcoming. In England, both David Brewster and John Herschel pursued the analogous subject of the selective absorption of light by coloured liquids. By 1833, Brewster had drawn attention to 'a remarkable series of dark lines and bands' which appeared when fuming nitric acid was interposed between the sun and the spectroscope.[1] To Brewster, the absorption lines so produced afforded a valuable new method of chemical analysis. He went further: 'When the sun descends towards the horizon and shines through a rapidly increasing depth of air, certain lines which before were little, if at all, visible, become black and well defined, and dark bands appear even in what were formerly the most luminous parts of the spectrum.'[2] Everything pointed to the fact that the variable lines, the so-called *telluric* lines, were produced by 'the absorptive effect of the earth's atmosphere.'

Brewster's countryman, Professor J. D. Forbes, strongly contested these views. From observations of the annular eclipse of 1836, Forbes concluded that the sun's atmosphere was not responsible for the solar lines.[3] He toyed with the idea of making what was, in principle, a prominence spectroscope but, lulled by the apparently decisive nature of his eclipse observations, pushed the idea aside.[4] In England, W. H. Fox Talbot confused certain bright-line spectra with dark-line spectra and, at one time, ascribed the D line to both sulphur and the salts of sodium.[5] In 1849, the coincidence of the bright lines of sodium with the dark D lines of the sun was first demonstrated by Léon Foucault at Paris[6] and, independently, by W. A. Miller, professor of chemistry at King's College, London. Miller stated that the coincidence was 'accurate to an astonishing degree of minuteness.'[7] Foucault's method was to pass sunlight through an electric arc charged with sodium salts and thence into the spectroscope. On its own, glowing sodium vapour gave two bright D lines. When sunlight was passed through this vapour, however, the black D lines of the solar spectrum were considerably reinforced. 'Thus the arc' Foucault wrote, 'presents us with a medium which emits the rays D on its own account, and which at the same time absorbs them when they come from another quarter.'[8]

Research in this subject was by no means confined to Europe. Between 1845 and 1855, David Alter of Pittsburgh used home-made spectroscopic apparatus to classify flame and spark spectra. In his first paper, published in 1854,[9] Alter described the appearance of the spark spectra of several metals, noted the appearance of lines which are not always seen and which he ascribed to impurities in the metal, and found that an alloy of two metals shows the lines of both constituents. In the following year he published a second paper[10] in which he described the spark spectra of hydrogen, air, and other gases. As before, he found that each gas has a characteristic spectrum. The colours of aurorae, he thought, ‘ probably indicate the elements involved in that [*sic*] phenomena . . . [while] the prism may also detect the elements in shooting stars, or luminous meteors.’ But a full consideration of Alter's work and the claims of other investigators of the forties and fifties such as Stokes, J. W. Draper, Becquerel, Wheatstone, Balfour Stewart, and Ångström, does not lie within the scope of this history.

Bunsen and Kirchhoff's revelation came in 1859. Using flames of different temperatures in which some common salt was placed, they were able to strengthen, weaken, and even reverse the dark absorption D lines observed when sunlight was fed via the flame to a slit-spectroscope. ‘ I conclude ’ Kirchhoff wrote,[11] ‘ that the dark lines of the solar spectrum which are not evoked by the atmosphere of the earth exist in consequence of the presence, in the incandescent atmosphere of the sun, of those substances which in the spectrum of a flame produce bright lines at the same place.’ More particularly, it was clear that, in order for the D lines to come out dark, a salt flame of lower temperature is required. The astrophysical implication was clear—the incandescent body of the sun is ‘ surrounded by a gaseous atmosphere of somewhat lower temperature. . . . From the occurrence of these lines (D), the presence of sodium in the atmosphere of the sun may therefore be concluded.’ What was true of the sun applied with equal force to the stars. A new era in astronomy had begun, to be compared only with that which followed the invention of the telescope.

The years which followed Kirchhoff's first announcement saw the careful mapping of the lines given by common elements. It was clear that great caution had to be exercised in deducing the chemical composition of a substance from its spectrum. In most cases the lines were numerous and easily confused. The smallest amount of foreign materials changed the spectrum of the body under examination—that is, new lines appeared. For comparison and identification purposes, the scrupulous measurement of line positions on as large a scale as possible was essential. Kirchhoff and his pupil Hoffmann, for example, made a drawing about eight feet long of the solar spectrum, using a millimetre reference scale which started from an arbitrary point.[12] Brewster, Gladstone, and Janssen worked assiduously at the delineation and identification of the telluric lines.[13] Ångström attained modern rigour in his now classic normal map of the solar spectrum. This was published at Upsala in 1868 and gave the position of some 1200 lines, of which 800 were identified with lines of common elements.[14] The unit he used, a ten-millionth part of a millimetre, is our modern *angstrom unit*.

Ångström's normal map was made possible by the use of a reflection grating instead of a prism. We have seen (p. 186) how Fraunhofer obtained small, bright spectra by setting fine parallel wires over the object-glass of a telescope. Using a similar principle, Fraunhofer measured, in 1821, the wavelength of the D line with considerable accuracy.[15] Whereas the spectrum produced by a prism is contracted at the red end and extended at the violet, that given by Fraunhofer's grating (i.e. by diffraction) is almost *normal*. More precisely, the angular deviation produced by the latter is nearly proportional to wavelength, so that both the distance between the spectrum lines and the wavelength can be measured on a uniform scale. Early gratings were comparatively coarse; Ångström's finest contained about 5000 lines to the inch, ruled on glass and used as a transmission grating.[16] In America, J. W. Draper possessed similar gratings, made for him by Saxton of the United States Mint in Philadelphia, as early as 1820.[17] He found that they gave brighter spectra when used as reflection gratings, that is, when silvered with mercury–tin amalgam.[17] Lewis M. Rutherfurd before 1850 produced gratings coated with silver and also diamond-ruled speculum-metal reflection gratings, some with nearly 20,000 lines to the inch.[18] They were small, but possessed quite high dispersive power and corresponding resolving power, factors which depend on the total number of lines. A good grating must contain some 10,000 to 15,000 lines per inch, all equally spaced on a highly polished and accurately figured reflecting surface. We see later how Henry Rowland met these exacting conditions.

Early spectroscopists did not stop at mere line plotting. In 1862, J. Plücker of Bonn pointed out that the same substance at different temperatures might give different spectra.[19] A. Wüllner found that both pressure and temperature changed the appearance of bright-line or emission spectra.[20] More extended research by Clifton, H. E. Roscoe, J. N. Lockyer, and others revealed that spectral changes also depended on molecular structure. No sooner was one problem solved than another took its place. A subject of apparent simplicity soon became one of surprising complexity.

Astronomers possessing fairly large equatorial telescopes found an exciting if laborious field of inquiry in solar and stellar spectroscopy. Rutherfurd, Young, and Langley in America, Huggins, Miller, and Lockyer in England, Respighi, Secchi, and Donati in Italy, Ångström in Sweden, S. Merz, Zöllner, Lamont, and Vogel in Germany, and Wolf and Rayet in France, all pioneered in astrophysics. To their own inventiveness was added the constructional skill of a number of opticians and instrument-makers. It is no exaggeration to say that the rapid rise of astrophysics was made possible by craftsmen like C. A. Steinheil and the Merz organization at Munich, John Browning, William Simms and, later, the Hilger brothers of London, and Howard Grubb of Dublin. These workers soon became acknowledged experts in the design and manufacture of high-grade prisms, spectroscopes, and auxiliary spectroscopic apparatus.

Investigators subsequent to Fraunhofer changed his method of observation of the solar spectrum by introducing the prism or prisms between two convex lenses. The slit was placed in the focal plane of one lens while the other projected the spectrum on a screen. This method was adopted independently by several investigators about the same time. We know that Simms, for one, constructed a spectroscope like this on Airy's suggestion as early as 1848.[21]

In practice, the spectroscope was so arranged that the slit was fixed in the focal plane of the object-glass. Light diverging from the slit was then rendered parallel by a collimating lens and, in this condition, was ready for dispersion. Some spectroscopes employed a train of three or five prisms arranged in such a way that dispersion without deviation was obtained. These compound prisms, first made by Amici, were introduced into astronomy by the French astronomer Janssen.[22] Alternatively, dispersion was produced by two or three separated dense flint-glass prisms. A third arrangement combined these two types of prism train, and a fourth, used by Lockyer and independently by Young, incorporated a right-angled prism.[23] When the light had passed through the lower part of Young's prism train it was totally reflected twice by the right-angled prism and sent back through the upper part of the prisms, thus being doubly dispersed. Trains of as many as fifteen prisms were used in solar spectroscopes but, for stellar studies, where little light was available, three to five prisms were at first used. To complete the spectroscope, astronomers added an observing telescope with cross-wires and devices for measuring line positions.

The highly original work of William Huggins from 1860 to 1869 was done at the focus of his Clark refractor with a two-prism spectroscope. Huggins conducted his researches from a private observatory which adjoined his house at 90 Upper Tulse Hill, London. His first telescope was a 5-inch Dollond equatorial refractor through which he made drawings of the planets.[24] Then came news from Dawes about the remarkable performance of Alvan Clark's objectives and, in 1858, he purchased an 8-inch glass which Cooke mounted equatorially and provided with clock motion.[25]

> I soon became a little dissatisfied with the routine character of ordinary astronomical work [he wrote later][26] and in a vague way sought about in my mind for the possibility of research upon the heavens in a new direction, or by new methods. . . . It was just at this time, when a vague longing after newer methods of observation for attacking many of the problems of the heavenly bodies filled my mind, that the news reached me of Kirchhoff's great discovery of the nature and the chemical constitution of the sun from his interpretation of the Fraunhofer lines. Here at last presented itself the very order of work for which in an indefinite way I was looking—namely, to extend his novel methods of research upon the sun to the other heavenly bodies.

Huggins' spectroscope contained a cylindrical lens, as first used by Fraunhofer,[27] to give width to a star's spectrum. To assist in the identification of the different spectral lines, Huggins used comparison spectra. These he formed by passing light from an induction coil, sparked between electrodes of different pairs of metals, into a small reflecting prism placed over part of the slit. Thus equipped, Huggins and his friend W. A. Miller examined the spectra of the brighter stars and of the sun, moon, and planets. Early in 1863, they sent a preliminary note to the Royal Society *On the Lines of the Spectra of some of the fixed Stars*.[28] On the same day as their paper was read, news arrived that similar observations had been made a month earlier by Rutherfurd.[29] At the same time, Secchi at Rome and a little later, Vogel at Potsdam entered into stellar spectroscopy.

In April of 1864, Huggins and Miller quite surpassed their contemporaries with their second paper on stellar spectra.[30] The latter were compared directly with the spectra of ten to twenty elements and many lines were identified with considerable certainty. It was clear that stars had elements in common with those of our own earth and had atmospheres like the sun. It was also clear that a complete study of even a single star would take many years. Photographic apparatus was installed

Fig. 116—William Huggins' 8-inch Clark-Cooke equatorial refractor

Huggins used the instrument from 1860 to 1869. The star spectrograph and the induction coil for comparison spectra can be seen.

in 1863, but Sirius and Capella gave only faint impressions; ' from want of steadiness and more perfect adjustment of the instrument, the spectra, though defined at the edges, did not show the dark lines, as we expected.'[31] The two astronomers used wet collodion plates in preference to the slower dry plates of the period, but they found the irregular drying and draining of the wet plates a great inconvenience.

In 1864, Huggins made his important discovery of the gaseous nature of the so-called ' unresolved ' nebulae, considered by many to be aggregations of stars too remote to be resolved by even the largest instruments.

> On the evening of August 29th [he writes][32] I directed the telescope for the first time to a planetary nebula in Draco. I looked into the spectroscope. No spectrum such as I expected! A single bright line only! At first I suspected some displacement of the

prism, and that I was looking at a reflection of the illuminated slit from one of its faces. This thought was scarcely more than momentary, then the true interpretation flashed upon me. The riddle of the nebulae was solved. The answer, which had come to us in the light itself, read: Not an aggregation of stars, but a luminous gas.

On May 18, 1866, Huggins observed the 'new' star in Corona Borealis discovered four days earlier by J. Birmingham. The nova was then just below mag-

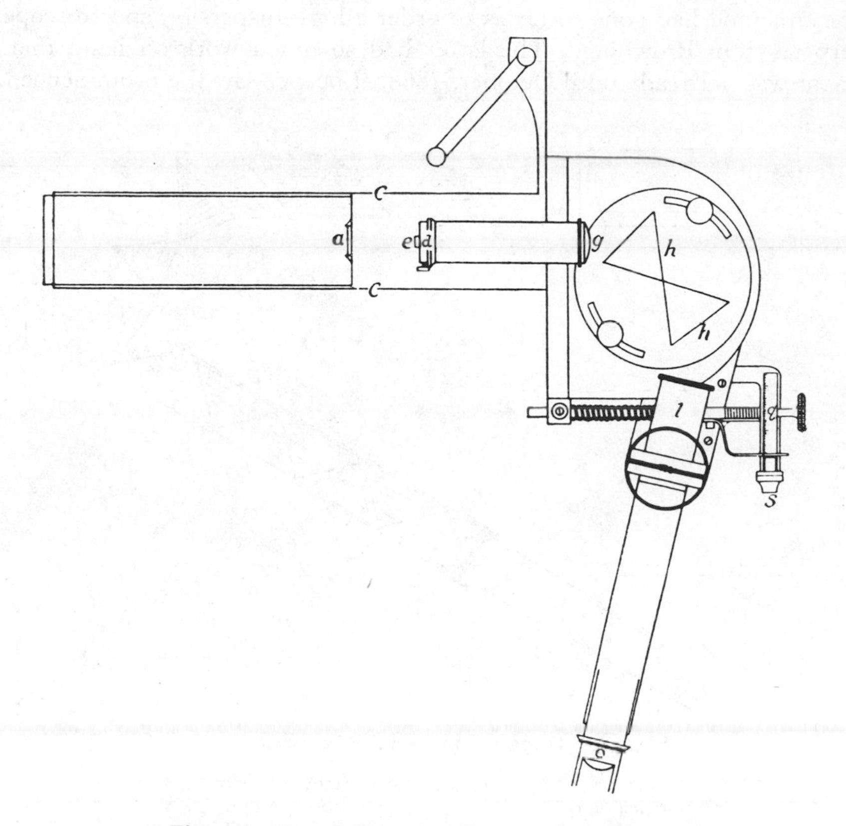

Fig. 117—Huggins' first stellar spectroscope

a	Cylindrical lens	*d*	Slit position
e	Slit prism	*h*	Flint-glass prisms
g	Collimator lens	*l*	Telescope

nitude 3·0 and 'the view in the spectroscope was strange, and up to that time unprecedented.'[33] Superimposed on a solar-type spectrum were a number of bright hydrogen lines, suggesting that the outburst was accompanied by the emission of a shell of gas at a temperature higher than that of the star's surface. The star waned rapidly and, within a few weeks, had sunk below naked-eye visibility. Speculations as to the nature of the outburst followed. One thing was clear—if our knowledge of novae was to be extended, the necessary instrument would be the spectroscope.

The year 1868 was a memorable one in astrophysics. Before then, De la Rue and Secchi had photographed solar prominences during total solar eclipses—the prominences were obviously solar appendages. During totality on August 18, 1868, Tennant, Pogson, Rayet, and Janssen in India saw the hydrogen lines in great brilliance. Inspired by the sight and in the hope of seeing the lines without an eclipse, Janssen placed the slit of his high-dispersion spectroscope tangentially to the sun's limb and found his wish granted.[34] Lockyer had entertained similar ideas for some time and had gone so far as to order a high-dispersion spectroscope for this purpose from Browning. The latter had so much work on hand that the instrument was not ready until October, 1868. Lockyer saw the prominence lines

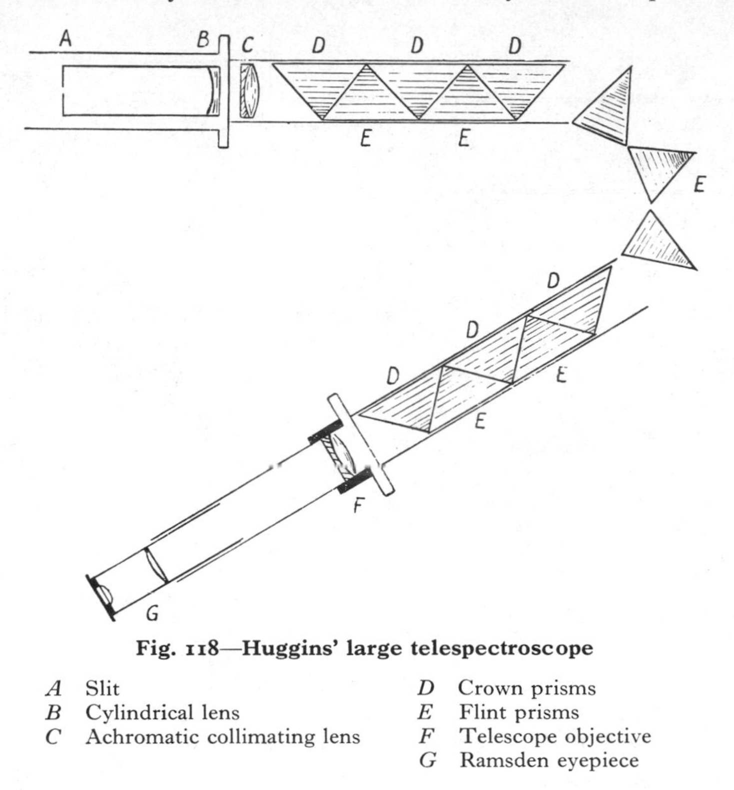

Fig. 118—Huggins' large telespectroscope

A	Slit	*D*	Crown prisms
B	Cylindrical lens	*E*	Flint prisms
C	Achromatic collimating lens	*F*	Telescope objective
		G	Ramsden eyepiece

on the 20th of that month, only to learn later of Janssen's prior discovery.[35] The incident reveals the dependence of the astronomer on the optician; Browning always insisted that Lockyer was forestalled solely because of his own delay in finishing the instrument.[36] No ill-feeling arose, however, and the merit of the discovery was shared equally between the two astronomers. It remained only to open the slit wide when the C line in the red was clearly in view to see the form and not merely a line section of the prominence. Solar physics leaped forward, as it were, overnight. Before long, astronomers had inaugurated the daily observation and delineation of prominences.

In 1868 also, Huggins read a paper to the Royal Society on his attempts to determine the radial velocity or relative motion in the line of sight of certain bright stars.[37]

Doppler had stated in 1841 that the wavelength of light or sound is altered by the motion of the source or the observer towards or away from each other.[38] He wrongly accounted for differences of colour of some binary stars in this way; red stars were receding, blue stars were approaching the observer. Subsequent investigators rightly criticized Doppler's interpretation on the ground that the eye would not be sensible of any change in colour—rather would the motion, as Fizeau pointed out in 1848,[39] produce a slight yet measurable shift of the spectral lines. After some difficulty and using a powerful spectroscope, Huggins observed this displacement for the F line of hydrogen in the spectrum of Sirius. The shift was only 0·109 millionth of a millimetre, that is, just over 1 angstrom, towards the red. When allowance was made for the earth's motion, this shift suggested a speed of recession for Sirius of 29·4 miles per second.[40] Further measures in 1872 yielded a lower velocity for Sirius of about 20 miles per second,[41] a value near the modern figure.

We must remember, however, that these radial-velocity determinations, in common with those which Huggins obtained for other stars, were all derived from line shifts too minute to be measured with accuracy by direct visual methods. In general, his values are subject to grave uncertainty and compare unfavourably with modern photographically determined velocities. Yet Huggins took every possible care and precaution when he made these extremely delicate observations. 'Unless the air is very steady' he writes, 'the lines are seen too fitfully to permit of any certainty in the determination of coincidences of the degree of delicacy which is attempted in the present investigation. I have passed hours in the attempt to determine the position of a single line.'[42] The sound workmanship of Cooke's mounting and clock-drive ensured almost strict coincidence of a vibrationless star image on a slit 1/300 inch wide, but small inequalities in the drive required continuous supervision of star–slit coincidence. This careful guiding, the interruption of direct observations in order to attend to the source of comparison spectra, the reading of the micrometer, and the strain on the eyes in repeating each observation several times, bespeak the great patience with which Huggins conducted his researches, at first ably assisted by Dr Miller and then by Mrs Huggins, later Lady Huggins.

As if this pioneer work in radial-velocity determination was not enough, Huggins subjected Comet 1868 II to careful direct and spectroscopic examination.[43] This comet was brighter than those of 1866 and 1867 and it enabled him to assert that the greater part of its light was due to luminous hydrocarbon vapour.

The eclipse of 1869, August 7, was visible in the United States and was assiduously observed by Young, Langley, Harkness and others. Harkness, and independently Young, observed the faint continuous spectrum of the corona with a brilliant line in the green.[44] Young's measurements seemed to indicate that this emission line coincided with one due to iron, but measurements at subsequent eclipses showed that it coincided with no solar line. The name *coronium* was given to the apparently new gas responsible for the line; this particular line is now known to be due to 13-fold ionized iron.

At the 1870 Spanish eclipse, Young placed the slit of his spectroscope tangentially to the sun's limb at the moment of totality and saw the fading absorption lines. Suddenly, 'the whole length of the spectrum was filled with brilliant coloured lines, which flashed out quickly and then faded away, disappearing in about two seconds—a most beautiful thing to see.'[45] This was the first observation of the flash spectrum, revealing the presence of the sun's reversing layer and chromosphere, which together normally occasion the dark absorption lines in the solar spectrum. Young identified many of the lines, among them the green coronal line and the D_3 line in the yellow seen in the 1868 eclipse prominence spectrum. This D_3 line was attributed to sodium until Lockyer showed that it did not correspond to any known element.[46] He called the 'new' element *helium*, a gas which Ramsey isolated from a terrestrial source in 1895.

Another highly significant development of the mid-nineteenth century was Lewis M. Rutherfurd's pioneer work in stellar photography. Hitherto the reflector, with its great light-grasp and comparative ease of manufacture, had taken all the credit in astronomical photography. Only in solar and eclipse work where exposure times were short and modest apertures no disadvantage, did the refractor find useful employment as a photographic instrument. In 1856, Rutherfurd built an observatory in the garden at the rear of his home in New York. The major instrument, housed beneath a 20-foot diameter dome, was an 11¼-inch visual equatorial refractor. Rutherfurd calls his telescope 'a very substantial instrument'; it was made for him by Henry Fitz, a progressive New York optician. Opening from the main building was a small transit apartment with a computing room attached. 'The transit' Rutherfurd writes, 'is 189 feet N.W. from Second Avenue and 76·3 feet N.E. from Eleventh Street.'[47]

Experiments in astronomical photography began in 1858. Images of the moon and of stars down to the fifth magnitude were obtained, but their definition was not good enough to satisfy Rutherfurd. He found the actinic or near ultra-violet focus to be 0·7 inch shorter than the visual and attempted to remedy this by inserting lens combinations between object-glass and photographic plates.[48] Meeting no success, he then attached a 13-inch silvered glass mirror to the tube but the device was cumbersome and lacked rigidity. He now returned to the refractor, this time computing a two-lens combination suitable solely for photographic work. Fitz ground and, with Rutherfurd's assistance, polished the glasses; in December, 1864, America had its first large photographic refractor.[49]

Fitz died at the early age of 55 years, just before the object-glass was finished. His work on refractors slightly predates that of his contemporary, Alvan Clark, but it did not receive anything like the same prominence. Fitz made several object-glasses and telescopes for Rutherfurd and has a number of 6- to 9-inch refractors to his credit. By 1862, when Clark finished his first 18½-inch telescope, Fitz had made a 12-inch for Vassar College, Poughkeepsie, New York, a 12½-inch for the University of Michigan, and a 13-inch for the Dudley Observatory, Albany, New York.[50] It appears also that he supplied a 16-inch or 18-inch refractor to a Mr Van der Zee of Buffalo, New York,[51] but the subsequent history of this telescope is obscure.*

* According to R. S. Bates (*Sky and Telescope*, **1**, p. 18, November, 1941), Fitz was born on December, 1808, at Newburyport, Massachusetts, and was first a printer and then a locksmith. He made five telescopes for Rutherfurd—of 4-, 5¾-, 6-, 9-, and 11¼-inch aperture. He died on October 31, 1863, just as he was about to sail for Europe to select glass disks for a 24-inch refractor.

With the $11\frac{1}{4}$-inch photographic Fitz refractor, Rutherfurd took numerous pictures of star groups, sunspots, and double stars and, with the wet plates of those days and exposures of three minutes, succeeded in recording stars to about the ninth magnitude.[52] He showed how micrometers, furnished with travelling microscopes, could be used to measure the relative positions of the stars recorded on these plates with ease and apparent accuracy.[53] B. A. Gould, later director of the Cordoba Observatory, Argentina, joined in this work, and from Rutherfurd's plates measured the positions of nearly fifty stars in the Pleiades cluster and a large number in Praesepe. When Rutherfurd made his next object-glass of 13 inches aperture, corrected for photography by the addition of a third (flint) lens, Gould bought the old one from him. To his bitter disappointment, the flint glass arrived in South America broken. Attempts at cementing the two pieces proved fruitless, but a specially designed frame eventually enabled him to get them nicely adjusted. With this arrangement, Gould obtained moderately good photographs of the moon and of a few star clusters.[54]

In 1873, Gould obtained another Rutherfurd–Fitz glass, but further disappointments were in store, for the photographer used up precious supplies by taking pictures of the moon and of terrestrial objects for his own amusement and profit. ' So that at the end of three years' struggle ' Gould writes, ' I found my resources diminished by several thousand dollars, with absolutely no return excepting a few indifferent photographs.'[55] The next photographer, J. A. Heard, proved more conscientious, and Gould began to see some return for his outlay. Heard obtained images of the moon $1\frac{1}{2}$ inches in diameter before enlargement, also impressions of double stars and various star clusters. Some of the latter contained nearly 200 measurable stars down to the ninth magnitude.[56] Altogether, measurements of these plates for thirty-seven clusters yielded the positions of 9144 stars. Most of the photographs were taken by the wet-plate collodion process, for Gould did not begin to use dry plates until 1881. Unfortunately, the plates soon deteriorated and those in store at Harvard College Observatory are now in a very bad condition.

While Rutherfurd was experimenting in stellar photography, Henry Draper at his Hastings-on-Hudson observatory succeeded in taking the first satisfactory photographs of stellar spectra. His first spectrogram, of the star α Lyrae, was obtained in May, 1872, with a quartz prism, placed just inside the focus of the secondary mirror of his 28-inch reflector.[57] In August of the same year, he obtained a better spectrogram of this star in which four lines were more distinctly shown. This was the first photograph to show the characteristic lines of a star's spectrum. Draper used neither slit nor camera lens.[58] On August 31, 1873, he obtained a spectrogram of α Aquilae (Altair) with the same apparatus, the photograph being $\frac{1}{2}$ inch long and $\frac{1}{32}$ inch wide. Although the exposure time was 10 minutes, it does not appear that this spectrogram showed any lines. In October, 1875, several spectrograms of Vega were taken with the 28-inch in conjunction with a Browning nine-prism direct-vision dispersion unit. Again no slit was used.[59] In the summer of the following year, however, Draper made a piece of apparatus which he called a ' spectrograph ' and which consisted of a slit, Browning direct-vision prism train, and a 7-inch focus Voigtländer portrait lens.[60] Yet no great improvement resulted and the apparatus proved cumbrous and awkward to adjust.

Late in 1876, we find Draper placing the direct-vision prism, without slit or camera lens, inside the focus of his 12-inch refractor.[61] All manner of arrangements and experiments followed until pressure of other work and a visit to Europe brought the investigations to a temporary halt.

In 1870, the Royal Society loaned Huggins a Howard Grubb equatorial reflector–refractor. Thomas Grubb had retired two years before this date and the original works, now under Grubb's son, Howard, were moved to larger premises at Rathmines, Dublin. The Royal Society telescope was, therefore, one of Howard Grubb's first instruments. It was furnished with two interchangeable telescopes—a 15-inch achromatic and an 18-inch Cassegrain reflector with metallic specula. Later, in 1882, both telescopes were mounted together on independent declination

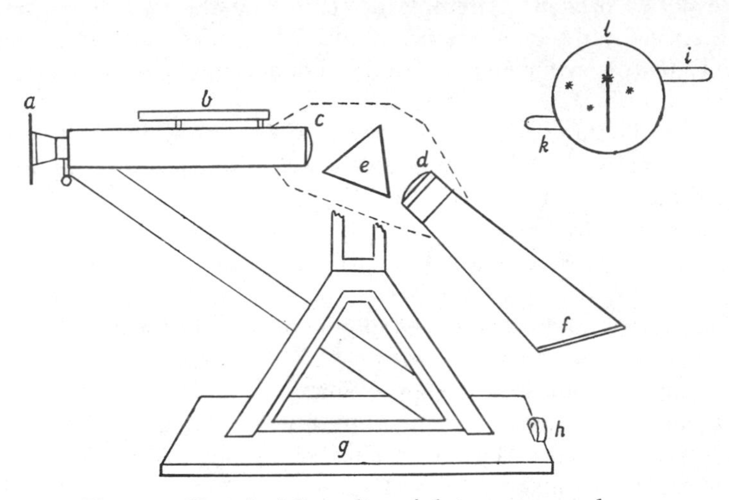

Fig. 119—Huggins' first ultra-violet spectrograph

a	Slit plate	*g*	Bevelled edge
b	Tube for collimation	*h*	Screw for adjustment in focus of mirror
c, *d*	Quartz lenses	*i*, *k*	Shutters of slit
e	Prism of Iceland spar	*l*	Silver plate with slit
f	Photographic plate		

axes, one on each side of the polar axis.* For the Cassegrain, Huggins designed the first ultra-violet spectrograph,[62] with a quartz collimating lens and an Iceland-spar prism made by Adam Hilger of London. A sensitive dry-plate replaced the observing telescope of the usual spectroscope.

With the slit of the new spectrograph at the primary focus, Huggins in 1875 obtained a promising spectrogram of Vega and followed this with spectrograms of many other bright stars, Venus, Jupiter, and the sun and moon. 'In the spectra of such stars as Sirius and Vega', he writes, 'there came out in the ultra-violet region, which up to that time had remained unexplored, the completion of a grand rhythmical group of strong, dark lines, of which the well-known hydrogen lines in the visible region form the lower members. Terrestrial chemistry became enriched with a more complete knowledge of the spectrum of hydrogen from the

* Until 1954 the instrument was installed at the Observatories, Cambridge, England.

stars. Shortly afterwards, Cornu succeeded in photographing a similar spectrum in his laboratory from earthly hydrogen.'[63] In the hands of Balmer and Rydberg, this pioneer photographic work laid the foundation of the study of line series in spectra.

By 1882, Huggins had photographed the spectrum of Comet 1881b[64] and, most difficult of all, had obtained impressions of a number of emission lines in the spectrum of the Orion nebula,[65] including a strong line in the ultra-violet. To secure these spectrograms, the object had to be kept on the narrow slit for an hour before the faint, disintegrated light made its impression on the plate. To assist in the guiding of the telescope, Huggins later made the slit plates of polished metal. They thus formed a bisected mirror in which the reflected image of the star field could be observed by means of a small Galilean telescope fixed in the central hole of the primary mirror.[66]

In 1880, Huggins discarded the first ultra-violet spectrograph for a more accessible instrument placed at the Cassegrain focus. Dispersion was effected by two Hilger 60° Iceland-spar prisms; two further prism arrangements were provided—a 60° Iceland-spar single prism and, for very faint objects, a compound prism composed of two quartz half-prisms.[67] The polished slit plates again sent a picture of slit and star field to the observer by means of a prismatic view tube. Since the star image on the slit was wider than the slit width (owing to diffraction and atmospheric effects), Huggins could set the latter in the direction of daily motion and, by making the clock control run a little slow, cause the star to drift between two marks on the slit. By doing this as often as possible, width was given to the spectrogram without using a cylindrical lens. To assist in this delicate guiding, the slit jaws were faintly illuminated by a small, red lamp placed inside the telescope.[68]

From the study of so many stellar spectra, Huggins could not fail to see that the stars could apparently be grouped according to the lines in their spectra. Rutherfurd had already told of similarities between certain stellar spectra and had attempted to classify them.[69] Secchi's more reliable observations involved the spectra of some 300 stars which, between 1863 and 1867, he sorted into three broad types.[70] The first type, the spectra of ' bluish ' stars, comprised more than half the stars observed. To the second belonged ' yellowish ' stars with solar-type spectra. The third embraced mostly red and variable stars. Later he added a fourth type: ' very deep red ' stars whose spectra showed heavy absorption bands. It was difficult to escape the conclusion that the stars in each specific spectral type differed physically from those of another type. H. C. Vogel, another champion of slit-spectroscopy, introduced in 1874 a simple classification of three main classes with subdivisions.[71] In his suggestion that thesequence was one of decreasing temperature from the white Sirian stars to yellow and red stars, he came part of the way towards the more modern Russell theory of stellar evolution. In conjunction with Dunér, who worked with the 9½-inch Merz refractor at Lund, Sweden, Vogel at Potsdam prepared catalogues of stellar spectra which involved the observation of tens of thousands of stars. Vogel's classification led to the idea of an orderly evolutionary sequence—that stars change in time from one spectral type to another.

Upon his return from Europe in July, 1879, Draper resumed his spectrographic investigations with new vigour. He brought back a two-prism Huggins star spectroscope and, this time, made more use of the appreciable light-grasp of the 28-inch reflector. A full account of the results of this second period—cut short by Draper's early death in November, 1882—appeared in 1883 when C. A. Young and E. C. Pickering presented his *Researches upon the Photography of Planetary and Stellar Spectra* to the American Academy.[72] In this paper, we find a list of 78 stellar and planetary spectrograms, all taken on dry plates with the Huggins star spectroscope, used in some cases with the reflector and in others with the refractor. E. C. Pickering measured line positions for many of these photographs and checked them with lines in Draper's 1872 normal reference map of the solar spectrum. Pickering's comparisons showed, among other things, that the constitution of α Boötis (Arcturus) and α Aurigae (Capella) was similar to that of the sun.[73]

To ensure that her husband's great work should continue, Mrs Draper established and maintained a department of stellar spectroscopy at the Harvard Observatory. Already E. C. Pickering, the director, assisted by W. H. Pickering, then instructor in photography at the Massachusetts Institute of Technology, were pioneering in stellar photography with a 7-inch, $f/5$ portrait lens and other lenses.[74] In 1885, the Bache fund of the National Academy enabled Pickering to purchase an 8-inch, $f/5{\cdot}6$ Voigtländer doublet, corrected and mounted by Alvan Clark & Sons and known as the 'Bache Telescope'.[75] Pickering was also anxious to speed up stellar spectrography and, to this end, pressed Fraunhofer's idea of the objective prism into service. With a 30° prism placed in front of a 2-inch, $f/3{\cdot}5$ photographic lens, he obtained spectrograms of Vega, Altair, and Polaris which showed many of their characteristic lines.[76] A 15° prism before the Bache object-glass gave him all that he needed for a complete spectrographic survey of the brighter stars in the northern hemisphere.

By the use of the objective prism, not only was a great increase in light secured, but all stars in the field of the telescope impressed their spectra simultaneously on the plate. Formerly, only one star spectrum could be photographed at a time, and this only when the star was of the first or second magnitude. By the new method, several hundred spectra could be recorded on one plate, some of them of stars of the seventh and eighth magnitude. Secchi used the objective prism for visual observations, but its addition to his 9½-inch Merz refractor reduced its aperture by almost a half and upset the smooth working of the clock-drive.[77] Merz specialized in the manufacture of objective prisms and, at one time, made a few compound direct-vision objective prisms, but their cost and limited size prevented their general adoption. Astronomers had no choice but to be content with a single objective prism or prism train; the deviation produced by them meant, of course, that the telescope could not be pointed direct to the stars being photographed. To give width to the spectra obtained in this way, Pickering arranged the dispersion to be in the declination direction. Slight over-weighting of the driving clock caused the spectrum to trail across the plate and so to gather breadth.

In 1886, Mrs Draper presented her husband's 11-inch Clark photographic telescope to Harvard College Observatory, together with sufficient funds for Pickering's proposed mass spectrography—the entire research to be called 'The Henry Draper Memorial'. Most of the spectrograms of this great project were taken with the

8-inch Bache telescope. For studying the spectra of the brighter stars, four prisms were placed in front of the object-glass, each prism having a refracting angle of 15°. All four prisms gave spectrograms about 4 inches long, but generally one prism was employed.[78]

In the three years from 1886 to 1889, Pickering and his co-workers made a complete spectrographic survey of the stars of the northern hemisphere. Before the end of the fourth year, the Bache 8-inch objective prism photographic doublet had been put to work at Arequipa, Peru, on the southern stars.[79] Here also, a 13-inch, *f*/15 Clark photographic refractor, the Boyden telescope, was installed together with three 13-inch objective prisms.[80] The classification of the many hundred thousand spectra obtained, a task in which Mrs W. P. Fleming, Miss A. C. Maury, and especially Miss A. J. Cannon, played a prominent part, took many years but resulted in the now classic Draper spectral classification. The first survey soon bore fruit. While examining plates in 1889, Miss Maury discovered that the brighter component of the well-known double star ζ Ursae Majoris (Mizar) has a variable spectrum—that the K and certain other lines become periodically double.[81] Further exposures indicated that the period of line doubling is about 20·5 days, that two stars are involved, each revolving about a common centre. The doubling is a Doppler effect and the star a spectroscopic binary with its orbital plane tilted 50° to the observer. Later in the same year, Miss Maury found that β Aurigae undergoes similar changes in a period of about four days. In 1889 also, Vogel found a similar orbital motion for the variable star β Persei (Algol)[82]—a star with a period of about 69 hours—thereby confirming Goodricke's earlier hypothesis that this star is an eclipsing binary. The discovery of some hundreds of new variable stars, several novae and many stars with peculiar spectra was an important by-product of the *Henry Draper Catalogue*.*

The addition of an 11-inch Merz refractor and more powerful spectrographic apparatus to the equipment at the Potsdam Observatory enabled Vogel, between the years 1887 and 1892, to determine the radial velocities of over 50 stars. These pioneer spectrograms gave an average probable error of 2·6 km/sec,[84] and added still further to Potsdam's claim to be the finest astrophysical observatory in the world. Like the Mount Wilson Observatory in modern times, Potsdam became a Mecca for astronomers from abroad.

E. C. Pickering pioneered in stellar photometry. He not only instituted at Harvard an ambitious scheme of brightness measurements with new and more accurate visual photometers, but greatly extended Bond's earlier photographic work. The latter half of the nineteenth century saw a rapid development in visual stellar photometers, which culminated in Pickering's famous meridian photometer.

We have seen how Argelander's work led to the publication of the great *Bonner Durchmusterung* between the years 1859 and 1862. His photometry was continued by Schönfeld, his successor at Bonn, and extended to the southern hemisphere by Gould and Thome at Cordoba. An alternative method of inquiry, begun in 1836 by John Herschel at the Cape with his astrometer, was refined by Steinheil's prism

*Draper's 11-inch refractor is now on long-term loan to the Sun Yat-sen University Observatory, Canton, China. The 8-inch Draper doublet, gift of Mrs Draper to Harvard, helped to resurrect astronomy at Torun, Poland, after the second world war.[83]

photometer. The latter consisted of a small divided object-glass, each half of which could be moved along the optical axis of the telescope, so spreading star images into blur circles of sensible area. The distance by which the semi-lenses were separated to produce diffused images equal in brightness became a measure of the difference in brightness of the two stars. In 1850, Arago suggested using the principle of the polarization of light in the design of an astronomical photometer and, a little later, Zöllner introduced his artificial star comparison photometer. In this instrument, the real star was viewed alongside the image of an artificial star, the brightness of which was controlled by the relative position angles of two Nicol prisms.[85]

E. C. Pickering exploited various possibilities of Zöllner's instrument. His first meridian photometer consisted of two equal but slightly inclined refracting telescopes placed horizontally in the east to west direction. Reflecting prisms before the object-glasses and polarizing prisms near the common focus enabled

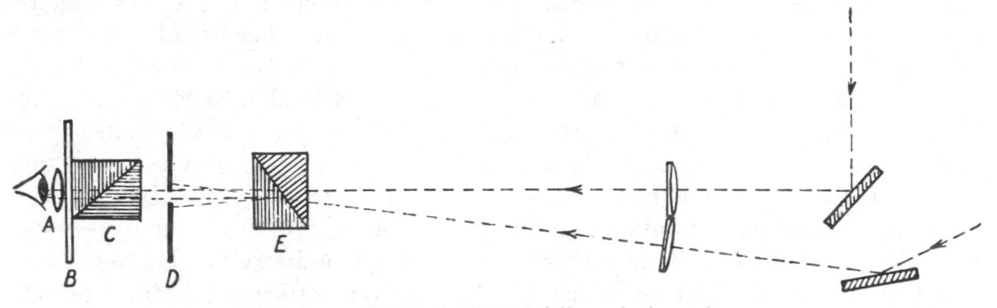

Fig. 120—Pickering's meridian photometer

The double-image Nicol prism *E* brings the two pencils of light from the object-glasses together. Each beam is separated into two by *E*, and the two inner beams are allowed by *D* to enter a second Nicol *C*. The amount of rotation of *C* is read from the scale *B*, behind which is the eyepiece *A*. For an account of the method of observation see Furness, C.E., 1915, *Introduction to the Study of Variable Stars*.

two star fields on the meridian to be brought into the same field of view and directly compared. The brightness of the image of Polaris, say, and that of the star being compared were equalized by rotating a Nicol prism, the position angle of which was a measure of the difference in brightness of the two stars.[86] In this way, using Polaris and, later, λ Ursae Minoris, Harvard astronomers determined the magnitudes of many thousand naked-eye stars with meridian photometers of only 2 and 4 inches aperture.

A further photometer of this period was the extinction wedge which C. Pritchard introduced to the University Observatory, Oxford, about 1881.[87] A thin prism of neutral glass was cemented to a similar wedge of white glass so that a plate of uniform thickness resulted. This could be slid in a groove in the cap of the eyepiece until the star under observation just disappeared from view. The point on an accurately calibrated wedge at which extinction occurred was a measure of the brightness of the star observed. In four years, Pritchard, W. Plummer, and B. C. Jenkins measured the brightnesses of 2784 stars, the substance of the now classic *Uranometria Nova Oxoniensis*.[88] Pritchard's wedge, however, had two important defects—selective absorption by the tinted glass and the tendency of the observer to exceed the actual extinction position, that is, to overestimate the apparent brightnesses of stars. For these and other reasons the extinction method was little used afterwards and is now obsolete.

A far greater contribution to the stellar photometry of this period was the carefully prepared and highly accurate *Potsdamer photometrische Durchmusterung* undertaken by G. Müller and P. Kempf with a modified Zöllner photometer. This great work, published in two sections (in 1894 and 1907), gave the magnitudes of 14,199 stars. The instrument was made by J. Wanschaff of Berlin and took the form of a large coudé theodolite of the Reichenbach type. A large right-angled prism reflected the light from the object-glass along the horizontal axis and thence into the photometer attachment. An oil lamp behind a pinhole gave rise to the comparison artificial star. Three alternative objectives, of 2·6-, 1·4- and 0·8-inch aperture and having different aperture ratios, enabled the observers to cover a brightness range which extended to stars of magnitude 7·5.

To extend his photometry, Pickering in 1898 installed at Harvard a single objective 12-inch meridian photometer. Comparisons were made against an optically produced comparison star, the brightness of which was varied by moving a neutral wedge in the optical train. Stars down to magnitude 13 could now be reached and the instrument proved invaluable for the photometry of the fainter comparison stars for variables.[89]

Pickering's experiments in photographic stellar photometry began at Harvard in 1882 and concerned the comparison of star trails taken with a stationary camera against a standard graded series of trails.[90] In 1885, the Bache 8-inch *f*/5·6 Voigtländer 'doublet' was used for the photometry of members of star groups, one of them the Pleiades, using the method of star trails.[91] The results were encouraging; Pickering estimated that his determinations had a probable error of less than 0·1 magnitude.[92] Of even greater value, although others were slow to appreciate it, was the insight the work afforded into both the permanent and incidental sources of error to be guarded against in photographic photometry.

We turn now to the Royal Observatory, Cape of Good Hope, where in 1882 David Gill, H.M. Astronomer, strapped an ordinary camera to his equatorial to take photographs of the great comet of that year. With exposures of one to two hours' duration, he obtained striking photographs not only of the comet but of the background of stars.[93] It occurred to him that here was a simple and quick way of constructing star maps free from personal errors and the laborious methods of eye observation. With different lenses he could obtain different scales, with different exposures he could reach stars of any desired magnitude. Gill thereupon ordered from T. R. Dallmeyer an *f*/9 'doublet' portrait lens of 6 inches aperture and with this began the photographic southern extension of Argelander's *Bonner Durchmusterung*. Gill exposed the plates, and J. C. Kapteyn of Leyden (later of Groningen) undertook the immense task of measuring the positions and photographic magnitudes of all the stars on them.[94]

The potentialities of the new photography for recording star positions were also seen by the brothers Paul[95] and Prosper[96] Henry of Paris, then engaged in the visual revision of Chacornac's charts of zodiacal stars. It was clear to them that the existing visual equipment at the Paris Observatory was unsuitable for this work, especially when it came to plotting some sections of the Milky Way. They decided to build a special photographic refractor, for, in those days, the method of correcting a visual objective by a yellow filter was unknown. They computed and made some small trial lenses, ultimately constructing a photographic refractor of 13·4 inches (33 cm) aperture and 11 feet 3 inches (343 cm) focus. This achievement was no less remarkable than that of Fitz and Rutherfurd for, not many years previously, the Henrys were only working opticians at Nancy. Leverrier, struck by their skill, invited them to the Paris Observatory, where they rigged up a modest workshop in a disused shed.[97] The success of the preliminary trials of the 13·4-inch refractor was announced to the Academy of Sciences in 1885 and, in the following year, their instrument was permanently installed in the observatory. It was mounted equatorially by P. Gautier, and was coupled with a guiding 9·4-inch visual refractor of nearly the same focal length.[98] The photographs obtained exceeded all expectations and, in 1887, an international congress, sponsored by Gill and Admiral Mouchez, director of the Paris Observatory, met in Paris to consider further extensions of this work.

The plan was to compile a great star map, the *Carte du Ciel*, which would cover the entire sky. Such a scheme, if completed, would require the exposure and reduction of over 10,000 plates and would give rise to a catalogue of over four million star positions. Eighteen observatories offered to share the work, among them Greenwich, Oxford, and the Cape, and their representatives then considered what instrument would be most suitable. Pritchard hoped to use the 13-inch De la Rue reflector at Oxford, and naturally advocated an instrument like it. Common, after his experience with large reflectors, which he ungallantly compared with the female sex, stressed the reliability and manageability of the refractor.[99] Pickering urged the adoption of a photographic doublet like the one he was then using and which covered a large area of sky at a time. None could deny, however, the outstanding merit of the Henrys' specimen photographs, and it was decided to make their telescope the standard. By doing this, the scheme was made as uniform as possible. The focal length of 343 cm meant that one minute of arc was represented by one millimetre on the photographic plate, a scale suitable for measuring minute changes in star positions over a period of years. Furthermore, given plates of equal sensitivity and fixed exposure times, stars could be photographed down to approximately the same order of magnitude.

The scheme was an ambitious one and took too much for granted. Most of the effort went into obtaining accurate star positions at the expense of the photometry. No account was taken of small differences in colour and aberration correction likely to arise in eighteen different objectives. Both the physical and chemical properties of photographic plates were not fully understood and the effects of seeing during exposures were ignored. The results of Pickering's important investigations with star trails had not yet been published and, as far as magnitude determination was concerned, rigorous standardization became impossible.[100] In short, astronomers lacked a fundamental scale of photographic photometry, a desideratum which only the most persistent work of many specialists over many years was to supply.

Fig. 121—Sir Howard Grubb 20-inch photographic reflector and 7 inch visual refractor

Made for Isaac Roberts, 1885

At this stage we again meet with the work of Howard Grubb, for he it was who supplied many of the astrographs for participating observatories. Grubb came face to face with the particular requirements of a photographic telescope when he supplied Dr Isaac Roberts with a 20-inch reflector. The instrument was designed for taking photographs at the primary focus and was installed in Roberts' observatory at Moghull* in April, 1885. ' It was soon discovered ' Roberts writes, that the instrument had several defects, though the works of the optician and mechanician were performed with all the care and skill consistent with a high standard in those days.'[101] Roberts was exacting in his demands; he placed them under three broad headings: First, the stand should be strong and so firmly constructed as not to vibrate when a moderately high wind is blowing or when tapped with the hand. Second, the tube should be rigid and free from measurable flexure to an extent greater than 2 seconds of arc in all positions. Third, ' The driving

* Moved, in 1890, to Crowborough, Sussex.

clock, sector and tangent screw, must be so perfectly made, that when a sixth magnitude star near the meridian is viewed on the cross-wires in the eyepiece of the guiding telescope, it shall be kept steadily bisected by the cross-wires for consecutive intervals of at least three minutes each, under a magnifying power of one hundred and fifty diameters.'[102] Grubb investigated the errors mentioned by Roberts and admitted their validity. The mounting was reconstructed and returned to Moghull, but star images were still elongated after long exposures. With the telescope in this condition, Roberts, on October 10, 1887, took a photograph of the Andromeda nebula which showed, for the first time, the spiral form. Many photographs of this and other nebulae followed and, in 1893, were published in Roberts' important *Photographs of Stars, Star Clusters and Nebulae*.

Grubb's finest work at this period was undoubtedly the design and construction of astrographic refractors. When the Henrys' instrument was taken as pattern for the *Carte du Ciel*, Continental observatories ordered theirs from the Henrys. British institutions preferred to patronize a British maker and unanimously chose Howard Grubb. Like everyone else, Grubb was new to the peculiar difficulties of photographic lenses and at first had to undertake much experimental work. To make matters worse, astrographic objectives had little in common with ordinary photovisual lenses, about which at least something was known. The standard photovisual is so computed that the brightest visual and the most effective photographic rays are brought to a common focus. This being so, the observer focuses the image on a ground-glass screen and knows that, when this is removed and replaced by a photographic plate, the image will still be in actinic focus. The photovisual objective, designed by H. Dennis Taylor and produced by T. Cooke and Sons in 1892 is, however, a compromise and not rigorous enough for photographic work in positional astronomy. Photography was then, of necessity, carried on in the violet and near ultra-violet region of the spectrum for it was to these rays that early dry plates were most sensitive.

Grubb's first photographic lenses were tested at Greenwich and from these he advanced by trial and error to larger sizes. The latter he sent to Pritchard at Oxford, who photographed stars with them and sent back lists of defects. These Grubb corrected as best he could, when the lenses were again sent to Oxford for further tests. So the work progressed, slowly but carefully until 1889, when a successful objective was considered worthy of a mounting.[103] The following year, Grubb despatched a completed instrument to Mexico and wrote to Pritchard to say that the telescope for Greenwich was finished and mounted. By 1896, Grubb had completed astrographs for Greenwich, Oxford, and the Cape, Tacubaya in Mexico, Melbourne, and Perth (Western Australia). He had, besides, just despatched a 12-inch refractor with a 10° objective prism to F. McClean of Tunbridge Wells, Kent, and was at work on a new 26-inch astrographic refractor for Greenwich—a gift from Sir Henry Thompson.

When E. C. Pickering's results in short-focus photography came to be fully appreciated, this department of practical astronomy reaped a rich harvest. Pickering at first worked mainly with the 8-inch Draper doublet, a duplicate of the 8-inch Bache telescope, and with this photographed objects invisible through the 15-inch Harvard refractor. With small portrait lenses, Dr Boeddicker at Lord Rosse's

Fig. 122—Sir Howard Grubb

observatory obtained photographs of the Milky Way; so also did Dr Sheldon with *f*/3 lenses of 6¼ inches focus and exposures of one hour and upwards.[104] On Mount Hamilton, California, E. E. Barnard photographed the entire constellation of Orion with a cheap magic lantern lens of 1½ inches diameter and using one to two hours' exposure. Much to his surprise, he obtained evidence of an extensive filamentous nebulosity which embraced the belt and great nebula of Orion.[105] Even more startling results were obtained with an *f*/5, 6-inch aperture Willard doublet strapped to a 12-inch equatorial telescope.[106] The small scale of these photographs was their drawback for the *Carte du Ciel*, but for the detection of variable stars, minor planets and faint nebulosity they proved invaluable. Barnard's photographs with the Willard lens showed both the star clouds of the Milky Way and dark markings which A. C. Ranyard, among others, thought were obscuring masses of dust.

With the early portrait lenses, only stars in the centre of the field gave circular images with anything near full aperture. Images near the edges of the field appeared more like parallelograms. Stopping down the aperture to *f*/9 or *f*/10 gave more uniform results but led to over-long exposure times and consequent difficulty in keeping the camera always trained on the same region of the sky. With the development of commercial photography, faster plates and new lens systems were introduced, so that these early defects were gradually eliminated. C. S.

Hastings in the United States and H. Dennis Taylor in England pioneered in the development of astrographic lenses, the first in collaboration with John A. Brashear at Pittsburgh and the second while manager of the optical workshops of T. Cooke and Sons of York.* Taylor's well-known 'Cooke' photographic lens (1894) consisted at first of three separated achromatic doublets, changed later to three spaced single lenses. With one of these astro-photographic lenses of 10 inches aperture and 45 inches focal length were taken the Franklin-Adams star charts.[107]

An important short-focus photographic refractor of the nineties was the $f/5 \cdot 5$ Bruce doublet of 24 inches aperture. This instrument, the munificent gift of Miss C. W. Bruce to Harvard College Observatory, was made in 1893 by A. Clark and Sons.[108] It remains one of the best examples of its type. From 1896 to 1927 it was used at the Boyden station of Harvard at Arequipa, Peru. With exposures of 2 hours, the plates showed stars down to magnitude 17.[109] Little wonder then, that with increasing exposures up to four hours, this telescope, with its wide field, photographed more nebulae at a time than any other single instrument. It was moved to Harvard's South African branch observatory at Harvard Kopje, near Bloemfontein, but the optics were replaced by a Baker-Schmidt system in 1950.

The Bruce telescope made possible the first detailed survey of the Magellanic Clouds. These two nearby galaxies contain a rich variety of objects, the study of which has provided the means for the exploration of our own and more distant galaxies. They are, moreover, well clear of bright foreground stars and clouds of obscuring matter in the Milky Way, yet near enough for their bright stars to be within the reach of present instruments. From plates taken with the Bruce telescope, Miss H. S. Leavitt discovered that, for certain variable stars in the Magellanic Clouds, there was a connection between luminosity and period.[110] The brighter variables had longer periods than the fainter, although all of them had periods of from one to about forty days. They are called Cepheids, after their prototype δ Cephei. Their great value as measuring tools came when Hertzsprung and Shapley recognized that the relationship noticed by Miss Leavitt was an intrinsic property of the stars themselves. Some years earlier, S. I. Bailey found on Bruce plates of globular clusters a number of variable stars with periods of less than one day—the so-called cluster-type Cepheids.[111] These stars made possible Shapley's classic work on the distances and space distribution of globular clusters.[112]

Short-focus cameras continued to play an important part in the observational programme at Harvard. A unique feature was the constant patrol of the sky, for the detection of novae and variable stars, by a number of camera lenses.[113] These were driven and controlled simultaneously by impulses from a single clock.† The comparison of photographs taken over many years has enabled astronomers to study the early history of novae and has resulted in the discovery of hundreds

* H. Dennis Taylor also introduced the first commercially made apochromatic objectives (*M.N.R.A.S.*, **54**, p. 67, 1894).

† Electrical methods for the driving of telescopes were introduced at Harvard College Observatory by Willard P. Gerrish. Instead of a clock mechanism actuated by a heavy weight, Gerrish used an electric motor synchronized by the pendulum of a standard clock. The method was applied to the 24-inch reflector at Harvard in 1907. It proved so successful that other Harvard telescopes, including the 60-inch reflector, were driven by modifications of this method. For photographic work with large instruments, irregularities of motion are corrected by the observer who maintains a visual check on image position through a large finder or guide telescope.

of variable stars. Another line of inquiry, begun by Harvard, is the counting of stars in the Milky Way down to the fifteenth magnitude. This piece of co-operative research now involves many American and other observatories.[114]

REFERENCES

[1] *Phil. Trans. Edin.*, **12**, p. 519, 1834.
[2] *Ibid.*, p. 528.
[3] *Phil. Trans.*, **126**, p. 450, 1836.
[4] *Ibid.*, p. 453.
[5] Roscoe, H. E., 1869, *Spectrum Analysis*, pp. 92, 118.
[6] Stokes, G., *Phil. Mag.*, **19** (4th series), pp. 194-196, 1860, translated from *L'Institut*, p. 45, 1849, Feb. 7.
[7] Lockyer, J. N., 1874, *Solar Physics*, p. 187.
[8] *Vide* Ref. 6.
[9] Alter, D., *Am. Jour. Sc.*, 1854.
[10] *Ibid.*, 1855.
[11] Read to Berlin Academy Oct. 27, 1859. Stokes, *op. cit.*, gives translation.
[12] *Vide* Facsimile in H. E. Roscoe's translation, 1862–63, of G. Kirchhoff's *Researches on the solar spectrum.*
[13] Roscoe, Ref. 5, pp. 204–205.
[14] Ångström, A. J., 1868, *Récherches sur le spectre solaire.*
[15] Fraunhofer, J., *Denk. d. K. Acad. d. Wiss. zu München*, **8**, pp. 28–34, 1821.
[16] Schellen, H., 1872, *Spectrum Analysis*, p. 237.
[17] Draper, J. W., 1878, *Scientific Memoirs*, p. 117.
[18] *Ibid.*
[19] Plücker and Hittorff, *Phil. Trans.*, **155**, i, pp. 1-29, 1865.
[20] Wüllner, A., *Pogg. Ann.*, **135**, p. 174, 1868; **137**, pp. 337-361, 1869.
[21] Schellen, *op. cit.*, p. 230. Lockyer, *op. cit.*, p. 152.
[22] Lockyer, *op. cit.*, p. 163.
[23] *Ibid.*, p. 167.
[24] Huggins, W., *Publications of Sir William Huggins's Observatory*, i, p. 6, 1899.
[25] *Ibid.*
[26] *Ibid.*, pp. 6–7.
[27] Lockyer, J. N., 1887, *Chemistry of the Sun*, p. 18, quotes *Edin. Phil. Jour.*, **10**, pp. 37–38, 1824.
[28] *Proc. Roy. Soc.*, **12**, pp. 444–445, 1863.
[29] Rutherfurd, L. M., *Am. Jour. Sc.*, **35** (2nd series), pp. 71, 407, 1863.
[30] *Phil. Trans.*, **154**, pp. 139–160, 1864.
[31] Huggins, *op. cit.*, p. 9.
[32] *Ibid.*, pp. 11–12.
[33] *Proc. Roy. Soc.*, **15**, p. 146, 1866.
[34] Lockyer, Ref. 7, pp. 127–128.
[35] *Ibid.*, pp. 125–126.
[36] *M.N.R.A.S.*, **90**, p. 361, 1930.
[37] *Phil. Trans.*, **158**, pp. 529–550, 1868.
[38] Doppler, C., 1842, *Ueber das farbige Licht der Doppelsterne.*
[39] Lockyer, Ref. 7, p. 200.
[40] *Phil. Trans.*, **158**, pp. 548–549, 1868.
[41] Huggins, W., *M.N.R.A.S.*, **32**, p. 361, 1872.
[42] *Phil. Trans.*, **158**, p. 546, 1868.
[43] *Ibid.*, pp. 555–564.
[44] Clerke, A., 1885, *History of Astronomy during the Nineteenth Century*, p. 219. Lane Hall, A. W., *J.B.A.A.*, **45**, p. 118, 1935.
[45] Young, C. A., 1891, *General Astronomy*, p. 200.
[46] Lockyer, T. M., and W. L., 1928, *Life and Work of J. N. Lockyer*, pp. 266–291.
[47] Rutherfurd, L. M., *Am. Jour. Sc.*, **39** (series 2), p. 304, 1865.
[48] *Ibid.*, pp. 304–305.
[49] *Ibid.*, p. 308.
[50] *Dictionary American Biography*, **6**, p. 433, 1931.
[51] Newcomb, S., 1881, *Populäre Astronomie*, Leipzig, p. 134.
[52] *Vide* Ref. 47, p. 308.
[53] *M.N.R.A.S.*, **53**, p. 230, 1893.
[54] Gould, B. A., *Observatory*, **2**, p. 18, 1878.
[55] *Ibid.*
[56] *Ibid.*, p. 19.
[57] Barker, G. F., 1888, *Memoir of Henry Draper*, p. 20.
[58] *Ibid.*
[59] *Ibid.*, p. 23.
[60] *Ibid.*
[61] *Ibid.*
[62] Huggins, Ref. 24, i, pp. 50–51, 1899.
[63] *Ibid.*, p. 22.
[64] *Proc. Roy. Soc.*, **33**, pp. 1-3, 1882.
[65] *Ibid.*, p. 425.
[66] *Vide* Ref. 62, p. 51.
[67] *Ibid.*, p. 55.
[68] *Ibid.*, p. 53.
[69] Rutherfurd, L. M., *Am. Jour. Sc.*, **35**, pp. 71, 407, 1863.
[70] *Compt. Rend.*, **64**, p. 774, 1867.
[71] *Astr. Nach.*, **84**, p. 113, 1874.
[72] Draper, H., *Proc. Am. Acad.*, **19**, pp. 231–261, 1884.
[73] Barker, *op. cit.*, pp. 26–27.
[74] Bailey, S. I., 1931, *History and Work of the Harvard Observatory*, p. 119.

[75] *Ibid.*, p. 42, p. 119.
[76] Barker, *op. cit.*, p. 28.
[77] Schellen, *op. cit.*, p. 465.
[78] Bailey, *op. cit.*, pp. 119–120, p. 150. Young, *op. cit.*, pp. 489–490.
[79] Bailey, *op. cit.*, p. 42.
[80] *Ibid.*, pp. 44–45, p. 119.
[81] *Ibid.*, p. 166.
[82] *Astr. Nach.*, **123**, p. 289, 1889.
[83] *Sky and Telescope*, **7**, p. 8, 1947.
[84] Merrill, P. W., *M.N.R.A.S.*, **99**, p. 318, 1939.
[85] Weaver, H. F., *Pop. Astr.*, **54**, p. 214, 1946. For constructional details of Zöllner-type photometers vide Ambronn, L., 1899, *Handbuch der Astronomischen Instrumentenkunde*, ii, pp. 701–710.
[86] Bailey, *op. cit.*, p. 128.
[87] Pritchard, C., *M.N.R.A.S.*, **42**, p. 68, 1881.
[88] Pritchard, C., *Oxford University Observatory Publications*, No. 2, 1886.
[89] Bailey, *op. cit.*, p. 134.
[90] *Ibid.*, p. 135; *Mem. Am. Acad.*, **11**, p. 179–226, 1888.
[91] Bailey, *op. cit.*, pp. 135–136.
[92] Weaver, *op. cit.*, p. 291.
[93] Gill, D., 1913, *History and Description Royal Observatory, Cape of Good Hope*, p. xlix.
[94] *Ibid.*, pp. xlix, ff.
[95] *M.N.R.A.S.*, **65**, p. 349, 1905.
[96] *Ibid.*, **64**, p. 296, 1904.
[97] *Bull. Soc. Astr. Fr.*, **17**, p. 420, 1903.
[98] *Astronomical Register*, **24**, p. 245, 1886.
[99] Turner, H. H., 1912, *The Great Star Map*, p. 25.
[100] Weaver, *op. cit.*, p. 292.
[101] Roberts, I., 1893, *Photographs of Stars, Star Clusters and Nebulae*, p. 21.
[102] *Ibid.*, p. 28.
[103] Pritchard, A., and Turner, H. H., 1897, *Charles Pritchard. Memoirs of his Life and of his Astronomical Work*, pp. 310–311.
[104] *M.N.R.A.S.*, **51**, p. 230, 1891; *J.B.A.A.*, **3**, p. 320, 1893; **5**, p. 397, 1895; **6**, p. 154, 1896.
[105] Barnard, E. E., *Ap. J.*, **2**, pp. 351–353, 1895.
[106] Hale, G. E., 1908, *The Study of Stellar Evolution*, pp. 30–31.
[107] Franklin-Adams, J., and Taylor, H. D., *M.N.R.A.S.*, **64**, p. 613, 1904; Obituary notice Franklin-Adams, *M.N.R.A.S.*, **73**, p. 210, 1913.
[108] Bailey, *op. cit.*, p. 43.
[109] *Ibid.*, p. 143.
[110] *Ibid.*, p. 202.
[111] *Ibid.*, p. 181.
[112] Shapley, H., Harvard Observatory Monograph No. 2, *Star Clusters*, 1930.
[113] Shapley, H., 1944, *Galaxies*, pp. 128, 144, 155.
[114] Bok, B. J. and P. F., 1945, *The Milky Way*, p. 47.

CHAPTER XV

> A hundred observers might have used the appliances of the Lick Observatory for a whole generation without finding the fifth satellite of Jupiter; without successfully photographing the cloud forms of the Milky Way; without discovering the extraordinary patches of nebulous light, nearly or quite invisible to the human eye, which fill some regions of the heavens.
>
> SIMON NEWCOMB

THE invasion of the dry plate into astronomy was, in general, a comparatively slow process. The surrender of stellar astronomy was completed in 1887, but it was some time before H. Grubb, Gautier, and the brothers Henry were able to supply the necessary astrographs. The reflecting telescope remained an almost untried instrument in America—its photographic possibilities were certainly not appreciated until James E. Keeler, at the close of the century, set to work with a somewhat antiquated Grubb–Common reflector. Interest in planetary studies and spectroscopy rose to a high pitch in the eighties and nineties and justified the continued erection of large visual refractors. During these two decades, the wave of refractor building in America rose to culminate in the erection of the great Lick and Yerkes refractors. European observatories found room for new and larger instruments which were both visual and, whenever possible, photographic. We shall see presently how the large refractors of Gautier and the Henrys threatened to wrest the lead in size from Alvan Clark and Sons.

For some years, therefore, large visual refractors were erected and abundantly justified their existence. Several devices were employed to make one mounting serve both visual and photographic purposes. When funds were generous, twin lenses and tubes were adopted. The 32·7-inch visual and 24·4-inch photographic twin telescope at Meudon is the largest and heaviest of this type. A less satisfactory alternative was to supply either the one tube with two interchangeable objectives, one photographic and one visual, or to add a third correcting lens to the visual objective. The Henry brothers furnished the 23·6-inch equatorial coudé at the Paris Observatory with two such alternative object-glasses. A 33-inch correcting lens made a new instrument of the Lick refractor and a similar lens enabled E. C. Slipher to take photographs of Mars with the 27-inch Lamont-Hussey refractor at Bloemfontein.[1] These lenses were little used after good yellow filters and orthochromatic plates became available. A fourth arrangement, a rare one, was to design the object-glass so that the front lens could be reversed. When the crown lenses of the 13-inch Boyden and 20-inch Van Vleck refractors are reversed and slightly separated from their flint components, the instruments are changed from visual to photographic. Finally, the telescopes could be Cooke photovisuals and serve a dual purpose or be supplied with slip-in visual correcting lenses. Brashear's 30-inch Thaw photographic refractor at the Allegheny Observatory is of the latter type; the 12-inch visual corrector is inserted near the middle of the length of the tube. It was not until the introduction of orthochromatic plates early in the present century that a visual refractor could be used photographically merely by the addition of a yellow filter.

A giant refractor, representative of the type we are now going to consider, was the 27-inch visual telescope which Howard Grubb made for the Observatory of Vienna in 1880. This was indeed a giant for those days—for five years it was the largest refracting telescope in the world. The instrument was ordered in 1875, and the mechanical parts were completed in the summer of 1878, but Feil of Paris had such difficulty in obtaining perfect glass blanks that work on the lenses did not start until late in 1879.[2] Grubb supplied the 45-foot diameter dome, still the central feature of the observatory, also three smaller domes and their revolving machinery. ' I have, with advantage, on splendidly clear evenings in September ' H. C. Vogel wrote in 1883, ' used a power of 1000 and even 1500 and perceived the fine details of planetary disks with admirable sharpness.'[3] Although the new observatory was erected some three miles from the centre of Vienna, the present size of that city makes a change in the location of the 27-inch highly desirable.

We left the Clarks busy with the mounting and 26-inch disks for the Washington telescope. A few years later, they were at work on the mechanical and optical components for a 15·6-inch, a 23-inch, a 16-inch, and another 26-inch refractor. The first was mounted in 1879 at the Washburn Observatory of the University of Wisconsin, where it is still the largest instrument. In recent years, this telescope figured in the pioneer experiments of J. Kunz, J. Stebbins, and A. E. Whitford which resulted in the modern highly sensitive photoelectric photometer.[4] The 23-inch was finished in 1882 for the Halstead Observatory, now the observatory of Princeton University, New Jersey. With this telescope, and using a Brashear spectroscope, C. A. Young observed, in 1892, many double lines in the spectra of sunspots.[5] These were thought to be the effect of ' reversal ' by superposed gaseous layers at different temperatures, but we now relate them to magnetic fields. The tube was remounted by J. W. Fecker, in 1932 and, during its active career, has been used by C. A. Young, Harlow Shapley, R. S. Dugan, and N. L. Pierce. The 16-inch found itself, in 1883, in the private observatory of H. H. Warner at Rochester, New York. Ten years later, Lewis Swift took it to its present site on Mount Lowe, California. Clark's second 26-inch was installed in 1883 as a gift from Leander J. McCormick to the observatory of the University of Virginia. It is at present engaged on the determination of stellar parallaxes by photography.

In 1884, the Clarks made a 30-inch diameter objective for the equatorial which the Repsolds were erecting at the Pulkowa Observatory. Otto Struve wrote to Newcomb for assistance in negotiating with the Clarks for the optical parts, and later travelled to Cambridge, Massachusetts, to see their workshops. He was surprised at the modest size and simplicity of the premises, but was ' so well pleased with what he saw that he decided to award them the contract for making the object-glass.'[6]

The Clarks followed the 30-inch objective with a pair of 20-inch object-glasses, one for the Van Vleck Observatory of Wesleyan University, Middletown, Connecticut, and the other, mounted by Saegmüller, for the Chamberlin Observatory of the University of Denver, Colorado. More important than these, however, was the 24-inch equatorial erected in 1896 near Flagstaff, Arizona, for Percival Lowell, a wealthy amateur astronomer. Lowell made the planet Mars his special study and the 24-inch supplemented the existing 18-inch used for this work. The results of his observations over many years were published in popular form and proved highly controversial. His interpretation of Schiaparelli's *canali* as lines of vegetation bordering aqueducts built by the inhabitants of Mars to convey water from the polar caps to equatorial desert wastes was a challenge to the thinking world. The rigid geometrical pattern of the *canali* is now generally considered to be an optical illusion—at least subsequent observers have failed to see the markings 'like the rails of a railway track'[7] and in such profusion as described by Lowell. In more recent years, V. M. Slipher pioneered in the determination of radial velocities of nearby galaxies with the 24-inch Lowell telescope.

The priority, in point of size, of the 30-inch Pulkowa telescope was shortlived, for, in 1887, a 30-inch visual refractor was erected at Nice for a banker named R. Bischoffsheim. This instrument was the joint work of P. Gautier and the brothers Henry; three years later, Bischoffsheim presented his observatory to the University of Paris. Early in its career, the Nice refractor was used by M. Perrotin for planetary work, but its large size, once considered so important, seems now to be its greatest drawback. This refractor was followed by an even larger Henry–Gautier visual telescope of 32·7-inch aperture for the new branch of the Paris Observatory at Meudon. The Meudon refractor, still the third largest in the world and the largest in Europe, was coupled in 1891 with a photographic refractor of 24·4 inches aperture.[8] Here again, interesting work on planetary detail was done with the visual object-glass. We see later that it was at Meudon that Deslandres carried out experiments which resulted in his independent invention of the spectroheliograph.[9] The Henrys could, of course, draw readily on the old-established Paris glassworks for their large disks and, in Gautier, they had an earnest collaborator and skilled engineer. Had the project been financially possible, there is no doubt that the Henry–Gautier liaison could have produced a 36-inch or 40-inch refractor equal in its performance to the giants of Lick and Yerkes. As it was, these French artists turned their attention to the construction of elbow or coudé refractors and to two reflectors—a 32½-inch for the University of Toulouse and a 39½-inch for Meudon.

Howard Grubb followed his 27-inch Vienna refractor with his largest, a 28-inch *f*/12, which in 1893 replaced the old 12¾-inch Merz at Greenwich. The new telescope was fixed to the mounting that had carried the smaller instrument and which, owing to the excellence of its design and construction, took the larger tube without tremor or strain.[10] The longer tube, however, meant that a special super-hemispherical dome had to be made to accommodate it. Reversing the front component of the objective converts a visual into a photographic instrument. This telescope has been used mainly for double-star measurements and is now at Herstmonceux, near Hastings, the new home of the Royal Observatory.

In 1896, the first of a number of large objectives left the workshop of C. A. Steinheil at Munich. This was one of 26·8 inches aperture and 68·9 feet focal length—an unusual aperture ratio of nearly *f*/31. Since the telescope was intended for instructional purposes at the Archenhold Observatory in Berlin-Treptow,

F. S. Archenhold was anxious to reduce eyepiece movement to a minimum. To meet this requirement and to accommodate so long a focal length, the engineering firm of C. Hoppe in Berlin turned to the unconventional. Residents near Treptow Park viewed with surprise the assembly of enormous castings which constitute the equatorial mounting. Surprise turned to excitement as section after section was added to the tube until, with fourteen sections in line, a 60-foot gun-like barrel stretched skywards. The counterpoised tube is almost entirely above the supporting trunnions (declination axis) and is braced by four adjustable steel tie-rods. The instrument has always been used in the open air, although the observer and the lower parts of the mounting are somewhat protected by a surrounding brick and stone platform. Bomb-damage during the second world war was fairly extensive around Treptow and, by 1948, the instrument was in a dilapidated condition. It has since been restored; the objective has been refigured by Carl Zeiss of Jena and the tube, still the longest telescope tube in the world, appears even more impressive than ever before in its new coat of aluminium-bronze paint. The refractor, along with several smaller instruments, is still primarily used for demonstration purposes, for the Archenhold Observatory is one of the most popular public centres for instruction in observational astronomy in Eastern Germany.

Also of this period, 1897, is the Grubb 26-inch refractor (the Thompson equatorial) of the Royal Observatory at Herstmonceux. It is a photographic telescope and has been used for many years (at Greenwich) on a programme of stellar parallax determination. Instead of the usual weight counterpoise, the tube was balanced by a Grubb reflector with 30-inch primary mirror by A. A. Common. The reflector has been used principally for the photography of comets and satellites. With this instrument, P. J. Melotte in 1908 discovered the eighth satellite of Jupiter.[11] It will have a separate mounting and dome at Herstmonceux.

The Lick telescope was made possible through the whim of James Lick, a California millionaire, who offered to finance the erection of memorials, to himself and to his wife, such as statuary, free baths, a home for old ladies, and a large mountain observatory. The main feature of the last project was to be 'a powerful telescope, superior and more powerful than any telescope yet made.'[12] The site chosen was on Mount Hamilton, in Santa Clara County, California, some 4250 feet above sea level and high above the mists and clouds of the lowlands. Lick offered to allocate $700,000 for the purchase of land, the construction of the telescope and the provision of suitable buildings.

> No one seems to know how or when the idea of a great telescope entered Mr. Lick's mind [M. W. Shinn writes].[13] It was there before he took anyone into his confidence. He had never seen a real telescope, so far as anyone knows. He was entirely ignorant of astronomy and not even in an unlearned way an observer of the skies. . . . In some way, however, through chance readings, a sense of the glory of astronomical discovery had laid hold upon him, and the first persons with whom he talked of the disposition of his fortune found the plan of the telescope firmly fixed in his mind.

To bring Lick to the astronomer's way of thinking was a difficult task. Indeed, to fix his ideas on the type of telescope and its suitable location required months of tactful guidance and repeated presentations of the facts.

At first, a large reflector was contemplated, and G. Calver nearly obtained the

Fig. 123—Worcester Reed Warner

(*Warner & Swasey Co.*)

order.[14] As this instrument was then little known in America, however, the authorities decided to ' play safe ' with a 36-inch refractor. The flint glass was cast by Feil and Co. of Paris, and arrived safely at the Clarks' factory in 1882, but the crown disk cracked during its packing.[15] For two years the Feil brothers tried in vain to produce another crown blank and succeeded only after nineteen failures. The great cost of this work so strained their resources that, soon afterwards, they went into bankruptcy. Thereupon, the elder Feil, who had retired some years previously, again took charge of the business and, in 1885, shipped a perfect disk to the Clarks.[16] During these years, Newcomb visited Europe and, in his report on the work, tells of the difficulties met with in getting so large a glass of uniform texture throughout.[17] The flint component alone weighed 375 lb and cost $10,000, while the complete objective cost over $50,000.

The mounting was the work of Messrs Warner and Swasey of Cleveland, Ohio, a firm with little experience in large telescope design, but with a growing reputation as machine-toolmakers. When tenders were invited by the Lick Trust, Warner and Swasey's bid was the highest of all but, so superior was the design submitted, that they were awarded the contract.[18]

To appreciate how rapidly Worcester R. Warner and Ambrose Swasey mastered the difficulties of large telescope design, it is necessary to retrace our steps. Of the

Fig. 124—Ambrose Swasey

(*Warner & Swasey Co.*)

two men, Warner had the greater interest in astronomy, although the characters and interests of both developed with a remarkable unity. They were both born in the same year, 1846, were both reared on New England farms and both responded to the call of the new mechanical age then growing in America. They first met in a machine works in Exeter, New Hampshire, at the age of nineteen, where both were apprentices. Upon completing their apprenticeship in 1870, both young men entered the employ of Pratt and Whitney in Hartford, Connecticut. Swasey was given charge of the gear-cutting shop, where he designed and constructed a machine for generating and, at the same time, cutting the teeth of spur gears. Warner also became a foreman and attracted attention by his devices for reducing production times. In his spare time, he constructed a mounting for a portable telescope and soon decided to set up in business with Swasey.[19]

In 1880, Warner and Swasey opened a small shop in Chicago, but the venture was a failure. They therefore moved to Cleveland, Ohio, there to be joined by four friends from Messrs Pratt and Whitney. The shop succeeded and, from that time, the design and manufacture of machine tools, turret lathes, and grinding machines became their principal work.[20] Their first telescope, a $9\frac{1}{2}$-inch equatorial, was made in 1881[21] for Beloit College, Wisconsin. The following year, they built a revolving dome for the 26-inch refractor at Leander McCormick. Five years later, Swasey was supervising the construction of the 36-inch Lick telescope.

The Clarks were delayed in their work some two years owing to the non-arrival of the crown disk.[22] On the night of January 3, 1888, the object-glass was directed to a few bright stars but rapidly thickening clouds terminated observations. It was apparent, however, that the focal length of the objective was a few inches shorter than the contract had stipulated; the difference was remedied by cutting off a short section of the lower end of the 57-foot tube.[23] On January 7, details of the central part of the Orion nebula were examined by Swasey, Alvan Clark junior, J. E. Keeler, and R. H. Floyd, chairman of the Lick Trust.[24] Two nights later and despite bad seeing, Keeler found the rings of Saturn ' very sharply defined ' with a power of 1000. ' The great telescope is equal in defining power to the smaller ones [6½-inch and 12-inch Clark refractors] ' Keeler writes, ' and has in addition the immense advantage of greater light-gathering power.'[25]

Doubts as to the suitability of the selected site had been dispelled by S. W. Burnham who, during a three months' stay on the mountain in 1879, recorded forty-two nights as ' first-class '.[26] During his visit, and with only a 6-inch Clark refractor, Burnham discovered forty-eight new double stars.[27]

In its early years, the Lick Observatory was not a rich institution. The cost of the telescope and building left only $125,000 for endowment, and staff had to be reduced to the minimum. The great amount of visual double-star work accomplished with the telescope during its infancy speaks highly for the industry of the small group of six resident astronomers. The light-grasp enabled J. E. Keeler, later director of the Allegheny Observatory, to use a Brashear spectroscope for his now classic studies of the radial velocities of nebulae.[28] In 1892, the instrument revealed to Barnard a faint fifth satellite of Jupiter,[29] a body about 100 miles in diameter and close to the planet. With the advance of photography, a 33-inch diameter correcting lens was added; by means of colour filters and special plates, planets were photographed in light of wavelengths from the infra-red to the ultra-violet. In this way, valuable knowledge about the surfaces and atmospheres of these bodies was obtained.

The 36-inch telescope made possible two important and extensive observational programmes. Early in the history of the observatory, an investigation was started of all stars down to the ninth magnitude with a view to determining the ratio between the number of visual stars and the number of visual double stars. R. G. Aitken and W. J. Hussey played an active part in the programme, which resulted in the publication, in 1932, of a catalogue of 17,180 double stars[30] and which was continued mainly by H. M. Jeffers. The second line of inquiry is the radial-velocity programme undertaken by W. W. Campbell, W. H. Wright, and J. H. Moore, in particular. This work is as remarkable for its magnitude as for the sustained accuracy of its results. Most members of the observatory staff have had some share in it. Since 1896, when the three-prism Mills spectrograph by Brashear was attached to the Lick telescope, more than 26,000 spectrograms of about 3000 stars have been obtained.[31] These stars are scattered in both hemispheres for, between 1903 and 1929, thanks to the financial assistance of D. O. Mills and his son, Campbell was able to install a 37-inch Brashear Cassegrain reflector and spectrograph at Santiago, Chile. Where Vogel and Scheiner had had to be content with an average probable error of 2·6 km/sec in their radial-velocity determinations, Campbell and his associates reduced this error, in the most favourable cases, to 0·5 km/sec and less.[32] Their work continues, as does also the study of spectroscopic binaries to which it leads, aided by new and more powerful spectrographs and highly sensitive Kron photoelectric photometers.[33] The 36-inch Lick refractor, more than any other telescope, has enabled us to determine the sun's motion and the average random velocity of the stars. A result of Campbell's researches was the discovery that, with certain exceptions, the corrected radial velocities of stars change steadily along the Harvard spectral sequence. This significant relationship, together with work in radial velocities in general, has received and continues to receive the close attention of the astronomers at Mount Wilson, at Victoria, B.C., and at Toronto.

In 1895, Edward Crossley of Halifax, England, presented the 36-inch Calver–Common telescope, which he purchased from Common, to the Lick Observatory. The mirror was refigured by Sir Howard Grubb, but Keeler found the original mounting unsuitable for his purpose. He wished to use the instrument for photographic studies and went to great pains to improve and strengthen the vital parts.[34] His work with America's first large reflecting telescope is a landmark in the history of this instrument. His photographs revealed, for the first time, the existence of an immense number of either very small or very distant nebulae.[35] Large nebulae like M31 in Andromeda and M33 in Triangulum were either rare or, if small image size was interpreted as large distance, the observable domain was much larger than hitherto realized. We know now that Keeler had penetrated into the realm of extra-galactic nebulae that the images on his plates were those of galaxies very remote yet extremely numerous. At the time, astronomers were obliged to consider both a revision of cosmology and a change in viewpoint with regard to the photographic potentialities of the reflecting telescope.

In conjunction with the Lick refractor, the Crossley reflector, again re-mounted, enabled Curtis, Campbell, Moore, Wilson, and Wright to extend our knowledge of planetary and irregular nebulae. Slitless quartz spectrographs are used in these investigations since these nebulae emit mostly short-wave radiations. Both instruments have likewise shared in the photography of star clusters. N. U. Mayall determined the radial velocities of a number of globular clusters, and R. Trumpler has made a special study of open clusters like the Pleiades.[36] A photographic search with the Crossley reflector added three satellites to Jupiter's system; No. VI and No. VII were found by Perrine, and No. IX by Nicholson.

Another large telescope at Lick is the 20-inch Carnegie double astrograph made by Warner and Swasey. The acquisition of this comparatively new instrument, planned by W. H. Wright, was made possible by the Carnegie Foundation. An ambitious programme is under way with this instrument—no less than the construction of a new fundamental reference system, a system independent of the proper motions of the stars. Such a system is to be framed by using distant galaxies as reference points and determining star positions relative to them. In this way, a complete study of stellar motions in our galaxy will be possible. To cover the part of the sky visible from Lick, over 1100 plates, each 17 × 17 inches, are required.[37]

Fig. 125—The 36-inch Lick refractor

(*Warner & Swasey Co.*)

James Lick died in 1879 and, consequently, never saw the completion of the project he had sponsored. In accordance with his wishes, his remains were carried to the mountain and rest in the brick pier which carries the mounting of the telescope. The success of the Lick undertaking proved, beyond all doubt, the practicability of a large telescope situated at a high elevation. Furthermore, it brought the firm of Warner and Swasey well into the limelight as telescope engineers and added further to the fame of Alvan Clark and Sons. When the University of Southern California decided to acquire a 40-inch refractor, therefore, Warner and Swasey received the contract for the mounting.

Clark obtained the disks from M. Mantois of Paris at a cost of $20,000 but, no sooner did they arrive, than he heard that the university's plans had failed and that the necessary funds were no longer available. With two unwanted glass disks on his hands and with Mantois pressing for payment, it was a lucky day for Clark when he met George Ellery Hale. Hale was then assistant professor of astrophysics at Chicago and Clark mentioned what a great acquisition a telescope of this size would be to the new university.[38] Hale was amenable to any plan that might further astrophysics and forthwith threw all his energy into the project. He persuaded Dr W. R. Harper, president of the University of Chicago, to join forces in trying to raise the necessary funds. Their efforts met with little success until they approached C. T. Yerkes, Chicago trolley-car magnate. Yerkes took some persuading but, at length, consented to pay for the lens and mounting. Further solicitations by Hale, and repeated visits to Yerkes' office, led to the promise of further grants, until the beleaguered magnate found that he had dispensed over $349,000—sufficient to provide for lenses, mounting, and complete observatory buildings.[39]

After he had inspected most of the large telescopes in American and European observatories, Hale wrote that he turned to Warner and Swasey ' not only because of the well-known excellence of their workmanship, but more especially on account of the valuable experience gained by the firm in designing and constructing the mounting of the 36-inch Lick refractor.'[40]

Warner and Swasey received the order for the Yerkes telescope in the autumn of 1892 and, by the following summer, the mounting was on exhibition at the Columbian Exposition in Chicago. Hale chose a site on the shores of Lake Geneva, some 80 miles north-west of Chicago. The observatory's elevation is only 240 feet—Hale's previous experience at the Bellini Observatory on Mount Etna and on top of Pike's Peak (where he encountered severe electrical and atmospheric storms), had temporarily cooled his ardour for mountain observatories (but *vide* page 323).[41] Even at this moderate elevation, the Wisconsin winters are sometimes severe and temperatures have been recorded below $-20°$F. In a long, cold season, the ice on Lake Geneva may reach a thickness of over 24 inches.[24] E. B. Frost, second director at Yerkes, comments on the loneliness and discomfort he experienced during long night vigils at the telescope in extreme cold.[43] When visitors to the observatory asked E. E. Barnard ' But how do you keep warm? ' he would answer ' We don't.' One night, when the temperature in the dome reached 6°F, Barnard stopped work and closed the dome, although the stars were still shining. He was worried lest, with everything so cold, something in the telescope might break or be injured.[44]

Fig. 126—Pier and mounting of the 40-inch Yerkes telescope

(*Warner & Swasey Co.*)

One morning, after a night of observing, Barnard appeared with a raw scratch on the side of his nose. In the intense cold, moisture on the rim of the eyepiece had frozen it to his nose and, when he moved his head away, the rim tore off a narrow strip of skin.[45] To work, as did Barnard, from dusk to dawn in the bitter cold of winter nights calls for the highest in enthusiasm, industry, and perseverance.

The 40-inch lenses were finished in October, 1895. Together, they weighed 500 lb and, with their mount, half a ton. They were naturally made as thin as possible to reduce the weight to a minimum. The crown lens is $2\frac{1}{2}$ inches thick at the centre while the flint, $8\frac{3}{8}$ inches behind, has a central thickness of $1\frac{1}{2}$ inches. The focal length is 62 feet.[46] With 500 lb at the end of a long tube, it follows that this had to be of exceptional rigidity. Together with the moving parts it weighs over 20 tons, yet it is so well balanced that the slightest pressure suffices to direct it to any part of the sky. This pressure is provided by electric motors acting through gear trains, and is controlled by the observer. A similar system controls the motion of the 90-foot dome. Tube and axes are carried by a massive cast iron column, 43 feet high, the top section of which houses the driving clock. The base of the column rests on a brick pier embedded in solid concrete.[47]

To reach the eyepiece in all tube positions, the 75-foot diameter floor has a vertical motion and is of the same pattern as that installed at Lick. The idea of a rising floor is due to Grubb-Parsons who estimated for supplying the floor of the Lick Observatory.[48] The firm lost both the contract and designs, for one condition for submitting a bid was that the winner, in this case Warner and Swasey, should have access to the designs submitted. Warner and Swasey also supplied the Yerkes rising floor and, when the 26-inch Clark refractor was remounted, another for Washington.

A few days after observations had started with the 40-inch telescope, an accident occurred which might have had disastrous consequences. One of the cables which supported the floor slipped from its fastenings, and the entire south side of the floor fell to the ground from a height of 45 feet. Fortunately, the floor was not in use when the accident happened, but the telescope was shaken and it was feared that the object-glass had been damaged. No permanent damage was done, although the date of the opening ceremony had to be postponed by over a fortnight.[49]

The Yerkes floor weighs $37\frac{1}{2}$ tons. The dome and the heavier parts of the floor were erected by the King Bridge Company of Cleveland, Ohio. In the basement of the brown Roman-brick main building are laboratories for figuring and testing mirrors, lathes and machinery for carrying out repairs and making auxiliary apparatus, and a spectrographic laboratory.

The advantages of the Yerkes refractor for certain types of work are now well-known. On May 21, 1897, only two days after the lenses arrived, the first observations were made by Hale, Barnard, and Ellerman. Despite poor seeing, Barnard saw the Ring nebula in Lyra and M13 in Hercules better than he had ever seen them from Mount Hamilton. He noted also a faint optical companion to Vega which had not been seen before.[50] Hale writes that the 40-inch showed him minute structures on the moon which were quite invisible through the 12-inch Kenwood telescope on the same night.[51] A European astronomer, who had spent many years studying Jupiter, remarked after looking at that planet through the

Fig. 127—The Bruce telescope of the Yerkes Observatory

40-inch glass that his studies seemed useless, so much more could he see at a single glance.[52] Burnham, Barnard, Ritchey, Frost, Adams, Schlesinger, Ellerman and others have done invaluable work with the telescope in many varied and interesting fields—procuring results unobtainable with a smaller instrument.

Since its foundation, the Yerkes Observatory has contributed to almost every branch of astronomy. At the turn of the century, Ritchey showed how the 40-inch telescope could be used as a photographic instrument by placing a yellow filter in front of an isochromatic plate.[53] With the introduction of orthochromatic plates, whose sensitivity extended from the green to the red parts of the spectrum, came ideas of determining the photovisual magnitudes of stars. The pioneer experiments of J. Parkhurst, R. Wallace, and F. Jordan[54] at Yerkes established this new method of photographic photometry. Here also, S. W. Burnham continued his work on double-star measurements, and E. E. Barnard pioneered in short-focus photography.

In the previous chapter, we referred to Barnard's work with the 6-inch Willard lens at the Lick Observatory. Barnard came to Yerkes in 1895 and, in 1904, began work with a 10-inch aperture doublet, the gift of Miss C. W. Bruce and made by J. Brashear, and Messrs Warner and Swasey. With this instrument Barnard took photographs of the Milky Way which were published in two volumes in 1927.[55]

At Yerkes, Dr F. E. Ross introduced his wide-angle lens, the power of which he demonstrated in his beautiful photographs of regions of the Milky Way.[56] Most significant of all, however, was F. Schlesinger's work on stellar parallax determination during the years 1903 and 1905. From photographs taken with the 40-inch, Schlesinger determined 26 new parallaxes with an average probable error of $\pm$ 0″·013.[57] The importance of this investigation will be appreciated when we realize that, with all the graduated instruments at many observatories to command, astronomers at that time knew only about 30 parallaxes to within 0″·05.[58]

REFERENCES

1 Whipple, F. L., 1947, *Earth, Moon and Planets*, pp. 213 ff.

2 Grubb, H., *Engineering*, 1881. 'Description of 27-inch Refractor.'

3 *M.N.R.A.S.*, **92**, p. 254, 1932.

4 Stebbins, J., *Publications Washburn Observatory*, **15**, Pt. I, 1928.

5 Adams, W. S., *Ap. J.*, **87**, p. 380, 1938.

6 Newcomb, S., 1903, *Reminiscences of an Astronomer*, p. 145.

7 Lowell, P., 1907, *Mars and its Canals*, p. 193.

8 Meudon *Annales*, **1**, 1896.

9 Stratton, F. J. M., *M.N.R.A.S.*, **109**, pp. 141–142, 1949.

10 Maunder, E. W., 1900, *The Royal Observatory, Greenwich*, pp. 272–273.

11 *M.N.R.A.S.*, **69**, p. 331, 1909.

12 Lick, J., Extract from Will of, quoted in Shapley, H., and Howarth, H. E., 1929, *Source Book of Astronomy*, p. 316.

13 Shinn, M. W., *Overland Monthly*, 'The Lick Observatory', p. 2, 1892.

14 Denning, W. F., 1891, *Telescopic Work for Starlight Evenings*, p. 37.

15 Shinn, *op. cit.*, p. 10.

16 *Ibid.*

17 *Ibid.*

18 Holden, E. S., 1897, *The Work of the Lick Observatory.*

19 *The Warner and Swasey Company*, 1880–1920 (Private Publication), pp. 11–14.

20 *Ibid.*, pp. 19–21.

21 *Ibid.*, p. 20.

22 Holden, E. S., *Observatory*, **8**, p. 85, 1885; *Sidereal Messenger*, **7**, pp. 61–62, 1888.

23 Swasey, A., *Publ. Astr. Soc. Pac.*, **31**, p. 58 1919.

24 Keeler, J. E., *Sidereal Messenger*, **7**, p. 79, 1888.

25 *Ibid.*, p. 83.

[26] Burnham, S. W., *Publications Lick Observatory*, **1**, p. 14, 1887.
[27] *Ibid.*, pp. 24–33.
[28] *Publications Lick Observatory*, **3**, pp. 165–229, 1894.
[29] *A. J.*, **12**, p. 81, 1893.
[30] *New General Catalogue of Double Stars within 120° of the North Pole*, 1932.
[31] Vinter Hansen, J. M., *Pop. Astr.*, **55**, p. 189, 1947.
[32] Merrill, P. W., *M.N.R.A.S.*, **99**, p. 318, 1939.
[33] Vinter Hansen, *op. cit.*, p. 192; Kron, G. E., *Ap. J.*, **103**, p. 326, 1946.
[34] Hale, G. E., *Ap. J.*, **11**, pp. 325–349, 1900; 1908, *The Study of Stellar Evolution*, p. 42. *Vide* also *Publ. Astr. Soc. Pac.*, **12**, pp. 146–167, 1900.
[35] Hale, G. E., 1908, *The Study of Stellar Evolution*, p. 45.
[36] Vinter Hansen, *op. cit.*, p. 193.
[37] *Ibid.*, *pp.* 194–195.
[38] Frost, E. B., 1933, *An Astronomer's Life*, p. 97.
[39] Struve, O., *Pop. Astr.*, **55**, p. 230, 1947; Bates, R. S., *The Telescope*, **7**, p. 80, 1940.
[40] Hale, G. E., *Ap. J.*, **6**, p. 41, 1897.
[41] Hale, Ref. 35, pp. 117–119.
[42] Frost, *op. cit.*, p. 203.
[43] *Ibid.*, p. 205.
[44] *Jour. Tennessee Acad. Sc.*, **3**, p. 29, 1928.
[45] *Ibid.*, p. 13.
[46] Hale, G. E., Ref. 40, p. 43.
[47] *Ibid.*, pp. 41–42.
[48] *Proc. Roy. Soc.*, **135**, p. vii, 1932 (Obituary notice H. Grubb).
[49] Struve Ref. 39, p. 232.
[50] *Ibid.*, p. 232.
[51] Hale, Ref. 35, p. 26.
[52] *Ibid.*, p. 25.
[53] Ritchey, G. W., *Ap. J.*, **12**, p. 352, 1900.
[54] Parkhurst, J. A., and Jordan, F. C., *Ap. J.*, **23**, p. 79, 1906. Parkhurst, J. A., *Ap. J.*, **17**, p. 373, 1903. Wallace, R. J., *Ap. J.*, **24**, p. 268, 1906; **25**, p. 116, 1907.
[55] Barnard, E. E., 1927, *Atlas of Selected Regions of the Milky Way*. *Ap. J.*, **49**, p. 1, 1919.
[56] Struve, Ref. 39, p. 237.
[57] Spencer Jones, H., *M.N.R.A.S.*, **104**, p. 95, 1944.
[58] *Ibid.*, p. 94.

CHAPTER XVI

The only remedy is a most serene and quiet Air, such as may be found on the tops of the highest Mountains above the grosser Clouds.

ISAAC NEWTON

EARLY in the present century, general interest in astrophysics in America ran high, and good reasons were not wanting for the establishment of new observatories. Those already in existence generally had just one large refractor and were now ready to receive more up-to-date equipment. In all cases, public munificence played a large part, especially when controlled by an organizer as far-sighted and as resolved as George Ellery Hale. Four times over, Hale succeeded in raising sufficient money for the erection of the largest telescopes of modern times—the 40-inch Yerkes refractor, the 60-inch and 100-inch Mount Wilson reflectors, and the 200-inch Hale telescope.

Hale was born in Chicago on June 29, 1868. At an early age he became acquainted with S. W. Burnham, then a stenographer in the Chicago law courts by day and a double-star observer by night. Through Burnham, Hale purchased a second-hand 4-inch Clark refractor which he mounted on the roof of his father's house and used for observations of a general nature.

At this period [Hale writes],[1] the $18\frac{1}{2}$-inch Dearborn refractor stood at the summit of a lofty tower forming part of the . . . old University of Chicago (long since demolished, after the removal of the telescope to the Northwestern University at Evanston). Professor G. W. Hough, known for his observations of Jupiter and double stars, who was in charge of the instrument, was kind enough to let me look through it frequently. I was also permitted to aid the feeble gas-engine in turning the dome, or rather drum, of the observatory. Naturally, my own ambitions were thus stimulated, but neither Hough nor Burnham had the slightest interest in astrophysical research, and I could not have devoted my life to such work as they were doing, valuable as it was to science. The reason lay in the fact that I was born an experimentalist, and I was bound to find the way of combining physics and chemistry with astronomy. Fortunately, it was not far to seek.

In Cassell's book [Hale continues], is a description of a spectroscope, with an account of the construction of a simple instrument, using either a luster prism or a hollow one filled with carbon disulphide. I built both forms, and the odor of the disulphide abides with me after a lapse of fifty years. Then my father, always ready to encourage serious efforts, enabled me to buy a small spectrometer. This had a single prism, and I lost no time in fitting it with a small plane grating. For by this time I had learned of the work of Rutherfurd and Rowland and had acquired some slight conception of the possibilities of high dispersion. Nothing could exceed my enthusiasm in observing the solar spectrum and in measuring the principal lines. I bought Lockyer's *Studies in Spectrum Analysis* and began the observation of flame and spark spectra and their comparison with the spectrum of the sun. At last I had found my true course, and I have held to it ever since.

Meanwhile, Hale prepared himself for his career. He received his engineering training at the Massachusetts Institute of Technology in Cambridge, at the end of which period he took his degree. While a student there, he observed solar prominences with a spectroscope and conceived the idea of the spectroheliograph.[2] During this period also, he met E. C. Pickering who, impressed by the possibilities of the spectroheliograph, gave Hale opportunities for trying out his experimental models.[3] Many reasons prevented Hale from fixing these to his 4-inch telescope and his father therefore bought him a 12-inch Warner and Swasey equatorial refractor and paid the expenses for a solar laboratory at Kenwood, Chicago. Between 1891 and 1895, assisted by Ellerman, Hale devoted himself to the steady development of his invention.[4]

The spectroheliograph was independently Hale's own idea.

> As he [the astronomer] sits in his laboratory [he wrote later][5], surrounded by lenses and prisms, gratings and mirrors, and the other elementary apparatus of a science that subsists on light, he cannot fail to entertain the alluring thought that the intelligent recognition of some well-known principle of optics might suffice to construct, from these very elements, new instruments of enormous power.

One feels that Hale is here describing the circumstances of earlier meditations, for his invention was little more than a particular arrangement of lenses and prisms. The method embodied in the apparatus was known to Janssen as early as 1869, and Braun of Kalocsa, Hungary, thought along similar lines.[6] Young at Princeton, New Jersey, in 1870[7] and Lohse at Potsdam[8] both tried the instrument in practice, but without success. The French astronomer, Henri Deslandres, of the Meudon Astrophysical Observatory, was more successful, and, in 1890, produced a spectroheliograph by modifying an instrument which he called a *spectrographe des vitesses*.[9] Hale was unaware of this earlier work and, incidentally, his first trials at Harvard Observatory, using a line of hydrogen, were unsuccessful.

In principle, the spectroheliograph consists of two moving slits, a collimator, prism system or grating, and a photographic plate. An image of the sun is formed on the slit of a spectroscope and a second slit is so arranged as to isolate a narrow band of wavelengths. If the solar image now traverses the first slit at a uniform rate and a photographic plate behind the second slit moves at the same rate, an image of the sun in almost monochromatic light is built up. In the Kenwood spectroheliograph, the solar image and photographic plate were fixed, while the first and second slits moved across them by means of a system of levers actuated by hydraulic power. Hale first tried the narrow band centred at $\lambda 6563$ where the prominent C hydrogen line occurs; but this, being in the red region of the spectrum, failed to affect the early red-insensitive plates.[10] At Kenwood he turned to the H and K lines in the extreme violet. These lines are difficult to perceive visually but register well on the photographic plate. A further advantage is that they are bordered by broad, dark wings which appreciably reduce the brightness of the continuous spectrum. Using these lines, Hale procured in 1892 highly successful pictures of calcium prominences 'with the sharpness and precision characteristic of the best eclipse photographs.'[11] More particularly, the instrument revealed a new class of phenomena—the bright calcium flocculi. The plates suggested some association between faculae and the clouds of incandescent calcium vapour, but further work at Yerkes revealed differences, and the term *flocculi* was applied to the calcium clouds in view of their fleecy appearance.

The great sunspot of February, 1892, afforded by its exceptional activity 'the very conditions required to bring out the peculiar advantages of the spectroheliograph.'[12] Hourly, indeed minute-to-minute, records could be taken of the rapid movements and consequent changes in form of the associated flocculi. During the years at Kenwood, and until the instruments were transferred from there to the Yerkes Observatory, Hale obtained over 3000 photographs of solar phenomena.[13]

The outstanding success of this work, coupled with its great interest and value in investigating the nature of the sun's atmosphere, determined Hale to apply the spectroheliograph to the Yerkes refractor. The 7-inch solar image given by the

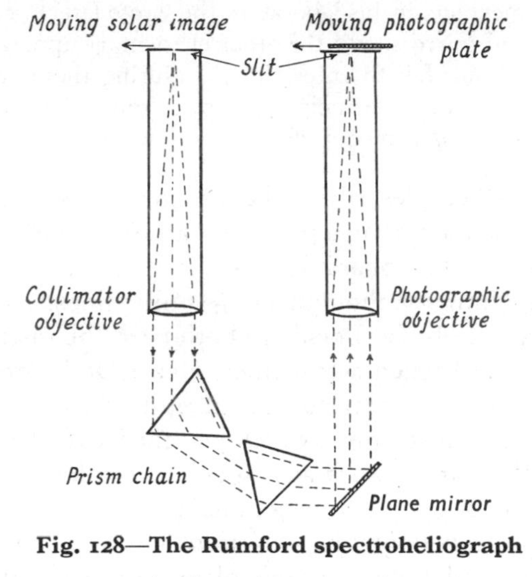

Fig. 128—The Rumford spectroheliograph

A schematic cross-section

40-inch objective, as compared with the 2-inch given by the Kenwood refractor, promised finer details of the calcium flocculi and the use of lines other than those of calcium. A special instrument known as the Rumford spectroheliograph was built in the workshop at Yerkes.

> In this instrument [Hale writes][14] the solar image is caused to move across the first slit by means of an electric motor, which gives the entire telescope a slow and uniform motion in declination. . . . While the solar image is moving across the first slit, the plate is moved at the same rate across the second slit, by a shaft leading down the tube from the electric motor, and connected, by means of belting, with screws that drive the plate-carriage.

With the Rumford spectroheliograph, Hale was able to study calcium flocculi at various levels corresponding to different points on the contours of the H and K lines. In May, 1903, he obtained the first photographs with the $H\gamma$ line of hydrogen, discovering thereby the presence of dark hydrogen flocculi.[15]

To procure these photographs, Hale employed a large diffraction grating and two prisms. The increased dispersion was necessary 'to exclude completely the light from the continuous spectrum on either side of the line employed. The admission of even a small quantity of this light might completely nullify the slight differences of intensity recorded by the aid of the comparatively faint light of the dark line.'[16]

From the hydrogen lines $H\alpha$, $H\beta$, and $H\gamma$, in the red, blue, and violet respectively, Hale and his colleagues turned to the strongest lines of iron and other elements. But even with the grating, only mediocre results were obtained. Hale had to admit that the Yerkes refractor, as a solar telescope, fell short of the ideal instrument. For obtaining sunspot spectra, a much larger image was desirable and, to produce the high dispersion necessary for his investigations, Hale found he would need a 10-foot spectrograph. This would be too long and too heavy for a moving tube. Fitting the cumbrous Rumford spectroheliograph to the tube took up valuable time and required the addition of balance weights.[17]

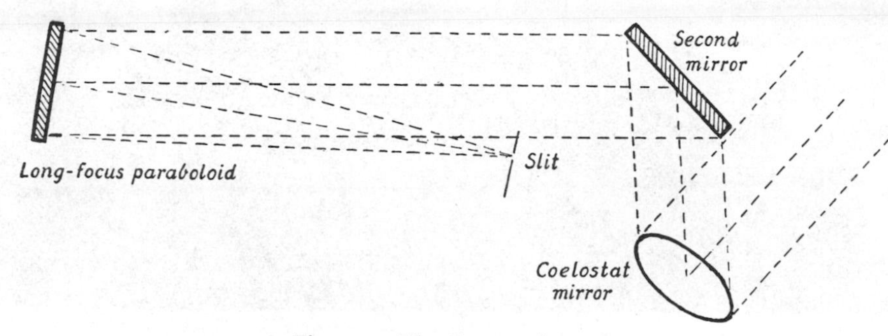

Fig. 129—The Snow telescope

A diagram showing the optical layout used to obtain a stationary image of the sun

The solution lay in a fixed, horizontal telescope, fed by a coelostat and secondary mirror and in which lenses, mirrors, and interchangeable spectrographic apparatus could be firmly mounted on stone or concrete piers. By this arrangement a large, well-defined solar image could be formed inside a laboratory. The first experimental telescope of this kind was built by Hale and G. W. Ritchey in the grounds of Yerkes Observatory. A long-focus objective and spectroheliograph were fed by a 30-inch coelostat and were housed in a temporary wooden structure some 80 feet long. Trials were well in progress when a spark from the induction coil set the housing on fire and, in a few minutes, the entire equipment was destroyed.[18] A second telescope, made possible by a $10,000 grant from Miss Helen Snow, was of a more solid nature.[19] It was completed in the autumn of 1903 but gave indifferent definition owing to currents of heated air in the optical path and to warping of the mirrors by the sun's heat.

These initial setbacks served only to strengthen Hale's resolve to establish a mountain observatory, devoted primarily to solar research and equipped with a large reflector for stellar photography. In 1902, he became member of and, later, secretary to the astronomy committee of the Carnegie Institution of Washington, an organization devoted to the advancement of scientific research. His proposals

Fig. 130—The Snow telescope building from the south-east

(Mount Wilson and Palomar Observatories)

to this committee met with warm support and $5000 were advanced to enable Professor Hussey of Ann Arbor to investigate observing conditions at potential sites in the United States.[20] Mount Wilson (5886 feet), was the committee's final choice, one probably influenced by the fact that W. H. Pickering had, some years before, spoken highly of the observing conditions there and had established a small Harvard substation on the top.[21]

In March, 1904, systematic meteorological observations were begun at Mount Wilson, together with frequent tests of solar definition with a $3\frac{1}{4}$-inch refractor.[22] How well Mount Wilson satisfied Hale's exacting demands we shall see later. To a summer visitor, unfamiliar with the niceties of atmospheric conditions, the location appears ideal. On all sides stretches a beautiful panorama, bathed in sunshine and reaching to islands nearly one hundred miles out in the Pacific. During summer months, the sun shines almost continually by day; the fog clouds and sea breezes of the valley fail to reach the clear, steady air of the mountain top. In winter, the weather is severe, with low temperatures and heavy falls of rain and snow.

In 1904 also, a further grant by the Carnegie Institution brought Ritchey, Adams, Pease, and Ellerman to Mount Wilson, together with the Snow telescope. Pasadena formed a good assembly base, and here the first workshop and laboratory

Fig. 131—The Snow coelostat and second flat from the north

(Mount Wilson and Palomar Observatories)

were established. In the summer of 1904 all was ready for hauling the instruments to the summit. Mount Wilson was best reached from the San Gabriel Valley by the 'New Trail', a steep and precipitous path about $9\frac{1}{4}$ miles long starting only $6\frac{1}{2}$ miles from Pasadena. At first, this track was only two feet wide in places, but sufficient for the relays of mules carrying the lighter building materials. Later, the trail was widened into a road to permit the passage of the heavy base castings and mirrors for a 60-inch reflector.

The Snow telescope was mounted on the south-east edge of the mountain top, with its axis as level and as close to the meridian as the configuration of the ground allowed. A carefully levelled pair of rails some 150 feet long were laid on concrete piers built on solid rock and, to the south, projected over a precipice, so that the coelostat was some 30 feet above ground. The coelostat and secondary mirrors, of 30 and 24 inches in diameter respectively, were mounted on individual carriages and fed the light to either one of two 24-inch mirrors housed in a large, white, well-ventilated wooden shed. One mirror had a focal length of 60 feet and formed an image of the sun 7 inches in diameter. The other, mounted on its own carriage and at the extreme northern end of the rails, had a focal length of 143 feet and gave a 16-inch disk. Slight adjustment of the mirrors sufficed to throw the solar image successively on the slit of a 5-foot spectroheliograph, an 18-foot Littrow or auto-collimating spectrograph (the largest then built), and bolographic apparatus, all securely fixed to concrete piers.[23]

For sunspot spectrography, light from only the umbra was allowed to enter the slit by closing all but a small portion in the centre. Comparison spectra were obtained by covering the centre of the slit and allowing two strips to pass light from the photosphere into the collimator. The plates so obtained underwent critical examination in the laboratory for the strengthening and widening of certain lines in the spot spectrum as compared with the same lines in the spectrum of the photosphere, two effects first noticed by Lockyer. Collaborative spectrographic work in the Pasadena laboratories left no doubt that sunspots are cooler than other parts of the sun's disk. This contribution to solar physics at once ruled out all theories on the nature of the sun which required very high-temperature sunspots, and Hale was able to attack afresh the problem of the physical nature of sunspots.[24]

The tube of the spectroheliograph stood immediately below the spectrograph, and rotation of the latter allowed light from the main mirror to enter the slit:[25, 26]

> On account of the large size of the solar image, which is about 6·7 inches in diameter, the slit is $8\frac{1}{2}$ inches long. After passing through the slit, the light falls upon a large collimating lens 8 inches in diameter, which renders the rays parallel. They then meet a silvered glass mirror, from which they are reflected to two prisms, of $63\frac{1}{2}°$ angle. After being dispersed by the prisms the rays strike the 8-inch camera lens, which forms an image of the spectrum on the camera slit. The optical train thus resembles that of the Rumford spectroheliograph, but the lenses and prisms are so much larger that no light is lost from the circumference of the solar image. . . . [Curvature of the spectral lines due to the prisms was overcome] by dividing the curvature evenly between the two slits.
>
> In the daily programme of observations [Hale continues][27] at least one photograph with the H_1 line of calcium, showing the faculae and low level calcium vapour; one with the H_2 line of calcium, showing the flocculi at a higher level; one with the $H\gamma$ line of hydrogen; and one with an iron line, are made in the early morning and again, if circumstances permit, in the late afternoon. Since the weather is clear day after day through the summer and autumn months, and not infrequently during the rainy season, the instrument thus yields a large number of plates, suitable for the comparative study of the flocculi. Photographs of the prominences are also made daily, when circumstances permit. These are used to determine the changes in number and total area of the prominences during the Sun-spot period.

With special red-sensitive plates in the spectroheliograph, Hale obtained photographs of the sun in $H\alpha$ radiation. He found that the hydrogen flocculi showed striking vorticity in the neighbourhood of sunspots, a vorticity which, if interpreted in terms of the rotation of charged particles, might give rise to the strong magnetic fields associated with sunspots.[28]

Not every observation with the Snow telescope was straightforward. Heating of the mirrors by the sun produced marked changes in focal length, together with astigmatism due to distortion of the coelostat mirrors. As the mirrors were exposed to the sun, so focal length increased—sometimes by as much as 12 inches in the case of the 60-foot mirror. The focal length then remained fairly constant, but as in time the silver tarnished, so the surface absorbed more heat and the lengthening increased. Disturbing air currents, so noticeable at Yerkes, were less marked but still prominent, and their elimination called for all Hale's ingenuity. He limited regular observations to the early morning and evening, he shielded the mirrors from sunlight until the moment of exposure arrived, he used electric fans to keep the mirrors cool, then heated them electrically at the back. He spread white cheese-cloth over the ground to minimize heating of the ground, and then stirred the air by means of electric fans. At one time, he had ideas of replacing glass mirrors by quartz disks, but these proved too porous for optical working.[29] The result of all these efforts was a series of spectroheliograms equal, if not superior, to those taken with the 40-inch Yerkes refractor and Rumford spectroheliograph.

We have seen that Hale's principal aim during the busy years at Kenwood and Yerkes was the extension, by spectrograph and spectroheliograph, of our knowledge of the sun. This earlier work was, however, part of a much larger and more ambitious scheme—the elucidation of the many problems of stellar evolution. After the completion of the Yerkes Observatory, we find Hale directing more and more of his energy to the next line of attack—the erection of a large reflecting telescope suitable for taking high-dispersion spectrograms of the brighter stars. The scheme for a large reflector had been in his mind for many years, for, soon after the building of the Yerkes Observatory, his father had acquired (for $25,000) a 60-inch mirror blank from the St Gobain Glassworks, Paris.[30] This he had offered to the University of Chicago, provided that the latter supplied the mounting and observatory. Ritchey partly ground the mirror at Yerkes, but lack of funds prevented its completion. In 1903, a grant from the Carnegie Institution, given as the direct result of Hale's untiring efforts, made the project possible. Ritchey began designing the mounting and optical system, while Hale contributed ideas and designed the observatory building.

George W. Ritchey figures so prominently in the development of Mount Wilson that some account of his previous work is necessary. He was born in 1864, at Tupper's Plains, Ohio, and, as a boy, caught his father's enthusiasm for astronomy. While a student at the University of Cincinnati, he made a 9-inch reflecting telescope and, in 1888, set up a small astrographic laboratory at his home in Chicago. Here he developed many of the ideas which were used later at Yerkes and Mount Wilson, and here he acquired great skill in figuring optical surfaces.[31] His first outstanding achievement was a 24-inch mirror of 8 feet focus and, when Hale engaged him at Yerkes, he was able to mount this for photographic work in the south-east dome. The improved method of mirror mounting, the skeleton tube, and the rigidity, yet with delicate control, of the entire instrument were particular features.[32] The series of photographs of clusters and nebulae, taken at the primary focus, bear witness to the excellent optical and mechanical properties of this telescope. In 1901, and with 40 minutes' exposure, Ritchey discovered the expanding nebula round Nova Persei.[33] Using colour filters and special plates, Ritchey also adapted the 40-inch refractor for photography and took photographs of the moon and of star clusters[34]—copies of which appeared in many popular astronomical books published between the two world wars.

At Yerkes, Ritchey had control of the optical and instrument shops, where he

Fig. 132—G. W. Ritchey

He is seen during work on the 40-inch reflector of the U.S. Naval Observatory, December 29, 1931

(*American Museum of Natural History. Photograph by Clyde Fisher*)

designed and constructed most of the auxiliary equipment for the 40-inch refractor. Later, he became instructor in practical astronomy at the University of Chicago and, finally, assistant professor there. With the establishment of the Pasadena optical and mechanical shops, Ritchey was appointed director and given the task of grinding and figuring the 60-inch disk.

His first step was to design and construct a grinding and polishing machine. The disk, about a ton in weight and 8 inches thick, was supported by a massive cast iron turntable and rested on two thicknesses of Brussels carpet, the looped threads of which acted like innumerable small springs. In this position the top face was ground flat and parallel to the lower face, already so treated.[35] The edges were ground by a flat plate which rotated at high speed and which was fed with abrasive and water. The grinding tools were cast iron, ribbed at the back and divided into a number of small squares on the grinding face.[36] They were suspended from a counterpoised lever arm by which they could be swung over and lowered on the mirror. Both disk and grinder rotated, the grinder also moving over the surface from edge to edge in a series of short strokes. The abrasive was carborundum, a carbide of silicon, invented in 1898 and first made at the Niagara Falls Elstree Works.[37] Its cutting power is about six times that of emery which it soon rivalled in optical work. As grinding proceeded, finer and finer grades of carborundum were used, and weights added to the lever arm reduced the pressure of the grinder until the glass had a smooth and almost polished surface. The back of the disk then received similar treatment and was silvered to prevent rapid temperature changes.[38] The top face was now fine-ground with a slightly convex tool and the curvature tested from time to time with a spherometer. Ritchey made the polisher of narrow strips of pinewood, saturated with paraffin so as to retain their form. It was thus both rigid and light in weight. On the convex face were fitted rosin squares lightly covered with beeswax; these formed the polishing face, jeweller's rouge the abrasive.[39]

Ritchey took every possible precaution to exclude dust from the polishing shop. Walls and ceiling were varnished while the painted cement floor was kept wet during polishing operations. Windows were double and carefully sealed, and incoming air was filtered. Over the mirror, Ritchey suspended a canvas screen to protect the surface from falling particles. The temperature of the shop was kept as constant as possible and only the optician, dressed in surgeon's cap and gown, was allowed to pass through the door.[40] From a perfect spherical surface, the mirror passed through the figuring stages, being turned on edge at intervals for the Foucault test.* This test was carried out at the focus and required a well-figured plane mirror of 36 inches diameter, together with a small, diagonal plane mirror.[41]

The mounting is of the fork type and takes the form of a cast iron fork attached to the end of a hollow nickel-steel polar axis. To relieve friction on the upper and lower bearings, a hollow steel float, 10 feet in diameter, is fitted just below the fork and floats in a tank filled with mercury. The tube has an open framework and the mirror is so mounted as almost to ' float ' in its housing.[42]

Four optical arrangements are possible—one Newtonian, two modified Cassegrains, and a polar or coudé Cassegrain.[43] As a Newtonian, the focal length is 25 feet; this focus is used for direct photography and low-dispersion spectrography. Here a 16-inch spectrograph enables the spectra of all stars within a field of about 30 minutes of arc to be photographed at once. With this spectrograph, the spectrum of an eighth-magnitude star can be photographed in 70 minutes. In the modified Cassegrain, the light meets a convex hyperboloidal mirror which reflects it, not through a hole in the primary mirror, but to a comparatively small plane mirror at the lower end of the tube. This last mirror passes the rays to the side of the tube and to a double-slide plate-holder; the equivalent focal length is then 100 feet. A spectrograph can be mounted in place of the photographic plate, in which case a convex secondary giving an equivalent focal length of 80 feet is used. In the coudé form,[44] a third alternative convex secondary gives an equivalent focal length of 150 feet. A plane mirror is then so geared that the light is sent down the hollow polar axis for all tube positions. At this focus is the slit of a 13-foot spectrograph fixed to a solid concrete pier and housed in a constant-temperature chamber. By means of a further plane mirror at the lower end of the polar axis, the light

* The method of testing a paraboloid at its focus is described by Ritchey in *Ap. J.*, **14**, pp. 218–220, 1901.

can alternatively be fed to a vertical 18-foot Littrow spectrograph mounted 14 feet below ground level. For all three spectrographs, electric slow motions under the observer's control keep the star image always on the slit aperture, a condition achieved by watching the reflection of part of the image from the polished slit jaws.[45]

The fork and polar axis forgings, made by the Union Iron Works of San Francisco, together with all the other large parts, were first assembled in an erecting shop in Pasadena. Here also the 10-foot driving wheel and its worm gear, driving clock, motors, and divided circles were made, fitted, and adjusted. Each tooth of the driving wheel was spaced off by a finely graduated circle fixed to the polar axis and was formed by a special cutter. Teeth and worm gear were then ground together for several hours and the teeth finally polished with jeweller's rouge.[46] As soon as Ritchey had assembled and checked the motions, the set-up was dismantled and sent in convenient sections to the mountain top.

Hale's chief aim with the 60-inch reflector was to obtain spectrograms of the brighter stars comparable in scale with the negatives used in making Rowland's map of the solar spectrum.[47] While the new instrument was building, therefore, Hale tried a 13-foot grating spectrograph with the Snow telescope on Arcturus. The small aperture ratio of the Snow telescope necessitated long exposures and these, in turn, required not only that the spectrograph should be rigidly mounted but also that its temperature, and particularly that of the grating, should be constant. By an ingenious device, Hale restricted the latter to within $\pm 1\,^{\circ}C$ and, by other precautions, obtained promising spectra. One exposure lasted for five nights in succession and aggregated twenty-three hours.[48] Great care had to be taken to ensure that, after the day's solar work, the mirrors had settled down to their normal figures. Unless this time interval was observed—and it took many hours—star images were distorted and often multiple in character.[49] The collected results, although not as good as Hale could have wished, showed that, with about four hours' exposure, the same spectrograph on the 60-inch reflector should give critical spectra of first-magnitude stars. Subsequent experiments showed that, for stellar spectra, large prisms could be used to greater advantage than gratings, and the next spectrograms of α Orionis (Betelgeuze) were taken with the aid of a dense flint prism belonging to the spectroheliograph. Adams made the exposures on two successive nights with an aggregate time of seven hours, but bad weather cut short the exposure planned. The result, however, was good enough to show the close resemblance of the spectrum of this star (a low-temperature giant) to those of sunspots.[50]

The first visual and photographic observations with the 60-inch were made in December, 1908. Following the experience with the mirrors of the Snow telescope, provision had to be made for the mirror's protection against changes in temperature. Ritchey covered the dome with a canvas screen supported on a skeleton framework and encased the greater part of the mounting during the day with a canopy of blankets. The almost airtight shutters were kept closed during the day and were not opened until shortly before sunset. After these precautions, it was found that the mirror's figure about midnight was as perfect as when tested in the optical shop.

Fig. 133—The 60-inch reflector from the north-west

Mount Wilson and Palomar Observatories)

Visually [Hale writes in his 1909 Report][51] the images of stars, planets and nebulae obtained with it on a good night are excellent. The star images are very small and sharp and can be observed with great precision. Such an object as the Great Nebula in Orion shows a bewildering variety of detail. Globular clusters are specially remarkable because of the large number of stars made visible by the great light-gathering power.

In 1911, Barnard spent a month at Mount Wilson making photographic and visual observations of Mars and Saturn with the 60-inch.[52] Most of this work was done at the 100-foot focus, with a few visual observations at the Newtonian focus. The Saturn negatives show belts, polar caps, rings, and the Cassini division well defined, also traces of the Encke division. Exposures were made through a pale yellow filter on Isochrome plates and were of 10 to 12 seconds' duration. Photographs of Mars were less successful. They were taken through red filters on Seed Process plates, stained with a red-sensitizing dye. Hale reports on the visual observations as follows:[53]

Compared with the images of Saturn and Mars in the 60-inch, those in a refracting telescope have a muddy or dirty look. This was perhaps more striking in the case of Mars than in that of Saturn. Mars was almost colourless. There was a slight pinkish colour on which the dark details had been painted with a grayish coloured paint, supplied with a very poor brush, producing a shredded or streaky and wispy effect in the darker regions. In the visual observations there was a better chance than in the case of the photographic work, for one could pick out moments of steadiness that would be lost to him during the exposures. The results in the case of Mars are said, by Prof. Barnard, to be quite beyond his power of either describing or depicting. An artist could have studied the planet and given a representation of the structural details shown in the dark regions that would have conveyed some idea of what was seen, but no one could accurately delineate the remarkable complexity of detail of the features which were visible in moments of greatest steadiness. Some sketches were made in an effort to show these features, but Prof. Barnard considered his artistic skill entirely inadequate to the effort. In spite of the immense complexity of detail visible, he was unable to see any trace of the system of fine lines as drawn by Lowell.

At the principal focus, Barnard obtained impressive views of parts of the Milky Way.

The stars looked like jewels on black velvet. The sky was rich and dark, and every star was a glowing, living point of light. . . . Given an equal facility for handling the large reflector [Hale concludes], Prof. Barnard would prefer it for visual work on the planets to any of the large refracting telescopes with which he is familiar.

Photographic tests were conducted by Ritchey, who took nebulae and star-clusters as his subjects, both at the principal focus and 100-foot Cassegrain focus. The results, using exposures much longer than those necessary for the moon or planets, were encouraging, despite changes in focal length due to expansion and contraction of the tube. With a knife-edge, Ritchey found no difficulty in determining the focal plane to within 0·001 inch and, with this device, checked the focus between exposures.[54] With double-slide plate-holders of improved design, he obtained with Seed 23 plates, 'perfectly round star-images 1″·03 (of arc) in diameter after an exposure of 11 hours.'[55] In 1910, with exposures of 4 hours, E. A. Fath obtained images of stars of the twentieth magnitude.

With the 60-inch reflector well in hand, the Mount Wilson astronomers heard, with some surprise, that J. D. Hooker, a Los Angeles business man, had agreed to provide $45,000 for the main mirror of a second large reflector on the mountain. Hooker had in mind, at first, an 84-inch reflector but, feeling that this was likely to be surpassed in a few years, increased his grant to permit the making of a 100-inch.*

Only the St Gobain glassworks in France agreed to undertake the casting and annealing of so large a disk. After preliminary trials, they produced a disk 101 inches across, 13 inches thick and over 4½ tons in weight.[56] Three melts were required to fill the mould, each one containing 1½ tons and all poured into the mould in quick succession. This procedure caused innumerable tiny air-bubbles to become trapped in the interior, while the long annealing process gave rise to the partial devitrification of the glass and consequent loss of strength and rigidity.[57]

When the disk arrived at Pasadena in 1908, Hale and Ritchey thought it unlikely that it would take, let alone retain, a good surface figure. The management of

* The remainder of the $600,000, the cost of the complete installation, was granted by the Carnegie Institution.

St Gobain thereupon signified their willingness to bear the loss and to construct another disk. Ritchey went to Paris, and there discussed the best arrangements for completing the work, and a new glass furnace and an annealing oven were constructed.[58] In the meantime, smaller disks up to 60 inches diameter arrived at Pasadena (for use in the telescope and for testing the 100-inch mirror) and a 100-inch grinding machine approached completion. But all the costly efforts of the St Gobain glassworks failed to secure a disk better than the first. Finally, European war intervened and put a stop to their experiments.

The succession of failures led Hale to return to the rejected glass and he noticed that the bubbles did not approach the surface close enough to interfere with the formation of an optical surface. A. L. Day made tests to determine the chances of using the disk and declared that the layer of bubbles strengthened rather than weakened it.[59] In 1910, therefore, Ritchey went ahead and, for five years, he and two assistants made the work of grinding and polishing some 7800 square inches of glass their main concern.

As with the 60-inch mirror, every precaution was taken to ensure the constancy of the temperature of optical room and testing hall. To prevent the air stratifying into layers of different temperature, electric fans were distributed at various points to stir the filtered air. Woollen wadding enclosed in canvas covers was used to protect the back and sides of the mirror.[60] Parabolization was done on a large polishing machine with polishers of 90°-sector form, one of 850 and the other of 415 square inches area. As this work proceeded, the mirror was brought to the vertical and tested at the centre of curvature, about 84 feet away, by the Foucault method.[61] Tests were made every morning following a day's figuring and were limited to summer months. Winter required the use of artificial heat in the optical shops and tests became difficult and uncertain; this time of the year was spent in rough-grinding the two auxiliary mirrors. The work of changing the figure from spherical to paraboloidal took nearly a year. With the completion of a 60-inch plane silvered mirror, tests were made at the focus and proved invaluable for detecting and correcting zonal errors in the general curvature. At the last stages, small figuring hand-tools were used to 'soften-down' several slightly prominent zones too narrow for treatment by the machine. Photographic Hartmann tests* were made both to check the visual Foucault pictures and to provide a permanent record of the state of the surface.[62]

The Hooker telescope is mounted English fashion in a rectangular steel cradle which forms most of the polar axis. This type of mounting prevents observation of circumpolar stars but, at the time, this restriction was not considered a great disadvantage. More important is the rigidity provided by this design (the choice of F. G. Pease), and the smooth motion obtained by relieving the bearings in the north and south pedestals by a mercury flotation system.

The complete instrument weighs 100 tons yet it can be made to follow stars with great accuracy. As with the 60-inch, the smaller parts and accessories were made in the Pasadena shops. The driving clock alone required more than a ton of bronze

* Hartmann, J., *Zeit. f. Instrument.* **24**, p. 1, 1904, described his method of test for the correction of the 80-cm objective at Potsdam. By this method residual aberrations can be assessed quantitatively.

castings and nearly $1\frac{1}{2}$ tons of iron castings in addition to the 2-ton driving weight.[63] This clock is mounted in a room below the south pier and actuates a 17-foot driving wheel.

The optical design embodies both Newtonian and Cassegrain forms. The Newtonian focus is 42 feet and is used for direct photography and spectrographic work. A special observing platform permits observations at this focus. Two Cassegrain forms are available. In the first, a diagonal plane mirror reflects the convergent light to focus at a point 11 feet from the base of the tube and gives an equivalent

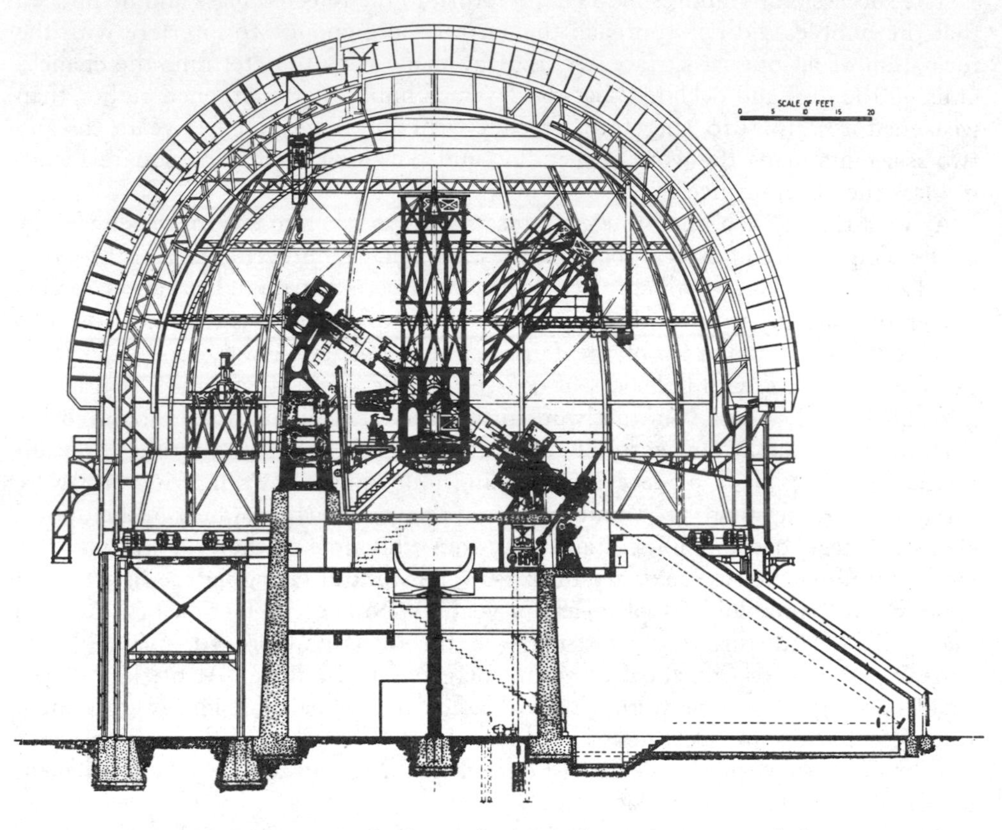

Fig. 134—Diagram of the 100-inch Hooker telescope and dome

(Mount Wilson and Palomar Observatories)

focus of 134 feet. In the second, a polar coudé, the beam is reflected down the polar axis to give an equivalent focus of 250 feet.[64]

All motions of telescope, dome and shutters are electrically controlled and involve over 30 electric motors. The dome is of thin sheet-steel and 100 feet in diameter. Like the walls, also of sheet steel, it is of double thickness, with a generous air-space in between.[65] One early difficulty was the slow rate at which the 100-inch mirror adjusted itself to changes in observatory temperature. The mirror required some 24 hours to cool 10°C, during which period it was useless for

Fig. 135—The dome of the 100-inch telescope from the south

(*Mount Wilson and Palomar Observatories*)

accurate observations. The addition of a cold-water pipe system behind the disk and maintained at night temperature proved of little benefit owing, chiefly, to the uncertainty as to what the night temperature would be.[66] Present precautions are to keep the dome closed all day and to house the mirror in a cork-lined insulating chamber. A slight falling off in definition was at one time noticed when the tube was inclined in certain directions and, at first, the mirror was blamed. Pease suspected the mirror's flotation system and eventually eliminated the trouble, which he traced to friction between the back of the disk and its supports.[67]

Preliminary tests took the form of comparisons between the Hooker and 60-inch telescopes, and mainly at the 134-foot Cassegrain focus of the former, by photographing stellar spectra on the same nights with similar spectrographs.[68] These tests, and others which followed, revealed the great gain in light-grasp due to a nearly three-fold increase in mirror area. Regular research with the 100-inch began in 1918 and, since then, has covered a wide field. The instrument brought faint stars, nebulae, and star clusters within photometric and spectrographic range. Direct photography of star clusters by Harlow Shapley has brought new knowledge as to their size, distance, and space distribution. In the hands of Russell,

Pease, Merrill, and Joy—to mention but a few—and supplemented by spectrographs and photocells, the 100-inch has provided a mass of information concerning the temperatures, composition, motions, intrinsic brightnesses, distribution, and distances of stars. The Cassegrain foci have been utilized for radiometric observations of the moon and planets, and the great light-grasp for the photography of very distant nebulae. The work of Humason, Hubble, and Baade in the realm of galaxies has yielded surprising results. Observations with both instruments enabled detailed surveys of neighbouring galaxies to be made and later their distances became known with greater certainty than before. Red shifts in the spectra of extra-galactic nebulae were recorded which, if ascribed to the Doppler effect, denote large velocities of recession. Hubble showed the direct proportionality of the apparent velocity and the distance, a relationship which, if correct in interpretation, has important cosmological implications.

Many auxiliary instruments have been developed from the experimental stage at Mount Wilson in consequence of these researches. The bolometer, invented by S. P. Langley in 1880[69] and improved by C. G. Abbot,[70] has allowed the critical study of the distribution of energy in the solar spectrum, while the perfection of the thermocouple by S. B. Nicholson and E. Pettit has made possible similar work on stellar spectra. J. Stebbins and A. E. Whitford pioneered in the improvement of potassium, caesium and other photocells at Mount Wilson, and T. Dunham developed a photoelectric microphotometer for solar spectrum studies. An important outcome of some of the photocell investigations of Stebbins and Whitford was a photometry of the Andromeda nebula.[71] They showed that this object is far more extensive than was formerly realized—that it is, in fact, comparable in size to our own galaxy.

Another important auxiliary employed at Mount Wilson is the Rayton lens. By the 1930s the 100-inch had reached, for galaxies, the limits of its spectrographic range. It became necessary to consider faster spectrograph camera lenses, the aperture ratio of which determines the speed of the telescope-spectrograph combination. Exploitation of this principle by V. M. Slipher had already led to the first radial-velocity determination for the galaxy M31 in Andromeda.[72] Slipher attached a short-focus camera to the low-dispersion spectrographs of the 24-inch Lowell refractor and, by 1925, had determined the velocities of forty-one galaxies.[73] To extend this work to fainter systems, W. B. Rayton of the Bausch and Lomb Optical Company undertook the design of suitably modified 16-mm and 4-mm microscope objectives. Enlarged eight-fold, the second of these objectives had a numerical aperture of 0·85, equivalent of *f*/0·6.[74] It was used dry and enabled Humason to obtain spectrograms of nebulae distant some 140 million light-years, the red-displacements in which indicated velocities of recession of over 14,000 miles per second.

Since the second Rayton lens represented the limit for aperture ratio for a dry lens, attention turned to the adaptation of a basically homogeneous 2-mm microscope objective. In 1935, R. J. Bracey of the British Scientific Instrument Research Association designed an enlarged type of 2-mm microscope objective.[75] Messrs R. and J. Beck Ltd made the objective, and Messrs Ross Ltd the appropriate prism and collimator system. After satisfactory laboratory tests in England, the completed spectrograph was shipped to the Mount Wilson Observatory. The camera lens was of numerical aperture 1·4, equivalent to f/0·36, and represented a gain in rapidity over the 4-mm Rayton system of approximately 25 . 9. This increase in aperture ratio required the emulsion to be in oil-immersion contact with the rear lens surface. But for the deterioration in the purity and darkness of the night sky at Mount Wilson (due to industrial developments near Los Angeles), the Bracey lens would double the range for spectrographic work on distant galaxies.

The effective field at the primary focus of the 60-inch and 100-inch mirrors is comparatively small owing to coma. To enlarge the field, F. E. Ross designed

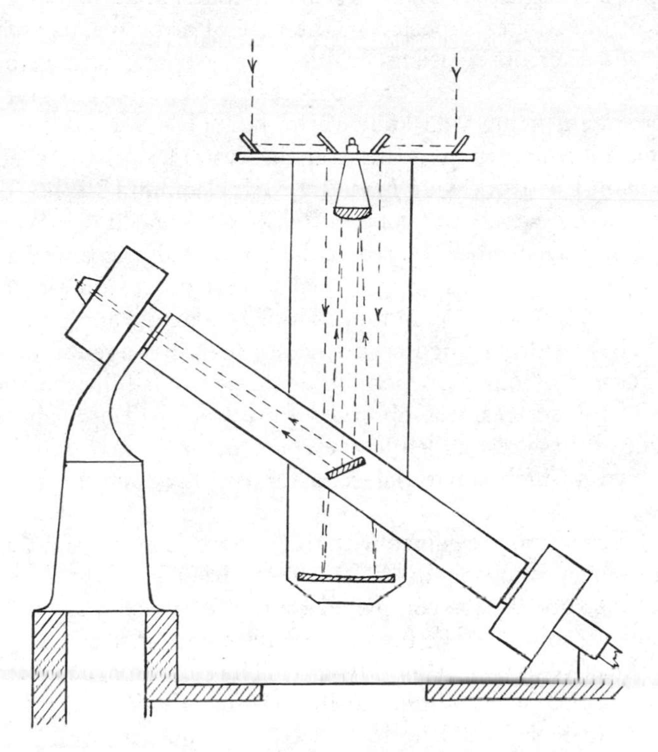

Fig. 136—Optical arrangement of the 20 foot interferometer

Diagram showing the interferometer attached to the 100-inch Hooker telescope

a new-type zero-power correcting lens which is placed in front of the photographic plate. This corrector has negative coma equal to the positive coma of the paraboloid, but its insertion causes some loss of light and introduces a little spherical aberration. The Mount Wilson optical shop has provided the 60-inch telescope with an 8-inch zero-power corrector and the 100-inch with a 7·5-inch triple- and 12-inch single-lens corrector.[76] Similar lenses are used at the primary focus of the 200-inch Hale telescope.

The 100-inch telescope has allowed the measurement of stellar diameters by interferometers. The interferometer method grew out of a suggestion by A. A.

Michelson in 1890[77] to use the interference of light for measuring the diameters of distant objects. He used it to determine the diameters of Jupiter's satellites and further trials suggested that it could be applied to the 100-inch telescope to measure stellar diameters. In 1920, an interferometer consisting of two plane mirrors, separated by a distance of 20 feet, was attached to the open end of the tube. From these mirrors, starlight was reflected to two similar mirrors 4 feet apart and thence to the main mirror, to be eventually reunited in the eyepiece in the form of interference fringes. During preliminary tests, fringes were obtained for the spectroscopic binary *α* Aurigae (Capella). It was found that Capella gave rise to the disappearance of the fringes expected in the case of a double star of nearly equal components.[78] Further observations enabled both separation and position angle to be determined to an accuracy of 1 per cent.[79] The spectrographic elements of the Capella system being well known, the new observations gave a good idea of the orbit, distance, and mass of the Capella system. All the preliminary tests and adjustments in this work were made by Michelson and Pease. Upon the former's return to the University of Chicago, Pease commenced regular work with the interferometer. On December 13, 1920, he measured the angular diameter of *α* Orionis (Betelgeuze). The fringes vanished at a mirror separation of 10 feet while settings of 21 and 12 feet were found for *α* Boötis (Arcturus) and *α* Scorpii (Antares) respectively. The resulting angular diameters are: Betelgeuze 0″·047, Antares 0″·040, and Arcturus 0″·022.[80] A definite decrease in visibility, without complete disappearance of the fringes, was observed for *o* Ceti, *α* Tauri, and several other stars. The value of these results led to the construction of a 50-foot interferometer carried on its own 36-inch equatorial reflector with 15-inch interferometer plane mirrors.

As a result of the experience gained with the Snow telescope, Hale, in the summer of 1906, aided by suggestions from C. G. Abbot, drew up plans for a vertical solar telescope. The instrument was completed by the following year and consists of a 65-foot vertical steel tower, at the top of which is a coelostat and secondary mirror and a 12-inch objective of 60 feet focal length. The solar image is formed at a short distance above ground level, where the light enters the slit of a Littrow grating spectrograph, of 30 feet focal length, housed in a concrete well nearly 9 feet in diameter.[81] There are several novel features. The glass mirrors are thick so as to reduce distortion when heated by the sun, the subterranean chamber ensures nearly constant temperature for the spectrograph, and the coelostat is well above disturbing air currents found near the ground. With these refinements, the photographs obtained are superior to those taken with the Snow telescope and 18-foot Littrow spectrograph. 'Double lines, which appear single in the previous photographs' Hale wrote, 'are now clearly resolved, and a great number of additional faint lines, particularly those of flutings, are recorded.'[82]

To obtain the longer focal length that would provide a solar image large enough for the close study of sunspots, Hale designed a 150-foot tower telescope.[83] At this height, the question of stability had to receive close attention. To ensure protection from the wind, the only source of vibration, each of the steel members of the tower is enclosed inside the corresponding hollow member of a second skeleton tower mounted on independent foundations. The journey to the top is accomplished

Fig. 137—The 150-foot tower telescope at Mount Wilson Observatory

(Mount Wilson and Palomar Observatories)

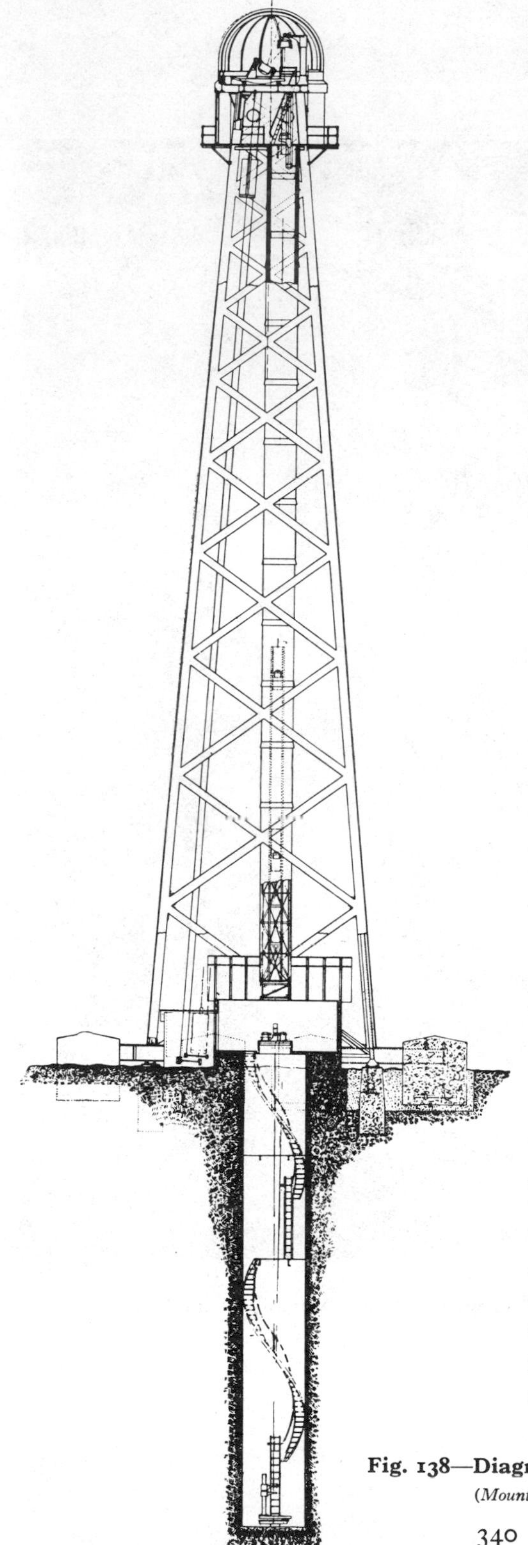

by riding a small electric elevator. The combined spectrograph and spectroheliograph, of 75 feet focal length, are mounted inside a 10-foot diameter concrete-lined well. Nearly all the optical parts of this unique telescope were made in the Pasadena optical shops under Ritchey's supervision. The mirrors are 12 inches thick, and to prevent distortion by heating, the mirror sides are enclosed in a closely fitting water-jacket through which a stream of water is kept circulating. The lens is a triple objective of 12 inches aperture designed by Hastings to reduce secondary spectrum to a minimum.[84] The 75-foot focus spectrograph lens is of the same form. From telescope lens to spectroheliograph extends a 5½-foot square-sectioned tube, protected from the sun by canvas louvres. Its object is to protect the light path from disturbing air currents. The final solar image is 17 inches in diameter at the 150-foot focus, but can be reduced by swinging in objectives of 30 and 60 feet focus situated further down the tower.

Preliminary tests were satisfactory, save for variations in the focal length of the objective during the day and signs of astigmatism about noon. Increase in focal length amounted, in some instances, to 5 feet in a few hours. To remedy these changes, water-jackets fed by running water were extended round the sides and back of the coelostat mirror and its secondary. As soon as thermal equilibrium was established, the astigmatism vanished and the change in focal length was reduced to a fraction of its former extent.[85] Fears as to the instability of

Fig. 138—Diagram of the 150-foot tower telescope

(Mount Wilson and Palomar Observatories)

Fig. 139—Mount Wilson Observatory from the air

(*Mount Wilson and Palomar Observatories*)

the tower proved groundless. In a wind of 20 miles per hour, photographs of the flash spectrum were obtained. Definition proved superior to that of the Snow telescope, and no convection currents of any consequence appeared in the vertical tube. With the 75-foot spectrograph and a Michelson grating, the distance between the D lines in the third-order solar spectrum is over an inch.[86]

> The combined spectrograph and spectroheliograph [Hale reported in 1912][87] is also designed for the photography of the spectrum of the chromosphere ('flash spectrum'), the comparative study of the spectra of centre and limb, the measurement of the solar rotation, the determination of the pressure at different levels in the solar atmosphere, and for various other purposes. The auxiliary apparatus includes a reflecting slit, especially adapted for the photography of the spectra of pores and other minute regions; a device for securing uniform density of the spectra of umbra, penumbra, and photosphere in a single exposure; . . . a parallel-motion apparatus, to facilitate accurate orientation of the instrument; polarizing apparatus, with simple and compound quarter and half wave plates for various wave-lengths; and an electric arc for comparison spectra.

At present, both tower telescopes and the Snow instrument are in regular use, for the latter still possesses advantages over the former. Due to the exclusive use of mirrors, the Snow telescope has perfect achromatism and its angular aperture is double that of the 60-foot tower object-glass. In 1917, it was equipped with a vertical spectrograph mounted in an underground concrete-lined chamber for investigations on the relative positions of solar and arc lines. With all three solar telescopes in constant use, it also became desirable to complete the 60-foot tower along the same lines as those of the 150-foot. In 1914, an outer tower, dome and vertical tube were provided[88] and the permanent concrete laboratory below now houses a 30-foot spectrograph and 13-foot spectroheliograph. The telescope was again remodelled in 1939–1940.

One further invention of Hale's remains to be described—the spectrohelioscope, designed and perfected after his resignation from the directorship of Mount Wilson Observatory in 1923.[89] A series of nervous breakdowns, due to years of overwork of body and mind, forced this ' retirement ' upon him. Rest he could not and, in Pasadena, he had built, largely at his own expense, a solar observatory for the continuation of his researches. The spectrohelioscope, born in this laboratory, is an ingenious offspring of the spectroheliograph. $H\alpha$ radiation is isolated and both slits are vibrated in unison relative to the solar image on the first slit. Behind the second slit, an eyepiece replaces the photographic plate. Owing to persistence of vision and the small amplitude of vibration, successive images produced by the motion of the first slit are blended into one composed of monochromatic light. For the arrangement to be effective, high dispersion is necessary, so that the $H\alpha$ line is at least 0·008 inch wide.

In Hale's original instrument, the slits were carried by a light bar which oscillated about a pivot at its centre. Most instruments now have fixed slits and eyepiece, while a portion of the solar image is made to vibrate across the slits by refraction through square, rotating glass prisms. In this modification, due to J. A. Anderson, the prisms are mounted in line on a spindle, one immediately in front of each slit, while spindle and slits are rapidly rotated by an electric motor. This arrangement enables the measurement of different velocities in the line of sight and so gives a picture of the movement of hydrogen flocculi.

While in Pasadena, Hale encouraged international co-operation in the regular observation of solar phenomena. He always held amateur work in high esteem and, in various articles, showed how the average amateur could make his own spectroheliograph or spectrohelioscope.

Hale died in his Pasadena home on February 21, 1938, following many years of illness.

> A natural leader of men [his friend, Dr S. B. Nicholson, writes][90] his eagerness and enthusiasm never failed to make those with whom he talked wish to contribute something to his projects. Men of means were inspired to co-operate financially in the promotion of cultural achievement and scientists in all fields were enlisted under his leadership in great co-operative organisations. The Yerkes Observatory of the University of Chicago, the Mount Wilson Observatory of the Carnegie Institution of Washington, the California Institute of Technology, the *Astrophysical Journal*, the International Astronomical Union and the National Research Council are direct results of his ability to organise and to stimulate others to work for the advancement

Fig. 140—George Ellery Hale

(*Mount Wilson and Palomar Observatories*)

of knowledge. He was indeed a true citizen of the world, yet, with all his world-wide activity, found time to take an active, although usually a silent, part in the cultural development of the city where he lived.

REFERENCES

1 Adams, W. S., *Ap. J.*, **87**, p. 371, 1938.
2 Dunham, T., *M.N.R.A.S.*, **99**, p. 323, 1939.
3 Adams, *op. cit.*, p. 373.
4 *Vide* Ref. 2.
5 Hale, G. E., 1908, *The Study of Stellar Evolution*, p. 13.
6 *Ibid.*, p. 82, footnote.
7 Young, C., *Journal Franklin Institute*, **60**, p. 334, 1870.
8 Hale, G. E., *Ap. J.*, **23**, p. 93, 1906.
9 Meudon *Annales*, iv, p. 19, 1910; *M.N.R.A.S.*, **73**, p. 327, 1913.
10 Hale, Ref. 5, p. 83. For an account of Hale's early spectroheliographs *vide Astronomy and Astrophysics*, **12**, p. 241, 1893, also *Ap. J.*, **23**, p. 54, 1906.
11 Hale, Ref. 5, p. 84.
12 *Ibid.*, p. 86.
13 *Ibid.*, pp. 86–87.
14 *Ibid.*, pp. 88–89.
15 *Ibid.*, pp. 94–95.
16 *Ibid.*, p. 94.
17 *Ibid.*, p. 104.
18 *Ibid.*, p. 109.
19 Hale, G. E., *Ap. J.*, **17**, p. 314, 1903.
20 Adams, *op. cit.*, p. 374.
21 During 1889–1890. *Vide* Bailey, S. I., 1931, *History and Work of Harvard Observatory*, pp. 56–58.
22 Hale, Ref. 5, p. 126.
23 *Ibid.*, pp. 131–138. Hale and Ellerman, F., *Ap. J.*, **23**, p. 54, 1906. Hale, *Ap. J.*, **24**, p. 61, 1906.
24 Ref. 5, p. 163.
25 *Ibid.*, p. 140.
26 *Ibid.*
27 *Ibid.*, pp. 141–142.
28 Hale, G. E., *Ap. J.*, **28**, p. 315, 1908. Carnegie Institution, Washington, Mount Wilson Observatory *Year Book*, **7**, p. 150, 1908.
29 Mount Wilson Observatory *Year Book*, **4**, pp. 60–61, 1905.
30 Dunham, *op. cit.*, p. 325.
31 Hargreaves, F. J., *M.N.R.A.S.*, **107**, pp. 36–37, 1947; Antoniadi, E. M., *J.B.A.A.*, **39**, pp. 326–328, 1929.
32 Ritchey, G. W., *Ap. J.*, **14**, p. 217, 1901.
33 *Ibid.*, p. 167.
34 *Ap. J.* **12**, p. 352, 1900.
35 Ritchey, G. W., *Smithsonian Contributions to Knowledge*, reprint of part of Vol. 34, 'On Modern Reflecting Telescopes', 1904, pp. 10–11.
36 *Ibid.*, p. 6.
37 *Amateur Telescope Making*, 1933, p. 74.
38 Ritchey, Ref. 35, p. 2.
39 *Ibid.*, pp. 7–10.
40 Hale, Ref. 5, pp. 223–224; Mount Wilson Observatory *Year Book*, **5**, p. 83, 1906.
41 Hale, Ref. 5, p. 226.
42 *Ibid.*, pp. 226–227.
43 *Ibid.*, pp. 227–228; Mount Wilson Observatory *Year Book*, **5**, pp. 83–84, 1906; **6**, p. 148, 1907.
44 For early account of the coudé form *vide* Turner, H. H., 1900, *Modern Astronomy*, pp. 125–131.
45 Mount Wilson Observatory *Year Book*, **9**, p. 157, 1910.
46 Hale, Ref. 5, p. 228.
47 Mount Wilson Observatory *Year Book*, **5**, p. 75, 1906.
48 *Ibid.*, **4**, p. 68, 1905.
49 *Ibid.*, **5**, pp. 66–67, 1906. Hale, Ref. 5, p. 137.
50 Mount Wilson Observatory *Year Book*, **5**, p. 75, 1906.
51 *Ibid.*, **8**, p. 170, 1909.
52 *Ibid.*, **11**, pp. 196–197, 1912.
53 *Ibid.*, pp. 197–198.
54 *Ibid.*, **8**, p. 172, 1909.
55 *Ibid.*, **9**, p. 165, 1910.
56 Joy, A. H., *Publ. Astr. Soc. Pac.*, **39**, p. 14, 1927.
57 Ritchey, G. W., *Trans. Opt. Soc.*, **29**, p. 197, 1928.
58 Mount Wilson Observatory *Year Book*, **8**, p. 179, 1909.
59 *Ibid.*, **10**, pp. 181–182, 1911.
60 *Ibid.*, **13**, pp. 281–282, 1914.
61 *Ibid.*, **14**, p. 284, 1915.
62 *Ibid.*, **12**, p. 238, 1913.
63 *Ibid.*, **13**, p. 282, 1914.
64 Joy, *op. cit.*, pp. 14–16.
65 Mount Wilson Observatory *Year Book*, **12**, pp. 239–240, 1913.
66 *Ibid.*, **18**, p. 260, 1919.
67 Pease, F. G., *Publ. Astr. Soc. Pac.*, **44**, pp. 308–312, 1932.
68 Mount Wilson Observatory *Year Book*, **17**, p. 217, 1918; **18**, pp. 261–263, 1919.
69 *Proc. Am. Acad.*, **16**, p. 342, 1881.
70 Abbot, C. G., *Ap. J.*, **18**, p. 1, 1903.
71 *Proc. Nat. Acad. Sc.*, **20**, p. 93, 1934.
72 *Lowell Observatory Bulletin*, No. 58, 1914.
73 *Pop. Astr.*, **23**, p. 21, 1915; *Ap. J.*, **61**, p. 353, 1925.
74 *Ap. J.*, **72**, p. 59, 1930.

[75] *Nature*, p. 227, pp. 643–5, 1936. Hale, G. E., *Ap. J.*, **82**, p. 111, 1935. Bracey, W. D., *Ap. J.*, **83**, pp. 179–186, 1936.

[76] Mount Wilson Observatory *Year Book*, **35**, p. 191, 1936.

[77] Michelson, A. A., *Phil. Mag.*, series 5, **30**, p. 1, 1890. *Ap. J.*, **51**, p. 263, 1920; **53**, p. 249, 1921.

[78] Mount Wilson Observatory *Year Book*, **19**, p. 252, 1920.

[79] *Ibid.*, **20**, p. 280, 1921.

[80] *Ibid.*, p. 278.

[81] *Ibid.*, **6**, p. 149, 1907; *Mount Wilson Observatory Contributions*, **1**, pp. 283–291, 1905–1908.

[82] Mount Wilson Observatory *Year Book*, **7**, p. 130, 1908.

[83] *Ibid.*, **8**, pp. 179–180, 1909.

[84] *Ibid.*, **9**, p. 176, 1910.

[85] *Ibid.*, **11**, p. 177, 1912.

[86] *Ibid.*, p. 178.

[87] *Ibid.*, p. 179.

[88] *Ibid.*, **13**, p. 283, 1914.

[89] Hale, G. E., *Ap. J.*, **70**, pp. 265–311, 1929. Dunham, *op. cit.*, p. 327.

[90] Nicholson, S. B., *J.B.A.A.*, **48**, pp. 318–319, 1938.

CHAPTER XVII

> Much is to be hoped from the new kinds of glass now being made for optical purposes at Jena, Germany, as the result of experiments conducted by Professor Abbe at the expense of the German Government. Cooke & Son, English opticians, since 1894 advertise 'Photovisual lenses which are practically aplanatic' and offer to make them as large as twenty inches in diameter. . . . Possibly a new era of telescope-making will open with the coming century.
>
> C. A. YOUNG

EARLY in the present century, astronomical telescopes of increasing size began to emanate from the old German university town of Jena in Thuringia, where the Zeiss Foundation was working in close collaboration with Schott's glassworks. Under the progressive leadership of the younger Zeiss and Ernst Abbe, the Carl Zeiss Foundation concentrated on the manufacture of precision optics, while the Otto Schott glassworks, well subsidized by the Prussian government, produced optical glass in great quantity. In 1919, Schott surrendered his ownership of the glassworks and became a member of the Zeiss organization which, in those days and owing to World War I, was the largest European centre for optical instruments. Between the two world wars, the foundation became unique in the history of business and technical organization. Its early success was due, not only to the receipt of generous state subsidies, but also to the liberal and self-sacrificing spirit of its founders. Here were business interests controlled by expert financiers, technical controls exercised by scientists, and prominence given to research by leading authorities in lens-design and glass manufacture. Here was an organization aiming to flood the world's instrument market with telescopes and optical apparatus which would surpass all others in design, finish, and performance.

The Zeiss concern was founded by Carl Zeiss, a toymaker, who set up his first small workshop in 1846. With only 100 thaler, borrowed from his father, a master-toymaker, Zeiss purchased a small stock of magnifiers and optical accessories. In 1847, he moved to larger premises in the same town, and repaired microscopes and laboratory instruments for Jena University. His assistant during these early years was A. Löber, who introduced interferometric methods into the workshop and, by his industry and regard for precision, did much to make the Zeiss venture a success.[1]

In 1861, Zeiss made his first compound microscope, but this and his subsequent instruments offered no advantages over the standard product of the time. He felt the need for a collaborator who understood optical computation and who could thereby introduce new and better lens systems. Such a partner was Ernst Abbe, doctor of philosophy of Göttingen University and son of a foreman spinner. Zeiss met Abbe in 1875,[2] when the latter possessed no particular optical knowledge, but his scientific training and interest in research enabled him to master Zeiss's immediate problems. He went further. He increased the accuracy of the work, developed methods of testing optical elements, became master of a valuable body of theoretical knowledge, and sought for improvements in optical glass. So urgent indeed became the demand for a wider variety of glass that, at one time, Abbe experimented with fluid media, but without practical success.

> For years [he writes][3] we combined with sober optics a species of dream optics, in which combinations made of hypothetical glass, existing only in our imaginations, were employed to discuss the progress which might be achieved if the glass makers could only be induced to adapt themselves to the advancing requirements of practical optics.

In 1879, however, Abbe met Otto Schott. Schott's father owned a plate-glass factory in Westphalia and his interest in the chemistry of glass led him to introduce lithium glass. This new glass had no optical use but, under Abbe's influence, Schott undertook fresh experiments which led, eventually, to the introduction of borate and phosphate glasses. In 1884, the glassworks of Schott and Sons was established by Abbe, Schott, and Zeiss (father and son).[4]

Before World War I, the name of Zeiss on an optical instrument was generally taken as a hall-mark of optical and mechanical perfection. Already the Germans had begun their attack on the world market. Single, double, and triple photographic refractors on original multi-counterpoised mountings were well advertised. These were followed by the manufacture of three 25·5-inch refractors, a visual for the Berlin-Babelsberg Observatory and one for the University of Belgrade, while the third, a photographic with 15-inch guide, went to the Imperial University Observatory near Tokyo. A 23·6-inch visual–photographic twin refractor for the Bosscha Observatory at Lembang, Java, followed, while telescopes of 15·7-inch diameter and under had world-wide distribution. Practically every German observatory, university, museum, and college benefited by the outpouring of Zeiss instruments.

M. Wolf's work in the discovery of nebulae was extended in 1905 by the addition of a 28-inch Zeiss reflector on the pattern of the 60-inch Mount Wilson telescope. In 1913, a 39·4 inch reflector was installed at the Hamburg Observatory, Bergedorf, while a similar instrument went to the Royal Observatory at Uccle, near Brussels. A third reflector of this size, but mounted on a polar axis, was installed at the Royal Observatory at Merate, near Como, while in 1927, a 49-inch on a weight-stress compensated equatorial mounting went to the University Observatory at Berlin–Babelsberg. This instrument, a Newtonian–Cassegrain, is still the largest reflector in Europe.[5]

The mounting of refractor tubes so that the declination axis comes near the eyepiece, a design made possible by weight-stress compensating systems, means that the movement of the eyepiece is small compared with its sweep on standard instruments. This mounting, introduced in 1906 with a 12-inch refractor for Urania, Zurich's public observatory, was found particularly suitable for telescopes set aside for public use. Refractors of this type are now installed at the Deutsches Museum, Munich, the Planetarium and Hall of Science in Griffith Park, Los Angeles, and the Franklin Institute, Philadelphia.

Zeiss specialized in the construction of large and small astrographs with four-lens objectives mounted singly and in duplicate. Sizes ranged from star cameras

Fig. 141—Zeiss 28-inch reflector of the observatory at Heidelberg

of 2·4 inches (60 mm) aperture, to giants of 15·7 inches (400 mm) aperture. Great use was made of the bent or offset column which afforded free mobility to the tube around the polar and declination axes. Zeiss astrographs, spectrographs, and other instruments found ready welcome in observatories as far apart as Uccle, Belgrade, Leningrad, and Tokyo. The tower telescopes at Potsdam (the Einstein tower) and Tokyo are of Zeiss construction, also, during World War II, the four smaller tower telescopes at stations in the Alps.

In his survey of German astronomy during the war,[6] G. P. Kuiper tells of the observatory which Hitler promised to present to Mussolini. As usual, Hitler's promise meant nothing. Most of the equipment, among it a 24-inch refractor and an astrograph, was completed in 1944, but never left Germany. The domes went to Italy in 1942, only to be brought back two years later. The most interesting item, an unfinished 1·5-metre Schmidt camera, was finished at Jena by the Russians for the new Pulkowa Observatory.

Important Zeiss-equipped observatories are at Bergedorf, near Hamburg, Potsdam and Neubabelsberg, both on the outskirts of Berlin, and Heidelberg. Both the Hamburg and Potsdam observatories have concentrated for many years on stellar photometry and on the classification of the spectra of stars in special regions of the sky. Hamburg, like Bonn, has taken its full share in the repetition of the *Astronomische Gesellschaft* star catalogue begun in 1863. A 6-inch Zeiss astrograph

Fig. 142—Zeiss $15\frac{3}{4}$-inch reflector at the Innsbrück Observatory

Notice the two 3-inch ultra-violet cameras, and the 7-inch coudé guide telescope and a 3·4-inch Zeiss *A.G.* zone camera are used for this work. The 39 4-inch reflector is still in use, together with a 23·6-inch reflector and the Zeiss 13·4-inch Lippert astrograph. The observatory also possesses a 23·6-inch Repsold-Steinheil refractor and several Schmidt-type cameras, the largest being a 32-inch × 48-inch *f*/3 Schmidt installed in 1954. As we shall see directly, Hamburg was the scene of the first practical development of this remarkable camera. Reference has already been made to the work done at Potsdam under Vogel's leadership. The observatory's largest instrument is still the 31·6-inch Steinheil-Repsold astrograph, with 19·7-inch visual guide, orginally installed in 1899. There is also a smaller version of this astrograph—an 11·8-inch with 9·4-inch visual guide. The Berlin–Babelsberg Observatory at Neubabelsberg operates the 49-inch Zeiss reflector, a 25·5-inch Zeiss refractor and a 15·8-inch astrograph by Toepfer. All three observatories suffered little damage during the last war. Both Potsdam and Berlin–Babelsberg, in common with the original Zeiss works at Jena, are in the Soviet zone.

Heidelberg Observatory was the scene of Wolf's important pioneer work in asteroid discovery and stellar and nebular photography. Using small portrait lenses, Wolf took photographs which delineated faint and extended nebulosity hitherto unrecorded except by E. E. Barnard. Like Barnard, he demonstrated the existence of dark patches in the Milky Way, now known to represent clouds of obscuring matter. Wolf also discovered the North America nebula[7] and extensive nebulosities near the Pleiades.[8] His discovery of about 5000 nuclear nebulae,[9] now known to be distant galaxies, was recognized by the award of the Gold Medal of the Royal Astronomical Society. With the 28-inch reflector, Wolf pioneered in nebular spectrography, using at times exposures of nearly eighty hours spread over several weeks. His spectrograms of M31 and other nebulae now known to be extra-galactic revealed a solar-type spectrum and indicated, for the first time, that these objects were great but remote systems of stars.

The early years of the present century also saw the continued erection of large telescopes by Sir Howard Grubb, four of which had an aperture of 24 inches. The first, a photographic refractor with 18-inch guide telescope and 24-inch objective prism, was presented to the Royal Observatory, Cape of Good Hope, by Frank McClean, a London surgeon and an able amateur astronomer.[10] Although ordered in 1894, difficulties over the mounting and electrical attachments delayed the installation of the instrument. Even the object-glass had to be returned for refiguring, and the telescope was not ready for use until 1901. Gill proved as capable an engineer as he was astronomer and personally supervised every detail of the erection of the 'Victoria telescope', for so McClean had christened his munificent gift. The telescope has played, and continues to play, an important part in the determination of radial velocities, proper motions, and parallaxes of southern stars. Grubb followed it with a similar telescope for the Radcliffe Observatory, Oxford, a 24-inch reflector for W. E. Wilson of Westmeath, and a 24-inch visual refractor for the National Observatory, Santiago, Chile. The Radcliffe and Wilson telescopes are now both at the Mill Hill Observatory of the University of London.

A larger undertaking by Grubb at this period was a 26½-inch visual refractor for the Union Observatory, Johannesburg, but war delayed its construction and shipment, and it was not erected at Johannesburg until 1925. With this instrument R. T. A. Innes, an accomplished double star observer, initiated a programme of micrometrical measures of double stars, a programme which continues under the direction of the present Union astronomer, W. H. van den Bos.

One of Grubb's last projects, and certainly the largest, was the 40-inch reflector and 32-foot dome for the southern station of the Pulkowa Observatory at Simeis in the Crimea. This instrument was not erected until 1928, three years after Grubb's retirement at the age of eighty-one, by which time the firm had passed into the hands of Sir Charles Parsons. When the Nazis overran the Crimea they shipped the Simeis reflector and other equipment to Potsdam—an act of wanton piracy unmatched in the history of astronomy.

In his search for minor planets and faint variable stars, Wolf used a triple photographic refractor with two 16-inch Brashear objectives and a 10-inch objective. The Brashear plant was then a flourishing concern. Through the generosity of the philanthropist William Thaw, Brashear left the rolling mill in 1881, to set up in business as instrument-maker.[11] So many orders arrived that the new Pittsburgh workshop was hard-pressed from the start. Brashear enlisted the services of a mechanic together with those of his son-in-law, J. B. McDowell, an expert in glass manufacture. In 1887, Brashear and McDowell secured the collaboration of Charles Hastings of Yale.[12] The concern made both reflectors and refractors for American observatories. Large Brashear reflectors are the (1904) 37-inch for the Lick Observatory's station at Santiago, Chile (now the Catholic University Observatory), and a similar instrument for the University of Michigan Observatory. In 1906 they also installed a 30-inch reflector in the Allegheny Observatory. A Brashear reflector of the same size is at the observatory of the University of Illinois, and the Steward Observatory at Tucson, Arizona, has a 36-inch with mounting by Warner and Swasey. Work on refractors included a number of 12-inch object-glasses and several up to 24 inches diameter (the 24-inch at the Sproul Observatory, Swarthmore, Pennsylvania).[13] The largest and finest example of the Brashear concern's work is the Thaw memorial photographic telescope of the Allegheny Observatory at Riverview Park, Pittsburgh.

It was with the Thaw refractor that Frank Schlesinger, appointed director of the Allegheny Observatory in 1905, continued his pioneer work in parallax determination. For this investigation, the Thaw telescope was a more convenient instrument than the 40-inch Yerkes refractor, and Schlesinger's first 50 determinations had a probable error of only $\pm 0''\cdot0085$.[14] Altogether, he determined at Allegheny the parallaxes of 365 stars, took plates for the determination of several hundred more and arranged a scheme of co-operation between the seven observatories of Allegheny, Dearborn, Greenwich, Leander McCormick, Mount Wilson, Sproul, and Yerkes. After taking up the directorship of the Yale Observatory in 1920, Schlesinger began an intensive programme of stellar-parallax determination for southern-hemisphere stars with a 26-inch McDowell photographic refractor erected at Johannesburg for the purpose.[15]

For his attacks on the problems of positional astronomy, Schlesinger also pressed wide-angle cameras into service. The time had arrived when many of the northern zones of the *Astronomische Gesellschaft* star catalogue were ready for re-observation.

This work normally required accurate measurements with meridian circles. Schlesinger demonstrated how the work could be done more quickly and more accurately by small doublets each covering 25 square degrees of the sky. He did not stop at this, however, but soon had a 5·6-inch $f/14·3$ lens in operation which covered 140 square degrees on plates 6 mm thick and measuring 48 cm × 58 cm.[16] A similar 5-inch camera lens was mounted on the Yale telescope at Johannesburg, so that the work was extended to southern skies. A special machine was employed for the measurement of the hundreds of stars on these plates; some idea of the immensity of Schlesinger's work will be seen from his catalogues published between 1925 and 1943—they give the positions of over 92,000 stars.[17] Schlesinger died in 1943 but the programme he initiated continues.

The quality of Brashear's optical surfaces assisted the production of many present diffraction gratings. Henry Rowland, professor of physics at Johns Hopkins University, spent many years perfecting machines for ruling fine and large reflection gratings. He finally evolved an automatic ruling engine which was kept in a basement free from vibration and temperature changes and with which he was able to rule 6-inch gratings having 14,438 lines to the inch. To Rowland is due the invention of the concave reflection grating which focuses its spectra and so dispenses with lenses. Slit and eyepiece (or photographic plate) are all that are necessary and, in the absence of the absorptive effects of lenses, spectra can be examined much further into the ultra-violet. The different orders of spectra overlap—a property which may be used to measure wavelengths in the infra-red and ultra-violet. Since the focal curve is the same for all wavelengths, an adjustment made for visible radiations is valid in the infra-red and ultra-violet. From photographs of the solar spectrum made with the aid of a concave grating, Rowland, and later, Higgs, extended line measurements far into the ultra-violet. Rowland's now classic table of solar spectrum wavelengths records the position of about 20,000 lines and, for many years, constituted the spectroscopist's standard of reference.[18] Glass gratings proved inferior to metal ones, and Rowland found that they wore down the point of the diamond. He gave his attention to metal gratings, therefore, the specula for which were provided by Brashear.

Brashear's greatest optical undertaking was to construct the first large Canadian reflector, then the second largest telescope in the world.* We refer to the 72-inch Victoria reflector,[19] completed in 1918 for the Dominion Department of Mines and Resources. The instrument is at the Dominion Astrophysical Observatory, some 7 miles north of Victoria, B.C., and 730 feet above sea level.

> The observatory [J. S. Plaskett writes][20] is much indebted to the Warner and Swasey Co., who have made most of the large mountings in America, for the spirit in which they undertook and carried through the work. Their sole object was to produce the best possible mounting regardless of cost and no suggestion of the writer looking to improvement was refused.

Elsewhere, Plaskett states:[21]

> The symmetry and the beautiful lines of the mounting were due to the artistry of Mr. Swasey, but the details of the design were beautifully worked out by Mr. Burrell,

* For a few months, before the 100-inch went into operation, the Victoria reflector was the largest.

who showed no less than genius in developing the mechanism required for the operation of the telescope in the most suitable and efficient, and at the same time, in the simplest possible form. It was the first telescope in which the polar and declination axes were wholly carried by self-aligning ball bearings and in which the motions were electrically operated and controlled. The completed telescope, in simplicity and beauty of design, in accuracy of construction, and in speed and convenience of operation, as yet (1937) unsurpassed by any working telescope, forms a great tribute to Mr. Burrell's engineering ability and to his skill in design.

The optical components were made by the Brashear Company.

Mr. J. B. McDowell, head of the firm since Dr. Brashear's death [Plaskett continues][22] and Mr. Fred Hegemann, his chief optician, are to be highly congratulated on the perfection of figure obtained under specially difficult circumstances.

As World War II delayed the mirror of the Pretoria telescope, so World War I almost delayed the 72-inch by many years. It left Antwerp from the St Gobain glassworks only a week before war was declared. The 55-inch disk for the testing flat never left France, for the works at St Gobain were destroyed by fire only a few days later. Brashear, although in his seventy-third year and immersed in public duties, himself took part in working the 72-inch disk, his last great optical undertaking. He died in 1920, two years after the completion of the telescope.

The 72-inch Victoria telescope allowed the determination of radial velocities, spectrographic magnitudes, and parallaxes of much fainter stars than those reached by the 36-inch Lick refractor. Using aluminium-on-glass gratings and Hilger and Warner and Swasey spectrographs, much valuable work has been done on stars with 'peculiar' spectra and on the orbital elements of eclipsing variables and spectroscopic binaries. At Victoria, J. S. Plaskett pioneered in the investigation of the absorptive effects of interstellar gas, a line of inquiry continued by J. A. Pearce, C. S. Beals, and A. McKellar.[23]

In 1943, an 84-inch aluminizing chamber was installed at Victoria, and McKellar successfully aluminized the primary and 20-inch secondary mirrors. Tests with light of wavelength 3900 A showed that the aluminium coat reflected 82·6 per cent of the incident light as compared with 65 per cent from the original silver coat when freshly deposited. It was considered that the photographic limits of the instrument had thereby been extended by a further magnitude and that spectra of eleventh magnitude stars could be recorded in reasonable exposure times.

In the previous chapter, we left Ritchey working on the final stages of the optical parts of the 100-inch Hooker reflector. This task he completed in the early months of World War I and then temporarily gave his services to war production. In the large optical shop of the U.S. Army Ordnance Department at Pasadena, he supervised the training and working of men and women in making optical parts for gunsights. At the end of the war, he set up his own private laboratory in Pasadena and brought to maturity several new ideas in telescope design.

Ritchey developed an entirely new telescope, the aplanatic Ritchey–Chrétien photographic reflector. The optical system, devised jointly by Ritchey and Henri Chrétien (a French optician who spent some time at Mount Wilson studying telescope design), gives a field both larger and flatter than that produced by an ordinary reflector.[24] The arrangement is a Cassegrain, working at about *f*/7, with near paraboloidal and hyperboloidal mirror surfaces and a slightly concave photographic plate. The latter is necessary to allow for astigmatism and curvature of field; coma and spherical aberration are absent. In practice, an optically flat plate is pressed by atmospheric air pressure against a curved plate-holder by which it is distorted to the required amount. After exposing the plate, air pressure behind it is restored and the plate returns to its original flat form. The small, round images which this design is capable of giving, however, demand refinements in guiding and focusing, and attention to flexure and temperature changes in the tube and mirrors.

To ensure that his mirrors for the Chrétien telescope retained their figures, and these were by no means easy to produce, Ritchey suggested the use of cellular mirrors.[25] These consisted of a number of comparatively small plates of quartz or Pyrex glass fitted together by fine-grinding and then cemented by an extremely thin layer of Bakelite cement to form an open, square latticework. The front and back circular glass plates were cemented to this rib system, the front plate being ground and polished to the required form. Holes cut in the side of each rib allowed air to circulate within the mirror so that changes in temperature had no effect on the figure. Further advantages were lightness and ease of mounting in the telescope tube. The composite disk weighed only about one-fifth of the weight of a solid glass disk, and both tube and mounting could be made smaller and lighter than those required for a Newtonian–Cassegrain of the same size. Disadvantages arise from the difficulty of figuring the necessary surfaces on comparatively thin plates and the uncertainty of their subsequent behaviour when cemented on a lattice base. To assemble the mirror and to figure the top plate in position with reasonable accuracy is impossible owing to the varying resistance to pressure offered by the supported and unsupported areas.

In 1923, Ritchey was invited to France by the National Observatory Council at Paris. He established an astrographic laboratory at the observatory and, with the assistance of the Paris Optical Institute and the St Gobain glassworks, so developed the cellular mirror technique as to construct a 20-inch Ritchey–Chrétien reflector.

In 1930, Ritchey returned to America and, in the following year, was in charge of the construction of a 40-inch Ritchey–Chrétien telescope for the U.S. Naval Observatory, Washington. For some years, he was director of photographic and telescopic research at the observatory but, in 1936, he retired from active scientific work. He died nine years later, at the age of eighty-one.[26]

Ritchey also worked at the Schwarzschild photographic reflector, the design of which was introduced in 1905 by the German astronomer, Karl Schwarzschild.[27] This telescope has a concave mirror in place of the Chrétien convex and is best suited to aperture ratios of about *f*/3. It is, consequently, photographically faster than the *f*/7 Chrétien, but its focal images are smaller. The figures of both mirrors depart slightly from true conicoids and only the defect of astigmatism appears in the focal plane. Compared with the Ritchey–Chrétien of the same aperture ratio, Schwarzschild's accepted design is longer. The photographic plate, moreover, is situated about midway between the two mirrors and, to prevent its exposure to direct light from the sky, a sky-fog baffle or tube extension is necessary. This

Fig. 143—The 72-inch Victoria reflector

(*Warner & Swasey Co.*)

addition need be made of only light-weight material, but it requires a dome larger than is usual. Only two examples of Schwarzschild's design appear to have been attempted. The larger, of 24 inches aperture, is at the University of Indiana and was constructed by W. A. Cogshall. The other Schwarzschild mirror, of 12 inches aperture and made and mounted by C. H. Smiley and his associates, is at the Ladd Observatory of Brown University, Providence, Rhode Island.

An alternative to the Schwarzschild system is that proposed by A. Couder of the Paris Observatory. Couder computed the curves of the primary and secondary concave mirrors to give zero astigmatism on a curved focal surface.[28] This means that the plate must either be distorted to this curve during exposure or preceded by a ' field-flattener '. The latter takes the form of a weak plano-convex lens, with its plane side just in front of the plate emulsion.* A sky-fog baffle is also necessary, since the plate-holder is in the upper part of the tube.

By far the most successful short-focus photographic telescope is the Schmidt camera, first described in 1932 in a paper by Bernhard Schmidt of the Hamburg Observatory, Bergedorf.[29] Schmidt's life is as interesting as the novel optical system he introduced. He was born in 1879 on the island of Nargen, Esthonia. As a boy, he made a convex lens by grinding the bottom of a bottle in a saucer containing fine sea sand. Another boyhood experiment was to make gunpowder, which he rammed into a piece of metal pipe. One Sunday morning, with his elders conveniently at church, Schmidt applied a match to his missile which exploded with such violence as to carry off his right hand and forearm. Schmidt studied engineering at the Institute of Technology at Gothenburg. He then proceeded to Mittweida, near Jena, where, for twenty-five years, he supported himself by making mirrors and lenses for amateur astronomers. In 1905, he made his largest mirror, of 40-cm aperture and $f/2{\cdot}26$, for the Potsdam Astrophysical Observatory. This task, performed with simple means and executed by the left hand only, he completed in the remarkably short space of three months.

The excellence of Schmidt's optical work became known to professional astronomers and the leading optical houses. Yet, individualistic and independent to a degree, Schmidt resisted all offers of full-time employment. It was not until 1926 that Dr Schorr, director of the Hamburg Observatory, prevailed upon him to live at the observatory and to participate in its work on a voluntary basis.[30] Here Schmidt worked at new types of telescope mountings and drives and, in 1930, constructed his first camera.

The system introduced by Schmidt, generally referred to as the basic Schmidt, consists of a concave spherical mirror in front of which is a thin and optically weak corrector plate. If a circular stop is placed at the centre of curvature of a concave mirror and if every ray of an incident parallel bundle passing through the stop aperture falls on the mirror, the reflected cone of rays will be symmetrical (Fig. 144). The straight line from the centre of curvature and drawn parallel to any given ray of light may be regarded as an optical axis of the mirror, an axis of symmetry. There is no coma, no astigmatism, no distortion—only spherical

* First used for this purpose by C. Piazzi-Smyth in 1874. He placed a simple lens just in front of the photographic plate of his Petzval portrait camera, and thereby improved the definition (*Brit. Jour. Phot. Almanack*, pp. 43–47, 1874).

aberration and curvature remain. If spherical aberration can be eliminated we have, in a spherical concave mirror, means of producing point images of axial and extra-axial point objects. Schmidt effected this by placing an aspherical correcting plate at the centre of curvature, a plate which at once acted as a stop and corrected the spherical aberration (Fig. 145). Curvature of field remains—the focal surface follows a curve convex to the mirror and of radius of curvature equal to the mirror's focal length. This necessitates the use of film or thin plates sprung in a holder to follow the curve. Photographs distorted in this way do not lend themselves to direct measurement and, to flatten the field, Schmidt suggested that a thin plano-convex lens be placed before a flat photographic plate with the convex lens surface facing the mirror.

Schmidt did not divulge his method of figuring the aspherical surface of the correcting plate. His first camera, made in 1930, was an $f/1 \cdot 7$ system with a corrector of 36 cm aperture. The mirror was 44 cm in diameter. We shall, in future, refer to the dimensions of such a system as 36-cm × 44-cm. With this instrument,

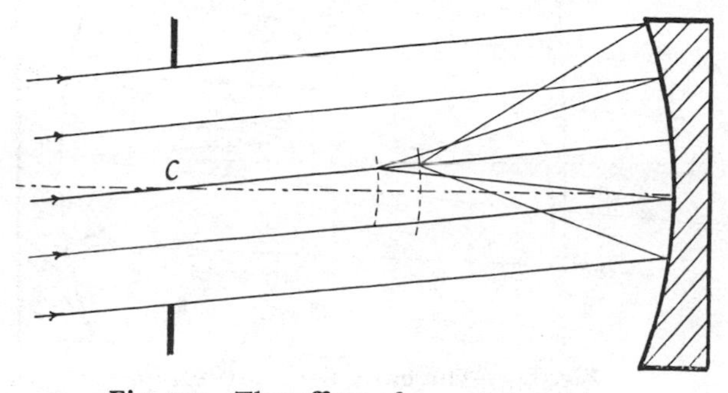

Fig. 144—The effect of an aperture stop

If an aperture stop is placed at C, the centre of curvature of a concave spherical mirror, the reflected beam retains its symmetry, but the system suffers from spherical aberration and curvature of field

Schmidt amused himself and his colleagues by reading the inscriptions on tombstones in a nearby cemetery. On a moonless night he also photographed a distant windmill. Applied to the stars, the camera gave round stellar images on curved film over a field as large as 16° angular diameter.

Schmidt spent several weeks working at the figure of the correcting plate. Professor Baade recalls[31] that he visited Schmidt one morning after the latter had spent some thirty-six hours of uninterrupted work on the glass disk. Schmidt was dozing but, on being awakened, accepted cigars but not food, since he was anxious to proceed at once with the drudgery of twelve hours' polishing.

Schmidt made and mounted a 60-cm × 60-cm system, but he died in 1935, before making the final tests. In the following year, Schorr divulged how the correcting plate was made.[32] Schmidt had placed an optical flat at the end of a cylinder, of nearly the same diameter, from which the air was pumped. When the glass disk had been bent by what was considered to be the required amount, Schmidt polished the outer surface to a perfect plane, let in the air, and had straightway a plate of the required figure.

The first Schmidt cameras were to Schmidt's original design. For moderate apertures, the amount of chromatic aberration introduced by the correcting plate was found to be negligible. The first Schmidt made outside Germany appears to be the $f/2 \cdot 4$ system which Dr Page Bailey of Riverside, California, attached in 1932 to his 15-inch Cassegrain reflector.[33] The 9-inch mirror has a focal length of 19 inches and the 9-inch correcting plate is stopped down to 8 inches. This was followed by the 8-inch $f/1$ Schmidt which C. A. and H. A. Lower of San Diego mounted independently on a polar axis with a 7-inch guide telescope.[34] The optical elements were mounted in a sturdy, square-sectioned tube, and gave a workable field of 20°. Star images at the edge of the field showed traces of astigmatism, an unavoidable error with standard Schmidts of large aperture ratio.

In England, the first Schmidt camera was made in 1936 by H. W. and L. A. Cox.[35] It was an $f/1$, $6\frac{1}{2}$-inch × 9-inch and, fitted with curved film-holder, was fixed to the Cox's home-made 12-inch equatorial reflector. A second Schmidt,

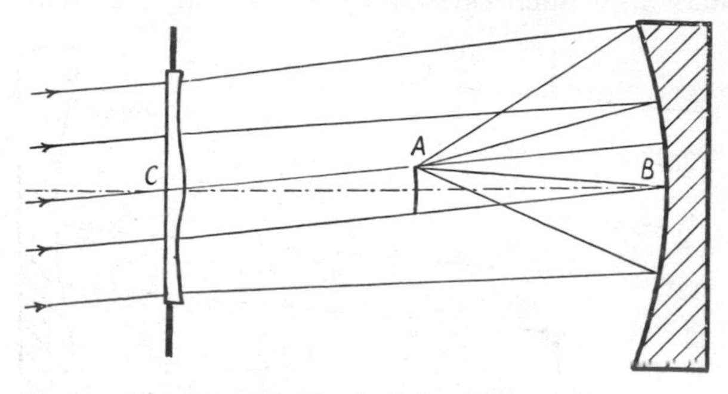

Fig. 145—The basic Schmidt system

When an aspherical corrector is placed in the aperture stop, the system is rendered free from spherical aberration for direct and oblique rays, but curvature of field remains. The aspherical surface of the correcting plate is purposely exaggerated.

$6\frac{1}{2}$-inch × 10-inch and working at $f/1 \cdot 5$, was made by H. W. Cox for J. P. M. Prentice in 1939. This camera covers a field of 20° and, with 45 seconds' exposure on Ilford Hypersensitive film, records stars to the thirteenth magnitude.

Both amateur and professional opticians have given attention to the two main defects of the standard Schmidt system. First, the tube length is double that of an ordinary reflector of equal focal length. This is not a serious problem and is mainly one of increased cost. Second, the curvature of the field and residual errors become more pronounced when the aperture ratio is increased. Y. Väisälä of the Turku University, Finland, used a plano-convex flattening lens with success.[36] F. B. Wright of Berkeley, California, showed in theory[37] that an ellipsoidal mirror with an altered form of correcting plate placed just behind the primary focus would not only give a flat field and short tube length, but permit the plate-holder to be attached to the rear face of the corrector. In comparison with the standard Schmidt, the field is limited owing to greater astigmatism and chromatic aberration. Väisälä, apparently, invented the Schmidt-Wright camera independently and, in a second paper,[38] proposed making a system of 11·8 inches aperture and 39·4 inches focus.

Dimitroff and Baker list[39] three Wright-type cameras in use in America—two 8-inch × 10-inch working at *f*/4, one by R. T. Smith of the Lick Observatory and the other by C. E. Wells of Roseville, California, and the third, an *f*/4, 5-inch × 8-inch, at Harvard. The last was built for flat-field spectrography. J. G. Baker has considered[40] further alternative forms which involve a convex mirror of special figure instead of a flattening lens; these would appear to be of particular value in spectrography.

Considerable theoretical work has been done on Schmidt-type systems, notably by C. R. Burch[41] and E. H. Linfoot[42] in England and by Wright,[43] Baker,[44] Smiley[45] and others in the United States. The result has been a variety of

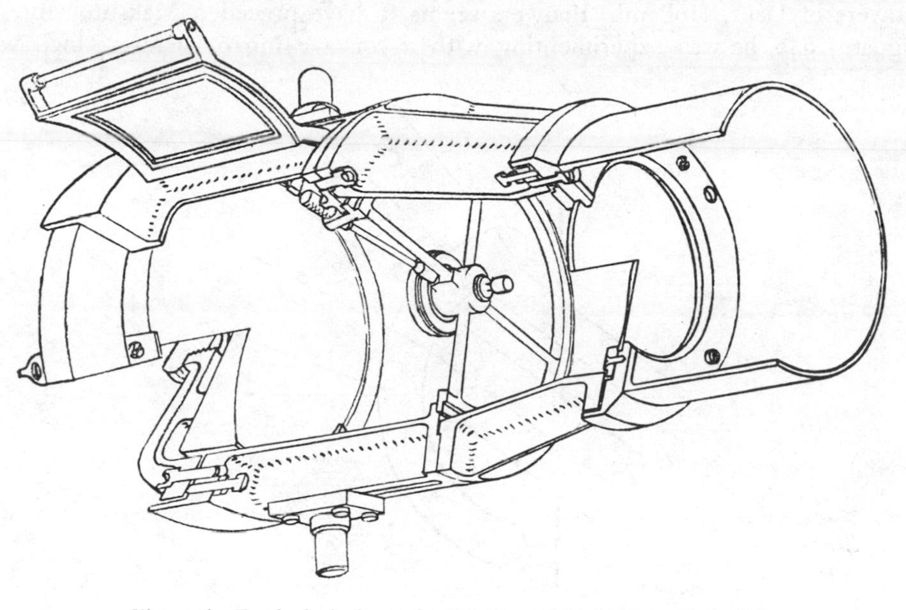

Fig. 146—Exploded view of a 6-inch × 8-inch f 1·25 Schmidt camera

(Optical Works Ltd, Ealing, London)

Schmidt–Cassegrain and other forms, many of which are still in the theoretical stage owing to constructional difficulties.

An alternative to the aspherical corrector of the basic Schmidt is to use a corrector with spherical surfaces, an idea which occurred to two quite independent investigators. In 1941, D. D. Maksutov of the State Optical Institute, Moscow, considered the effects obtainable by placing a weak negative lens or ' shell ' before a spherical mirror. For a lens which has concentric or near concentric spherical surfaces, any line drawn through the centre of curvature can be taken as an axis of symmetry. If, moreover, a circular aperture is placed at the centre of curvature, groups of incident parallel rays inclined up to a certain limiting angle will be refracted symmetrically (Fig. 147). The only monochromatic aberration will be spherical aberration and image curvature. The amount of spherical aberration compensates that due to the concave mirror and the final image will be located on an approximately spherical surface, the radius of curvature of which will be equal to the focal length of the system. One advantage of this arrangement over the Schmidt is now evident—the corrector is always perpendicular to principal oblique rays, rays which pass through the stop centre.

In May, 1944, in a paper to the Optical Society of America,[46] Maksutov developed this idea in a number of forms, all basically the same. He also referred to the possibility of placing a meniscus in the convergent beam in order to correct the spherical aberration of the mirror. In one form, the meniscus surfaces are worked on the faces of a 90° prism so that a Newtonian-type telescope results.

Investigations with meniscus correctors were made quite independently by A. Bouwers of Delft, Holland. Bouwers seems to have preceded Maksutov for, in August, 1940, he was experimenting with a concave mirror before which was

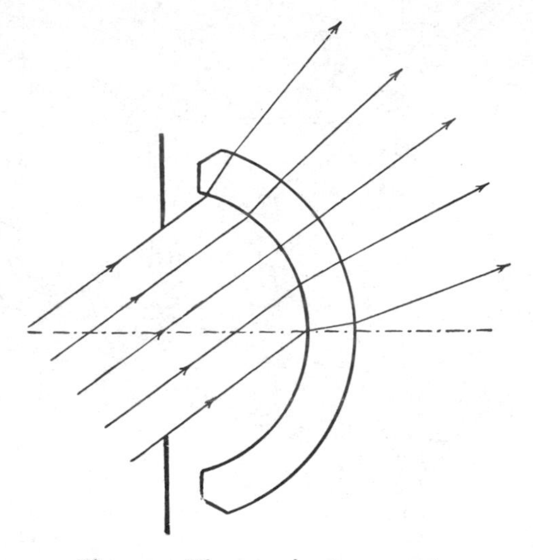

Fig. 147—The meniscus corrector

A parallel beam retains its symmetry after refraction by a meniscus corrector. The aperture stop is shown placed at the centre of curvature of the refracting surfaces.

placed a meniscus corrector. His work is summarized in his *Achievements in Optics* (1950), a most readable and valuable little book on catadioptric systems. Several degrees of freedom are available from the design viewpoint—the power, refractive index and thickness of the correcting meniscus, its distance and orientation convex or concave towards the mirror, the departure from sphericity of one face. The results achieved by the exploitation of some of these features are given by Bouwers in his book.

Many of the disadvantages of the Schmidt system are encountered in the Maksutov. The central position of the plate-holder, the obscuration it produces and the necessary curvature of field are features common to both. The Maksutov gains in the ease of its construction for low aperture ratios and in the shortness of tube length. It loses when the aperture ratio is above $f/1$ for then the Schmidt gives the better relative imagery. For *f*-values like *f*/7 and *f*/8, the Maksutov design offers possibilities as a visual instrument. Only three spherical surfaces have to be worked and the system is entirely closed. Only a few small Maksutov telescopes have been made, but the design appears to be one suitable for amateur construction. In his 1944 paper, Maksutov gives a photograph of a working model, whilst Bouwers has constructed both Cassegrain- and Gregorian-meniscus types (Fig. 149) in the form of compact monoculars.

Now that faint stars and nebulae are being examined in increasing numbers by slit and slitless spectrographs, rapid Schmidt-type cameras are essential. We have already seen, in connection with the Bracey lens (page 336), that a telescope with a slitless spectrograph is optically equivalent to a system having the focal length of the telescope and the aperture ratio of the spectrograph camera. If the Bracey lens is replaced by a faster Schmidt system it becomes possible to obtain spectrograms of fainter nebulae. A limit to exposure times is set by background illumination of the night sky which fogs the plate—hence the importance of a suitable telescope location.

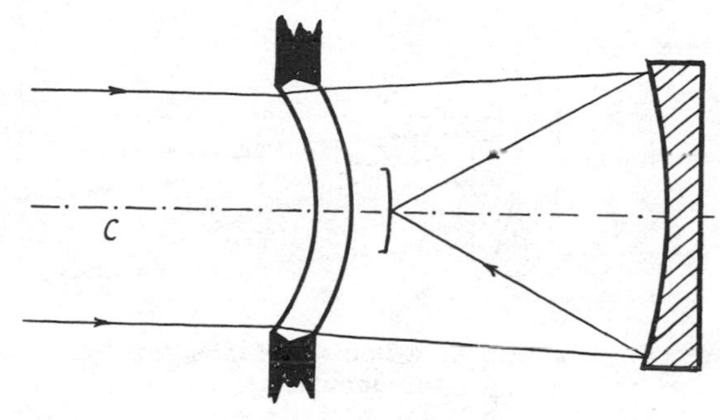

Fig. 148—The Maksutov system

Introduced in 1944, this system employs only spherical surfaces. The surfaces of the correcting meniscus lens are almost concentric with that of the spherical mirror.

The McCarthy slitless spectrograph used with the 82-inch McDonald reflector employs an *f*/2 Schmidt camera and specially made masks to reduce sky fogging.[47] One of the spectrograph cameras of the 200-inch Hale telescope operates at *f*/0·6 for an aperture of 13·8 inches.[48] The system is of the type known as a thick-mirror Schmidt and incorporates a fused quartz sphere. Apart from the corrector plate this camera consists solely of spherical surfaces. In the *Astrophysical Journal* for 1937,[49] there is an account of a Schmidt slit spectrograph which was attached to the Yerkes refractor for obtaining spectra of faint, diffuse galactic nebulae. This unit consisted of a wide slit attached to the upper end of the tube, two 60° quartz prisms and an *f*/1 Schmidt camera of 94 mm aperture focused on the slit. Resulting photographs showed nebular emission lines and weak continuous spectrum. 'The method is by far the most powerful in existence for the discovery of faint emission nebulae' Otto Struve wrote. 'With long exposures we can detect emission lines of nebulae which are too faint to be photographed directly on the background of the night sky.' A 150-foot nebular spectrograph was used on Mount Locke by Struve and Elvey, who established that material similar to that found in diffuse nebulae seems to extend throughout the galaxy.[50] This material requires excitation by strong ultra-violet radiation to make it luminous. In 1950, similar studies for the southern skies were undertaken on Mount Stromlo, Canberra, with a 130-foot nebular spectrograph. An $f/1{\cdot}0$ Schmidt camera of 10 cm aperture was incorporated but the spectrograph was unfortunately put out of action by the Stromlo fire of February 5, 1952.

The design of rapid Schmidt-type cameras for astronomical spectrography is proving a promising field for optical computers. On the practical side, D. O. Hendrix, director of the optical shops at Mount Wilson, has experimented with both solid and thick-mirror Schmidts.[51] The solid Schmidt is a cylinder of glass or quartz with worked end-faces (Fig. 151). If made of material of refractive index n this system becomes, theoretically, n^2 times faster than a basic Schmidt of similar dimensions. Effective speeds of $f/0{\cdot}35$ are practicable, whilst a solid Schmidt of diamond could have an aperture ratio of $f/0{\cdot}2$. As with the majority of Schmidt

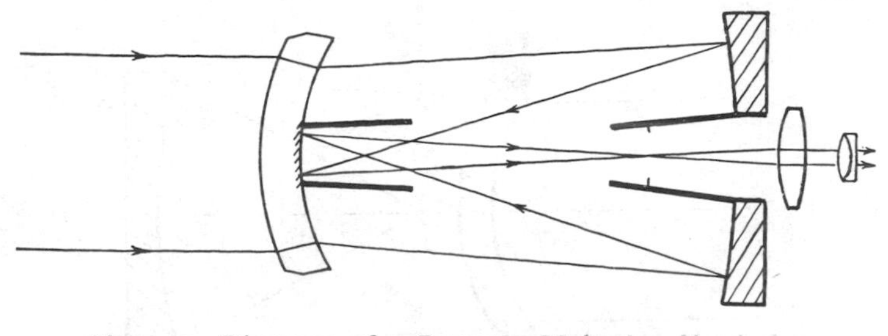

Fig. 149—Diagram of a Bouwers-Maksutov Gregorian monocular

(*After Bouwers, 'Achievements in Optics'*, 1950)

forms, the central obscuration (film-holder) remains. In the solid form due consideration must be given to absorption by the glass or quartz used. The thick-mirror Schmidt likewise ensures that the final image is formed in glass. A suitable arrangement is outlined in Fig. 152. With air only between the corrector plate and mirror surface the semi-image size would be as shown by the full arrow of size h. If a thick glass mirror of axial thickness $r/2$ is used, rays incident obliquely to the plane surface are refracted towards the normal and appear, as seen from the mirror surface, to emanate from a smaller object. The final image is correspondingly diminished to size h' (dotted arrow). The same amount of initial luminous flux is concentrated into a smaller image area, with consequent increase in illumination by the factor n^2, n being the refractive index of the material of the thick mirror. By 1939, Hendrix had constructed, among other systems, a thick-mirror Schmidt of 2·125 inches focus and 2 inches aperture. The corrector plate was separated by a distance of 1·34 inches from the mirror's plane surface, an arrangement which, in glass of refractive index 1·52, yielded $f/0{\cdot}62$. A small, circular, plane-parallel glass disk cemented to the upper (plano) mirror surface defined the focal plane and to this $\frac{1}{2}$-inch square plates were oiled.[52]

Schmidt cameras are no longer the rarities they were a decade ago. Progress has

Fig. 150—The 40-inch Yerkes telescope with nebular spectrograph

been most rapid in the United States where professional astronomers and amateur telescope-makers have taken the lead and where the professional optician can now undertake the construction of large instruments. The optical and mechanical shops of the Mount Wilson Observatory have played a significant part in Schmidt development. Here, in the mid-thirties, T. Dunham junior made an *f*/1·8 of 5 inches focal length and another of *f*/1·0.[53] In one year alone (1937–1938), W. S. Adams recorded that ' much work has been done on Schmidt cameras, including three with focal ratios of 0·66, 1·0 and 1·0 and foci of 2, 3 and 9 inches respectively. Other Schmidt mirrors of 17, 20, 26 and 30 inches were also finished.'[54] The 18-inch × 26-inch *f*/2 Schmidt at Palomar Mountain, designed by Russell W. Porter, originated from the Mount Wilson shops.[55] With this instrument, F. Zwicky obtained unique photographs of large regions of the northern sky. Here also J. S. Dalton figured the 48-inch corrector plate and Hendrix and Dietz ground and figured the 72-inch spherical mirror for the 48-inch *f*/2·5 Schmidt at Palomar Mountain.[56]

Many amateurs followed in the steps of Bailey and Lower and have made Schmidt systems which professionals have often been glad to use. Thus C. H. Nicholson, of the Chicago Amateur Astronomical Association, made elements for a 12·2-cm *f*/2·0 Schmidt and a 9·4-cm × 11·0-cm *f*/1·0 Schmidt for the quartz spectrograph of the 82-inch McDonald telescope.[57] The 20-inch × 20-inch *f*/2·25 Schmidt which Mellish of Escondido, California, made and mounted, appears to be the largest system yet made by an amateur.[58]

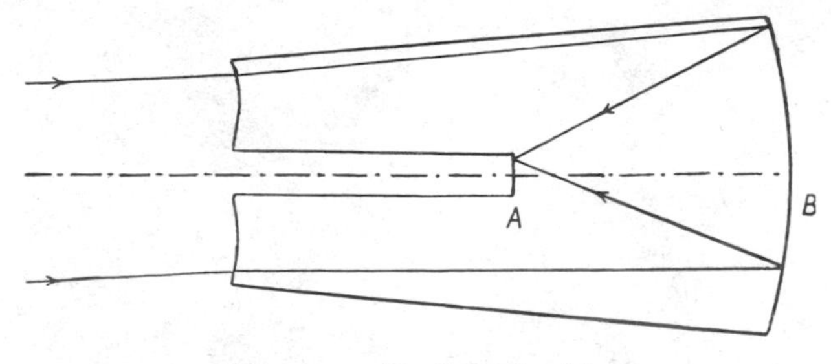

Fig. 151—A solid Schmidt

Surface *B* is aluminized. A plunger is necessary to position and to withdraw the plate at *A*, the plate being in oil contact with the glass or quartz face.

The Perkin-Elmer Corporation, founded in 1939 at Norwalk, Connecticut, now has the optics of several large Schmidt cameras to its credit. Its 26-inch × 31-inch *f*/3·5 Schmidt, with mounting made in the Harvard Observatory workshop, forms the main instrument at the Tonanzintla Observatory, Mexico. For this Schmidt, Perkin-Elmer made a 27½-inch objective prism of 4° apical angle. Another large Perkin-Elmer Schmidt-type telescope is the 32-inch × 35·6-inch *f*/3·75 Cassegrain-Schmidt recently completed for the Armagh, Dunsink and Harvard observatories and known as the A.D.H. telescope. These three observatories joined in the building and operation of this instrument which, in 1951, replaced the 24-inch Bruce objective on its Fecker mounting at the Boyden station of the Harvard Observatory in Bloemfontein, South Africa. The optical system is of the form designated ' C4 ' by J. G. Baker, chief optical consultant to Perkin-Elmer, in his important paper on flat field cameras.[59] The two mirror and corrector system gives a flat field and reduced tube length (168 inches) compared with the classic Schmidt. The field-flattening lens often used with the latter is here replaced by a slightly convex mirror of 16·8 inches diameter. With the 35·6-inch spherical concave which it faces, this secondary mirror covers afield of 4°·56 on a plate 9½ inches in diameter. Perkin-Elmer has also made a full aperture 33-inch objective prism for this telescope.[60]

The A.D.H. photographs are proving excellent for the more detailed study of globular star clusters—for stellar distribution within the clusters, for comparative studies of photographs of cluster stars in blue and red light, for the discovery and study of cluster Cepheids. Bart J. Bok, formerly associate director of the Harvard College Observatory, writes[61] that the A.D.H. plates ' open whole new areas of research on the Magellanic Clouds. . . . At the moment the greatest interest attaches to a precise knowledge of the varieties of stars, clusters and nebulae in the Clouds.'

Perkin-Elmer has also constructed, again to the design of Baker, six Schmidt-type or super-Schmidt cameras for meteor photography. Initially, the U.S. Navy

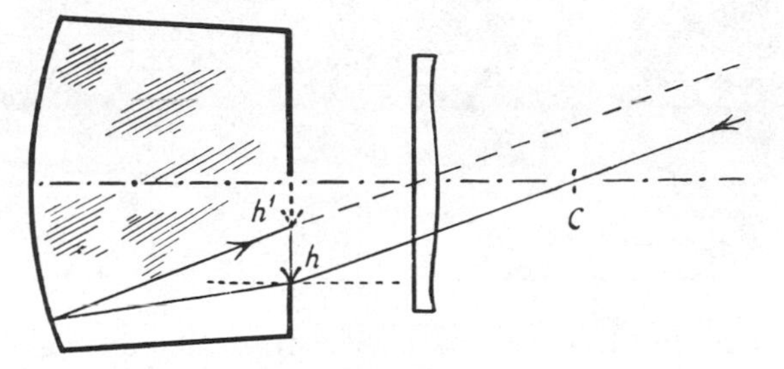

Fig. 152—The Hendrix thick-mirror Schmidt

Bureau of Ordnance purchased two of these cameras for use at Harvard's two meteor-observing stations near Las Cruces, New Mexico. Since these stations are 18 miles apart, simultaneous meteor photographs from them permit meteor height determinations. The cameras embody several most interesting optical and mechanical features. They have an effective aperture of 12¼ inches, an effective aperture ratio of *f*/0·82, and can cover a field of 55° angular diameter (52° diameter for meteor work). A single film at the focal surface covers nearly one-tenth of the area of the night sky and can record meteors down to the fourth magnitude.[62]

The corrector unit is compound and consists of two 18-inch nearly concentric meniscus lenses or ' shells ' of negative equivalent power and a two-lens achromatizing plate. The outer lens is convex towards the incident light, the inner lens is convex towards the mirror. These lenses enclose the corrector plate and are balanced about the centre of curvature of the 23-inch spherical mirror. The curved film-holder is located near the concave front surface of the inner meniscus lens—incident rays traverse this second lens twice. The film is first moulded under heat and pressure into a spherical cap of 8 inches radius of curvature and is held to the holder by vacuum control. As will be seen from Fig. 154, the focal surface is inaccessible from the side. Hence, to load the camera, the corrector unit has to be

Fig. 153—The Armagh-Dunsink-Harvard telescope

On Harvard Kopje, Bloemfontein, South Africa

hinged apart, an operation which, thanks to the careful mechanical design and construction, in no way affects the relative positioning of the optical elements.[63]

Only 2 mm in front of the film-holder is a shutter which rotates at 1800 r.p.m. The operating spindle passes through the centre of the spherical mirror and through the rear lens. Stars at the centre of the field are thereby obscured but this is of no consequence in view of the remarkably large angular field. By using the shutter, recorded meteor trails have a ' chopped ' appearance. The exposed and developed film is copied by a lens system, designed by Baker, which gives gnomonic projection on a flat glass photographic plate. The meteor trail sections are then measured, the measurements leading directly to an estimate of the meteor's deceleration and hence of air density at high altitudes.

At one period, the first camera photographed fifty-six meteors, only one of which was recorded by a 3-inch aperture *f*/2·5 aerial camera previously used for this work. From March to August, 1952, the two cameras simultaneously photographed more than 300 meteors, an average yield of one meteor per 30 minutes' simultaneous exposure time for the two cameras. In the early days of the Harvard meteor programme the photographic yield was only 5 or 6 meteors per year.[64] Effective

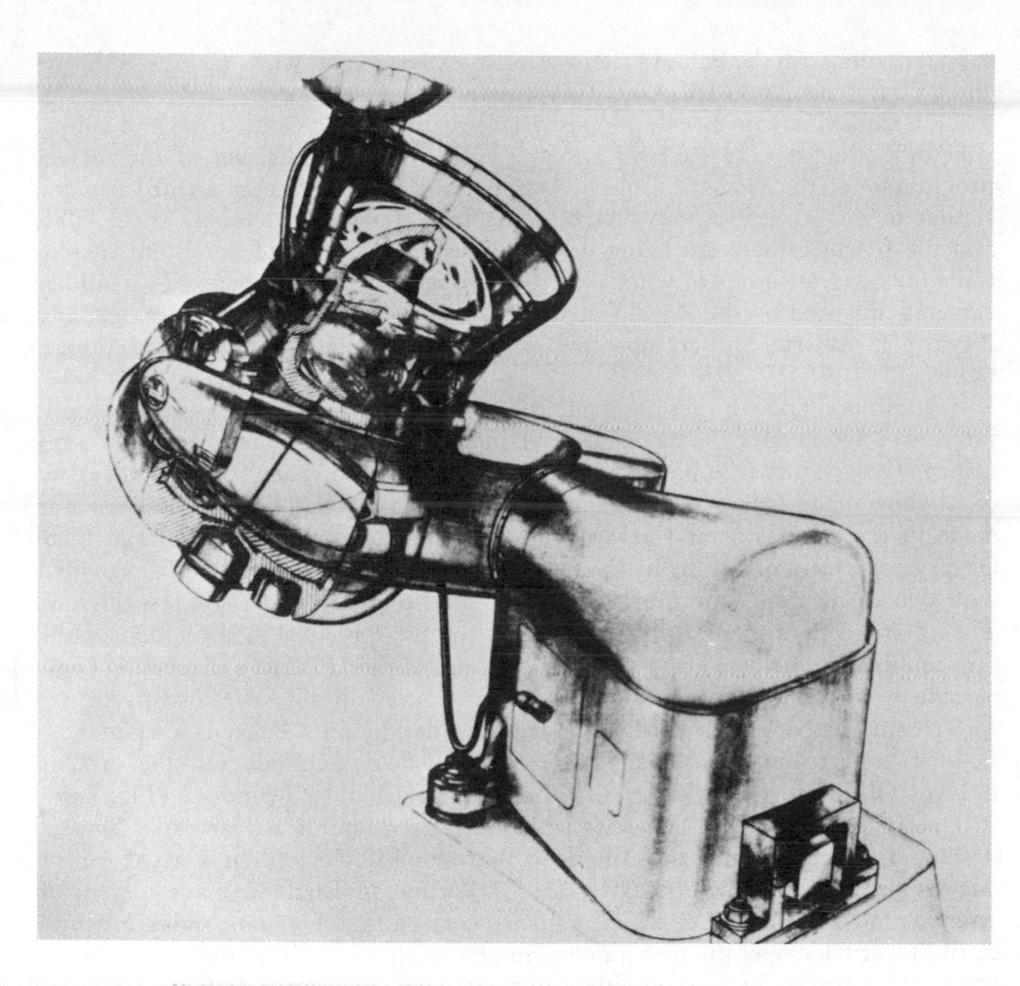

Fig. 154—Exploded view of the super-Schmidt meteor camera

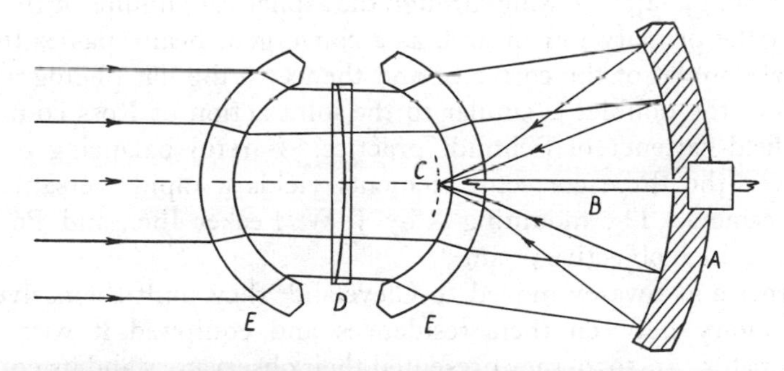

Fig. 155—Optical system of the Baker super-Schmidt

A Primary spherical mirror   *C* Focal surface
B Shaft of rotating shutter   *D* Achromatizing corrector
E Meniscus shell lenses

exposure time with the Schmidts on a clear moonless night, when using the shutter and a film of maximum sensitivity, is approximately 12 minutes.

Each camera is equatorially mounted and, with electric clock-drive and robust yoke mounting, weighs just over 2 tons. F. L. Whipple, director of the meteor programme of the Harvard College Observatory, reports [65] that a third camera, similar to the other two, was delivered in New Mexico in October, 1952. This and the fourth camera are being used for the photography of persistent meteor trails for determinations of wind velocities in the upper atmosphere. Two similar cameras, financed by the Royal Canadian Air Force, are located at Meanook and Newbrook, Alberta, and are operated by the staff of the Dominion Observatory, Ottawa.

In addition, Perkin-Elmer has made the optics for a 25-inch × 30-inch Schmidt for Lowell Observatory, a 24-inch × 31-inch for the Agassiz station of Harvard College Observatory (the Jewett telescope), and a similar system for the University of Michigan (the Curtis telescope). The mounting for the Jewett telescope was made in the workshops at Harvard. The Curtis telescope, at the Portage Lake Observatory, has a mounting by Warner and Swasey. This instrument is provided with two 24-inch aperture flint prisms of 4° and 6° refracting angle respectively. The 4° prism gives the low dispersion necessary for statistical investigations while the combination yields a dispersion of 110 A/mm at the $H\gamma$ line, a dispersion comparable with the lower dispersions customarily used in slit spectroscopy.[66]

A recent telescope with Perkin-Elmer optics designed by Baker is a 24-inch × 24-inch convertible Newtonian–Cassegrain flat-field Schmidt for the Arthur J. Dyer Observatory of Vanderbilt University, Nashville, Tennessee (Fig. 157). In December, 1947, following tests with the *f*/3·5 Schmidt at Harvard's Agassiz station, Baker reported to the American Astronomical Society on a new form of mirror-corrector. Two years later, C. G. Wynne in England made a general investigation of three-element, near-focus correctors for paraboloidal mirrors. In the Dyer telescope, the perforated primary paraboloid allows the use of interchangeable mirrors for work at either an *f*/4·5 Newtonian or an *f*/16·5 Cassegrain focus. To convert the telescope into an *f*/3·5 Schmidt-type system, a compound corrector is swung into a position near that occupied by the secondary Cassegrain mirror. Incident light after passing through the aspherical annulus of this corrector is reflected by the primary mirror and, as a convergent beam, passes through the convex central doublet of the corrector and thence to the flat photographic plate. The function of the doublet is similar to the joint action of Ross coma-corrector and convex field-flattener of Schmidt practice. Careful balancing of the aberrations between the three corrector elements yields a rapid, versatile, flat-field astronomical camera. The mounting is by J. W. Fecker Inc., and Perkin-Elmer is providing a 24-inch objective prism.*

When Warner and Swasey moved to Cleveland, they built themselves a small, private observatory between their residences and equipped it with a 9½-inch equatorial refractor. In 1919, they presented their observatory and its equipment to the Case School of Applied Science, of which college they were trustees. The

* For further information on the three-element corrector the reader should consult Dr Baker's article 'Optical Systems for Astronomical Photography' in *Amateur Telescope Making*, iii, 1953.

Fig. 156—The Baker-Perkin-Elmer super-Schmidt meteor camera

One of the instruments near Las Cruces, New Mexico.

observatory, situated some 300 feet above Lake Erie, is now known as the Warner and Swasey Observatory. In 1939, the authorities decided on the addition of larger equipment, and funds were provided by Katherine Burrell in memory of her husband, E. P. Burrell, for many years director of engineering at the Warner and Swasey Company. The resulting instrument, the 'Burrell telescope', is a 24-inch × 36-inch Schmidt.[67] The spherical mirror is aluminized Pyrex and the Vita-glass[68] corrector transmits more of the shorter photographic rays than does ordinary glass. All the optical components were figured by C. A. Lundin,[69] optician to the company and son of Carl Lundin, successor to Alvan G. Clark. Working at *f*/2·3, the system photographs stars fainter than magnitude 18 in ten minutes[70] and covers a field 5° in angular diameter. With a 24-inch objective prism of 4° angle before the corrector plate, J. J. Nassau has obtained spectra of twelfth-magnitude stars in only 30 minutes' exposure.[71] This significant re-

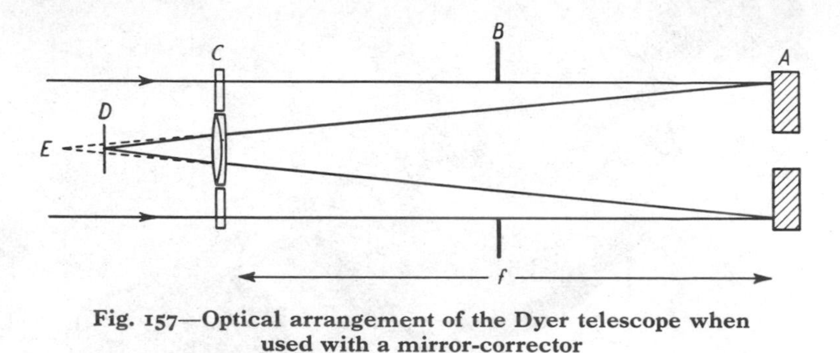

Fig. 157—Optical arrangement of the Dyer telescope when used with a mirror-corrector

A 24-inch diameter primary paraboloid
B Stop, 21·5-inch diameter, used for photographic purposes only
C 24-inch annular corrector with central doublet
D Focal plane, 102 inches from vertex of *A*
E Focus of *A*, 107·25 inches from mirror vertex
f Effective focal length of system, 81·8 inches

duction in exposure time, coupled with the sharpness of the images produced, demonstrates the great importance of Schmidt telescopes for mass spectrography.

The late J. W. Fecker was one of the first telescope-makers to master the constructional difficulties of the Schmidt camera. For many years Fecker worked under his father, Gottlieb L. Fecker, who joined Warner and Swasey in 1895 and became superintendent of the instrument department. J. W. Fecker set up on his own account in Cleveland and, on the death of McDowell in 1926, took over the Brashear-McDowell business.[72] Fecker made a 10-inch × 15-inch *f*/2·5 Schmidt for G. W. Cook's Roslyn House Observatory, now the Cook Observatory of the University of Pennsylvania, also a 16-inch × 25-inch *f*/2·7, and an 8-inch × 12½-inch *f*/1·5, which he mounted together on his favourite fork-type mounting for the private observatory of M. R. Schottland. Before his early death in 1945, Fecker had started work on a 60-inch × 60-inch *f*/2·5 Schmidt for Harvard.

The Penn Optical Company of Pasadena is responsible for the optics of a 24-inch × 36-inch *f*/3 Schmidt which, thanks to American financial aid, working through

Fig. 158—The 24-inch × 36-inch Burrell Schmidt

Warner and Swasey Observatory of the Case Institute of Technology, East Cleveland, Ohio

(*Warner & Swasey Co.*)

the European Recovery Programme, will be installed in the Gran Sasso Observatory. This observatory is located in the central Apennines, near the summit of Gran Sasso d'Italia, and is a station of the Observatory of Rome. Hendrix supervised the completion of the optical elements but the fork-type mounting is of Italian construction.[73]

The 48-inch × 72-inch *f*/2·5 Schmidt at the Palomar Observatory is now the largest instrument of its type. Reference has already been made to the working of the optical elements under the supervision of Hendrix. The corrector is figured on both sides and has a clear aperture of 49·5 inches. Behind it is a duralumin shutter of the 'clam shell' type, motor-driven and operated from either the central control desk or from the guiding head of the instrument. Two 10-inch *f*/15·6 guide telescopes are provided. The spherical mirror, of focal length 121 inches, weighs over a ton and is supported by thirty-six weighted compound levers. A further eighteen equally spaced peripheral supports share the load when the mirror is tilted.[74] The 20-foot long tube is carried by a fork-type mounting and, with other moving parts, weighs approximately 12 tons. This great mass is operated by remote control, as is also the steel and concrete dome. Like the Hale reflector (Chapter XVIII), the instrument has no need of graduated circles for right ascension and declination indications. This service is provided by the indicators of a series of Selsyn motors all operated from the control desk on the main observing floor.

The Palomar Schmidt is employed on the formation of a sky atlas, sponsored and financed by the National Geographic Society of Washington, D.C. It is estimated that some 2000 photographs will be necessary since each survey field covers an area of 6° × 6° with a margin for overlapping and is photographed through both red and blue filters for comparison purposes. Nebulae and stars down to about magnitude 20 can be recorded, so that the atlas should provide valuable information about clusters of nebulae and their relative frequency in the general field. The very thin glass plates, 14 inches square, are held in the focal surface by vacuum control and before use are each subjected to a bend test. The dark slide is brought to and away from the focal surface by a semi-automatic loading mechanism.

Schmidt-type systems are of especial value in Great Britain since they enable astronomers to obtain a wealth of information in the all-too-short periods of first-class seeing. A few photographs secured during good observing conditions provide material for many weeks' measurement and reduction. It is therefore considered advisable to concentrate on high optical performance and especially to investigate to the full the photometric problems to which Schmidt star images give rise. The images of faint stars (or of bright stars after short exposure) are so small as to be commensurable with the size of grains in the photographic emulsion. Because of this, and sometimes because of residual extra-axial aberrations, they differ in shape over the extent of the field. Light reflected between the surfaces of the corrector plate gives rise to ghosts on the film—bright stars are encircled by a halo. The intensity of the light reflected by the plate can be reduced by 'blooming' the plate surfaces (page 384) but a fainter halo persists. Furthermore, aberrations of the system affect the light distribution over this halo. Present Schmidts in Great Britain are therefore largely experimental. Each instrument is

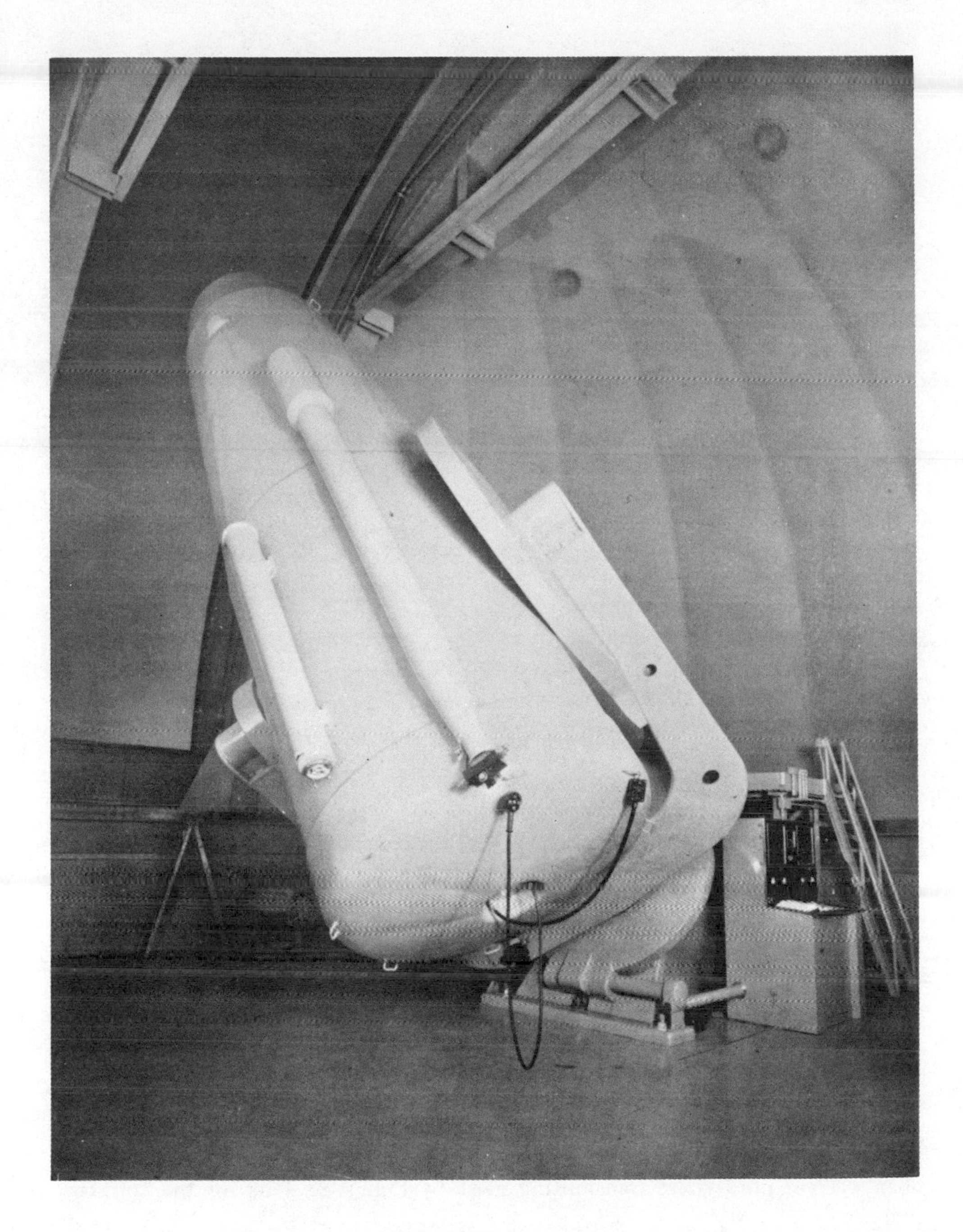

Fig. 159—The Palomar 48-inch Schmidt telescope

(*Mount Wilson and Palomar Observatories*)

Fig. 160—The 48-inch Schmidt telescope at Palomar Observatory

A drawing by Russell W. Porter

(California Institute of Technology)

considered capable of improvement and each can be altered without the cost and labour which attends the replacement, figuring, or adjustment of large elements.

Scotland led the way in 1946 when it was decided to install a Cassegrain-Schmidt in the observatory of the University of St Andrews, Dundee. The project came under the direction of E. Finlay-Freundlich, with R. L. Waland in charge of the mechanical and optical work. Before proceeding to the contemplated 30-inch × 38-inch Schmidt, Waland completed a 15-inch × 19-inch *f*/3 (effective *f*/3·9) pilot model[75] designed by E. H. Linfoot of Cambridge. It was felt that the experience gained by constructing and using this half-scale model would pay rich dividends when it came to making the final instrument. Waland now has the mechanical parts of the fork-type compensated mounting well under way and has figured the 38-inch primary and 19-inch secondary mirrors. He is also working on a disk of ultra-violet transmitting glass by Schott of Jena for the corrector plate.

Another Scottish Schmidt is the 16-inch × 24-inch *f*/2·5 camera recently installed at the Royal Observatory, Edinburgh. Tube and optics are by the London firm of Cox, Hargreaves and Thomson and replace a 15-inch Grubb visual refractor although using its drive and mounting. In the absence of film in the film-holder a built-in 90° prism sends the light to a Newtonian focus. Preliminary visual knife-edge tests at this focus followed by a series of photographs in the primary focal surface ensure the best possible imagery and check the optical stability of the system. As with the Palomar Schmidts, a constant mirror-to-film distance is maintained by an adjustable arrangement of invar and brass rods. These compensate changes in tube length and in mirror focal length due to changes in temperature. The corrector plate has surfaces coated with magnesium fluoride. To secure a homogeneous low-expansion disk, a number of selected thin plates were successfully fused and moulded to form the meniscus-shaped mirror blank. By keeping the mirror back convex and concentric with the aluminized optical surface, any slight lateral shift of the disk during exposure does not appreciably displace images in the focal surface.[76]

Financial considerations often restrict the establishment and operation of large Schmidt-type telescopes. At the Observatories, Cambridge University, it has been decided to adapt existing buildings to take a 36-inch Newtonian and a

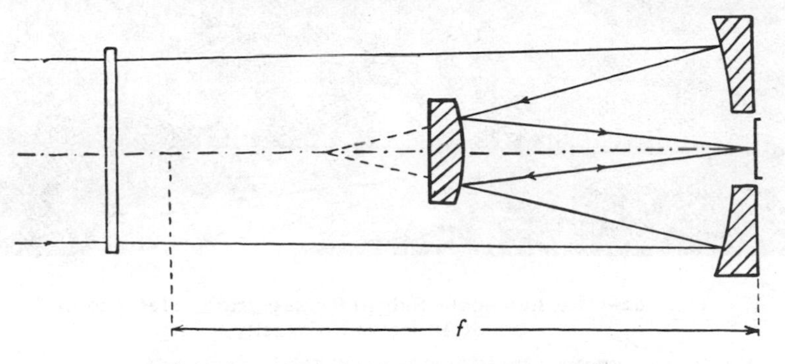

Fig. 161—A flat-field Schmidt-Cassegrain

The system has a correcting plate and two spherical reflecting surfaces.

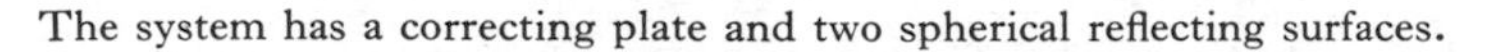

separate 17-inch × 24-inch $f/4$ Schmidt. Both instruments are made by Grubb-Parsons (pages 387 ff.). The Schmidt, in a fork mounting, was designed by Linfoot and will be used primarily for investigating the problems of stellar photometry outlined on page 372. To avoid reflections at the surfaces of the corrector plate giving rise to ghost images, the plate is made slightly convex and mounted with the convex side facing the mirror. A brass-invar compensating system is incorporated and fitments for a field-flattening lens, filters, and an objective prism indicate the scope of future investigations.[77]

A large meteor super-Schmidt is soon to be used in conjunction with the radar equipment at the University of Manchester's research station at Jodrell Bank, Cheshire (Chapter XIX). A team of workers under N. F. Mott at the University of Bristol have already progressed through a small experimental model to a pilot model of $f/0{\cdot}63$ (nominal) or $f/0{\cdot}75$ (effective) with a field of 56 degrees. The mirror diameter is 17 inches and a 28-inch mirror is planned for the full-scale model. The design is optically similar to that of the six Perkin-Elmer meteor cameras previously mentioned. Spherical cap glass plates will be used, and an

Fig. 162—The half-scale Schmidt-Cassegrain telescope of the St Andrews University

(Reproduced by kind permission of the University authorities)

original shutter design avoids having to drill a central hole in the mirror and rear meniscus lens of the corrector system.

Observatories about to modernize their equipment have to decide on either a convertible Schmidt or, like Cambridge University, on a separate Schmidt and large reflector. In some European observatories the problem has yet to arise owing to present lack of funds. The Royal Observatory at Herstmonceux, however, has decided on a giant convertible Schmidt, the 98-inch spherical mirror of which has been worked by Grubb-Parsons. This instrument receives our further attention in Chapter XIX.

With all the powerful equipment at Mount Wilson at his disposal, there was still one outstanding solar problem which Hale was unable to solve—the nature of the corona. Eclipse work had shown a connection between the general form of the outer regions of the corona and sunspot activity and prominences, but more information was required as to its composition and structural details. Research had always been limited to the few minutes of total eclipse, and the most assiduous observer, taking advantage of every total eclipse, could not hope for more than an hour's observation during his lifetime. As early as 1882, Huggins attempted to photograph the corona in full sunshine, but without success.[78] Others tried

Fig. 163—The 38-inch spherical primary mirror for the full-scale Schmidt-Cassegrain telescope of the St Andrews University

The mirror is shown on the polishing machine. R. Waland is seen in the background.

(Reproduced by kind permission of the University authorities)

photographic methods, with different colour filters and specially sensitized plates, but these also failed. Hale tried two methods. In the first,[79] he set the second slit of a spectroheliograph on the centre of a coronal line. This reduced the brightness of the sky without affecting the light from the corona, but the latter is so intrinsically faint that it failed to shine in contrast with the enfeebled background. The method failed both on Pike's Peak in 1893 and (in company with Deslandres), on Mount Etna in 1894. The second method utilized a sensitive bolometer[80] but was no more successful; Abbot showed later that corona radiations are too feeble for bolometric measurement.

During these early experiments, the apparatus was generally taken to places of high altitude to escape atmospheric haze, for it was appreciated that scatter by the earth's lower atmosphere was sufficient to obliterate the corona. Photographic methods indicated a halo round the sun, but this was due to atmospheric scatter and glare produced by the instrument. Exposure times were, moreover, too short to register even the brighter inner corona.

With the intention of eliminating instrumental scatter, Bernard Lyot, of the Meudon astrophysical observatory, investigated the brightness of the halo produced by an apparently first-class lens of 8 cm diameter and 2 metres focal length. He found the brightness to be everywhere more than two hundred times stronger than that of the corona itself.[81] It was apparent that, if he could obtain a lens free from scratches and small bubbles, with highly polished surfaces free from dust particles, much of the glare would vanish. From a selected disk of borosilicate crown glass, a 15-cm diameter lens was made, but veins still gave rise to considerable glare.[82]

Further glare was contributed by light diffracted by the edge of the lens and internally reflected by its surfaces. To eliminate this, Lyot designed his first coronagraph,[83] using the above-mentioned 8-cm lens. The direct solar image was formed on a black screen made slightly larger than the image of the disk. Behind the screen, a field-lens produced an image of the object-glass on a small diaphragm. The latter cut off light diffracted by the first lens while a small central screen blocked out the light of the image produced by reflection on the surfaces of the plano-convex objective. Achromatism was ignored at this stage but, in later instruments, was effected by a specially corrected lens placed behind the diaphragm and so shielded from stray light. Using an eyepiece and with the aperture reduced to 3 cm to hide the principal defects of the object-glass, Lyot saw prominences ' with a violet-pink tint '.

> The use of a red glass [he writes][84] greatly increased the contrasts and allowed of good observations. The Sun was surrounded by a slight halo without details. The polarimeter showed that this halo was polarized in a radial plane like the corona, but more faintly. The polarization appeared at 6′ from the limb; it increased rapidly toward the Sun and then remained almost constant under 3′. The more transparent the sky, the stronger it was.

Lyot concluded, from these and other observations, that four-fifths of the halo was due to diffusion. With a spectrograph, illumined by light from this halo, Lyot obtained the first spectrogram of the corona without the condition of a total solar eclipse.[85]

These observations were made in the summer of 1930, at the Pic-du-Midi Observatory, a site on the French side of the Pyrenees, over 9000 feet above sea level. This observatory, the highest permanently-manned coronagraph station in Europe, often enjoys periods of excellent seeing. The Baillaud telescope, brought to the station in 1904, offered a sturdy, cradle-type English mounting suitable for carrying the 8-cm or larger coronagraph.

In 1931 a 13-cm aperture coronagraph and improved spectrograph were taken to the Pic-du-Midi. Lyot resumed his polarimeter studies with a diffraction polarimeter capable of detecting a polarization of one part in a thousand, and several more coronal lines were photographed,[86] but with difficulty, since the plates used were not sensitive enough.

In 1931 also, Lyot installed his largest coronagraph.[87] The telescope lens is a plano-convex of borosilicate crown glass, 20 cm in diameter and of 4 metres focal length. It was ground and polished at the Paris Observatory by A. Couder from a special disk cast by the firm of Parra-Mantois. The tube is of duralumin and the portion which extends well beyond the object-glass is lined with grease to catch falling dust particles. The aperture is always closed except during moments of observation and the front surface of the first screen is silvered, and so inclined that the sun's light and heat are reflected out of the tube. To the coronagraph is attached a single spectrograph with a 4-inch Rowland grating; the latter can be placed in two alternative positions to produce either large- or small-scale spectra. With this instrument, mounted on the Baillaud equatorial at the Pic-du-Midi, Lyot, in July, 1931, photographed the inner corona without an eclipse.[88] Scores of photographs followed and revealed many of the phenomena so eagerly studied during the few minutes of a total solar eclipse. Spectra of corona and prominences were photographed, curtailed in extent only by the limit of transparency of the coronagraph and atmosphere (3300 A) and the limit of sensitivity of the plates available (12000 A). In the infra-red, Lyot obtained, with a three-hour exposure, record of a faint coronal line 8024 A and of two strong coronal lines, 10747 A and 10798 A, never before photographed during eclipses.[89] In 1935, he obtained the first

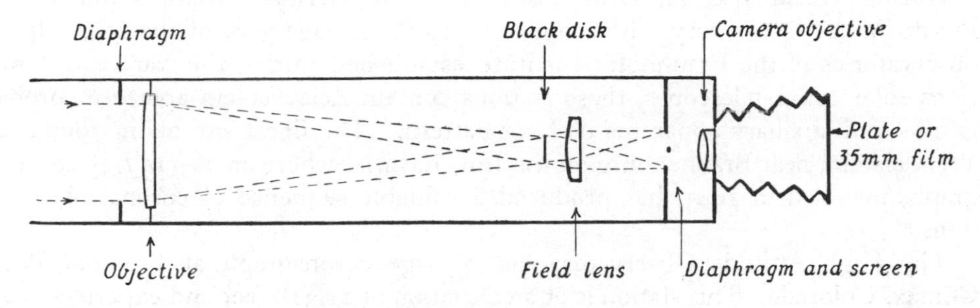

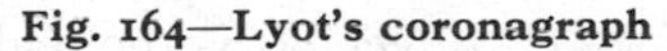
Fig. 164—Lyot's coronagraph

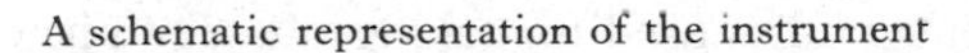
A schematic representation of the instrument

cinematograph film of prominences. Pictures were taken on 35-mm film at the rate of two a minute, sometimes automatically, sometimes choosing through an eyepiece the moments of best seeing. They revealed, on projection, the complex movements of the prominences and detail invisible in the spectroheliscope.[90]

For isolating the light of the corona and prominences, Lyot made use of another device, first proposed for laboratory work by R. W. Wood.[91] We refer to the quartz-polaroid monochromator, a practically monochromatic filter system. This device is so attached to the coronagraph that light from the occulting disk passes first through a coloured filter and then through a series of polarizing quartz crystals and polaroid filters. A collimator lens renders the emergent beam parallel and a camera lens forms an image of the prominences, say, on the photographic plate. The use of this principle is by no means restricted to the Pic-du-Midi Observatory. J. W. Evans, when at the Chabot Observatory, Oakland, California, used it with singular success.

Further proof of the fine seeing conditions at the Pic-du-Midi station is provided by recent photographs taken there by H. Camichel.[92] The Baillaud telescope, originally a twin 20-inch reflector and 10-inch refractor, now has its reflector replaced by the $23\frac{1}{2}$-inch photographic object-glass of the Loewy coudé telescope at Paris Observatory. This glass has a focal length of 59 feet which, in the Baillaud telescope tube, is 'folded' by the incorporation of two plane mirrors.* Loewy found that, during four years at Paris, only ten nights could be termed first-class—even then his plates failed to show lunar detail now recorded by Camichel, whose photographs equal those taken with larger American refractors and plead strongly for the proposed establishment at the Pic-du-Midi of a 60-inch aperture reflector.

Lyot found that the best observing times followed heavy falls of snow, for the falling flakes carried with them dust particles, fine flower seeds and vegetable fibres. Coronagraphs work effectively only at selected high-altitude stations, preferably above the tree line and where the annual snowfall is heavy. M. Waldmeier of Zurich has a Lyot-type coronagraph at Arosa, 6200 feet above sea level, where atmospheric purity has enabled him to follow α Leonis (Regulus) to within 12·4 minutes of arc from the sun's limb.[93] Waldmeier's work at Arosa has added further evidence for the intimate relationship between sunspot and coronal activity.

During World War II, Lyot forestalled weak German attempts to use the Pic-du-Midi Observatory. Mention was made, on page 348, of the four Alpine observatories of the Fraunhofer Institute established during the war years. Besides solar tower-telescopes, these stations contain Zeiss 11-cm aperture coronagraphs and auxiliary apparatus of Lyot pattern. The finest site of the four is at Wendelstein, near Brannenburg on the Inn, Bavaria, where an 11-cm $f/15$ coronagraph, installed in 1943, has produced a valuable sequence of coronal observations.[94]

The High Altitude Observatory has a large coronagraph at Fremont Pass, Climax, Colorado. This station is at an elevation of 11,318 feet and experiences an average annual snowfall of some 20 feet. It is the highest coronagraph station in the world. The coronagraph is used for taking photographs of the sun at intervals of 10 seconds for many hours. A photoelectric guiding device, attached to the telescope, keeps the sun's disk centred in the field of view so that both long and short periods of guiding are quite automatic.[95] At another high-altitude station at Sacramento Peak, near Alamogordo, New Mexico, the American Air Force has financed the installation of a 16-inch aperture, 26-foot coronagraph of advanced design. A duplicate coronagraph, made possible by a grant from the Office of Naval Research, is being installed at Climax.[96]

Cinematography of the solar prominences has been taken up in an ambitious way at the McMath-Hulbert Observatory of the University of Michigan at Lake Angelus, 30 miles north of Detroit. This unique observatory is the culmination of a series of projects started in 1926 by F. C. and R. R. McMath and H. S. Hulbert, pioneers in the application of motion-picture technique to astronomical photography. These three, two engineers and a judge respectively, experimented until they developed a technique and an instrument by which the motion of astronomical

* The Baillaud telescope has at various times had several different optical systems installed in it. The 20-inch reflector never gave critical definition and was adapted to take a 15-inch object-glass from Toulouse. The instrument now carries both the folded $23\frac{1}{2}$-inch refractor and the 20-cm Lyot coronagraph.

Fig. 165—The Climax station of the High Altitude Observatory, Fremont Pass, Colorado

An artist's impression of the station as originally planned. The conical 'dome' is designed to shed the heavy snows of winter.

objects could be viewed on the projection screen. Through the introduction of frequency-controlled synchronous motors, they achieved the necessary accuracy in guiding. Experimental automatic cameras, attached to a 10·5-inch equatorial reflector, and operated over a wide range of exposure times and running rates, eventually enabled them to project motion pictures of sun, moon, and planets on a cinematograph screen.[97] In 1936, with the erection of a 50-foot tower telescope, emphasis was placed on solar research, and a programme of recording prominence motions perpendicular to the line of sight was undertaken.[98] In 1939, the horizontal Stone spectroheliograph was installed[99] in the 50-foot tower so that movements along the line of sight could be continuously recorded. In the following year, a 24-inch Cassegrain reflector with optical parts by Perkin-Elmer[100] replaced the original 10·5-inch reflector, while a second tower telescope, the 70-foot McGregor tower,[101] began to supplement the work of the 50-foot tower. In the 50-foot tower telescope, a range of focal lengths is made possible by the use of off-axis paraboloidal mirrors which can be introduced to reflect light from the coelostat up to a flat mirror and down again to the spectroheliograph head. In the

McGregor tower, light from the coelostat is passed through a long-focus objective. Quartz and Pyrex mirrors are used throughout, and the double walls of the towers, designed for adequate ventilation and insulation, reduce disturbing air currents to a minimum.

With the equipment in the McGregor tower, invaluable direct motion pictures of sunspots, faculae, and fine surface details can be obtained. A powerful Littrow-type grating spectrograph, used in conjunction with this telescope, permits the study of high-dispersion solar spectra. An all-reflecting Pfund-type spectrometer makes it possible to extend these studies far into the infra-red. Complex motions and striking forms of prominences are shown on the 35-mm spectroheliokinematograph records obtained at this observatory. On these films are recorded (in $H\alpha$ light) arched phenomena, eruptive prominences sending material away from the disk with velocities well above those of escape, bright surges of matter subsiding as rapidly as they rise, and slender faint streamers pouring downwards with little sign of ascending material.

Another important development in telescope construction in the 1930s was the introduction of the aluminizing process for forming reflecting films on optical surfaces. The process grew out of attempts to deposit a thin protecting layer on the faces of rock-salt prisms, the surfaces of which deteriorate when exposed to air. Silver coats, protected by a thin film of quartz—both applied by evaporation techniques—were then formed on glass surfaces. Ideas for coating astronomical mirrors with deposits other than silver soon matured and, by 1932, technical difficulties were so far overcome as to make possible the aluminizing of a 12-inch mirror.

Aluminium coats are, in many ways, superior to silver films. Aluminium reflects 89 per cent of visible light, and its reflectivity drops only to 85 per cent for the ultra-violet (3150 A). Silver, freshly applied and burnished, reflects 95 per cent of visible light, but falls to 4 per cent for the ultra-violet. Tests at the Lick Observatory showed that, for stellar photography, an aluminized mirror reflects about 50 per cent more light than does one of silver.[102] An aluminium film is more durable than one of silver. It need not be burnished after deposition and, on exposure to air, a hard, transparent oxide coat makes it possible to remove dust and marks with a moist, soft cloth, or even with soap and water, an operation which would entirely remove a silver film. Aluminium is by no means an ideal metal for coating astronomical mirrors, but it has the highest reflectivity of all the metals so far tried. Aluminium films are softer than those of chromium, which are so hard as to require the use of abrasives for their removal; but the reflectivity of chromium is 65 per cent in the ultra-violet and only 69 per cent in the visible part of the spectrum. Apart from silver, chromium was the first metal to be used (in 1932, on a 15-inch mirror at the Lowell Observatory) for coating astronomical mirrors.[103] Rhodium provides hard films, but it is an expensive metal and has a reflectivity slightly less than that of silver and aluminium. Physical and chemical ageing affect the properties of all thin metallic coats. The former gives rise to a granular structure which makes the film act like a reflection grating for short wavelengths, with consequent lack of definition under high powers. Chemical ageing gives rise to selective absorption by the film. Of all metals used in coating mirrors, rhodium is least affected by chemical and physical action.

Fig. 166—The McMath-Hulbert Observatory, University of Michigan

R. C. Williams[104] and J. D. Strong[105] were the first to discover a practical technique for coating mirrors with aluminium. Strong's account of the process is simply but directly told in the following extract from *The Telescope* for August, 1934.[106]

> Aluminum is heated in tungsten coils to such a temperature that the metal evaporates. This occurs in a large bell jar which is evacuated, so that an aluminum molecule, evaporating from the tungsten coil, travels in a straight line without interruption by collision with other molecules until it strikes some object, either the walls of the container or the mirror face to be coated. In order to obtain this collision-free path it is necessary that the container be evacuated to a pressure of 1/10,000 of a millimeter of mercury or less. . . . The container, in which such a pressure is possible, must be free from leaks to the extent that, if it were evacuated and sealed off, it would attain ½ atmospheric pressure only after the lapse of 15 years' time. . . . The pumps used in maintaining this high vacuum are of the oil diffusion type.
>
> In the process of coating an astronomical mirror the films are formed by evaporation from twelve coils arranged in a circle above the mirror. This arrangement of coils forms a film, on mirrors as large as 40 inches in diameter, 1/1000 of a millimeter (1/25,000 of an inch) in thickness, and uniform to four or five per-cent.

Another coating process applied to optical surfaces aims at making them reflect as *little* light as possible. In 1892, H. Dennis Taylor of the Cooke concern noticed that lenses which had weathered, and had thereby acquired a purple tarnish or bloom, transmitted more light.[107] He thereupon attempted to tarnish glass artificially by immersing crown-glass lenses in an aqueous solution of sulphuretted hydrogen. Surfaces ' leached ' in this way acquired a ' slaty brown colour ' and Taylor estimated that the transparency of a single surface was increased from 96·5 per cent to 97·5 per cent.[108] This meant that the reflectivity was reduced by about one-third from 3·5 per cent to 2·5 per cent. The chemical method of blooming was subsequently developed by Kollmorgen (1916),[109] Jones and Homer (1941)[110] and Nicoll (1942),[111] but for high-class optical work evaporated films are now preferred.

The evaporation process was introduced by Strong in 1936,[112] although Zeiss claimed to have used it in 1935.[113] The process aims at coating the surfaces of a lens with a thin, permanent, and transparent film. Ideally, the perfect coating for a surface in air should have a refractive index equal to the square root of that of the glass to which it is applied. The optical thickness should then be one-quarter wavelength of the incident light. Under these conditions, light reflected from the air-film boundary is 180° out of phase with light reflected from the film-glass boundary. By the principle of interference the intensity of the reflected light is zero. These requirements cannot be met in practice owing to the absence of suitable coating media and the usual wide range of wavelengths in the incident light. Using fluorite, Strong was able to decrease the intensity of reflected light at normal incidence by 54 per cent. Films of magnesium fluoride and cryolite, first tried out by Strong, are even more effective. A film of magnesium fluoride on crown glass of refractive index 1.52 reduces the reflectivity from 4.2 per cent to 1.3 per cent. For a similarly coated flint-glass surface of index 1.65, the reflectivity is reduced from 6 per cent to 0·6 per cent.[114] The value of coated elements, whether to increase image brightness or to reduce the intensity of ghost images, particularly when the optical system has several surfaces, is therefore appreciable.

REFERENCES

[1] *Optician*, **93**, 1937, May 7.
[2] Czapski, S., *Ap. J.*, **21**, p. 379, 1905.
[3] *Vide* Ref. 1.
[4] Czapski, *op. cit.*, p. 379.
[5] Zeiss, *Catalogue of Astronomical Instruments* No. 516e.
[6] Kuiper, G. P., *Pop. Astr.*, **54**, p. 269, 1946.
[7] Wolf, M., *Astr. Nach.*, **127**, p. 428, 1891; *Knowledge*, **14**, pp. 188, 230.
[8] Wolf, M., *Astr. Nach.*, **137**, p. 175, 1895; *Sirius*, p. 106, 1891.
[9] Wolf, M., *Sitzungsberichte d. kgl. bayer. Akad. d. Wiss.*, *Bd. xxxi*, Heft ii, p. 111, 1901.
[10] Gill, D., 1913, *History and Description of the Royal Observatory, Cape of Good Hope*, i, pp. 1–15.
[11] Brashear, J., 1925, *Autobiography*, pp. 70–72.
[12] *Ibid.*, p. 85.
[13] *Ibid.*, pp. 246–248.
[14] Spencer-Jones, H., *M.N.R.A.S.*, **104**, p. 95, 1944.
[15] *Ibid.*, p. 96.
[16] *Ibid.*, p. 97.
[17] *Ibid.*, p. 98.
[18] Rowland, H. A., Spectrum Wave-Lengths, *Ap. J.*, **1–5**, 1895–97.
[19] Plaskett, J. S., *Publ. Dominion Astrophysical Observatory*, **1**, No. I, 1922.
[20] Plaskett, J. S., *Handbook, Dominion Astrophysical Observatory*, p. 29, 1923.
[21] Plaskett, J. S., *Publ. Astr. Soc. Pac.*, **49**, p. 142, 1937.
[22] *Vide* Ref. 20.
[23] Plaskett, J. S., and Pearce, J. A., *Publ. Dominion Astrophysical Observatory*, **5**, pp. 168–235, 1935. Beals, C. S., *Ibid.*, **6**, pp. 333–337, 1938. McKellar, A., *Ibid.*, **7**, pp. 251–272, 1949.
[24] Ritchey, G. W., and Chrétien, H., *Compt. Rend.*, **185**, p. 266, 1927. Ritchey, G. W., *Revue d'Optique*, **1**, pp. 1–28, 1922.
[25] Ritchey, G. W., *Trans. Opt. Soc.*, **29**, pp. 198–200, 1928.
[26] Hargreaves, F. J., *M.N.R.A.S.*, **107**, pp. 36–38, 1947.
[27] Schwarzschild, K., *Mitt. der k. Stern. z. Göttingen*, ii, 1905.
[28] Couder, A., *Compt. Rend.*, **183**, ii, p. 1276, 1926.
[29] Schmidt, B., *Mitt. d. Hamb. Stern. i. Bergedorf*, **7**, p. 15, 1932.
[30] Schorr, R., *Astr. Nach.*, **258**, p. 45, 1936.
[31] Brockman, R., cites in *J.B.A.A.*, **59**, p. 112, 1949.
[32] *Vide* Ref. 30. For notes on the working and testing of Schmidt plates *vide* Twyman, F., 1952, *Prism and Lens Making*, pp. 347–353.
[33] *Amateur Telescope Making*, p. 397, 1944; *Publ. Am. Astr. Soc.*, **8**, pp. 110–111, 1935.
[34] *Amateur Telescope Making*, pp. 410–416, 1944.
[35] Cox, H., *J.B.A.A.*, **48**, pp. 308–313, 1938; **50**, pp. 61–68 1939.
[36] Väisälä, Y., *Astr. Nach.*, **254**, p. 361, 1935.
[37] Wright, F. B., *Publ. Astr. Soc. Pac.*, **47**, p. 300, 1936.
[38] Väisälä, Y., *Astr. Nach.*, **259**, p. 197, 1936.
[39] Dimitroff, G. Z., and Baker, J. G., 1946, *Telescopes and Accessories*, p. 294.
[40] Baker, J. G., *Proc. Amer. Phil. Soc.*, **82**, pp. 323–338, 1940.
[41] Burch, C. R., *M.N.R.A.S.*, **102**, p. 159, 1942; **103**, pp. 159–165, 1943.
[42] Linfoot, E. H., *M.N.R.A.S.*, **103**, p. 210. 1943; **104**, p. 48, 1944; **109**, p. 279, 1949, *Proc. Phys. Soc.*, **57**, p. 199, 1945; **58**, p. 65, 1946. *Proc. Roy. Soc.*, **186**a, p. 72, 1946. Linfoot, E. H., and Wayman, P. A., *M.N.R.A.S.*, **109**, p. 535, 1949. Linfoot, E. H., and Wolf, E., *Jour. Opt. Soc. Amer.*, **39**, p. 752, 1949.
[43] Wright, F. B., *Publ. Astr. Soc. Pac.*, **47**, p. 300, 1935.
[44] Baker, *op. cit.*
[45] Smiley, C. H., *Jour. Opt. Soc. Amer.*, **28**, p. 130, 1938.
[46] Maksutov, D. D., *ibid.*, **34**, pp. 270–284,
[47] *Vide* Ref. 34 pp. 399–400.
[48] Bowen, Ira S., *Jour. Opt. Soc. Amer.*, **42**, pp. 795–800, 1952.
[49] Struve, O., *Ap. J.*, **86**, pp. 613–619, 1937.
[50] *Ibid.*, p. 617.
[51] Hendrix, D. O., and Christie, W. H., *Scientific American*, Aug., 1939.
[52] Hendrix, D. O., *Publ. Astr. Soc. Pac.*, **51**, p. 158, 1939.
[53] *Publ. Amer. Astr. Soc.*, **8**, pp. 110–111, 1935.
[54] Carnegie Institute Washington, Mount Wilson Observatory, *Year Book*, **37**, p. 36, 1938.
[55] *Vide* Ref. 34, p. 398.
[56] *Vide* Ref. 54, **39**, p. 23, 1940; **40**, p. 27, 1941; **43**, p. 15, 1944.
[57] *Vide* Ref. 34, pp. 399–400; Ref. 49, p. 614.
[58] *Vide* Ref. 39, p. 292.
[59] Baker, *op. cit.*, pp. 339-349.

[60] Bok, Bart J., *Scientific American*, p. 46, July, 1952.
[61] *Ibid.*, p. 48.
[62] Whipple, F., *Observatory*, **72**, pp. 175–176, 1952.
[63] *Sky and Telescope*, **10**, p. 219, 1951.
[64] Whipple, F., *Perkin-Elmer Instrument News*, **4**, No. 1, p. 3, 1952.
[65] *Vide* Ref. 62, p. 176.
[66] Miller, F. D., *Perkin-Elmer Instrument News*. **3**, No. 2, p. 4.
[67] Nassau, J. J., *Ap. J.*, **101**, pp. 275–279, 1945.
[68] *Ibid.*, p. 278.
[69] *Ibid.*, p. 277.
[70] *Ibid.*, p. 279.
[71] *Ibid.*
[72] Fisher, C., *Pop. Astr.*, **54**, pp. 17–18, 1946.
[73] Cimino, Massimo, *Sky and Telescope*, **10**, pp. 263–264, 1951.
[74] Harrington, R. G., *Publ. Astr. Soc. Pac.*, **64**, pp. 275–281, 1952.
[75] Finlay-Freundlich, E., *Nature*, **165**, p. 703, 1950. Finlay-Freundlich, E., and Waland, R. L., *Sky and Telescope*, **12**, pp. 176–177, 1953.
[76] *Observatory*, **72**, pp. 60–62, 1952; **73**, pp. 182–183, 1953. *M.N.R.A.S.*, **113**, p. 314, 1953.
[77] *Observatory*, **73**, p. 181, 1953.
[78] Huggins, W., *Proc. Roy. Soc.*, **34**, p. 409, 1882–83.
[79] Hale, G. E., *Ap. J.*, **1**, p. 318, p. 438, 1895.
[80] Hale, G. E., *Ap. J.*, **12**, p. 372, 1900.
[81] Lyot, B., *M.N.R.A.S.*, **99**, p. 581, 1938–39.
[82] *Ibid.*
[83] *Ibid.*, pp. 581–582, 583–584.
[84] *Ibid.*, p. 583.
[85] Lyot, B., *L'Astron.*, **45**, p. 248, 1931.
[86] *M.N.R.A.S.*, **99**, pp. 584–585, 1938–39.
[87] *Ibid.*, p. 586.
[88] Lyot, B., *L'Astron.*, **46**, p. 272, 1932.
[89] *Vide* Ref. 86, p. 590.
[90] *Ibid.*, pp. 592–593; *L'Astron.*, **51**, p. 208, 1937; **52**, p. 204, 1938.
[91] *Vide* Ref. 39, p. 218. Evans, J. W., *Publ. Astr. Soc. Pac.*, **52**, pp. 305–311, 1940. Pettit, Edison, *Ibid.*, **53**, pp. 171–181, 1941. Lyot, B., *Annales d'Astrophysique*, **7**, pp. 31–79, 1944.
[92] Camichel, H., *L'Astron.*, **60**, pp. 161–162, Pl. 1 and 2, 1946.
[93] Waldmeier, M., *Zeit. f. Astr.*, **19**, p. 21, 1939.
[94] *Pop. Astr.*, **54**, pp. 271–272, 1946.
[95] *Electronics*, 1946, June.
[96] Menzel, D. H., *Sky and Telescope*, **10**, pp. 187–189, 1951.
[97] *Vide* Ref. 39, p. 199.
[98] *Ap. J.*, **85**, p. 4, 1937.
[99] *Publ. Observatory of the University of Michigan*, **8**, pp. 57–59, 1943.
[100] *Ibid.*, pp. 95–101.
[101] Sellers, F. J., *J.B.A.A.*, **51**, p. 187, 1941.
[102] Strong, J. D., *The Telescope*, **1**, p. 66, 1934.
[103] Strong, J. D., *Ap. J.*, **83**, p. 423, 1936.
[104] Williams, R. C., *Physical Review*, **41**, p. 255, 1932; **46**, p. 146, 1934.
[105] Cartwright, C. H., and Strong, J. D., *Review of Scientific Instruments*, **2**, p. 189, 1931. Strong, J. D., *Ap. J.*, **83**, pp. 401–423, 1936.
[106] *Vide* Ref. 102, p. 67.
[107] Taylor, H. Dennis, 1896, *The Adjustment and Testing of Telescope Objectives.*
[108] Taylor, H. Dennis, *Brit. Patent* 29561, 1904.
[109] Kollmorgen, *Trans. Soc. Ill. Eng.*, **11**, p. 220, 1916.
[110] Jones, F. L., and Homer, H. J., *Jour. Opt. Soc. Amer.*, **31**, p. 34, 1941.
[111] Nicoll, F. H., *R.C.A. Review*, **6**, p. 287, 1942.
[112] Strong, J., *Jour. Opt. Soc. Amer.*, **26**, p. 73, 1936.
[113] Twyman, *op. cit.*, p. 483.
[114] Candler, C., 1951, *Modern Interferometers*, p. 75.

CHAPTER XVIII

I believe that a 200-inch or even a 300-inch telescope could now be built and used to the great advantage of astronomy.

G. E. HALE

IN 1918, the Dublin works of Sir Howard Grubb were moved to St Albans, Hertfordshire, England, and, upon Grubb's retirement in 1925, the entire business was acquired by the Hon. Sir Charles Parsons, F.R.S., son of the third Earl of Rosse. Parsons, at this time, was owner of Messrs C. A. Parsons & Co., electrical engineers, and of the Parsons Optical Glass Company of Derby. In 1894, his invention of the steam turbine was applied to the yacht *Turbinia* and, four years later, he established a new company—the Parsons Marine Steam Turbine Co., with works at Wallsend, Newcastle. In 1925, Parsons erected new works at Walkergate, Newcastle, for the business Sir Howard Grubb, Parsons and Company, manufacturers of large astronomical telescopes.

The first large Grubb-Parsons telescope* was the 74-inch David Dunlap reflector, situated on a site on Richmond Hill, about 12 miles north of Toronto. Grubb-Parsons undertook all the work except the casting of the mirror which was finally ordered from the Corning Glass Works, Corning, New York.

> When the telescope was ordered in 1930 [Prof. R. K. Young wrote][1] we knew that the portion which would probably take the longest to complete was the big mirror. At that time, the Grubb-Parsons Company controlled the Parsons Optical Glass Works at Derby and Sir Charles Parsons, head of C. A. Parsons & Co. of which three other companies were subsidiaries, was confident that they could manufacture a suitable disk of glass for the telescope mirror. But Sir Charles was in his 76th year when the order for our telescope was placed, and unfortunately he did not live to see the disk made.

In 1932, after Parsons' death, the disk was not forthcoming.

> In 1932 [Young continues] unexpected help arrived in connection with the manufacture of telescope mirrors which was not available in 1930. In the latter year the only firms which could undertake the manufacture of large disks were Carl Zeiss and the Glass Works at Derby in England.

News reached Young that the Corning Glass Works were making plans for casting a 200-inch disk of Pyrex glass (for the Hale telescope). The Toronto authorities thereupon prevailed upon Corning to make them a 74-inch disk first. The mirror was cast on June 21, 1933, and came out of the annealing oven in September of that year.[2] Upon its arrival in England, the working of the mirror and subsequent

* If we overlook the 40-inch Simeis reflector, erected in 1928, but largely the work of Sir Howard Grubb (*vide* page 351), and the 40-inch reflector erected in 1931 at the Stockholm Observatory, Saltsjöbaden, Sweden.

testing by the Hartmann photographic method were entrusted to Mr Armstrong of Grubb-Parsons.

C. A. Chant and R. K. Young of Toronto University collaborated with C. Young, manager of Grubb-Parsons, in the design of both telescope and dome. A novel feature of the telescope is the large iris diaphragm fixed just in front of the primary mirror. This iris can be closed to form a circle 12 inches in diameter (the size of the central hole in the mirror) and, when closed, forms a chamber over the reflecting surface. The weather is changeable at Toronto, with variations in temperature and humidity which cause the telescope to ' sweat '. By keeping the chamber dry by small electric heaters, and by packing the mirror sides with absorbent cotton, the metal film is well protected and yet, by means of the iris diaphragm, is soon uncovered for use.[3] Two alternative optical arrangements are possible. As a Newtonian, the instrument operates at $f/4{\cdot}9$ (nominal) with a focal length of 30 feet. As a Cassegrain at $f/18$, the focal length is 111 feet.

The mounting is of the conventional cross axis English type, with polar axis 22 feet long and 9 tons in weight. The axis is built in three sections, with two conical steel sleeves bolted to a central cubical box and two steel pivots shrunk into the extreme ends. This axis turns on self-aligning ball bearings and carries on one side the lattice tube and, on the other, the 13-foot declination axis and massive counterpoises. The entire instrument is electrically operated and is covered by an 80-ton dome. Both dome and circular observatory walls are of steel, double-walled to allow free circulation of air. Inside and outside surfaces are covered with agasote, a hard paper product, while the outside is further protected by copper sheeting.[4]

Like the 72-inch Victoria telescope, the 74-inch has made extensive contributions to astrophysics. Its main employment is radial-velocity determinations, an extension of the Lick radial-velocity programme to fainter stars. An unusual but important activity at David Dunlap was the meteor observing programme started in 1933 by P. M. Millman. Prof. J. F. Heard is now director of the observatory in succession to Prof. F. S. Hogg.

Other large Grubb-Parsons instruments of the thirties are the 36-inch reflectors of Edinburgh and Greenwich (the Yapp reflector, now moved to Herstmonceux, near Hastings), a 40-inch reflector, and a 24-inch and 20-inch twin refractor for Saltsjöbaden, Sweden, a 33-cm astrographic refractor for Rozspienwany, Poland, and a 40-cm twin astrographic refractor for the Leyden Rockefeller foundation for use at the Union Observatory at Johannesburg.

In 1934, Grubb-Parsons built their second solar telescope, the first (1927) being a 12-inch fixed refractor for the Commonwealth Observatory, Canberra, Australia. The second instrument, the Oxford solar telescope,[5] is of interest because it is the only vertical solar telescope in England and because it shows what could be done with only £4000. The instrument is housed in the old De la Rue dome and is supported by the tapered brick column which previously carried the 13-inch De la Rue reflector. A limit of £1800 was imposed for the cost of the mechanical and optical parts. The telescope takes the form of a vertical Cassegrain reflector with a $12\frac{1}{2}$-inch primary mirror and 6-inch diameter convex of equivalent aperture ratio $f/62$. Beneath the original dome is the 16-inch coelostat and secondary mirror, both of silvered quartz. This telescope, together with the auxiliary Casella-Hilger

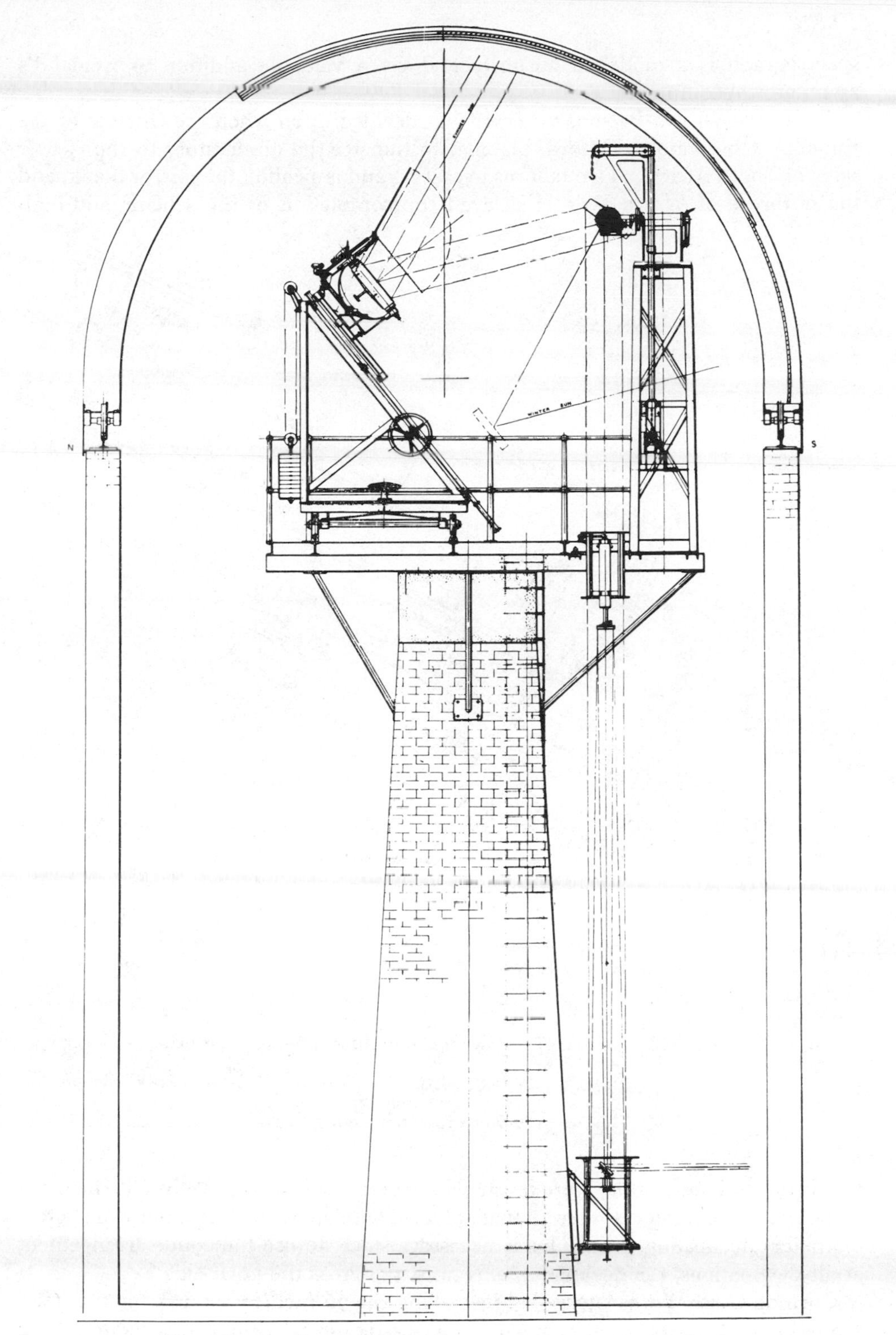

Fig. 167—Section of the solar telescope at the Oxford University Observatory

(Sir Howard Grubb, Parsons & Co.)

spectrograph is a model of austerity and yet a valuable addition to England's astronomical equipment.

A 74-inch Grubb-Parsons reflector was decided upon when the trustees of the Radcliffe Observatory, Oxford, planned to transfer the observatory to the clearer skies of South Africa. Two failures in casting and annealing the mirror blank, and the outbreak of World War II delayed the completion of the scheme and both

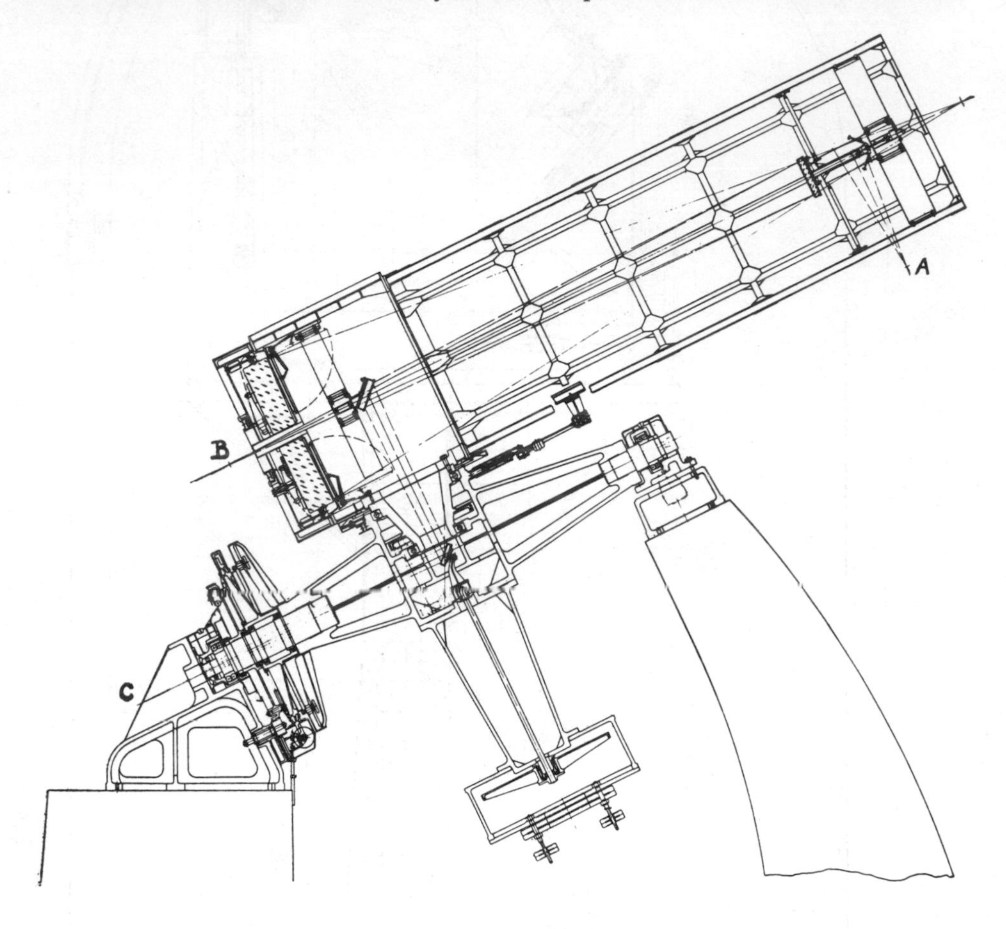

Fig. 168—Section of the 74-inch Radcliffe reflector, Pretoria

A 30-foot Newtonian focus
B 111-foot Cassegrain focus
C 173-foot Cassegrain-coudé focus

mounting and observatory were finished long before the mirror received its final figuring. The mounting is similar to that at David Dunlap, with a short but wide (4 feet diameter) declination axis. This is necessary since, to use the coudé arrangement at all declinations, the declination axis must not cross the polar axis.[6]

Corning Glass Works succeeded in making the 76-inch Pyrex disk for this telescope in 1938. It was then shipped to Newcastle for grinding and polishing. It is

Fig. 169—The 74-inch Radcliffe reflector, Pretoria

(Sir Howard Grubb, Parsons & Co.)

11 inches thick and has a central aperture 7 inches in diameter. Nine pads on double-ring ball bearings support the disk in its cell; the bearings, like those constructed for the 100-inch Hooker telescope, prevent friction between the back of the mirror and the supports. To keep the temperature of the mirror as even as possible and to reduce convection currents near its surface, the cell is packed in asbestos. When not in use, the mirror is protected by a cover formed of twelve sectors, counterbalanced and readily removed by turning a handwheel.

There are three foci. The Newtonian focus is 30 feet from the mirror and, with a 9-inch correcting lens of the kind introduced by F. E. Ross,[7] is used for direct photography. Two alternative Cassegrain convex mirrors give foci of 111 and 173 feet at the central aperture of the main mirror and north end of the polar axis, respectively. All secondary mirrors are of aluminized fused silica, cast by the Thermal Syndicate of Wallsend, Newcastle. The spectrograph prisms and lenses for use at the Cassegrain foci (as with the Toronto telescope) are by Messrs Hilger and Watts Ltd, with mechanical parts by Messrs C. F. Casella & Co.

The centrepiece of the tube and the mirror cell are of fabricated steel, while the open lattice tube is composed of duralumin bars and tie rods. This design leads to lightness in weight and low heat capacity.

The telescope drive departs from weight-driven clock and governor for a variable-frequency drive of the type developed at the McMath-Hulbert Observatory.[8] The frequency of the A.C. supply was not considered constant enough for use with a synchronous motor and the McMath-Hulbert drive was adopted with, however, a low-frequency tuning fork in place of a thermionic oscillator.

Instead of the more usual dome, the lower brick wall is surmounted by a cylindrical double-walled turret. This enables the observer, by special platform, to work at the Newtonian focus in all positions of the tube. In designing the building, every precaution was taken so that the telescope should remain at almost constant temperature. At telescope level, the temperature range inside the turret is less than 7° F, despite the large diurnal range on the veldt outside.[9]

Another Grubb-Parsons reflector is now installed at the Commonwealth Observatory, Mount Stromlo, Canberra. Before being shipped to Australia, the instrument was a major exhibit under the Dome of Discovery at the South Bank Exhibition of the 1951 Festival of Britain. The glass disk of 76 inches diameter, made by Messrs Pilkington Bros Ltd, is of 74 inches clear aperture. The telescope is of the same size and design as the Pretoria reflector. At the time of writing, November, 1953, Grubb-Parsons are well advanced in work on yet another 74-inch, this time for the Helwân Observatory, Egypt. A further addition to the equipment at Canberra is a large Gregorian-Schmidt, made possible by reconstructing the old Melbourne reflector (p. 267). The 50-inch spherical mirror, also by Pilkington, was worked by the London firm of Cox, Hargreaves and Thomson.[10]

Messrs Grubb-Parsons have now taken over the astronomical instrument side of Messrs Cooke, Troughton and Simms Ltd. Up to the death of James Simms in 1915, Troughton and Simms produced many large astronomical refractors and developed a world-wide reputation for surveying instruments. Similar work, notably equatorial refractors of 6 inches aperture and over, emanated from the workshops of T. Cooke and Sons, and it was not surprising that, in 1922, the two firms should combine to form Messrs Cooke, Troughton and Simms Ltd.

To undertake the various programmes of meridian observations at the Royal Observatory, Herstmonceux, two transit circles are necessary. The government of Victoria, N.S.W., have recently made the observatory a gift of the 8-inch transit circle by Troughton and Simms, installed at Melbourne Observatory in 1884.[11] The main instrument, however, is the reversible Cooke, Troughton and Simms transit circle installed at Greenwich in 1936. This comparatively new instrument is similar in design to the circle already built by the same firm for the Cape Observatory. But whereas the graduations of the fixed circle of the Cape instrument are made on an inlaid band of iridio-platinum and, on the movable circle, on inlaid silver, the 28-inch diameter circles of the Greenwich instrument are of glass with etched graduations.[12] The circles are thus free from the tarnishing propensities of bands of silver. Each glass disk weighs 16 lb and required careful design and construction. It became necessary to evolve a composite mount in which the differential expansion of zinc and gunmetal was used to ensure that the circles were perfect fits in their respective spigots over the range of temperature likely to be encountered. Flexure in the plane of the disks is minimized by a system of counterpoised levers, and cover disks protect all but the 24-inch diameter graduated annules. The 7-inch object-glass, together with the tube and moving parts, weighs about 1200 lb—a weight so relieved by counterpoises as to leave only 30 lb weight effective on each pivot bearing. Interesting features are the hollow, cast iron pillars, filled with a solution of ethylene-glycol in water to prevent sudden expansion with rise in temperature. Reversal of the axis is made possible by a carriage which runs on rails beneath the telescope.

Some idea of the accuracy of the mechanical parts can be seen by the fact that the $4\frac{1}{2}$-inch diameter pivots differ from true cylinders by not more than 0·00005 inch. The total range in the division errors of the fixed circle is about 0·7 second of arc and, with the travelling wire of the impersonal micrometer (driven by a synchronous motor), the probable error of a single transit is reduced to less than 0·01 second of time.

Since 1931, an ingenious method of photographic registration applicable to existing transit circles has been developed at the U.S. Naval Observatory. By this method the conventional microscope micrometers are replaced by a set of 35-mm-film cameras which automatically record the circle readings. A similar conversion is now in progress for the reversible transit circles at Herstmonceux and the Cape. A later addition at Washington is an automatic photoelectric measuring machine designed to measure the circle microscope photographs of the 6-inch transit circle. The machine measures some 300 photographs on a 35-mm film-strip and preliminary tests showed that it can read the graduated scale on a single photograph with at least the same accuracy as a practised observer, that is, within about 0·20 second of arc.[13] Since 1938, C. B. Watts of Washington has also experimented with another interesting photographic device. While the transit is in progress, a standard clock initiates periodical photographs of the divided head of the screw of the moving-wire micrometer. Watts applied the idea to the 6-inch transit circle and elaborated a method whereby any systematic displacements of the travelling wire from its indicated position while in motion could be detected and evaluated.[14]

Attention is also being given to designs of meridian instruments which promise to be less affected by systematic errors than are the most accurately-made transit circles. One proposal, due originally to H. H. Turner and developed by R. d'E. Atkinson,[15] is the mirror transit circle. In this a plane mirror replaces the usual moving telescope which is now fixed in the meridian. Effects of flexure are considerably reduced, and small displacements of the object-glass and parts of the micrometer eyepiece are eliminated. Also, with a fixed telescope a longer focal length can be used with advantage. A mirror transit circle has been constructed at Pulkowa and another is under construction at Oporto.

Whilst on the subject of high-precision instruments at the Royal Observatory, we are reminded of the Cookson floating photographic zenith telescope.[16] In 1885, S. C. Chandler of Boston, Massachusetts, found an apparent periodical variation in the latitude which Küstner of Berlin confirmed independently a year or two later. Chandler used an instrument which he had developed and which he called an *almucantar*. This was essentially a telescope fixed to a base which floated in liquid and which enabled the tube to rotate about a vertical axis. As a meridian transit circle is restricted to observations in the meridian, so the almucantar is similarly restricted to a horizontal circle. About the year 1900, Bryan Cookson designed a floating photographic zenith telescope for the purpose of investigating the constant of aberration and the variation of latitude by an adaptation of the Talcott method.[17] In Cookson's instrument, the Y's of the telescope are carried by an iron annulus which floats in an annular trough containing mercury. As with the almucantar, this arrangement dispenses with the usual levelling arrangements which form so important a part of transit instruments. The 6½-inch aperture, *f*/10 Cooke objective with the tube and iron float weighs some 350 lb and requires about 142 lb of mercury in the iron trough. N. Cookson and C. A. Parsons cast and turned the float and trough, but H. Darwin and technicians of the Cambridge Scientific Instrument Company made all the other mechanical parts. Cookson used his zenith telescope at the Cambridge Observatory between 1903 and 1908 and, after his early death in 1910,[18] the instrument passed to the Royal Observatory, Greenwich, and has since been in regular use. A photographic zenith telescope has also been used with success at Washington for the determination of time, and an improved form, for the same purpose, is being installed at Herstmonceux.

After the death of J. Brashear, his business was taken over by McDowell, who continued to manufacture large refracting telescopes. McDowell's two largest refractors have done useful work in the southern hemisphere; both were erected in the 1920s. The first, a 26-inch photographic refractor, with mechanical parts made in the local shops of the observatory, was at the Yale-Columbia southern station at the University of the Witwatersrand, Johannesburg.[19] As mentioned previously (page 351), this telescope played an important part in Schlesinger's programme of photographic stellar parallax determination. It is now at a new site at Mount Stromlo, Canberra—a move necessitated by the hindrances of city lights and smoke from industrial developments at Johannesburg. The second instrument, a 27-inch designed by Hussey, is at the Lamont-Hussey Observatory, Bloemfontein,

Fig. 170—J. W. Fecker

He is examining a quartz gravity pendulum in his workshop on February 3, 1931

(American Museum of Natural History. Photograph by Clyde Fisher)

South Africa.[20] This telescope made possible an intensive programme of double star discovery and measurement; up to 1944, January, 5179 new pairs were discovered.[21] The object-glass can be corrected for photography by an auxiliary lens; with the latter in position, E. C. Slipher, in 1939, took some of his finest photographs of Mars. In 1954 he again observed Mars with this instrument.

J. W. Fecker did not take over the Brashear concern until 1926, although McDowell invited him to join forces in 1923. Fecker's first large optical undertaking was to figure the 69-inch $f/4{\cdot}3$ mirror and smaller mirrors for the Perkins reflector, then under construction for Ohio Wesleyan University, Delaware, Ohio, by Warner and Swasey.[22] For a few years, this telescope was the third largest in the world.

In 1933, Fecker refigured the old Common 60-inch $f/5{\cdot}1$ mirror, the original mounting for which had stood for many years in the grounds of the Harvard College Observatory, Cambridge. Fecker made a new mounting and the telescope,[23] named the Rockefeller telescope and the third largest in the southern hemisphere, is now in use at Harvard's Bloemfontein station. The late J. S. Paraskevopoulos designed and constructed a set of control rods for this instrument, whereby he could adjust the figure of the primary mirror for different inclinations of the tube.[24] The large-scale images given by this telescope have proved invaluable for the detailed photographic study of southern nebulae and star clusters. A further example of Fecker's work at Bloemfontein is the new mounting which he provided for Clark's 24-inch Bruce photographic doublet. As mentioned on page 302, this mounting now carries a 33-inch × 35·6-inch Cassegrain-Schmidt in place of the Bruce lens.

In 1937, Fecker made his first large telescope, the 61-inch *f*/5·1 Wyeth reflector[25] for Harvard's Agassiz station, some 25 miles from Cambridge, Massachusetts. This telescope is a Cassegrain-Newtonian on Fecker's favourite fork-type mounting. A similar but smaller telescope was the 28-inch which he made for G. W. Cook's Roslyn House Observatory at Wynnewood, Pennsylvania. In 1940, we again find Fecker putting the finishing touches to a large mirror, this time the 60-inch *f*/5·0 for Warner and Swasey's reflector[26] at the Cordoba Observatory, Argentina. A 30-inch reflector, constructed in the observatory shops, a 13-inch Henry and Gautier astrograph, and a 12-inch Clark, with some smaller instruments, complete the equipment of this observatory.

Fecker was also responsible for the remounting of several of America's early reflectors and refractors, the construction of the 20-inch aperture, *f*/7 Ross astrographic lens at Lick (the largest of its type in the world), the design and erection of the Copernican Planetarium for the American Museum of Natural History, and the construction of the mounting for G. W. Cook's horizontal 15-inch siderostat.[27] The comparatively early death, in 1945, of so able and so versatile an engineer was a great loss for American astronomy.

The firm now operates as J. W. Fecker Inc.; J. Kalla, R. Kraus, and C. Binder comprise the present board of directors. Since 1945, and in addition to many smaller instruments, J. W. Fecker Inc. has constructed two 10-inch refractors, an 8-inch Maksutov telescope for the City College of New York, a 16-inch coelostat and 24-inch reflector for the Arizona State College, Flagstaff, Arizona, and the mounting for the Dyer telescope of Vanderbilt University, already referred to in Chapter XVII. The firm is now constructing a 38-inch Cassegrain for the observatory of Butler University, Indiana.

Worcester Warner died in 1929[28] while travelling in Germany with his family. Swasey lived eight years longer than his partner and died in June, 1937,[29] in Exeter, New Hampshire. Both men were widely travelled and both took an active interest in higher American education. Warner became President of the American Society of Mechanical Engineers in 1897 and, among other honours, received the degree of Doctor of Mechanical Science. Swasey received the decoration of the Legion of Honour in 1900 and, later, the Case School of Applied Science at Cleveland conferred upon him the honorary degree of Doctor of Engineering. In 1910, he received the degree of Doctor of Science from Denison University, Ohio.

The latest and largest undertaking of the Warner and Swasey Company was the construction of the 82-inch McDonald telescope on Mount Locke, Texas.[30] This reflector, third largest in the world, was completed in 1939. The project started when W. J. McDonald, a banker of Paris, Texas, left some $1,000,000 to the University of Texas for the purpose of constructing and maintaining a large astronomical observatory. By itself, the sum was insufficient to build and maintain a large observatory, but as the University of Chicago was anxious to have a large reflector at its disposal, its offer to staff an observatory and share in the costs of operation was accepted.

Fig. 171—The 69-inch reflector of the Perkins Observatory

Ohio Wesleyan University, Delaware, Ohio

(*Warner & Swasey Co.*)

The McDonald mounting is similar to that of the other large polar-axis type reflectors of this century but it incorporates several new features due to E. P. Burrell. J. S. Plaskett tells us:[31]

The conditions and specifications laid down for this instrument were most exacting and difficult of fulfilment and required the exercise of the utmost skill and ingenuity, as well as practically unending patience and perseverance on Mr. Burrell's part, to satisfy. I had the great pleasure and privilege, during visits to Cleveland, of looking over and discussing with him the drawings and various details in partial and completed form and my respect and admiration for his skill, ingenuity and perseverance could not help but be increased by this masterly example of the designer's art. The working out of the coudé form in a telescope of this type, the development of duplicate elevating platforms for reaching the Cassegrain focus in every position, as well as

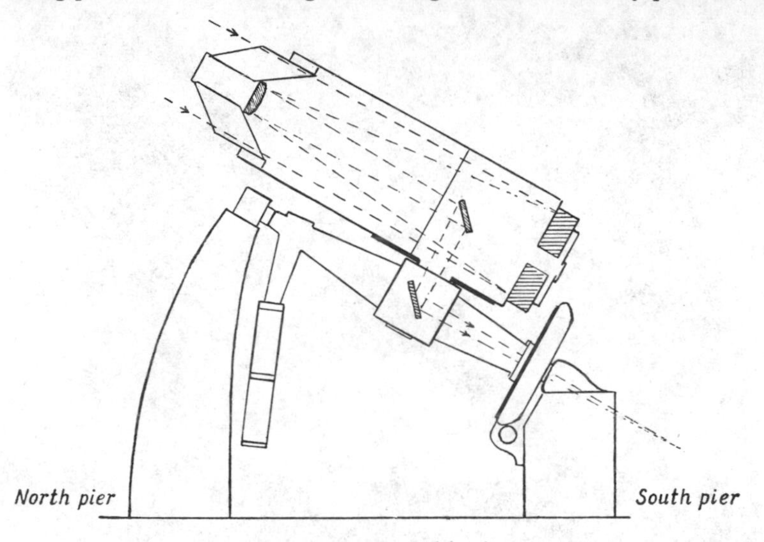

Fig. 172—The McDonald telescope

A simplified diagram showing the mirror arrangement for the coudé focus

for resilvering and changing accessories, the simplified arrangements for changing mirrors at the primary focus and the vacuum tube type of electric drive are some of the novel and ingenious features developed by Mr. Burrell in this instrument.

As if initial design difficulties were not great enough, Prof. Otto Struve informs us that one ' well-wisher ' wrote to the authorities to say that the 72-inch Victoria telescope had such tube flexure that the tube could not be used at zenith distances greater than 30°. Consequently, any design based upon that instrument would be a costly failure![32]

A novel feature of the McDonald reflector is the position of the counterpoise and the large disk bearing which replaces the declination axis. Instead of being attached to a declination axis, the lead and steel counterpoise is attached to the polar axis near the north pier end. This arrangement permits a low mounting, more freedom for the observer at the Cassegrain focus and unrestricted optical

Fig. 173—The 82-inch McDonald telescope

(*Warner & Swasey Co.*)

paths. It certainly disposes quite effectively of some 13 tons or more which would otherwise swing about midway on the polar axis.

The four radial fins which extend from the open end of the tube are unique in telescope design. They support the secondary convex mirror and give easy access to auxiliary apparatus at the primary focus. Altogether, the moving parts weigh 45 tons, yet a $\frac{3}{4}$-h.p. motor is sufficient to turn the declination axis, while an even smaller motor operates the polar axis.

The primary mirror was cast by the Corning Glass Works and took C. A. R. Lundin, at the Warner and Swasey Company, six years to bring to a paraboloidal figure.[33] It weighs $2\frac{1}{2}$ tons, was given a coat of aluminium-chrome by R. C. Williams at Cleveland and was ready for use early in March, 1939. G. P. Kuiper, the second director of the Yerkes and McDonald Observatories, and G. Van Biesbroeck conducted the first night tests and secured a number of spectrograms before May 5, the day of official dedication.

> Those of us who have had occasion to make visual observations with the 82-inch McDonald telescope [Otto Struve writes][34] will agree with me that on nights of good seeing the gain in visibility of details on the Moon and planets over the 40-inch [Yerkes refractor] is at least as great as the gain described by Mr. Hale of the 40-inch over the Kenwood 12-inch. [And later][35] I have estimated that with the 82-inch reflector I quite regularly secure in one month of concentrated work a larger amount of observational material, and material of far better quality, than I had secured with the 40-inch in 20 years of work!

On photographs taken with the McDonald reflector in February, 1948, Kuiper discovered Miranda, an additional moon for Uranus.[36] In the following year, Kuiper was able to confirm the existence of a second moon for Neptune. This addition to Neptune's family is of magnitude 19·5 and has a diameter of about 200 miles.[37]

Following the successful completion of the 100-inch Hooker telescope, it was natural that astronomers should consider the possibility of building a larger instrument. In the years just previous to the development of the 200-inch project, all manner of suggestions were made for telescopes of 200 and 300 inches aperture. Among these should be mentioned Ritchey's design for a vertical 300-inch reflector, with interchangeable cellular mirrors in Schwarzschild-Chrétien form, fed by a great coelostat and secondary mirror.[38]

> The observers work always in a stationary, comfortable position, as at their desks. The mirrors are light, cellular and ventilated, their optical form unaffected by temperature changes; they are supported without sensible flexure, by mechanical flotation. The weight of the coelostat mirror and its cell, the only parts which move during an exposure, is entirely floated in mercury; their rotating-mechanism is fine-ground and polished with optical accuracy. The stationary, vertical, square telescope-tube is near the centre of the massive, inner, concrete building; its flexures are immeasurably small; it is insulated with thick layers of cork; its expansion and contraction are very slow and small. Permanence of alignment of all optical mirrors is therefore practically perfect.

These and other ideals envisaged by Ritchey, to which he devoted much of his life's work, are obtainable only at enormous expense. The constructional difficulties of the 200-inch Hale telescope, overcome after what was tantamount to a national effort, would be small compared with those inherent in Ritchey's vast design.

Many substitutes for glass astronomical mirrors have been advocated, ranging from disks of cement and marble to composite mirrors of invar or stainless steel. Of these, obsidian held out attractive possibilities and is still under study. Obsidian is a black, volcanic glass which often occurs in large masses in the natural state; small disks take a good polish. Compound metal mirrors are difficult to make and are subject to the diffraction effects experienced by Lord Rosse. Telescopes consisting of several independent paraboloidal mirrors, all focusing to one common focal point, give poor images and have a resolving power equal to that of only a single mirror (page 435).

To escape from these difficulties, R. W. Wood, professor of experimental physics at Johns Hopkins University, attempted in 1908 to make a mirror by rotating a shallow pan of mercury at constant speed.[39] By centrifugal action, the mercury surface assumes a paraboloidal form, of focal length varying with speed of rotation. Wood used a 20-inch diameter pan, driven by a rubber thread transmission and magnetic clutch but, despite the velvety drive, disturbances appeared on the mercury surface. The ' mirror ' was placed at the bottom of a well and, in the summer of 1909 and in conjunction with a 20-inch flat, Wood obtained ' excellent views of the moon.' By altering the speed of rotation, he could vary the focal length from three to over twenty feet. ' The instrument resolves stars three seconds apart, shows the smallest craterlets on the moon, and yields wonderfully bright images of nebulae when running at short focus.' But the limitations imposed by the fixed direction, coupled with changes in focus and mirror performance, decided Wood against the scheme and, after 1909, he discontinued his experiments.

Ideas of a telescope larger than the Hooker reflector had occupied Hale's thoughts for a number of years, but it was not until after his ' retirement ' that he was able to give them serious attention.

> Starlight is falling on every square mile of the Earth's surface [he wrote in *Harper's Magazine* in 1928] and the best we can do is to gather up and concentrate the rays that strike an area 100 inches in diameter. I have never liked to predict the possibilities of large telescopes, but the present circumstances are so different from those of the past that less caution seems necessary. The question remains whether we could not safely advance to an aperture of 200 inches or, better still, to twenty-five feet.

A little later, Hale had conversations about large telescopes with Dr Wickliffe Rose, president of the Rockefeller International and General Education Boards, with the result that Dr Rose visited Pasadena and Mount Wilson to see the 100-inch reflector for himself. If a larger telescope were made, astronomers and engineers agreed that it should be limited to 200 inches aperture. A sudden advance from 100 inches to 300 inches was considered too uncertain for safety. Hale's intention, should a dream become reality, was to establish the new telescope on Mount Wilson, but the particular work awaiting the 200-inch, the study of faint extra-galactic nebulae, seemed to indicate a new observatory site. We have mentioned (page 337) that Mount Wilson no longer enjoys the original purity of night sky owing to industrial developments in the vicinity of Los Angeles, the effect of which became appreciable during the long exposures required for photographing faint extra-galactic nebulae.

Dr Rose seemed set on giving the telescope to the California Institute of Technology but, at the same time, did not wish the new organization to conflict in any way with the interests of the Carnegie Institution.

> As soon as the attitude of the authorities of the Carnegie Institution and the California Institute could be learned [Hale wrote in *Nature*][40] a plan was prepared in New York for Dr. Rose, embodying the following features: (1) Close co-operation between these institutions in the design and operation of the new observatory; (2) the laboratory, instruments and shops to supplement those of the Mount Wilson Observatory, thus providing new and very desirable facilities for joint research; (3) special stress to be laid on the development of new auxiliary instruments, so as to increase the efficiency of the 200-inch and other telescopes. This meant the construction of a very complete astrophysical laboratory, a well-equipped instrument shop and an optical shop large enough for the grinding, figuring and testing of the 200-inch mirror, all on the grounds of the California Institute in Pasadena, in close touch with the various laboratories of the Institute and the Pasadena headquarters of the Mount Wilson Observatory.

Following these plans, and after more money-raising efforts by Hale, came a grant of $6,000,000 to the California Institute of Technology, with promises of several more millions as the work proceeded.

The observatory council comprised Hale, R. A. Millikan, A. A. Noyes, and the banker, H. M. Robinson. Its first move was to collect recommendations from leading astronomers, technicians, and engineers. For the post of executive and liaison officer in the discussion of technical points, Hale chose Dr John Anderson. Dr Anderson was, for fifteen years, chief optical expert at Mount Wilson. To him fell eventually the tasks of choosing the observatory site and the organization of the optical work in the shops of the California Institute of Technology, Pasadena. Another valuable addition to the project was the late Russell W. Porter, tool designer, artist, draughtsman, and telescope-maker. Porter had a genius for visualizing complicated machinery in three dimensions. His rapid three-dimensional pencil sketches proved invaluable during the long design sessions. From Porter's office and drawing-boards in Pasadena originated designs for small telescopes for making preliminary seeing tests on the selected site and large numbers of drawings of the astrophysical laboratory, observatory, and telescope. His contributions, too numerous and too indefinite to list here, soon became necessary to the success of the scheme.

Mention of Anderson and Porter reminds us of the impossibility of apportioning credit with accuracy—someone would assuredly be overlooked. The Hale telescope and observatory resulted from the co-operative effort of a legion of experts. While we shall endeavour to be as objective as possible in the following pages, the critic will undoubtedly find omissions.

There were many suggestions for the composition of the mirror, among them invar, stellite nd stainless steel, composite metal and glass mirrors, Ritchey's cellular mirror, fused quartz, and Pyrex. From these, the observatory council and its advisory committee selected fused quartz, with Pyrex as alternative. Quartz is abundant in the natural state and fused quartz or silica has an expansion coefficient about one-fifteenth that of plate glass and a higher thermal diffusivity. It is harder than glass and less easily scratched; it takes a fine polished surface. Elihu Thomson, director of the Thomson Laboratory of the General Electric Company at Lynn, Massachusetts, had, moreover, substantial practical experience with quartz.

Thomson had previously made fused silica disks of small diameter by covering a quartz blank with slabs of clear quartz and melting them down to a smooth surface. Now, in conjunction with A. L. Ellis, assistant director, and with the resources of an industrial laboratory behind him, he resumed his experiments. Quartz fuses at 1650° C but vaporizes appreciably at a slightly lower temperature and, in order to eliminate air-bubbles from the surface, pulverized quartz was sprayed on with an oxy-hydrogen flame-gun. In this way, disks of 25 inches diameter were obtained. Further developments along these lines gave Thomson a 66-inch disk, but this and a second of the same size proved unsatisfactory. In 1931, after an expenditure of $600,000 and with little hope of a 200-inch disk being produced, the long series of experiments was abandoned.

> It is hardly necessary to tell you [Hale wrote to Thomson some years later][41] that we turned very reluctantly indeed from your bold and very ingenious work with fused silica (quartz) to Pyrex glass. I had been extremely anxious to give your great ability and skill their fullest scope, but the circumstances were such that we could find no way that seemed financially possible. There seems to be no doubt about getting a good Pyrex mirror which will perform well in the 200-inch telescope. But I shall always regret that we felt compelled, decidedly against our will, to stop the work at Lynn.

The direct suggestion of using Pyrex came from A. L. Day of the Geophysical Laboratory of the Carnegie Institution,[42] who had assisted in its original development. Hale had also been thinking of Pyrex from the very beginning of the scheme. Pyrex, a name coined to cover a number of borosilicate glasses possessing heat-resisting properties, was being manufactured at the Corning Glass Works in the form of cooking utensils and chemical ware. The coefficient of expansion of ordinary Pyrex is about five times that of fused quartz but one third that of plate glass. This coefficient can be reduced by increasing the quartz content and the material chosen for the 200-inch blank had a coefficient three times that of fused quartz.

Beginning with small disks, J. C. Hostetter and G. V. McCauley developed new methods of casting and annealing, while Day succeeded in winning Hale over to a mirror deeply ribbed at the back. This reduced the weight of the 200-inch disk by more than a half and, at the same time, allowed the use of internal and compensated supports. A 26-inch disk was produced and a 30-inch (the coudé flat) followed, then the 60-inch Cassegrain blank and a 120-inch testing blank, all with ribbed backs to Pease's design. At the outset, the cores fixed to the bottom of the mould, and about which the molten Pyrex flowed to give the pockets and ridges required, gave trouble. They would break loose from time to time as the high temperature melted the cement and weakened the fixing pins. Two attempts preceded the successful casting of the 30-inch disk, one attempt the 60-inch disk, and then neither disk was perfect, although usable.

The first 200-inch disk was poured on March 25, 1934,[43] in the presence of many observers. Owing to the rapid chilling of Pyrex glass when it is poured into the customary open mould, the glass was placed within its own temperature-controlled furnace. The latter took the form of a large, round dome, built with silicon bricks and in which the mould was suspended. Numerous flame jets maintained the high internal temperature. The tank containing 65 tons of molten Pyrex took 15 days to fill, and a further 16 days to heat to a temperature of 1575° C, when melting began to occur. The melt was transferred from tank to mould igloo by means of ladles holding about 700 lb of Pyrex at a time. These were suspended from overhead mono-rail trolleys and were guided with long handles by a crew of workmen. Not more than half of each ladle-full of molten Pyrex was poured; the remainder, clinging to the sides of the ladle, was broken away by workmen and returned to the tank. The mould had 114 cores, each bolted to the base and, once again, several cores came adrift and rose to the surface of the melt. There were anxious moments as the workmen, under McCauley's direction, jabbed at the floating cores with iron rods in the endeavour to break them. This set-back led to no change in the carefully planned casting procedure, however, and when full, the igloo's temperature was kept at approximately 1350°C for some hours to eliminate any large bubbles introduced during the pouring operations. Disk and mould were then allowed to cool to about 800° C, when they were lowered from the igloo to the annealing kiln directly below. The slow, regular cooling took four weeks—a rate ten times faster than that considered to be safe—but all went well. The mirror blank came out whole but polarized light tests revealed strains and the broken cores were embedded in the Pyrex. It was agreed, therefore, to use the disk as a spare and to attempt a second casting.

For the second disk, McCauley fixed the cores with bolts of chrome-nickel steel and introduced a special air-cooling system beneath the mould so that the interior of every core was below the danger temperature. The second disk was cast on December 2, 1934, this time with only a few visitors present. Ten months were considered necessary for satisfactory annealing but, three months before the process was complete, an unexpected flooding of the Chemung River occurred. The mirror was out of direct danger, but all attempts to stem the rising water by walls and dykes failed to prevent its reaching the temperature control equipment. This had to be moved to a higher level and necessitated switching-off the annealer for 72 hours, but with no harm to the disk. During the final three months an earthquake tremor passed through New York State, but with no deleterious effects to the mirror plant.[44]

For its journey from Corning to Pasadena, the disk was mounted vertically in a heavy steel case built in a special, low-slung railway truck. The operation of packing and transportation was a delicate one as the base of the steel case cleared the sleepers by only a few inches and the top came within a few inches of the tunnels and bridges on the route. The train moved only by day, at a speed of 25 miles per hour, and all vibrations were automatically recorded. The disk arrived at Pasadena in perfect condition.

> To an old-timer like myself [Hale wrote][45] it is difficult to realise, when looking at the new disc, that the central hole has an aperture equal to that of the 40-inch Yerkes refractor. No other scale gauge could be more striking to me, as I recall so vividly the arrival of the 40-inch objective at the Yerkes Observatory in 1897.

Fig. 174—The 200-inch mirror on its grinding and polishing machine

The machine weighed 160 tons and had three parts: the base frame, the turntable on which the mirror lay, and the bridge supporting the travelling vertical spindle to which the grinding and polishing tools were attached (an 8-inch tool is shown in position). The bridge ran on parallel rails at the sides of the machine. The turntable could be tilted to raise the mirror into a vertical position for testing (Fig. 175). Behind the machine to the left can be seen part of the full-sized grinding and polishing tool, and lower down a 50-inch and a 68-inch tool, all covered as a protection against dust.

(Photograph by J. V. Thomsen)

Fig. 175—The 200-inch mirror in testing position

The cellular structure of the back of the mirror is clearly shown

(Photograph by J. V. Thomson)

Meanwhile, the special optical shop had been built and equipped at the California Institute of Technology in Pasadena. Here J. A. Anderson, assisted by Marcus Brown, had trained some twenty workers who were engaged on the secondary mirrors. The front surface of the 200-inch disk was first ground flat. The disk was then turned over so that the ribbed surface could be similarly treated, wooden plugs and plaster of paris being used to fill the central aperture and other cavities so as to prevent chipping the rib edges. Before further work could begin on the front surface it was essential, in view of the flexibility of the disk,[46] to float the latter on its final bed of supports. The thirty-six pockets were therefore carefully ground cylindrical to receive the steel sleeves of the elaborate system of balances which occupied the mirror cell.

The first major task was to bring the front surface to approximately the requisite curvature, an operation which meant deepening the centre to nearly 4 inches, thereby removing 5 tons of glass. The surface had then to be made truly spherical and given a fine-ground finish preparatory to polishing. This was done by means of a hollow, full-sized, spherical tool made of welded sheet-steel, surfaced by numerous Pyrex blocks and fed by ever finer and finer grades of carborundum.

Fig. 176—Pressing the 50-inch polisher

The photograph records the method used for ensuring good contact between the mirror and the polishing tool

(Photograph by J. V. Thomson)

During these operations a horizontal beam or bridge, mounted on parallel rails, moved to and fro over the mirror. The spindle carrying the tool, and later a reciprocating machine with comparatively small polishers, moved back and forth along the bridge. In addition, mirror and cell were mounted on a slowly rotating turntable. In this way, a variety of polishing motions was obtained.[47]

Over three months were spent prior to the polishing operation in removing all traces of grit from the polishing machine and workshop. Squares of pitch were cemented to the Pyrex squares of the 200-inch grinder which now became the polishing tool. Each pitch square was divided into smaller squares so that the channels forming about 8000 facets distributed the rouge evenly over the mirror. The development of the paraboloidal figure meant a further deepening of the surface by 1/200th inch, a relatively large amount. Work on parabolization therefore began as soon as the mirror had taken its initial polish, grinding and polishing periods alternately following each other. Later stages were done with local polishers which ranged in diameter from 68 inches down to 8 inches. Altogether, 31 tons of abrasives, including rouge, were required, 10 tons of coarse abrasive being necessary for the initial rough grinding.[48]

As the work progressed the mirror was tested at its centre of curvature, the disk, cell, and turntable being first tilted into a vertical position. A large Hartmann screen pierced with radially-spaced apertures of 6 inches diameter was wheeled in front of the mirror so that a determination of zonal aberrations could be made. The 120-inch blank was not used, since it was estimated that its working into a testing flat would involve undue labour and time. Although rejected at this stage the blank is now the primary mirror of the new Lick reflector (page 434). By November, 1947, the 200-inch mirror was considered ready for final testing and figuring in the tube.

The final U-type mounting, a compromise between the fork and English types, was arrived at only after many discussions, experiments, and models. The idea appears to have come from Pease, although it had been known in modified form since 1918 when Porter introduced the split-ring equatorial for which he later took out a patent. As soon as drawings were ready, the Council appointed Captain C. S. McDowell to allocate the constructional work and to supervise the telescope's

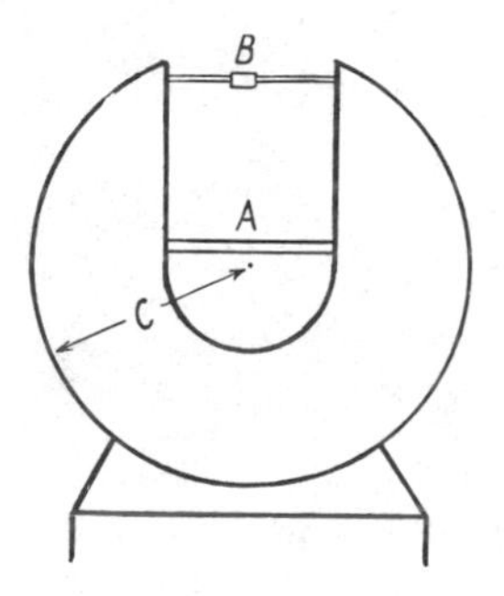

Fig. 177—The horseshoe mount of the 200-inch telescope

(*From 'The Telescope'*, 1940)

erection on the site. McDowell was a naval captain and, for some thirty years, had been engaged on ship-building work. He was a man of great energy and had useful scientific and industrial connections.

In 1935, the 19-ton, 20-foot diameter mirror cell (to Pease's design) was contracted to the Babcock and Wilcox Company, largest boiler makers in the world. The 48-foot diameter horseshoe, the yoke and tube were made by the Westinghouse Electric and Manufacturing Company (now the Westinghouse Electric Corporation), makers of large steam turbines and parts of the Boulder Dam electric generating plant. Hodgkinson, Ormondroyd, Froebel, Kroon and other Westinghouse technicians gave freely of their skill and energy.

> One interesting feature [Anderson and Porter write][49] somewhat disturbing at first, developed with the horseshoe. It was calculated that the horseshoe would not be sufficiently rigid to remain circular when the telescope tube moved over to the east or west. In other words, the upper horn of the shoe would sag and thus shift the polar axis of the instrument slightly. The cure for this was to deform the horseshoe by inserting a compression member at A and a tension member with turnbuckle at B, and drawing

Fig. 178—The 200-inch Hale telescope

The telescope is pointing to the zenith and is seen from the south. The 'slot' along the length of the southern side of the tube prevents obstruction of the beam from the coudé mirror in all positions of the telescope.

(Mount Wilson and Palomar Observatories)

the horns together by an amount equal to the deformation due to the weight of the tube. The horseshoe was then placed on the boring mill and its edge ground circular. When the bars A and B are removed, the horns spring apart, the horseshoe no longer being cylindrical. But, when the horseshoe is installed in the instrument the radius C will remain nearly constant for any position of the telescope tube.

A further difficult point was how best to overcome the friction due to some 500 tons weight on the polar axis bearings. Mercury flotation would have required the use of several tons of mercury and very large containing drums, all at great expense. Pease suggested mounting the horseshoe on rollers, but it was calculated that their use would require a torque of 22,000 lb-ft to overcome friction. The final arrangement, following up the suggestion of F. Hodgkinson of Westinghouse and worked out in detail by G. B. Karelitz, is to force oil through bearing pads at a pressure sufficient to float the instrument on a film of liquid. The film requires the use of only a few gallons of oil under a pressure of 210 to 385 lb per square inch. The two oilpads under the horseshoe are each 28 inches square and faced with Babbitt metal, while the south bearing takes the form of a 7-foot diameter hemisphere supported on three oil pads. A torque of only 50 lb-ft is sufficient to turn the complete telescope on these five pads.[50]

In its design, the tube departs from all previous forms. This is made necessary to satisfy the rigid specification laid down to prevent the tube from flexing under its own weight. The problem of flexure was solved by M. Serrurier, and Westinghouse's construction from his designs is one of the greatest achievements of the whole project. Tests show that, for all positions of the tube, the focus of the primary mirror does not depart from its mean position on the plate in the observer's cage by more than 0·01 inch. The central portion of the tube is 22 feet square, 44 feet long and weighs 75 tons. It contains the declination trunnions on which the tube rotates and which are themselves supported by radiating spokes like those of a wired car wheel but staggered and inclined differently. This arrangement provides great rigidity in planes at right angles to the declination axis, and leaves sufficient lengthwise freedom of the axis to accommodate any tendency of the tube to bind. From the corners of the centrepiece, 20-inch-wide I-section girders converge to the lower ring (supporting the mirror cell) and the upper ring (supporting a special observing cage). The tube does, of course, flex slightly under its own weight but within the very narrow limit stated above. With observing cage and mirrors, the tube weighs 140 tons.

Three alternative aperture ratios are provided—*f*/3·3 at the primary focus, *f*/16 as a Cassegrain, and *f*/30 as a Cassegrain-coudé. A distinctive feature is the cylindrical 'cage' which is supported within the tube and which allows an astronomer to work at the primary focus. The comparatively short focal length of 55 feet gives rise to bright images and enables faint objects to be photographed direct or through a spectrograph with short exposure times. The effective field at the primary focus of an *f*/3·3 mirror is only about half an inch in diameter owing to coma. To enlarge the field, F. E. Ross designed a series of two and three component correcting lenses of a type previously tried out at Mount Wilson. These lenses yield well-corrected fields up to 6 inches in diameter and can be made to give a telephoto effect. They then provide a choice of aperture ratios from *f*/3·6 to

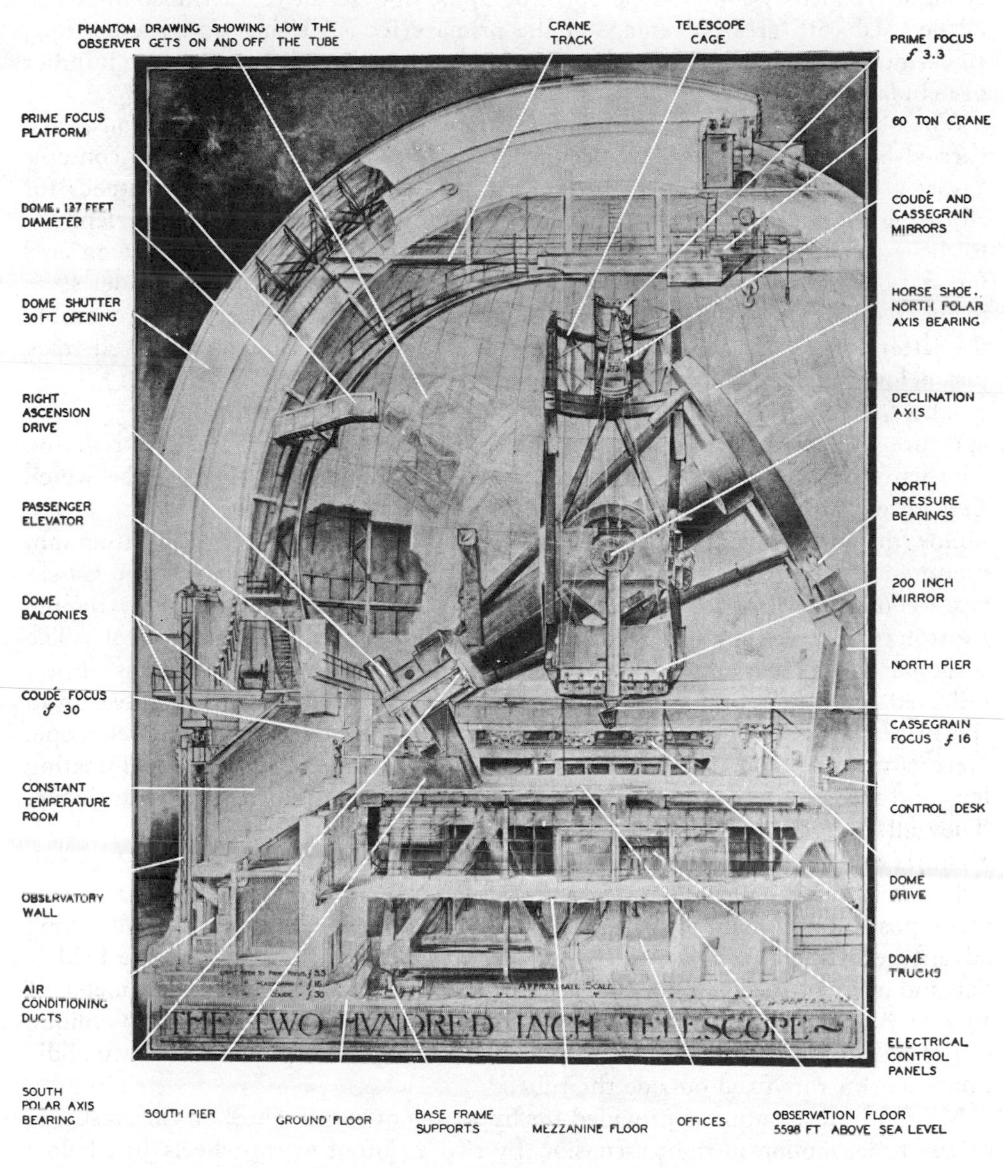

Fig. 179—Cross-section of the 200-inch Hale telescope installation

A drawing by Russell W. Porter

(California Institute of Technology)

f/6·0, a choice influenced by the nature of the object and the seeing conditions prevailing at the time of observation.[51] Double-slide plate-holders, photoelectric cells, short-focus Schmidt-type spectrographs, and sensitive thermocouples can all be used with great advantage at the primary focus. Using *f*/0·47 and *f*/0·95 thick-mirror Schmidts at this focus, M. L. Humason has obtained a large number of nebular spectrograms.

The observing cage or cylinder is 6 feet in diameter and is in two sections, each carried by its own set of radial knife-edge spiders. The upper cylinder contains a special observing chair, controls for the motion of the telescope, switches for adjusting the focus, indicators of right ascension and declination, and a telephone to the assistant on the main floor. These features are common to all three observing stations. The double-slide plate-holder is carried in the lower cylinder so as to be free from vibrations produced by movements of the observer. To enable the latter to enter or leave the tube, a movable platform runs on a curved track just below the main shutter arches.

The *f*/16 Cassegrain focus falls immediately below the centre of the 40-inch aperture in the primary mirror on to the slit of a 4-foot prism spectrograph. Alternatively, a 53½-inch × 36-inch diagonal flat, supported upon a tube which rises through the 40-inch hole, passes the convergent beam through one of the hollow declination trunnions. At this alternative focus is an 8-foot spectrograph, mounted within one of the 10-foot diameter yoke tubes. To use the *f*/30 Cassegrain-coudé, the first convex mirror is swung to the side of the tube by remote control, to be replaced in a few moments by another convex secondary mirror. The diagonal flat is then rotated and so geared that the convergent beam is always reflected down the polar axis. The focus, at an effective distance of 500 feet from the primary mirror, falls in a constant-temperature room south of the telescope. Here a 12-inch, 30-foot focus off-axis paraboloidal mirror acts as a collimating lens and feeds four gratings in composite. Five alternative cameras are available. They all embody the Schmidt principle, have corrector plates worked by D. O. Hendrix and focal lengths which range from 12 feet to 8·4 inches. The 8·4-inch Schmidt is of particular interest in view of its low aperture ratio of *f*/0·7. This is made possible by mounting a quartz sphere in the convergent beam and taking advantage of the aplanatic points of a spherical refracting surface. The field is flat and well corrected for violet and ultra-violet radiations over a diameter of 17 mm.[52] For stars with declinations greater than 50°, and to avoid very oblique angles of incidence at the flat mirror, an alternative path is provided via two additional flat mirrors fixed outside the tube.[53]

Motion in declination is provided for by a 14-foot worm wheel on the west side of the tube, motion in right ascension by two 14½-foot worm wheels just below the south bearing. One of these wheels gives rapid motion in right ascension, the other gives slow motion. These gears were made in the Pasadena instrument-shops under rigid temperature control conditions. Angular errors from tooth to tooth do not exceed 1 second of arc anywhere on the circumference.[54] The telescope is driven by a synchronous motor controlled by a quartz crystal oscillator; varying atmospheric refraction at different tube elevations and residual errors, due to changes in the mounting, are allowed for by a mechanical computer. To save

Fig. 180—The exterior of the dome of the 200-inch telescope

(California Institute of Technology)

time in directing the tube, a control desk between the north piers has indicators giving right ascension and declination, also dials on which can be set the next star position.[55] It is claimed that, having set the dials to the required values, the telescope can be re-directed in right ascension and declination to an accuracy of 5 seconds of arc.

The dome is 137 feet in diameter and weighs 1000 tons. Its position and the aperture of the twin shutters are continually controlled by a small dummy telescope which also controls the position of the wind screen. Both dome and building are made almost entirely of steel, with $\frac{3}{8}$-inch butt-welded covering plates. The precautions against rapid temperature changes at Mount Wilson apply equally well on Palomar Mountain.

> There are three floors [Anderson and Porter write]:[56] the ground floor contains offices, dark rooms, air conditioning equipment, library, lunch room, and a large storage area at the service entrance; the second, or mezzanine floor, is used for public rest rooms, electric panel boards, time-standard equipment, and the oil-pumping equipment; the third, or observation floor, contains the instrument itself. A glassed-in area for the public, at the head of the stairway, permits visitors to view the telescope without transferring their body heat (really a sizable number of calories) to the

observing room itself. A 60-ton crane travels between the shutter arches, far up in the dome. Freight and passenger elevators are provided.

Palomar Mountain, an isolated summit in the San Jacinto Mountains of southern California, is 125 miles south-east of Pasadena and 50 miles north of San Diego. It is a table top, thirty miles long and ten miles wide, covered with a heavy growth of trees and thick brush. For many miles around, the population is sparse. Hussey thought the site too isolated, but this fact should be an obstacle to any contemplated future industrial development. The mountain is in an earthquake area, as is Mount Wilson, but a study of the local geology showed that as the telescope would be anchored, as it were, to the mountain, there would be no danger from this source.

Dr Anderson spent five years supervising the seeing tests on Palomar Mountain. Many of these tests were made by Ellerman and other Mount Wilson astronomers. Two 12-inch Cassegrain reflectors, designed by Porter, were used for this work, also ten 4-inch polar refractors set on Polaris and charged with a magnification of 700. The state of the air was judged by the amplitude of lateral shifts of the star images relative to fixed vertical webs. From these observations it was found that seeing conditions on the mountain were particularly favourable and, in general, superior to those on Mount Wilson.

Even without unforeseen hitches, the work of testing and adjusting so large a telescope as the Hale reflector takes many months. The first visual and photographic test observations, preliminary to the work of adjustments, were made in January, 1948[57]. With his eye behind a small reading lens, Anderson was the first person to peer at the image formed by the primary mirror. He was followed by I. S. Bowen, E. P. Hubble, M. L. Humason, Russell Porter, Marcus Brown, and others connected with the project, also by J. H. Oort, visiting Dutch astronomer. The photographs were of stars taken with the tube vertical and then at 60° and 30° elevations north and south.

On June 3, 1948, an assemblage of astronomers and state officials witnessed the dedication ceremony. To L. A. DuBridge, president of the California Institute of Technology, fell the formal dedication of the telescope and observatory and the presentation to Mrs Hale of the printed resolution calling the telescope the *Hale telescope*. The unveiling of a bronze plaque below the bust of Hale in the foyer revealed that:

> The 200-inch telescope is named in honor of George Ellery Hale
> whose vision and leadership made it a reality.

> The Unity, the singleness of purpose, the faith, and the spirit to which this plan bears witness [said Dr V. Bush on this occasion][58] lie at the heart of any undertaking in which men or institutions associate together.

Final tests on the 200-inch mirror in the optical shop at Pasadena revealed a slightly raised outer zone or 'turned-up' edge. Since the amount was small (about one wavelength of green light), the mirror was sent to the observatory in this condition in the hope that orientations other than the vertical and modifications of the support system would cause the mirror edge to sag by this amount.

This hope did not mature, and attempts to cure the defect by fan-induced air circulation, changes in support pressure, and thermal insulation at the edge were unsuccessful. It was therefore decided to refigure the affected zone on the mountain, a task performed by Hendrix in only nine hours of polishing.[59] Further adjustments and photographic knife-edge and Hartmann tests took a further five months. By the end of October, 1949, the mirror was ready for a new aluminium coating.

Following the completion of the first Ross field lenses, several photographs of nebulae under various atmospheric conditions were taken. One of these, of NGC 147 in Cassiopeia, appeared in *Sky and Telescope* for February, 1950. The galaxy is clearly resolved, the smallest stellar image being only slightly greater than 0″·5 in diameter. The 100-inch Hooker telescope could resolve this object only with the use of red-sensitive plates.

The Mount Wilson and Palomar Observatories come under the joint directorship of Dr Bowen, who is directly responsible to the California Institute of Technology. Dr Anderson remains executive officer to the observatory council, whilst Dr Max Mason, since the death of Hale in 1938, is chairman of the observatory council. An immense programme of work now confronts Dr Bowen and his colleagues. Priority is being given to nebular astronomy.

> For the past two and one half years [Dr Bowen stated in his Halley lecture of May 13, 1952][60] the Hale telescope has been in use on every clear moonless night on a program which is planned to make a systematic step by step attack on the uncertainties that were inherent in the earlier determination of the distances, sizes and distribution in space of these extragalactic nebulae. Other studies are being initiated to investigate the structures and in particular the distribution over these structures of various stellar types.
>
> The most distant nebula observed [Dr Bowen continued], located in the Hydra cluster, showed a shift of the spectrum corresponding to a velocity of 38,000 miles per second or over 1/5 the velocity of light. Finally, the installation of the large coudé spectrograph in July 1950 made possible the start of observations for the second of the observatory's broad programmes. This programme has for its purpose the investigation of the chemical composition, the temperature, the pressure, the motions and other physical conditions in the atmospheres of stars including the planetary nebulae.

REFERENCES

1 Young, R. K., *Jour. R.A.S. Canada*, **29**, p. 303, 1935.

2 *Ibid.*, p. 305.

3 *Engineering*, 1934, March 9 and 30; April 20.

4 Young, *op. cit.*, p. 302.

5 *Engineering*, 1939, March 10 and 24.

6 *The Engineer*, 1938, Sept. 16, 23 and 30. Knox-Shaw, H., *Occ. Not. R.A.S.*, **1**, pp. 45–53, 1939.

7 Ross, F. E., *Ap. J.*, **81**, p. 156, 1935.

8 *Publ. Observatory of the University of Michigan*, **7**, p. 44, 1939.

9 Knox-Shaw, *op. cit.*, p. 47.

10 *Nature*, **166**, p. 144, 1950.

11 *J.B.A.A.*, **58**, p. 111, 1948.

12 Spencer Jones, H., *M.N.R.A.S.*, **104**, pp. 146–162, 1944.

13 Watts, C. B., *A. J.*, **53**, p. 152, 1948.

14 *Ibid.*, **50**, pp. 179–182, 1943.

15 For early reference to moving mirror *vide* Turner, H. H., *M.N.R.A.S.*, **54**, p. 412, 1894. *Vide* also *Observatory*, **66**, pp. 365–367, 1946; **68**, p. 98, 1948 for modern developments.

16 Cookson, B., *M.N.R.A.S.*, **61**, pp. 315–334, 1901.

[17] For Talcott's zenith telescope and his method of finding latitude *vide* Chauvenet, W., 1906, *A Manual of Spherical and Practical Astronomy*, ii, pp. 340–367.

[18] *M.N.R.A.S.*, **70**, p. 298, 1910.

[19] *Trans. Astronomical Observatory Yale University*, **8**, pp. 5–10, 1936.

[20] Dimitroff, G. Z., and Baker, J. G., 1946, *Telescopes and Accessories*, pp. 77, 283.

[21] Olivier, C. P., *Pop. Astr.*, **52**, p. 420, 1944.

[22] Crump, C. C., *Pop. Astr.*, **37**, pp. 553–559, 1929.

[23] Bailey, S. I., 1931, *History and Work of the Harvard Observatory*, pp. 47–48.

[24] From information received from Dr Roy K. Marshall, formerly at Fels Planetarium, Philadelphia, U.S.A.

[25] *Pop. Astr.*, **54**, p. 18, 1946.

[26] *Ibid.*, **30**, pp. 593–597, 1922.

[27] *Ibid.*, **54**, p. 19, 1946.

[28] Fisher, C., *Pop. Astr.*, **38**, pp. 253–259, 1930.

[29] Nassau, J. J., *Pop. Astr.*, **45**, pp. 407–418, 1937.

[30] Struve, O., *Publ. Astr. Soc. Pac.*, **55**, pp. 123–135, 1943.

[31] Plaskett, J. S., *Publ. Astr. Soc., Pac.*, **49**, pp. 142–143, 1937.

[32] Struve, O., *Pop. Astr.*, **55**, p. 284, 1947.

[33] *Ibid.*, p. 286.

[34] *Ibid.*, p. 236.

[35] *Ibid.*, p. 289.

[36] *Pop. Astr.*, **56**, p. 231, 1948.

[37] *J.B.A.A.*, **60**, p. 40, 1950.

[38] Ritchey, G. W., 1929, *Development of Astro-Photography and the Giant Telescopes of the Future*, Plate 33.

[39] *Amateur Telescope Making*, p. 323, 1933.

[40] *Nature*, **137**, p. 223, 1936.

[41] Woodbury, D. O., 1940, *The Glass Giant of Palomar*, p. 135.

[42] Hale, G. E., *Nature*, **137**, p. 225, 1936; Woodbury, *op. cit.*, p. 131.

[43] For account of pouring of first disk *vide* Woodbury, *op. cit.*, pp. 169–174. Spencer Jones, H., *Proc. Phys. Soc.*, **53**, p. 504, 1941.

[44] Spencer Jones, *op. cit.*, p. 505; Woodbury, *op. cit.*, p. 179.

[45] *Nature*, **137**, p. 1017, 1936.

[46] Bowen, Ira S., *Jour. Opt. Soc. Amer.*, **42**, p. 796, 1952.

[47] Twyman, F., 1952, *Prism and Lens Making*, pp. 568–576. Notes by J. V. Thomson.

[48] *Ibid.*, pp. 570–571.

[49] Anderson, J. A., and Porter, R. W., *The Telescope*, **7**, p. 35, 1940.

[50] Spencer Jones, *op. cit.*, p. 509.

[51] Bowen, Ira S., *Jour. Opt. Soc. Amer.*, **42**, p. 797, 1952.

[52] *Ibid.*, pp. 798–800.

[53] Anderson and Porter, *op. cit.*, p. 33.

[54] *Ibid.*, p. 36.

[55] *Ibid.*, pp. 36–37. Spencer Jones, *op. cit.*, pp. 510–511.

[56] Anderson, and Porter, *op. cit.*, p. 38.

[57] *Pop. Astr.* **56**, p. 172, 1948.

[58] Cal. Inst. Tech., 1948, *Dedication of the Palomar Observatory and Telescope*, p. 27.

[59] Bowen, Ira S., *Publ. Astr. Soc. Pac.*, **62**, p. 91, 1950. *Vide* also Ref. 51.

[60] *The Observatory*, **72**, p. 129, 1952.

CHAPTER XIX

In whatever directions the professional of the future may be forced to develop his methods, we may be confident that the great company of amateurs will never allow visual observations to become a lost art.

W. H. STEAVENSON

IN our survey of four centuries' development of the telescope we have made repeated reference to amateur work in astronomy. Mention has been made of Herschel, Rosse, Nasmyth, Lassell, Rutherfurd, and Draper, all amateur telescope-makers. Others, like E. E. Barnard (who started as an amateur), S. W. Burnham, and T. E. R. Phillips, were telescope users rather than telescope-makers. We have, perforce, done the latter scant justice; the book has yet to be written which will tell of their early struggles and contributions to astronomy. Some amateurs have been blessed with ample means: Lassell, Huggins, W. H. Pickering, and Hale prosecuted their first studies without financial handicap. Others, like Rosse and G. W. Cook, employed astronomers to work in their own private observatories. We followed the Newall telescope to the Observatory of Cambridge University. The founders of the McMath-Hulbert Observatory gave their establishment to the University of Michigan and, in 1940, the private observatory of G. W. Cook (mentioned in our account of J. W. Fecker) was bequeathed to the University of Pennsylvania. The story of the professional astronomer is but a part of the story of astronomy, for astronomy is a science which offers a common meeting ground for both amateur and professional.

A large number of amateur-made instruments are now doing valuable work in established observatories. Mirrors by With, Common, Calver, Blacklock, Wassel, and Slade are still in regular use although made many years ago. The Rev. W. F. A. Ellison, early in the present century, progressed from 'a spectacle lens, a sixpenny microscope and a pasteboard tube' to the manufacture of 'more than 140 mirrors of apertures from 6 to 12 inches, and object-glasses of 4-, 4½-, 5- and 5¼-inch apertures.'[1] Ellison stated that his life was spent helping lame dogs over stiles.[2] He communicated many articles to the *English Mechanic* and the *Journal of the British Astronomical Association* and, in his own book, *The Amateur's Telescope*, dealt at length with the construction of mirrors and achromatic lenses. Such information was far more scarce than it is now and, in publishing his methods in detail, Ellison rendered great service to astronomy.

Large Cassegrain-Newtonian reflectors by J. H. Hindle, a Lancashire manufacturing engineer, are still in use. Dr W. H. Steavenson worked for some years with a 20-inch Hindle reflector at Norwood, in south-east London, and now uses a 30-inch by the same maker at Cambridge. H. G. Tomkins, an assiduous lunar observer, made himself a 24-inch Cassegrain reflector[3] and, just before his death in 1934, started work on a 30-inch.[4] J. H. Reynolds was the first to widen the scope of the coelostat for photographic work. He used a 24-inch horizontal reflector and 30-inch plane mirror, both made by himself. His arrangement was later used at the Hamburg Observatory and dignified with the name *Uranostat.*[5] Reynolds both designed and supervised the manufacture of the mounting for a 30-inch Common mirror which came into his possession. In 1905, this telescope was sent to Helwân, Egypt, where Reynolds and Knox-Shaw used it to photograph nebulae lying between the celestial equator and $-40°$ declination, many of them for the first time. For some years this instrument was the largest telescope on

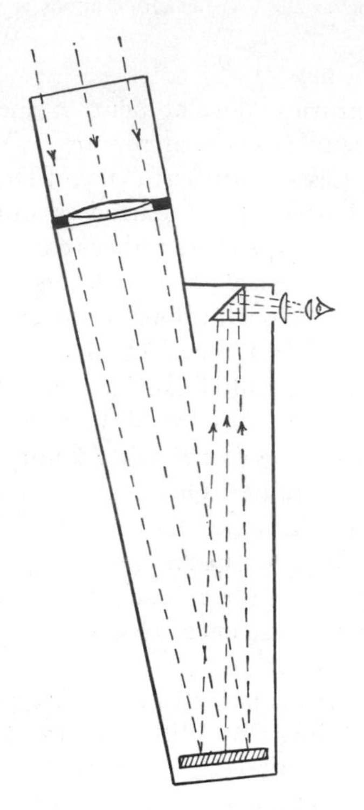

Fig. 181—A compact refractor

This unusual design allows an $f/14$ refractor to be used with a mounting normally suitable only for reflectors. An 8-inch Grubb objective has been used in this way by M. A. Ainslie with excellent results.

the African Continent. Reynolds also made a 28-inch reflector which he installed in his own observatory at Low Wood, Harborne, Birmingham, grinding and figuring the mirror himself.[6] Reynolds later donated this telescope, with a 30-inch mirror which belonged to C. R. D'Esterre, to the Commonwealth Observatory, Canberra. The 30-inch, with a new mirror by Ritchey (and with which Reynolds and Knox-Shaw took the above-mentioned photographs), remains at Helwân, a gift by Reynolds to the Egyptian government. Another British amateur responsible for figuring large mirrors is F. J. Hargreaves of Kingswood, Surrey. Hargreaves works with a 14-inch reflector made by himself and, during his active career as amateur astronomer, produced a variety of high-class optical elements. He is now in professional association with H. W. Cox and J. V. Thomson, with whom he has made the optics for the 16-inch × 24-inch Schmidt at Edinburgh (page 374) and the 50-inch mirror for the proposed Schmidt at Canberra (pages 267, 392). At the time of writing, the firm is refiguring a 28-inch mirror which formerly belonged to the above-mentioned Reynolds reflector at Harborne. This mirror is destined for a new Cassegrain reflector at the Dunsink Observatory, Ireland, and will be used primarily for photoelectric work.

Swiss telescope-making has been dominated by the work of Emile Schaer, optician and astronomer at the Geneva Observatory until 1926. Schaer's interest in optics took a practical turn in 1893 when he worked three 6-inch glass disks into convex lenses. He silvered one surface and refigured the other until the lens-mirror gave good images by reflection. Only after 160 hours of local polishing did a mirror of this type give satisfactory star images. In 1898, Schaer occupied a flat at the top of 2, rue de l'École de Chimie, Geneva. Here a small room contained turntables, tools and silvering trays while a wide balcony commanded a magnificent view which Schaer used as a testing ground. On one occasion a good mirror, complete with tube and most of the stand, slipped over the balcony and fell five storeys to destruction below. About this time Schaer experimented with silvering and photographic processes and introduced the Vautier-Schaer telephoto system. In 1908, he completed a 16-inch Cassegrain reflector for the Geneva Observatory and started work on a 39½-inch Cassegrain. All the optical components for these instruments were made in the small room on the fifth floor. The 39½-inch *f*/3·0 mirror turned out to be astigmatic owing, no doubt, to difficulties in manipulating and in testing so large a disk in so small a space. This defect Schaer remedied by a local polishing of the hyperboloidal secondary mirror—a remarkable achievement.

After his appointment in 1898 to the Geneva Observatory, Schaer spent many years working with the 39½-inch at temporary stations away from the lights of Geneva. In 1913 he received permission from the French to operate the telescope in a temporary solar observatory on Mont Salève. War broke out soon after the telescope was installed and the local inhabitants, mistakenly suspicious of the activities at the observatory, pillaged and sacked the buildings. Fortunately they spared the great mirror, but this was not brought back to Geneva until 1920. Schaer then advocated the establishment of a high-altitude Swiss observatory and, largely through his efforts, the 39½-inch, remounted and refigured to remove the astigmatism, was used between 1926 and 1930 at a temporary station at Jungfraujoch in the Bernese Oberland. This instrument, on a modern mounting and accompanied by a 24-inch Schaer Cassegrain, is now established in the permanent Jungfraujoch station of Geneva Observatory.

During the years 1913 to 1918, Schaer made a second 39½-inch Cassegrain, again supplied with an *f*/3·0 primary mirror. After four years' use at a provisional station at Petit-Saconnex, this instrument was installed in the Geneva Observatory.[7] Schaer's largest mirror, a 48·4-inch (Fig. 182), is now at Basle University, having spent the war period heavily sandbagged. Plans were made for its mounting but, as yet, nothing can be done owing to lack of funds. Schaer made a 16-inch

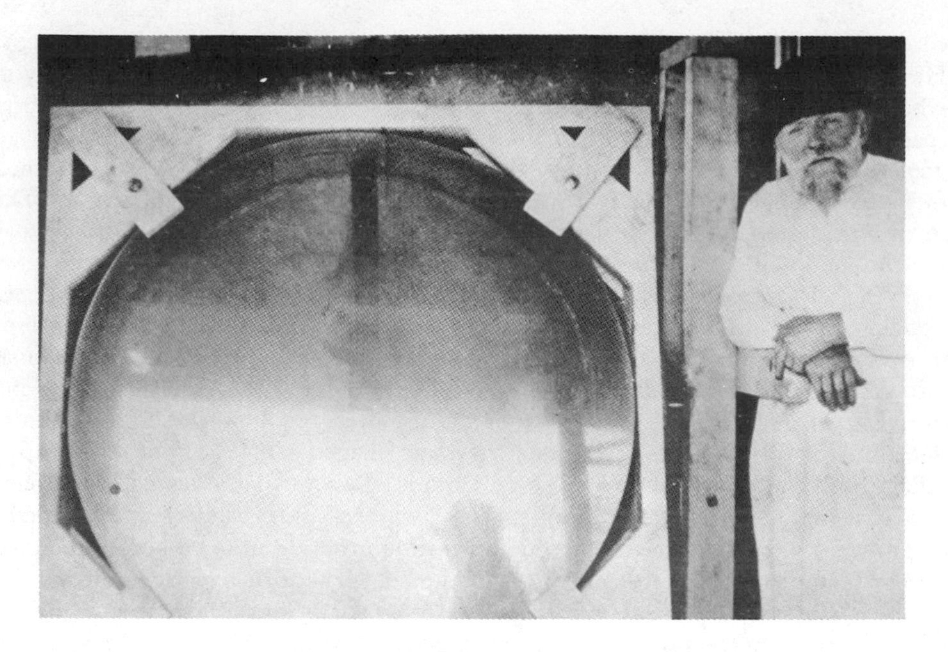

Fig. 182—Emile Schaer

Seen with his 48·4-inch mirror at the Basle Observatory, 1922

(Courtesy R. Schaer)

Newtonian and several 16-inch and 24-inch Cassegrains and invariably gave the primary mirrors aperture ratios of $f/2$ to $f/4$. A 24-inch, formerly made for Honegger-Cuchet of Conches, Geneva, and used later by Gentili di Giuseppe at Bue, near Versailles, is now at the Pic-du-Midi station. Another 24-inch, formerly at Meudon, went to the provisional Jungfraujoch station in 1924 (Fig. 183) for observations of Mars.

When Schaer took photographs at the primary focus of the 39½-inch telescope he also used the system as a visual Cassegrain for guiding purposes. The usual convex secondary was replaced by a small plate of optical glass of which the surface turned towards the primary mirror was worked convex and then figured hyperboloidal. This face was unsilvered and reflected sufficient light to permit good guiding without recourse to the usual auxiliary visual refractor. At one period, Schaer replaced the plate by a crown-flint doublet with the crown lens facing the primary mirror. The outer face of this lens was hyperbolized and silvered at the centre only; this central part then functioned as a small Cassegrain secondary mirror. Light reflected from the primary mirror thus traversed the doublet before it reached the photographic plate, the outer surface of the flint component being figured to compensate zonal effects produced by the hyperboloidal surface of the crown lens. The doublet was used in this way between 1913 and 1922 on both 39½-inch reflectors and also on the 24-inch at Conches. The device also gave a degree of field-flattening and coma-correction, properties

Fig. 183—The temporary Jungfraujoch station in the Bernese Oberland, 1924

Emile Schaer (in front) with two of his Cassegrain telescopes. The larger instrument is of 24-inch aperture. The station was 3,360 metres above sea level, close to the site of the permanent station erected later.

(*Courtesy R. Schaer*

which Ross subsequently (and independently) exploited and which Baker incorporated in the design of the Dyer telescope (page 368).

Schaer's largest refractors were visual instruments of 19·7 and 12·6 inches aperture for the private observatory of R. Jarry-Desloges at Sétif, Algeria. Smaller instruments, including several photographic objectives, were used at Geneva or were sent abroad, mainly to observatories in France. Schaer also introduced a number of 'réfracto-réflecteurs' or folded refractors. Two plane mirrors are so placed behind an objective that the refracted beam traces out the form of the letter N. The overall length of the instrument is thereby reduced to a third of its focal length and the observer continues to look in the direction of the object. The $23\frac{1}{2}$-inch Henry objective at the Pic-du-Midi station operates in this way.

Schaer died in 1931, aged 69 years, leaving behind him many finished instruments and a large number of unmounted prisms, mirrors, and objectives. Most of his optical work seems to have been done on a largely empirical basis and his methods of testing and mounting his elements were simple yet effective. Many of his telescopes are still in use—examples of a skill and industry which have imparted new life to telescope-making and observational astronomy in Switzerland.

American amateur telescope-making has been dominated by so many names that it is impossible to do adequate justice to all. For a fuller insight into the work of American amateurs, the reader is referred to the three volumes of *Amateur Telescope Making* and to the monthly journal, *Sky and Telescope*. We should mention, however, Joel H. Metcalf who died in 1923 and in whom we find the singular combination of a minister of the Unitarian Church and an expert optical computer.[8] Metcalf both designed and made his own photographic lens systems and, as a result of many years' work in astronomical photography, discovered six comets, forty-one minor planets, and many variable stars. Among other lenses, he made a 10-inch photographic triplet, a 16-inch $f/5{\cdot}25$ doublet, and, at the time of his death, was at work on a 13-inch triplet.[9] The last lens was later completed by Lundin for the Lowell Observatory and was used in the discovery of the planet Pluto. The 16-inch doublet, mounted in a fork equatorial, together with a 12-inch Metcalf doublet[10] and a 10-inch Metcalf triplet,[11] have done useful work at the Harvard College Observatory. In 1923, H. Shapley decided upon an extension of the *Henry Draper Catalogue* and the first region selected was covered by the 10-inch Metcalf triplet, which gave excellent definition.[12] The 16-inch doublet, the 'Metcalf telescope', is an unusual dioptric system since it requires the use of slightly curved photographic plates.

The largest amateur-made telescope in the United States would appear to be the 36-inch reflector which the Amateur Telescope Makers of Indianapolis, Indiana, made for the Goethe Link Observatory. The building was based on a sketch given to Dr Link by Russell W. Porter and was constructed by C. Bowers, an Indianapolis carpenter. C. D. Turner, an engineer, built the grinding-machine and was responsible for the design and construction of the cross-axis mounting. Two amateurs, V. Maier and C. Herman, undertook the optical work, making use of a 36-inch ribbed Pyrex disk cast at about the same time as the 200-inch.[13] This telescope, now supplied with a modern electronic drive and with the latest spectrographic equipment, comes under the control of the University of Indiana.

Several scale models of the 200-inch Hale telescope have appeared in the United States. Noteworthy are the two models made by C. E. Raible of Millvale, Pittsburgh.[14] The first is a 1/25 actual size model of the original design for the telescope. When the fork mounting was changed for the present yoke and horseshoe journal, Raible began, in 1939, a 1/16 steel and aluminium scale model, not completed until 1947. This second instrument stands nearly 6 feet high from the floor to the top of the tube (when placed vertical). The largest scale model, 1/10 actual size, is at Pasadena on the sun-roof of the Astrophysical Laboratory of the California Institute of Technology. It was made in the Pasadena optical shops and was used for mechanical tests during the design of the 200-inch. Another working model, made by S. Orkin at a cost of nearly $25,000 and complete with 6-foot diameter observatory, was placed on exhibition in the Hall of Science of the Griffith Observatory, Los Angeles.[15]

The interests of amateur astronomers are catered for by societies like the British Astronomical Association, the Société Astronomique de France, and the Astronomical Society of the Pacific, the American Association of Variable Star Observers and the Astronomical League in the United States. As far as telescope construction is concerned, recent years have witnessed considerable growth of mirror-making

Fig. 184—A 12-inch Newtonian reflector built by amateurs

An example of a simple, yet effective, instrument for amateur use. The major part of the construction was achieved with only woodworking facilities. The mirror was figured by F. J. Hargreaves.

(By courtesy of the Rev. R. C. Wood and the British Astronomical Association)

clubs and amateur telescope-making societies. These clubs are no longer confined to America although persistent co-operative effort by small groups, working together at regular intervals, is comparatively new elsewhere. Such a group was formed at Schaffhausen, Switzerland, in 1946 and another started at Cambridge, England. During their first three months, the Schaffhausen group produced twenty-one mirrors, mostly 6 inches in diameter.[16] In May, 1947, twenty-three finished reflectors were placed on top of the mediaeval town fortress and, in two nights, about a thousand people inspected the night sky through first-class telescopes.[17]. The Schaffhausen group aims at providing every school in the district with a 6-inch or 8-inch reflector, preferably figured by the science teacher himself. It also hopes to build and equip its own observatory on the heights above the town.

A rich harvest awaits the skilled amateur. Schmidt camera technique is in its infancy and many theoretical ideas await practical development. Lower, Cox and others have shown that the construction of the basic Schmidt system is within the capabilities of the amateur who possesses a workshop. Cassegrain and Gregorian Schmidt-type systems likewise await practical development. Maksutov's recent introduction of the meniscus system, with a steeply curved meniscus lens with spherical surfaces in place of the Schmidt correcting plate, is particularly suitable for amateur construction.[18] Amateurs should not hesitate to make their own telescope mirrors. Elihu Thomson's discovery in 1878[19] that if two disks of glass of equal diameter are ground together with abrasive, one will become convex and the other concave, renders the construction of special grinders and polishers unnecessary. Starting from small disks of 4 to 6 inches diameter, the amateur will gain sufficient confidence to undertake the manufacture of 8- and 12-inch mirrors and, with these, valuable astronomical work can be done. More attention should be given to fixed telescopes of long focal length, both for direct observation of the sun and for use in conjunction with a spectroheliograph. Hale has written enough to convince anyone with mechanical ability that the latter instrument and spectrohelioscope are within amateur reach. In England, A. M. Newbegin, F. J. Sellers, and M. A. Ellison have, between them, observed the sun over many years with spectrohelioscopes privately made and operated. One thing is certain—that if the amateur should attempt any of these projects and find himself in deep water, more experienced workers are always ready to assist and guide him to success. The various sections of the British Astronomical Association, to mention only one amateur organization, were formed with this object in view.

Having furnished himself with a suitable instrument, the amateur should endeavour to apply it in a direction most suited to its powers and to the furtherance of astronomy—if a Schmidt camera, to work on the detection of comets, variable stars and novae, if a reflector, of 6-inch aperture or larger, to work on variable stars, planetary markings, lunar features near the limb, and the more transient details of sunspots. With a spectrohelioscope and 'graph working on, say, a 2-inch disk, added value to direct visual observations is achieved, even if observations are limited to two or three days a week. The development of auxiliary apparatus is a fascinating and productive occupation. We refer to direct photography, photoelectric photometers for variable-star work, monochromators, automatic telescope

Fig. 185—A transportable 16-inch reflector

This instrument was built by W. A. Rhodes, an amateur astronomer of Phoenix, Arizona. It operates either as an $f/3 \cdot 3$ Newtonian or as an $f/18$ Gregorian. Despite a total weight of nearly a ton, the instrument and its cover may be transported from place to place on a trailer with heavy wheels. Special features include lead filled magnets for counterweights, and a selsyn control for focusing the secondary mirror. The roughing out and perforation of the primary mirror were performed with a sandblast gun

Fig. 186—Russell W. Porter

At Springfield, Vermont, September 1, 1929

(American Museum of Natural History. Photograph by Clyde Fisher)

drives, and devices for improving telescope performance. While useful work in astronomy can be achieved with a pair of binoculars or a table telescope, it is common experience that small telescopes bring a desire for something larger, while the more ambitious amateur usually carries off the best prizes.

In America, A. G. Ingalls, associate editor of *The Scientific American*, gave constant publicity to the pioneer work of the late Russell W. Porter. The three volumes of *Amateur Telescope Making*, edited by Ingalls and previously referred to, have had a wide circulation in England and in the United States.

Porter was born in 1871 and, as a young man, studied architecture. He became so enthusiastic about Arctic exploration that, for many years, he accompanied Peary, Fiala and others, on their trips north of the Arctic Circle. Before the first world war, he settled in his homeland and became interested in telescope-making, a knowledge of which he put to good account when war broke out by doing optical work at the National Bureau of Standards in Washington. Later, he returned to his birthplace at Springfield, Vermont, and became optical associate with the Jones and Lamson Machine Company, makers of high-precision machine-tools. Here he built fifty 6-inch reflecting telescopes, mounted on the split-ring equatorial which he had invented.[20]

To Porter we owe the invention of several types of fixed-eyepiece telescope. His aim was so to arrange the optical and mechanical layout of a telescope that the eyepiece remained fixed for all tube directions. This is an old problem—we saw that it was approached in the late seventeenth century (pages 60–61) by placing a mirror or mirrors before fixed telescopes. One major outcome, the coelostat (pages 278–279), made possible the modern vertical solar telescope. In the nineteenth century, Nasmyth's 20-inch altazimuth Cassegrain reflector (page 219) required only an additional plane mirror to send the light along the hollow horizontal axis. The eyepiece was fixed relative to the observer but only because

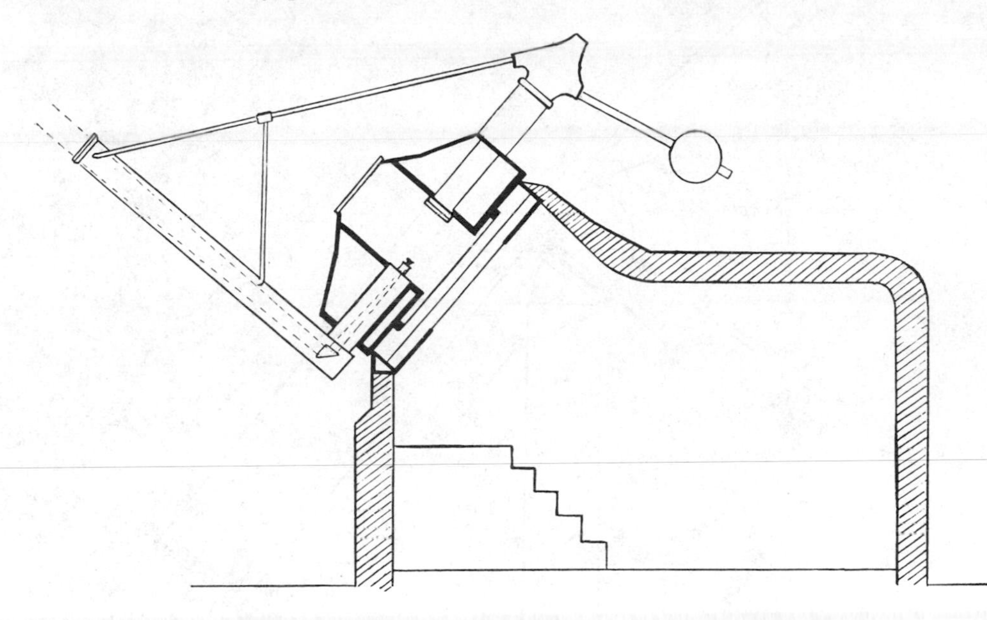

Fig. 187—Diagram of the Hartness turret telescope

Nasmyth's observing chair moved in azimuth with the telescope. Had Nasmyth's telescope been mounted in a yoke-type equatorial he could, still using a single mirror (but suitably geared), have sent the light down the (hollow) polar axis. This was done in the 60-inch Mount Wilson reflector and, as we have seen, in several later large reflectors where a stationary image is formed on a fixed spectrograph.

Porter took advantage of the fixed direction of the polar axis of the equatorial mount. At his former home on the Maine coast he designed and erected a polar Newtonian telescope (about 1917) but dismantled it later owing to the inconvenience of looking up the polar axis. With the development of Stellafane, the Vermont clubhouse of the Springfield Telescope Makers (a club initiated by Porter), came the erection of his turret reflector. The turret telescope was the invention of Governor J. E. Hartness, head of the Jones and Lamson Machine Company, who erected a 10-inch aperture turret refractor at Springfield.[21] In this design the observer is comfortably housed in the turret-shaped extension of the polar axis, protected from the rigours of the cold Vermont winter outside. The optical layout is simple and the single reflection is produced by a small prism, but the structure is not an easy one for amateur construction. The dome of the Porter turret at Springfield (Fig. 189) is made of concrete and once carried two telescopes, a 16-inch Newtonian of 17 feet focal length, and a Cassegrain of 12 inches aperture and 16 feet equivalent focal length;[22] the latter is now replaced by a counterweight. In 1920, Porter invented the first Springfield mounting which was assembled

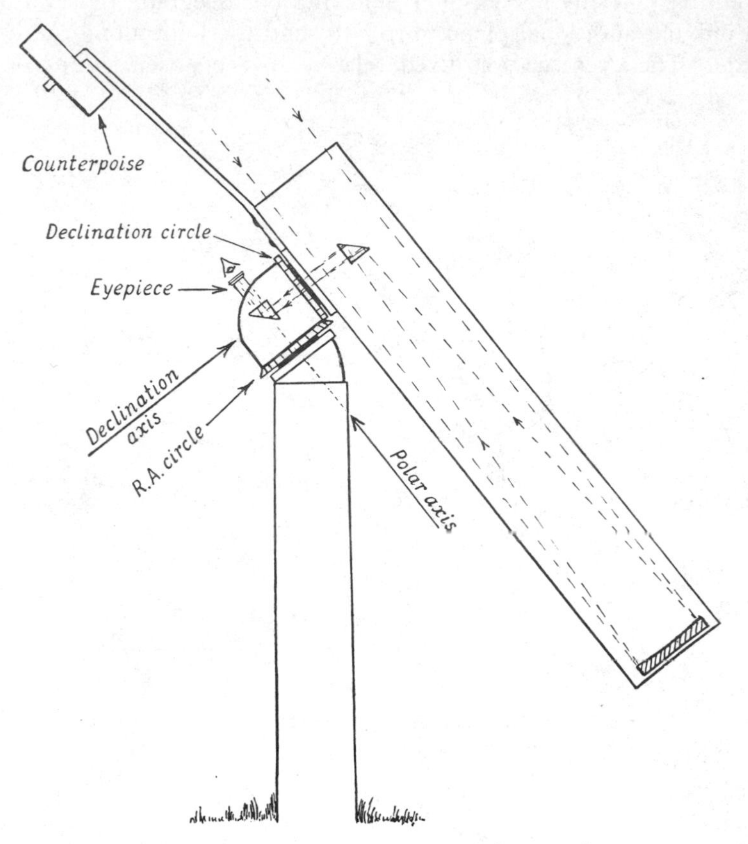

Fig. 188—The Springfield equatorial mounting

One of Russell W. Porter's fixed-eyepiece designs.

and tried out at Stellafane.[23] This mount and its offspring, the Pasadena, employ two reflections by small prisms and bring both right ascension and declination circles to within a few inches of the observer's face (Fig. 188).

Most astronomers prefer stiff necks and body contortions to loss of valuable light by additional reflections. Amateurs seem to endure long vigils at the telescope with remarkable fortitude. England, for all its reputed bad weather, has produced some of the finest amateur astronomers in the world—Dawes, Webb, Denning, Williams, and Phillips deserve special mention in this respect. Most of them worked

Fig. 189—Porter's turret telescope

The instrument is at Stellafane, near Springfield, Vermont. The Newtonian telescope is in place; a Cassegrain form was added later—hence the hole in the front of the turret

(Photograph by Robert E. Cox)

for long periods in the open, at all hours and in all temperatures, and never expressed any particular wish for the comforts of closed observatories and fixed eyepieces.

> Persistency in observation [Denning writes][24] apart from the value derived from cumulative results, increases the powers of an observer to a considerable degree.

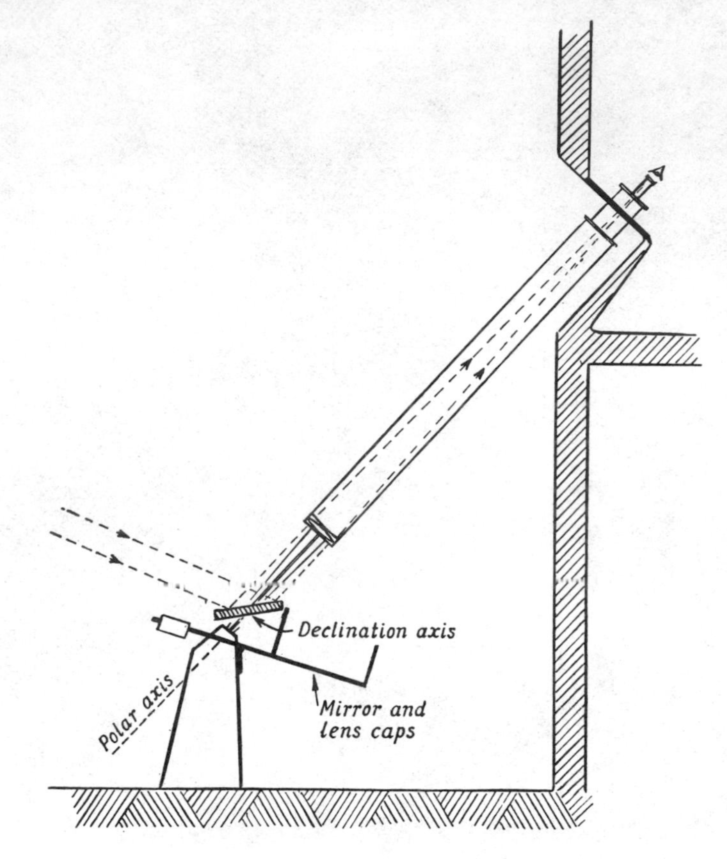

Fig. 190—Gerrish-type polar telescope

Motion in right ascension and inclination of the declination mirror were controlled from the eyepiece end. H. Grubb's form was similar, except that the entire instrument was mounted on a wheeled carriage.

Observations must be accurate and regularly made, to achieve which the amateur observer must sacrificc much of his time, energy, and home-comforts. It is one thing to possess a telešcope and quite another to use it to advantage. It is significant that, of all the thousands of telescopes in existence, comparatively few are actively engaged on regular astronomical work. The remark that we made in the case of the elder Herschel, whose telescopes were sent all over western Europe, is unfortunately true today.

A few further forms of fixed-eyepiece telescope should be mentioned for completeness. Where the restrictions caused by the mounting, size of mirrors, and height of the supporting building are of no consequence, the luxury of a stationary eyepiece cannot be denied. The 12-inch Gerrish polar telescope, mounted at Harvard in 1900, had a range of 80° in declination.[25] Pickering's photographic lunar atlas was made in 1900 with a polar telescope. In this case, the tube was nearly horizontal since the latitude of the observing site, a Harvard station at Mandeville, Jamaica, is only 18° 01′N. The objective, of 12 inches aperture and 135 feet focus, was fed by a clock-driven 18-inch heliostat.[26]

Another American fixed instrument is the 15-inch horizontal siderostat of the Cook Observatory, now part of the University of Pennsylvania. The Brashear

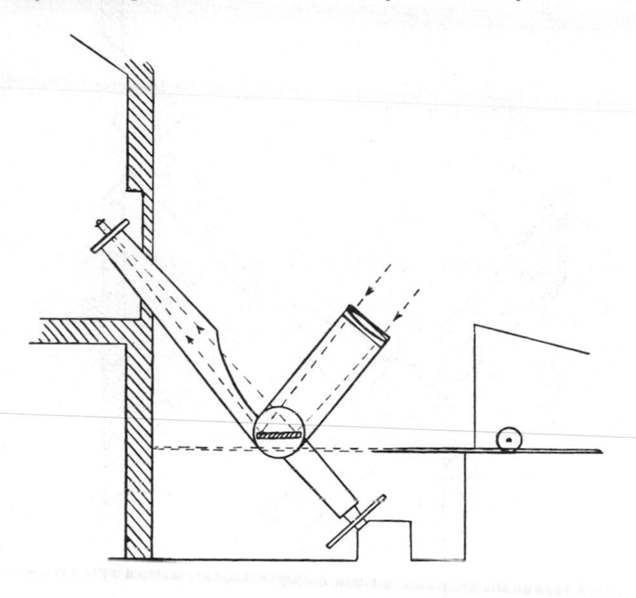

Fig. 191—Grubb modified coudé telescope

The Sheepshanks astrographic refractor of 12½-inch aperture at Cambridge, England, is a fine example of this type.

objective, originally made for the Philadelphia Central High School, was later purchased by G. W. Cook of Wynnewood, who got J. W. Fecker to set it up as a siderostat.[27] In 1934, A. M. Skellett of the Bell Telephone Laboratories used Cook's telescope for his experiments on the coronaviser (page 442). Yale Observatory possesses a 40-foot polar telescope with a 15-inch McDowell photographic object-glass and 9-inch visual guiding telescope.

Howard Grubb at one time took an interest in fixed-eyepiece forms. The 12½-inch Sheepshanks polar coudé at the University Observatories, Cambridge, England, is of his design and, with the plane mirror placed between objective and eyepiece, permits a reduction in area of the reflecting surface. The objective is a photovisual triplet designed by H. Dennis Taylor.[28] Further polar axis telescopes designed by Grubb were installed in Ireland in the eighties of the last century.

French astronomers in the eighties and nineties of the same century looked with favour on Loewy's coudé telescope. The brothers Henry overcame the difficulties attendant on figuring the large flat reflecting surfaces required, and their first instrument of 10½ inches aperture was pronounced a success. With a larger Henry–Gautier coudé of 23½ inches aperture and 59 feet focus, the largest yet built, Loewy and Puiseux took their now classic series of lunar photographs.[29] The telescope is equipped with both a visual and a photographic object-glass. The latter is now on temporary loan to the Pic-du-Midi Observatory.

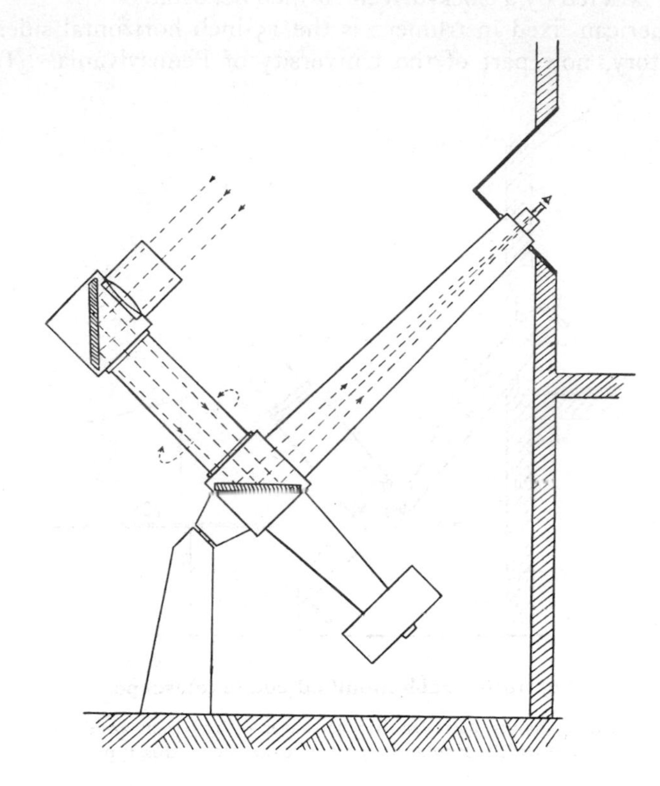

Fig. 192—Diagram of Loewy's equatorial coudé

While the age of giant photographic reflectors is only beginning, we can say with certainty that 40 inches is the limit for the useful size of a refractor. A larger glass, by reason of the increase in lens thickness and surface area, would probably absorb and reflect all, if not more than the extra light received by the increase in aperture. In addition, the cost and difficulty of obtaining suitable disks and compensating for their flexure would present great, if not insuperable, problems. Early in the present century, Grubb-Parsons made 41-inch diameter lenses for a large refractor intended for the Nikolaieff Observatory in southern Russia. Although, by 1929, work on the lenses, dome and rising floor was stated to be in its final stages, the project failed to materialize.

A telescope of 49·2 inches aperture was exhibited at the Paris Exhibition of 1900 and, for about a year, was the largest refractor in the world. To avoid the expense of a large dome and equatorial mounting, and to ensure adequate stability, the promoters, led by F. Deloncle, ordered from Gautier a fixed-tube instrument fed by a siderostat.[30] The Teumont glassworks cast the 80-inch diameter glass blank for the siderostat and Mantois provided 50-inch disks for two objectives, one photographic and one visual. The finished instrument had a focal length of 187 feet and consisted, apart from the siderostat, of a long, heavy sheet-steel tube mounted on stone pillars. At one end a carriage on rails carried the objective. At the other end a tube section, similarly mounted, carried arrangements for an eyepiece, micrometer, photographic plate, and projection lens. These devices were all fitted to an inner tube which could be rotated by clockwork to compensate field rotation. The siderostat mirror with its heavy iron cell, fork supports and base, floated in mercury.

The instrument was an impressive show-piece and, as such, attracted large numbers of visitors. It was mounted in its own building and visitors, under the glare of electric lights inside and of searchlights and illuminations outside, were invited to look into the eyepiece. After midnight, with the exhibition closed, E. M. Antoniadi of the Juvisy Observatory made the only observations of any consequence. Using the photographic objective, he made drawings of some of the more prominent nebulae, drawings reproduced in the *Bulletin* of the Société Astronomique de France for 1900.

The promoters held out the most extravagant hopes for this instrument. By its use the observer would see the moon as if it were only twenty-six miles distant. By projection methods, 12-foot diameter images of Mars would be formed on a giant translucent screen. On cloudy nights it was proposed to project photographs on the screen—the only sensible idea, seeing that the spectators would have been none the wiser. These and other plans failed to materialize. The French government refused to purchase the instrument and the promoters, faced with great financial loss, sold the parts as scrap metal. The optical elements were stored in the optical laboratory of the Paris Observatory where they apparently still remain.

Several factors will determine the erection of large Schmidt-type systems and further 100- or 200-inch telescopes and, maybe, a 300-inch telescope. The United States is younger and richer than European countries and the south-western States enjoy a climate against which Europe cannot, generally speaking, compete. Another 200-inch telescope will certainly cost up to $8,000,000 and a 300-inch much more. Wages, building materials, and maintenance costs will be high for the next few decades, and the expense of a telescope increases much faster than its diameter. Two great European wars and the threat of further war in such a short span of years have been, and continue to be, a strain on resources normally applied to the interests of art and science.

That the work of large telescope building will continue is shown by the proposed erection of a 98-inch reflector in southern England. In his presidential address to the Royal Astronomical Society on February 8, 1946,[31] Prof. H. H. Plaskett made a strong appeal for the revival of observational astronomy in England, and the council of the society subsequently resolved to support his idea of building a large reflector. British weather, however, dictates exposures during short and infrequent spells of good seeing, and the importance of high-dispersion spectrography bespeaks a large-aperture condensing system. To meet these two distinct requirements, an 80-inch × 98-inch Schmidt-type system[32] will be erected at the Royal Greenwich Observatory, Herstmonceux. With the corrector in position, direct photography and nebular spectrography can be undertaken at the primary focus. With the corrector dismounted, two aspherical concave secondaries will permit spectrographic observations at the coudé and Gregorian foci. The primary mirror is unperforated; one plane mirror will be required for observations at the Gregorian focus, and two for those at the coudé focus.

A large reflector was also planned for the University of Michigan Observatory. Heber D. Curtis was appointed director in 1930 and proceeded to design an 85-inch reflector, only to find later that the necessary funds would be delayed indefinitely. Nevertheless, a site was purchased and a Pyrex blank was ordered from the Corning Glass Company. The first disk proved defective but Corning offered one of 98-inch diameter at slightly increased cost. The offer was accepted and the larger disk was successfully cast, annealed, and delivered to Ann Arbor. In 1942 Curtis died, and the scheme which he had fostered for twelve years failed to mature. Through the generosity of the authorities of the University of Michigan and the Tracy McGregor Fund, this disk has been placed in the possession of the Royal Observatory, together with a 26½-inch Pyrex secondary disk. As Sir Harold Spencer Jones has pointed out, this gift should enable the new telescope —to be called the *Isaac Newton telescope*—to be in operation two or three years earlier than would otherwise have been possible. At the time of writing, Grubb-Parsons have figured the spherical primary mirror. An offset fork mounting will be adopted, with a polar axis in the form of an inverted and truncated cone.

Another large telescope is planned for erection at the Lick Observatory on Mount Hamilton. This will be a 120-inch reflector in a fork-type mounting designed by W. W. Baustian and chosen because it will enable observation of the entire sky except for a narrow zone near the horizon. The building which will house the telescope contains a basement optical shop in which mirrors can be worked, tested and aluminized. One section, a concrete-walled light tunnel, extends underground to 73 feet beyond the wall of the building. Building and 97-foot diameter dome were completed by Messrs Carrico and Gautier of San Francisco in March, 1952. They enclosed basement workshop, laboratories, darkrooms, staff-rooms and visitors' gallery. The optical work is under the supervision of Hendrix of the Mount Wilson and Palomar Observatories. The 120-inch ribbed blank, a proposed testing blank for the 200-inch Hale telescope (page 408), is now on the grinding machine and Messrs Pilkington Bros have supplied glass blanks for the five auxiliary mirrors. Work has also started on the mounting, which will be large enough to enable the observer to ride in the tube at the primary focus.[33]

The immense cost of large moving telescopes has encouraged at least two workers to explore the possibilities of a fixed vertical form provided with a mosaic or tessellated primary mirror. Mention has already been made of the promising results achieved by Lord Rosse with compound mirrors of speculum metal.

For many years G. Horn-d'Arturo, of the University of Bologna, experimented with this form and, during the period 1945 to 1953, assembled a mirror composed of sixty-one spherically worked tesserae, each of hexagonal outline and 8 inches across. Each mirror was placed on three adjustable screws which projected through a large marble slab. The tesserae were adjusted for coincidence of foci and together formed a mirror of about 70 inches aperture and 410 inches focal length. This was assembled and tested in the lofty top room of a high tower at the university, and a moving plate-holder was positioned at the centre of a hole in the ceiling. By giving the plate the requisite slow west-to-east motion, it was possible to counteract the effect due to the earth's rotation. The photographic performance was considered encouraging although, as expected, star images were enlarged by the combined diffraction effects of the edges of the tesserae, with consequent loss of resolving power. Analytical considerations, however, lead Horn-d'Arturo to state that, in theory, the form and dimensions of the photographic image formed by a tessellated mirror are practically identical with those generated by a uniform paraboloid.[34] Y. Väisälä of Turku, Finland, is also working with this type of mirror. In 1949, he began to build a vertical Schmidt of which the spherical mirror is composed of seven circular mirrors, each of 12½-inch diameter and giving an overall diameter of 33½ inches. The large corrector plate will be supported by a sturdily constructed wooden framework. Horn-d'Arturo, Väisälä and now J. P. Hamilton of Victoria, Australia, write enthusiastically about the merits of the idea and its application to reflectors which, in imagination, attain apertures of 500 and even 800 inches.

The present century has witnessed the transference of many observatories from towns, cities and universities to sites remote from the lights, vibration and haze of industrial and commercial areas. We have followed in these pages the removal of the Radcliffe Observatory to its station near Pretoria. Herstmonceux Castle, near Hastings, has become the new home of the Royal Observatory, for nearly 300 years at Greenwich. There is still scope for more large reflectors in the southern hemisphere, particularly in territory belonging to the Commonwealth. The Pretoria telescope and the 74-inch reflector at Mount Stromlo, Canberra, are already meeting this need. Harvard College Observatory has advanced astronomy by establishing a station at Arequipa, Peru, later at Bloemfontein, South Africa. Ritchey at one time visualized a series of five 315-inch vertical ' universal-telescopes ' under one management and spaced in different latitudes about twenty degrees apart. He admitted that his plan called for the institution of an international commission, untrammelled by considerations of boundary and the niceties of national prestige.

> These telescopes [he wrote][35] may well become the mighty guns of Peace. For inevitably, nations will soon decide, and gladly, that all such efforts for human betterment, for science and for education, are infinitely more interesting and more profitable than the wholesale mutilation and murder of our fellow-men called war.

The improvement in sensitivity and directivity of radio receivers and transmitters used for radar during the war facilitated post-war studies in the new subject of radio astronomy. As early as 1931 and 1932, K. G. Jansky at the Holmdel, New Jersey, branch of Bell Telephone Laboratories, was investigating a residual radio

Fig. 193—30-foot radio telescope

A small steerable radio telescope. Winter scene at the Jodrell Bank Experimental Station of the University of Manchester

noise which came, apparently, from some extra-terrestrial source.[36] Later investigations revealed a periodicity of 23 hours 56 minutes, maximum disturbance in approximately the direction of the galactic centre and distribution along the Milky Way.[37] Jansky was unable to detect any high-frequency radio radiation from the sun and concluded that the radiations possibly came from the stars themselves or from material excited by stellar radiation.

Radio noise was first picked up from the sun and associated with a solar source by J. S. Hey during the war years in England. His investigations for the Army Operational Research Group were made with Army radar equipments operating on wavelengths of 4–6 metres.[38] In 1943, Grote Reber used a 'radio telescope' with a 31-foot diameter paraboloidal mirror of sheet metal and 20 feet focus. Signals were received by a dipole at the focus while the mirror, fixed in azimuth, could be moved in altitude. This radio telescope gave better angular resolution than previous systems and Reber's work with it revealed a rough correlation between radio intensity and the visible distribution of matter in the Milky Way.

Fig. 194—Forty-eight helical antennae of the Ohio State University radio telescope

The array can be tilted around a horizontal axis. It has recently been enlarged to carry ninety-six helices.

(Ohio State University)

Using a wavelength of 1·85 metres, Reber found that the sun radiates with an intensity equal to that of the Milky Way region in Sagittarius.[39]

None of the radio telescopes at present in use has anything like the resolving power of an optical instrument. A resolving power of 1 minute of arc at 4 metres wavelength, for instance, would require an aerial system about two miles square.[40] Large paraboloidal antennae necessitate more stringent surface contour tolerances and their construction would be a large and expensive engineering feat. A move in this direction, however, has already been made. A fixed paraboloid at the University of Manchester's research station at Jodrell Bank, Cheshire, has a diameter of 218 feet and is used by R. Hanbury Brown, J. A. Clegg, C. Hazard and other members of Prof. A. C. B. Lovell's research team. The aerial was designed by Clegg and has a mirror contour formed by wires strung on steel cables. These cables are supported by concentric rings of posts and give a form paraboloidal to within five inches. With a wavelength of 1·9 metres the beam width is about 2°. The receiving dipoles at the focus are mounted on a 126-foot tower pivoted on an east to west axis at its lower end. By moving the tower in a meridional plane it is possible to cover regions within 15° of the zenith.[41] In the summer of 1950, using this instrument, Hanbury Brown and Hazard detected and plotted radio emission intensities for M31 in Andromeda, one of the nearest galaxies.[42] The same observers have recently (October, 1953) reported[43] the detection and measurement of radio emission from the late-type spiral M81 in Ursa Major.

Comparatively small paraboloids can be equatorially mounted as, for instance, is the 17-foot aperture equatorial radio telescope built in 1948 by engineers at Cornell University. The paraboloidal antenna covers a sky area of from 2° to 30° according to the wavelength used, and rotates about its axis for polarimetric studies. The entire unit, of some 8 tons weight, can be rotated in azimuth for calibration purposes.[44]

Work is now in progress at Jodrell Bank on the construction of a 250-foot steerable radio telescope. The financial outlay on this great project, involving some £450,000, is being met by the Department of Scientific and Industrial Research and the Nuffield Foundation. The aerial is a steel bowl covered with 2-inch mesh and mounted on a horizontal axis for movement in altitude. Axis, operating gears, and bowl are mounted on a 300-foot diameter track which will permit moving the instrument in azimuth. A control system will aim the telescope to any part of the sky and automatic scanning motions in altitude or in azimuth will be provided.[45] The reduction in beam width offered by an aerial of this great size, coupled with its directional versatility, will enable more accurate mapping of the contours of radio intensity levels (isophotes) within the galaxy and will permit the more precise localization of small-area radio sources.

By means of an interference technique it is now possible to overcome the resolving-power limitations of small paraboloids. Two methods are available. One is analogous to the optical interferometer known as Lloyd's mirror and consists essentially of a single aerial perched on a high cliff overlooking the sea.[46] The other is analogous to the Michelson stellar interferometer: two aerials are spaced laterally in an east to west direction.[47] Although the resolution is poor by optical standards, these methods enabled studies of radio-frequency emissions from large sunspot areas and led to the discovery of two major sources of radio frequency in Cygnus and Cassiopeia. Every effort has been made to determine the apparent angular size of these sources in the hope of identifying them with a visual object. In 1948, using the Lloyd's mirror technique, J. G. Bolton and G. J. Stanley of Sydney, Australia, showed that the Cygnus source has an apparent size of less than 8 minutes of arc.[48] In 1950, Stanley and O. B. Slee claimed that the same technique had enabled them to reduce this limit to 1′ 30″.[49] Using the Michelson method and a base line of 500 metres, M. Ryle and F. G. Smith of the Cavendish Laboratory, Cambridge, reduced the angular size of the Cassiopeia source to 6 minutes of arc.[50] Investigations at Jodrell Bank, Cambridge, and Sydney during 1950–52 reduced these limits still further and it is now evident that the radio source in Cygnus can be identified with what appears to be two extra-galactic nebulae in collision and the one in Cassiopeia with diffuse galactic nebulosity.[51]

Another comparatively intense source is in Taurus and in 1949, Bolton and Stanley suggested[52] its possible connection with M1, the well-known Crab nebula. Recent work at Sydney confirms the association and, moreover, locates the fourth major source, known as Centaurus A, with a nebula.[53] An increasing number of other discrete sources are being detected, some remote from the galactic plane, but none has yet been identified with an individual bright visual star.[54] With the 218-foot paraboloid at Jodrell Bank, Hanbury Brown and Hazard in August, 1952, received weak signals from a discrete source in Cassiopeia.[55] Its position agrees well with that of Tycho's supernova of 1572. Since the Crab nebula is believed to be the remnant of the supernova of 1054, it is not unlikely that other supernovae remnants are also sources of radio radiation. It is likely that the use of receivers of greater sensitivity and resolving power will reveal many more discrete sources of radio emission—that these sources are both numerous and well distributed in galactic latitude. They are sometimes referred to as 'radio stars' but with the realization that they may well prove to be extra-galactic nebulae and/or certain regions of interstellar material in our own galaxy. The galactic radio stars must clearly have their counterparts in M31, M81 and other external galaxies but, as yet, the origin of the radiation remains unexplained.

In one particular instance the origin and probable conditions of radio emission are known. We refer to radiation of 21-cm wavelength (1,420 Mc/sec) which H. C. van de Hulst predicted in 1944 as coming from low-temperature interstellar hydrogen. This radiation was first detected on the night of March 25, 1951, at Harvard University.[56] The signal came from an apparently extended source near the galactic plane in the direction of Ophiuchus. Van de Hulst was visiting Harvard at the time and at once notified the Radiophysics Laboratory at Sydney and his associates in Holland. In Holland, J. H. Oort and C. A. Muller first detected the hydrogen signal on May 11, using a 25-foot paraboloid at Kootwijk radio station.[57] They used a beam width of 2·8 degrees at half-power and, from the observed wide spread in galactic latitude, suggested that the responsible hydrogen clouds must be relatively close to the earth. In June, a cable from Sydney gave further confirmation. Since then, attention has also been given to differences in the frequency in different galactic directions which, interpreted as Doppler effects, add further to our knowledge of galactic structure. In the summer of 1953, F. J. Kerr and J. V. Hindman of Sydney detected 21-cm radiation from the Magellanic Clouds.[58] Once again, measurable frequency shifts could be interpreted in terms of the relative motions of the two clouds.

In January, 1946, radar contact was established with the moon by the Engineering Laboratories of the U.S. Army Signal Corps. They transmitted pulses of ¼-second duration with 3 kilowatts peak power at a wavelength of 2·7 metres. The aerial system consisted of an array of 64 dipoles, with a wire-netting reflector, supported on a 100-foot tower and directed horizontally. Reflections from the moon could be observed at moonrise and moonset.[59] This achievement raises interesting questions as to what extent this method can be used to check the distance of the sun by sending pulses to targets like Eros in opposition. Radar pulses are also reflected by the ionized gases in meteor trails as J. S. Hey and J. W. Phillips conclusively showed in 1946–1947.[60] This important method of meteor detection both supplements visual night-observations and extends meteor counts throughout the day. Furthermore, it provides important information about the physics of the upper atmosphere.

The cost of large mirrors and lenses has led many radio engineers to look for a substitute image-producer in the fields of electronics and television. Initial investigations have met with appreciable success. In 1933, V. K. Zworykin of Camden, New Jersey, introduced an instrument which he called an *iconoscope* whereby an image received on a mosaic of microscopic cesium-silver dots on a mica disk could be magnified electrically.[61] Each silver dot, due to the photoelectric action of light, receives a small individual charge proportional to the intensity of the light falling upon it. This charge is accumulated by depositing a continuous metallic film on the other side of the mica, so making the receiving screen into a kind of condenser. If a scanning beam of electrons now passes over the mosaic, the dots are discharged one by one in rapid succession, and a pulsating current is produced which is fed from the condenser plate through a suitable amplifier. This arrangement forms in television technique a well-known and sensitive type of analyser, and its application to the telescope has possibilities. If the mosaic and the electron beam can be made respectively fine and narrow enough (and mosaics giving almost photographic resolution are being developed) this electronic analyser could replace the conventional eyepiece. The reproduction of the image from the amplified

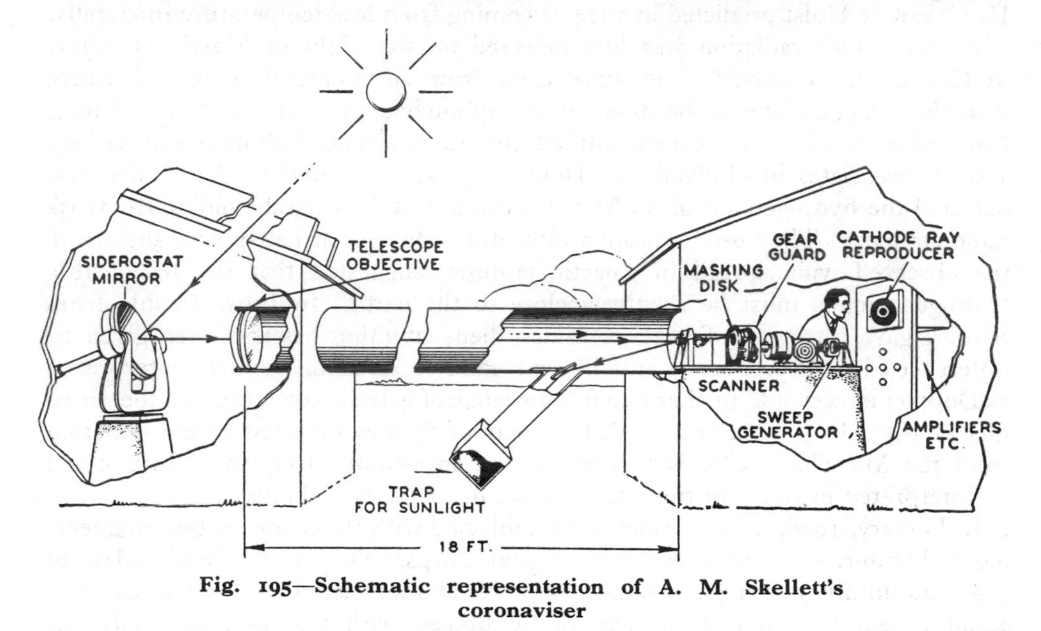

Fig. 195—Schematic representation of A. M. Skellett's coronaviser

current is a comparatively simple matter and could be effected with the aid of, say, a second electron beam and fluorescent screen. The resultant image would be both brighter and larger than the original.

At the Paris Observatory, A. Lallemand and M. Duchesne have recently had some success with an electrophotographic technique.[62] They made use of the 26-cm $f/40$ coudé refractor to focus images on a cathode plate. Electrons ejected from the plate were accelerated and refocused electrostatically on an ordinary photographic plate at the anode. To maintain the necessary high degree of vacuum in the electron tube, a special magnetic plate-changing mechanism was incorporated. The tube was also enclosed in a liquid-oxygen refrigerator. Records of Saturn, which appeared well-defined, were obtained in only one-fifth of a second. Ordinary photographic techniques would require ten seconds for comparable results. An image of θ Orionis was obtained in five seconds and showed the four component

Fig. 196—A versatile and well-equipped reflector

This instrument was designed and built by an American amateur, P. R. Engle, and is in operation at the New Mexico College of Agriculture and Mechanic Arts at Las Cruces, New Mexico. The optical system, with 17-inch primary mirror, operates either as an *f*/7·8 Newtonian or as an *f*/22·8 Cassegrain. Special features include electric controls, an eighteen-point mirror flotation system, a removable Newtonian nosepiece, single-strut swing-over spiders for the secondary mirror support, and a cork-lined aluminium tube which can be rotated in the cradle.

stars more clearly than would an ordinary photograph of three minutes' exposure.[63]

A device somewhat similar to Zworykin's and called the *coronaviser* was developed in 1939 by A. M. Skellett of the Bell Telephone Laboratories.[64] This instrument purported to render the corona visible in full sunlight. The solar image produced by a long-focus objective fed by a siderostat was scanned spirally. In practice, the solar disk was blocked out by a circular screen and scanning was limited to the annulus immediately round its edge. Behind the scanner was a photocell connected to amplifiers and thence to a cathode-ray reproducer. The outfit was given a practical trial at the late Dr Cook's observatory at Wynnewood, Pennsylvania, and images of solar prominences were obtained. Glare from the earth's atmosphere was practically uniform in the area scanned and was represented in the photoelectric current by a uniform direct current which was eliminated by the amplifier. To distinguish 'parasitic images' caused by instrumental defects and to check the true appearance of the prominences, a spectrohelioscope was at hand for direct comparisons. At the time, Skellett considered that the capabilities of the device could be realized only under the crystal-clear skies encountered at high altitudes.[65]

Thus we return again to the old problem of escaping from the damaging optical effects of the earth's atmosphere—and there is no escape as yet. The ocean of air in which we move, breathe, and have our being is at the same time the most troublesome factor in our attempts to survey the sidereal domain. In the future, when interplanetary travel becomes possible, observing stations and, ultimately, observatories may be established out in space. For the present, we must admit that we are impotent against the atmospheric restriction but, in the development of the telescope, we recognize no limitations

> Telescope building is at once an exact science and a fine art [the late J. W. Fecker once said].[66] While resting in the activity of telescope making, we grow naturally in the knowledge and expression of its art. The successful growth, development, design and building of telescopes has been made possible only through the complete and harmonious co-operation of astronomer and instrument designer, and with this continued co-operation, effectiveness, power and efficiency of astronomical instruments will continue to increase until eventually we will be limited only by atmospheric conditions on this earth.

REFERENCES

1. *Amateur Telescope Making*, p. 73, 1933.
2. *J.B.A.A.*, **47**, p. 189, 1937.
3. *M.N.R.A.S.*, **88**, p. 158, 1927–1928.
4. *J.B.A.A.*, **45**, pp. 80–82, 1934; *M.N.R.A.S.*, **95**, p. 332, 1935.
5. *M.N.R.A.S.*, **68**, p. 488, 1908.
6. *Observatory*, **70**, p. 30, 1950; *M.N.R.A.S.*, **110**, pp. 131–133, 1950.
7. Tiercy, G., *Publications de l'Observatoire de Genève*, **35**, pp. 25–32, 1940. Also from information received from Mme Renée Schaer, Lausanne.
8. Bailey, S. I., 1931, *History and Work of the Harvard Observatory*, p. 269.
9. Bailey, S. I., *Pop. Astr.*, **33**, p. 493, 1925.
10. Bailey, Ref. 8, p. 49.
11. *Ibid.*, p. 145.
12. *Ibid.*, p. 159.
13. Edmondson, F. K., *Sky and Telescope*, **8**, p. 34, 1948.
14. *Sky and Telescope*, **7**, p. 3, 1947.
15. *Ibid.*, p. 4.
16. Rohr, J., *Pop. Astr.*, **55**, p. 28, 1947.
17. *J.B.A.A.*, **58**, p. 110, 1948.
18. Maksutov, D. D., *Jour. Opt. Soc. Amer.*, **34**, pp. 270–283, 1944. Tenukest, C. F., Schaefer, R., and Pinnok, H., *J.B.A.A.* **56**, pp. 130–131, 1946.
19. Thomson, E., *Journal Franklin Institute*, **76**, 3rd series, pp. 117–121, 1878.

[20] *Amateur Telescope Making*, p. 485, 1933. *Pop. Astr.*, **26**, p. 147, 1918.

[21] Hartness, J. E., *Jour. Am. Soc. Mech. Eng.*, pp. 1511–37, 1911.

[22] Porter, R. W., *Pop. Astr.*, **29**, pp. 249–251, 1921.

[23] *Amateur Telescope Making*, pp. 333–375, 1944.

[24] Denning, W. F., 1891, *Telescopic Work for Starlight Evenings*, p. 79.

[25] Dailey, Ref. 8, p. 45. Bell, L., 1922, *The Telescope*, pp. 122–123.

[26] Pickering, W. H., 1903, *The Moon*, Preface. Bell, *op. cit.*, p. 123.

[27] *Pop. Astr.*, **54**, pp. 18–19, 1946.

[28] Bell, *op. cit.* ,pp. 124–125.

[29] Loewy, M., and Puiseux, P., *Atlas photographique de la Lune*, 1896, 1897, 1898. The instrument is described in *Compt. Rend.*, **96**, pp. 735–741, 1883.

[30] Lockyer, J. N., *Nature*, **61**, pp. 178–181, 1899. Butler, C. P., *ibid.*, **62**, pp. 574–576, 1900. Antoniadi, E., *Bull. Soc. Astr. de Fr.*, **14**, pp. 218, 375–376, 385–387, 459–460, 1900. *Vide* also *Jour. R.A.S. Canada*, **31**, p. 141, 1937.

[31] *Observatory*, **66**, pp. 234–238, 1946.

[32] Linfoot, E. H., *Proceedings London Conference on Optical Instruments*, 1951, p. 176.

[33] Baustian, W. W., *Publ. Astr. Soc. Pac.*, **58**, pp. 173, 349, 1946. **64**, pp. 122–127, 1952. *Sky and Telescope*, **10**, p. 248, 1951.

[34] *Scientific American*, pp. 60–63, Jan., 1951; pp. 99–102, May, 1954. *J.B.A.A.*, **63**, pp. 71–74, 1953.

[35] Ritchey, G. W., *Trans. Opt. Soc.*, **29**, pp. 213–218, 1928.

[36] Jansky, K. G., *Proc. I.R.E.*, **20**, p. 1920, 1932.

[37] *Ibid.*, **21**, p. 1387, 1933; **23**, p. 1158, 1935; **25**, p. 1517, 1937.

[38] *Nature*, **157**, p. 296, 1946.

[39] Reber, G., *Proc. I.R.E.*, **28**, p. 68, 1940; *Ap. J.*, **91**, p. 621, 1940.

[40] *Nature*, **164**, p. 816, 1949.

[41] *J.B.A.A.*, **61**, p. 181, 1951.

[42] Hanbury Brown, R., and Hazard, C., *Nature*, **166**, p. 901, 1950.

[43] *Ibid.*, **172**, p. 853, 1953.

[44] *Pop. Astr.*, **42**, pp. 71–74, 1949.

[45] *J.B.A.A.*, **62**, pp. 241–242, 1952.

[46] McCready, L. L., Pawsey, J. L., and Payne-Scott, R., *Proc. Roy. Soc.*, A, **190**, p. 357, 1947. Bolton, J. G., and Stanley, G. J., *Nature*, **161**, p. 312, 1948; *Aust. J. Sci. Res.*, A, **2**, p. 139, 1949.

[47] Ryle, M., and Smith, F. G., *Nature*, **162**, p. 462, 1948.

[48] Bolton, J. G., and Stanley, G. J., *Nature*, **161**, p. 312, 1948.

[49] Stanley, G. J., and Slee, O. B., *Aust. J. Sci. Res.*, A, **3**, p. 234, 1950.

[50] Ryle, M., and Smith, F. G., *Nature*, **162**, p. 462, 1948.

[51] *Nature*, **170**, pp. 1061–1065, 1952.

[52] Bolton, J. G., and Stanley, G. J., *Nature*, **164**, p. 101, 1949.

[53] Mills, B. Y., *Nature*, **170**, p. 1064, 1952.

[54] Smith, F. G., *M.N.R.A.S.*, **112**, pp. 497–513, 1952. Ryle, M., Smith, F. G., and Elsmore, B., *M.N.R.A.S.*, **110**, pp. 508–523, 1950.

[55] Hanbury Brown, R., and Hazard, C., *Nature*, **170**, pp. 364–365, 1952.

[56] Ewen, H. I., and Purcell, E. M., *Nature*, **168**, p. 356, 1951.

[57] Muller, C. A., and Oort, J. H., *Nature*, **168**, pp. 357–358, 1951.

[58] Kerr, F. J., and Hindman, J. V., 1953, *Prelim. Report Survey of 21-cm Radiation from the Magellanic Clouds.*

[59] Mofenson, J., *Electronics*, **19**, p. 92, 1946.

[60] Hey, J. S., and Stewart, G. S., *Nature*, **158**, p. 481, 1946. *Proc. Phys. Soc.*, **59**, p. 858, 1947.

[61] Zworykin, V. K,. *Journal Franklin Institute*, **215**, p. 535, 1933.

[62] Lallemand, A. and Duchesne, M., *Compt. Rend.*, **233**, p. 305, 1951; **235**, p. 503, 1952.

[63] *Sky and Telescope*, **12**, pp. 286 and 291, 1953.

[64] *Bell Laboratories Record*, **18**, p . 62, 1940.

[65] Skellett, A. M., *The Telescope*, **7**, p. 56, 1940.

[66] Fecker, J. W., *Jour. R.A.S. Canada*, **24**, p. 304, 1930.

INDEX